CHERRIES

CHERRIES
Crop Physiology, Production and Uses

Edited by

A.D. Webster

Horticulture Research International
East Malling
West Malling
Kent ME19 6BJ, UK

and

N.E. Looney

Agriculture and Agri-Food Canada
Research Centre
Summerland
BC V0H 1Z0, Canada

CAB INTERNATIONAL

CABI Publishing is a division of CAB International

CABI Publishing
CAB International
Wallingford
Oxon OX10 8DE
UK

CABI Publishing
44 Brattle Street
4th Floor
Cambridge, MA 02138
USA

Tel: +44 (0)1491 832111
Fax: +44 (0)1491 833508
Email: cabi@cabi.org
Web site: www.cabi-publishing.org

Tel: +1 617 395 4056
Fax: +1 617 354 6875
Email: cabi-nao@cabi.org

A catalogue record for this book is available from the British Library, London, UK

ISBN 0 85198 936 5

First published 1996
Transferred to print on demand 2004

Printed and bound in the UK by Antony Rowe Limited, Eastbourne.

Contents

Contributors

G. Bargioni, *Via G. Medici 4, 37126 Verona, Italy.*

G.K. Brown, *United States Department of Agriculture, Agricultural Research Service, Fruit and Vegetable Harvesting Investigations, Agricultural Engineering Department, Michigan State University, East Lansing, MI 48824-1323, USA.*

J.F. Brunner, *Washington State University, Tree Fruit Research and Extension Center, 1100 North Western Avenue, Wenatchee, WA 98801, USA.*

S.R. Drake, *United States Department of Agriculture – Agricultural Research Service Tree Fruit Research Laboratory, Wenatchee, WA 98801, USA.*

J.A. Flore, *Department of Horticulture, Michigan State University, East Lansing, MI 48824-1325, USA.*

E.J. Hanson, *Department of Horticulture, Michigan State University, East Lansing, MI 48824-1325, USA.*

E.J. Hogue, *Agriculture and Agri-Food Canada Research Centre, Summerland, BC V0H 1Z0, Canada.*

A.F. Iezzoni, *Department of Horticulture, Michigan State University, East Lansing, MI 48824-1325, USA.*

A.L. Jones, *Department of Botany and Plant Pathology, Michigan State University, East Lansing, MI 48824-1321, USA.*

K. Kaack, *Danish Institute of Plant and Soil Science, Department of Food Science and Technology, Horticultural Research Centre, DK-5792, Aarslev, Denmark.*

C.D. Kesner, *Department of Horticulture, Michigan State University, East Lansing, MI 48824-1325, USA.*

G. Kollár, *Department of Farm Management, University of Horticulture and Food Industry, PO Box 53, Budapest, Hungary.*

E.M. Kupferman, *Washington State University, Tree Fruit Research and Extension Center, 1100 North Western Avenue, Wenatchee, WA 98801, USA.*

M. Longstroth, *Department of Horticulture, Michigan State University, East Lansing, MI 48824-1325, USA.*

N.E. Looney, *Agriculture and Agri-Food Canada, Research Centre, Summerland, BC V0H 1Z0, Canada.*

G.I. Mink, *Irrigated Agriculture Research and Extension Center, Washington State University, Prosser, WA 99350-9687, USA.*

D. Pennell, *Brogdale Horticultural Trust, Faversham, Kent ME13 8XZ, UK.*

R.L. Perry, *Department of Horticulture, Michigan State University, East Lansing, MI 48824-1325, USA.*

E.L. Proebsting, *Irrigated Agriculture Research and Extension Center, Washington State University, Prosser, WA 99350-9687, USA.*

S.E. Spayd, *Department of Food Science and Human Nutrition, Washington State University, Prosser, WA 99350-9687, USA.*

H. Schmidt, *Bundesforschungsanstalt für Gartenbauliche Pflanzenzüchtung, Bornkampsweg 31, 2070 Ahrensburg/Holst, Germany.*

M. Thompson, *Department of Horticulture, Oregon State University, Corvallis, OR 97331-2911, USA.*

J. Vittrup Christensen, *Institute of Pomology, Kirstinebjergvej 12, 5792 Aarslev, Denmark.*

A.D. Webster, *Horticulture Research International, East Malling, West Malling, Kent ME19 6BJ, UK.*

Preface

Loveliest of trees, the cherry now
Is hung with bloom along the bough,
And stands about the woodland ride
Wearing white for Eastertide.

A.E. Housman 1859–1936

The sweet cherry is one of the most popular of temperate fruit crops with consumers, albeit often one of the more expensive. Its popularity is considered surprising by some connoisseurs of fruit, in that it is not the sweetest, the most flavoursome or the largest of fruits and much of each fruit is composed of an inedible stone (pit). Nevertheless, sweet cherries have, for several hundred years, epitomized luxury, affluence, even decadence. The fruits are undoubtedly attractive in appearance, on account of their bright, shiny skin colour, and their subtle flavour and sweetness is obviously of great appeal to most consumers of fruits.

The ancestors of the modern cultivated sweet cherries are believed to have originated around the Caspian and Black Seas on the borders of Europe and Asia. These bore much smaller fruits than the cultivated trees we are familiar with today. Nevertheless, the early migratory populations in Europe are believed to have cultivated them by planting occasional trees near their dwellings and they even carried out some early selection of superior types.

Sweet cherries are now cultivated commercially in more than 40 countries of the world, in temperate, Mediterranean, and even subtropical regions. They prefer regions experiencing warm, dry summers, but require adequate rainfall or irrigation during the growing season if fruit of marketable size is to be produced. Rainfall at harvest time may destroy the crop, however, by inducing the fruit to crack.

The sour or tart cherry, which has origins very close to those of the sweet cherry, is less widely cultivated. Nevertheless, large quantities of sour cherries are produced in many European countries and in the USA. Most of these are used in processing and a large range of processed cherry products are now on sale throughout the world.

Compared with other temperate fruits, such as the apple and peach, the horticultural improvement of cherries as a commercial crop has been slow. Until relatively recently, traditional scion varieties and vigorous rootstocks have persisted in commerce and many of the major production problems, such as bird damage, rain-induced skin cracking and bacterial diseases, have made fruit growers wary of investing in cherries.

For a crop with such great unfulfilled market potential, this previous neglect by horticultural researchers is surprising. Since the early 1970s, however, considerable progress has been made in the development of cherries as profitable crops for the fruit grower. New improved varieties of sweet cherry have been bred which have larger fruit and are more resistant to bacterial canker than traditional varieties. Also, self-fertile varieties have been developed which set fruits much more reliably in production areas experiencing more marginal spring and early summer climatic conditions. New high-yielding and high-quality sour cherry varieties have also been bred in Europe and mechanized systems of harvesting, largely developed in the USA, have revolutionized the production of this crop for processing. Dwarfing rootstocks are beginning to be released which should, in the future, enable sweet and sour cherries to be produced in high-density planting systems. These plantations will be cheaper and easier to manage and to protect from bird and rain damage.

Few books have been devoted to the cultivation of cherries, possibly an indication of the relative neglect the crop has suffered from biologists and horticulturists over the years. Apart from the comprehensive *Cherries of New York* by Hedrick *et al.*, published in 1915, and *Cherries and Cherry Products* (R.E. Marshall, 1954, Interscience, London), few authoritative texts written in English have emerged in the last 50 years.

The objective in compiling this book was to provide horticultural students, professional fruit growers, researchers and others with an interest in cherry culture with an informative, up-to-date textbook on cherries. The cherry species are still under-researched and it is hoped that the book may stimulate further research and development with these exciting crops.

A.D. Webster
N.E. Looney

I | INTRODUCTION

1 The Taxonomic Classification of Sweet and Sour Cherries and a Brief History of Their Cultivation

A.D. Webster

Horticulture Research International, East Malling, West Malling, Kent ME19 6BJ, UK

1.1 Introduction

Most botanists now classify all of the 'true' cherries within the genus *Prunus*, which is part of the family *Rosaceae*; very few now adhere to the older system of classification which considered the sweet and sour cherries to be within the genus *Cerasus*. Linnaeus (1810), whose system of plant classification hinges largely on floral characteristics in distinguishing species, recognized that the characteristics of those species previously grouped within *Cerasus* were very similar to those within *Prunus*. He suggested, therefore, that they should join the 'Laurel', 'Bird' and 'Apricot' cherries within the genus *Prunus*. Only those systems of classification which placed greater emphasis on fruit or stone, rather than flower characteristics continued to segregate cherries into the separate *Cerasus* genus. Most authorities now place the esculent or edible cherries within the genus *Prunus*, subgenus *Cerasus*; however, most Chinese literature still places cherries within the genus *Cerasus*.

Classification of the many species within the subgenus *Cerasus* into further subgroupings, or sections, has already occupied taxonomists for generations and will probably continue to do so for many years to come. As we still have very incomplete knowledge of the origin and evolution of the many cherry species, appropriate phylogenetic systems of classification cannot be constructed. Instead, most current systems depend largely upon observable similarities or differences in plant morphology or anatomy. In this sense they are 'artificial' – systems of classification constructed by humans, as an aid to communications concerning plants. That such systems of classification frequently differ is, therefore, not surprising. Where a particular species fits within a classification depends solely on the importance or emphasis the botanist chooses to give to particular plant features.

Cerasus, the earlier name of the genus containing both the sweet and sour cherries, is not only used to denote the subgenus but has also assumed the status

3

of section in some classifications of the genus *Prunus* (Koehne, 1912). Most modern classifications are based to a great extent on that proposed by Koehne or a variation of this, first proposed by Rehder in 1947 (Appendix B). The classification still most widely used in Britain is another variation of Koehne's system published by Collingwood Ingram in 1948 (see Appendix A). Ingram divides the subgenus *Cerasus* into eight separate sections; he places particular emphasis on the form of the calyx tube, the style pubescence and the dentation of mature leaves in making these subdivisions. Other taxonomists have, in contrast, placed more emphasis on the types of sepals – in particular, whether they are reflexed or spreading.

Although there is still disagreement concerning some of the taxonomic sections proposed by Collingwood Ingram, the section of most relevance to the edible cherries, namely Section 1 *Eucerasus*, which holds within it *Prunus avium* L. (the sweet cherry), *Prunus cerasus* L. (the sour cherry) and *Prunus fruticosa* Pallas (the Steppe cherry), is now generally accepted by most taxonomists. These species, which might be called the edible or esculent cherries, all originate from Europe, North Africa, Afghanistan and Iran. Although the numbers of species of *Prunus* originating in eastern Asia are much more numerous (more than 100 in the Himalayas and 75 in China), amongst these only *P. tomentosa* (the Nanking or Downy cherry) has any real value as an edible crop. Schmidt (1973) has argued that *P. canescens* (the Grey Leaf cherry) should also be included within the *Eucerasus* section of Collingwood Ingram's system or the *Cerasus* section of the Rehder system of classification.

For more recent studies of the taxonomy and nomenclature of *Prunus* the works of Hayashi (1971), Kubota (1978) and Kawasaki (1982) should be consulted.

1.2 Taxonomy and Brief Descriptions of Cherry Species and Types

A brief description of those species of *Prunus* which bear edible fruits and which are generally referred to as cherries follows.

1.2.1 The sweet cherry: *Prunus avium* L. (*P. sylvestris* Ray or *Cerasus avium* L. (Moench))

What is now referred to as the sweet cherry (or occasionally and wrongly the Bird cherry) has attracted a plethora of vernacular names. The French refer to them as *merisiers*, from which the occasionally used British term 'Merries' is undoubtedly derived. Merisier itself may have derived from a combination of *amère*, meaning bitter, and *cerise*, meaning cherry. Another, perhaps rather less plausible, explanation of its origin is that it derives from *merica*, a berry mentioned by Pliny (see translation 1793). Mazzards and Geans are alternative names also used to describe the group.

The sweet cherry is a deciduous tree of large stature, occasionally reaching almost 20 metres in height, with attractive peeling bark. Leaves are large, 7.5–

12.5 cm long and about half this width, with coarse and irregular toothing; the leaf stalk is long with reddish glands near the lamina. Flowers are pure white, approximately 2.5 cm in diameter, formed either singly in the axils of the previous year's wood or in groups of up to five surrounding a vegative bud in spurs on older wood. Fruits on the wild species are roundish, red or black in colour and small (approximately 2.0 cm in diameter).

The sweet cherry is indigenous to parts of Asia, particularly to what is now northern Iran, and also to the Ukraine and other countries to the south of the Caucasus mountains. It is also a native of Europe and is found from the south of Sweden to Greece, Italy and Spain. Wild populations are most frequently found on slightly calcareous soils. The species is thought to have originated somewhere close to the Caspian and Black Seas, from where it has slowly spread. This spread was brought about initially by birds eating the fruits and then depositing the stones in their faeces some distance away. But its distribution was slow, as its fruits ripen well after the migratory birds have passed over Europe and its spread would have mainly relied on the movements of local bird populations.

Its distribution and naturalization have proved extensive, for it is now found wild in areas as widely spaced as northern India, the southern plains of Europe and even on isolated islands such as Madeira. Whether this distribution is entirely attributable to birds is a subject of some dispute. Although its spread predates many of humankind's great movements across Europe and Asia, there is some evidence that its naturalization in many areas was post-Stone Age and after the Aryan migrations. Some authorities suggest that first attempts at cultivating cherries were made in the early Alpine lake-dwellings. What is certain is that the sweet cherry was known to the Albanians (as *kerasie*) long before the Hellenes founded their great civilization in Greece.

1.2.2 The sour or tart cherry: *Prunus cerasus* L. (*Cerasus vulgaris* Mill)

The Albanians called the sour cherry *vyssine*, from which, it is argued, the German name for sour cherries, *Wechsel*, and the Italian name, *visciolo*, both possibly derive. Some even suggest that the French for sour cherry, *guigne*, also originates from the same linguistic root. The sour cherry is thought to have first been cultivated at the beginning of the Greek–Hellenic civilization.

It forms a small tree, or more often a deciduous bush, which suckers profusely from the base, forming dense thickets in the wild. It has smaller leaves than the sweet cherry, oval or ovate with a pointed apex, which are 4–7 cm long and about half this in width. Leaf margins have double, gland-tipped teeth and glanded stalks. Flowers, which are formed abundantly on one-year-old wood, are white, 1.75–2.5 cm in diameter on stalks approximately 2.0 cm long. Fruits of the non-cultivated types are red or blackish and round with soft, acid flesh. The main distinguishing characters between the sour and sweet cherries are the leaves, which on the former are lacking in down, the acid or bitter flavour of sour cherry fruits and the campanulate shape of the sour cherry calyx tube (in sweet cherries it is urn-shaped and constricted at its mouth).

De Candolle (1884) deduced that the sour cherry originated from an area very similar to that of the sweet cherry, around the Caspian Sea and close to Istanbul. Other authorities have disputed this as being too restrictive, suggesting origins in an area stretching from Switzerland to the Adriatic Sea in one direction and the Caspian Sea to the far north of Europe in the other direction. It is not found is such abundance as the wild sweet cherry in any of its areas of origin and its natural spread is not nearly so extensive.

Some authorities have suggested that the sour cherry originated as a hybrid from crosses between the sweet and the Ground or Steppe cherry (Olden and Nybom, 1973). Kolesnikova (1975) has suggested that native sour cherries can be divided into two ecological groups:

1. Western Europe group – these have less hardiness than the other group but better eating quality and include varieties such as Griot Ostheim and Kentish.
2. Middle Russian group – these are much hardier than varieties in the first group but have poorer eating quality and are smaller with predominantly dark flesh and juice. This group includes Vladimirskaya and Lyubskaya. These cherry types spread as far west as the Rhine valley and the Balkans.

However, Hillig and Iezzoni (1988) argue that there are not two distinct groups but rather a continuous range of morphological variation between the two extremes.

The severe winter cold experienced in many parts of Europe meant that natural selection undoubtedly favoured those types within Kolesnikova's second group, i.e. those with more hardiness, possibly from *P. fruticosa* parentage. In contrast, in the parts of Europe experiencing milder climatic conditions, people's desire for improved fruit quality must have favoured selection of less hardy but superior-sized and flavoured types, possibly with more of the *P. avium* parent's traits in their blood.

The Maraschino cherry (*P. marasca* of some authorities), considered by many to be a race of the sour cherry, is found wild in the parts of Yugoslavia previously known as Dalmatia.

1.2.3 Subdivisions or groupings within sweet and sour cherries

In 1860, Robert Hogg in his *Fruit Manual* divided the cherry species into eight groups which he called races. These were as follows.

1. Black Geans – sweet, heart-shaped fruits with tender and dark-coloured flesh.
2. Amber Geans – pale-coloured sweet cherries with tender, light yellow and translucent flesh and skin.
3. Hearts – sweet cherries, dark in colour, with some Bigarreau characteristics, but flesh intermediate in texture between the latter and that of the Geans.
4. Bigarreaux – sweet cherries with light-coloured, mottled skin and hard, crackling flesh. These are the Duracines of earlier literature.
5. Dukes – dark, red-skinned fruits with semiacid juice. Possibly hybrids between sweet and sour cherries.
6. Red Dukes – differ from 5 in having uncoloured juice and palish red skin, which, like the flesh, is translucent.

7. Morellos – long, slender, pendant shoots, dark-coloured fruit with acid, dark-coloured juice.

8. Kentish – sour, acidic cherries, pale red in colour of which Kentish and Montmorency are the types.

Hogg's ranking is further explained in Table 1.1.

Table 1.1. Subdivisions within the cherries according to Hogg (1860).

Geans (including all of the sweet cherry types)

1. Fruit obtuse, heart-shaped; flesh tender and melting. These were divided into:
   (a) Black Geans – flesh dark, juice coloured. Examples are Knight's Early Black and Waterloo;
   (b) Amber Geans – flesh pale, juice uncoloured. Examples are Early Amber and Ohio Beauty.
2. Fruit heart-shaped, flesh half tender, firm or crackling:
   (a) Hearts – flesh dark, juice coloured. Examples are Black Tartarian and Büttners Black Heart;
   (b) Bigarreaux – flesh pale, juice uncoloured. Examples are Bigarreau Napoleon, Elton and Florence.

Griottes (including the sour and Duke cherries)

1. Branches upright, occasionally spreading, leaves large and broad:
   (a) Dukes – flesh dark, juice coloured. Examples are May Duke and Royal Duke;
   (b) Red Dukes – flesh pale, juice uncoloured. Examples are Belle de Choisy and Late Duke.
2. Branches long, slender and drooping, leaves small and narrow.
   (a) Morellos – flesh dark, juice coloured. Examples are Early May, Morello and Osteim;
   (b) Kentish – flesh pale, juice uncoloured. Examples are Kentish, Flemish and All Saints.

Most authorities now adopt a rather simpler scheme and divide sour cherries into two groups, the Amarelles and the Morellos. Amarelles are those cultivars with pale red fruits flattened at the ends and uncoloured juice (e.g. Montmorency); these are those also referred to as Kentish or *cerisier commun*. The Morellos or Griottes have, in contrast, dark spherical fruits and dark-coloured juice (e.g. Schattenmorelle). Linnaeus (1753) referred to these two groups as *P. cerasus capronia* (Amarelles) and *P. cerasus austera* (Morellos). The name Amarelle may have been derived from a similar word in Latin meaning bitter, while Morello is said to derive from the Italian for black and Griotte from the French, *aigre*, meaning sharp.

A further group of sour cherries are those with small fruits and strong-flavoured juice which are grown in Yugoslavia for maraschino production. In the past these have often warranted a separate specific grouping under *P. marasca* Jacq.

The sweet cherries are now usually split into three groups: the Mazzards, often wild types with small inferior fruits of various shapes and colours; the Hearts, Geans or Guignes, with soft-fleshed fruits (e.g. Governor Wood, Roundel and Black Tartarian); and the Bigarreaux with hard-fleshed, heart-shaped, light-coloured fruits (e.g. Windsor, Napoleon and Yellow Spanish).

Linnaeus (1753) previously gave the Hearts group the varietal name Juliana, and De Candolle called them *Cerasus juliana*. The Bigarreaux are similarly *P. avium duracina* of Linnaeus, *Cerasus duracina* of De Candolle, *P. avium decumana* of Koch and Roemer's *Cerasus bigarella*. Some suggest these Bigarreaux originated from a cross between the 'true sweet Gean' and the slightly sour 'Glass cherry', but there is no evidence to support this idea.

A fourth group, the Dukes, with heart-shaped, acid or subacid fruits (May Duke, Reine Hortense and Belle de Choisy), is occasionally classified with the sweet cherries. However, the Duke types are more often considered to be hybrids between sweet and sour cherries. They are tetraploids which are thought to have arisen following the pollination of sour cherry by an unreduced (2*n*) gamete of a sweet cherry. Such unreduced pollen is produced at low frequencies by a number of cherry cultivars. Known in France as Royales, the Duke cherries of England, named on account of one of the oldest varieties in the group, May Duke, have also been classified as *P. avium regalis* Bailey, *Cerasus regalis* Poiteau and Turpin or *P.* × *goudouinii* Rehd. Fruit set on Duke types is generally low in comparison with sweet or sour cultivars on account of meiotic disturbances.

1.2.4 Other cherry species cultivated for their fruits

Apart from the two principal cherries of commerce, *Prunus avium* and *P. cerasus*, several other species of cherries are occasionally cultivated for their fruits. Brief descriptions of these are given below.

More often, however, these and other *Prunus* species within the cherry groupings are grown for their ornamental value. Of more significance to the grower of cherries is that many other *Prunus* species and hybrids have been tested and used by horticulturists as rootstocks for sweet and sour cherries. This use is discussed fully in a subsequent chapter. Lists of those *Prunus* species which are usually classified as cherries are given in the Appendices.

The Mahaleb or Perfumed cherry: Prunus mahaleb L.

The Mahaleb is a native of all southern Europe as far north as central France, southern Germany, Hungary and the Caucasus region. The species is shallow-rooted and seems to prefer sandy or gravelly soils. It produces a tree of medium stature with white bark which, together with the leaves and wood, is strongly scented. The wood when dried is hard, brown-veined and takes a good polish. It is used for cabinet making in France and for manufacturing twisted stems for pipes in Austria. The tree bears black or yellow fruits, which ripen in July. Mahalebs are popular rootstocks for sweet and sour cherries in many parts of the world. Several subspecies, differing mainly in their leaf size, shape and degree of hairiness, have been distinguished in Hungary (Terpo, 1968). These are ssp. *simonkai*, ssp. *cupiana* and ssp. *mahaleb*.

The Siberian, Ground or Steppe cherry: Prunus fruticosa Pall. (Cerasus/Prunus chamaecerasus *Lois*)

The Steppe cherry is a small (1–3 m high) spreading and suckering shrub which originates in western and central Asia (Watkins, 1976), but which is also found in parts of northern Europe, including Germany. It bears umbels of white flowers, 2.0 cm in diameter, in May and its round, reddish, pea-sized fruits of extreme acidity ripen in August. This species is closely related to both the sweet and sour cherries. Kolesnikova (1975) has separated the Steppe cherries into three ecotypes; the middle Volga, Urals and western Siberian groups.

P. × *eminens* Beck (occasionally listed as P. *reflexa* or *Cerasus intermedia*) is believed by some authorities to be a hybrid between P. *fruticosa* and P. *cerasus*. It has leaves with deeper toothing, flowers with shorter stalks and forms trees of stronger habit than P. *fruticosa*.

Studies on the natural populations found in Hungary suggested that the pure species frequently grows in populations together with several different and related hybrids. These are generally believed to be hybrids between the Steppe and sweet cherries (similar to *Cerasus* × *mohacsyana* (Karp) Janch.). Other natural hybrids, with P. *fruticosa* as one parent, such as *Prunus* × *stacei* (= P. *fruticosa* Pall. × P. *cerasus* L. × P. *avium* L.) or P. × *javorkae* (Karp.) Janch. (= *Cerasus fruticosa* × *Cerasus mahaleb*), have also been described.

The Rocky Mountain cherry: Prunus besseyi *Bailey* (P. pumila *var.* besseyi *Waugh)*

This dwarf shrub is a native of the dry plains to the east of the Rocky Mountains in the USA. It has glabrous branches and grey/green oval leaves 2.5–6.0 cm long. The small white flowers are formed in clusters of two to four in the axils of the previous year's shoots. The fruits, which were a traditional food source for Indians and settlers in the USA, are oval, about 2.0 cm long and usually black, covered with a purplish bloom. It fruits only occasionally in northern Europe. Despite its vernacular name, the Rocky Mountain cherry is usually not graft-compatible with sweet or sour cherries and is usually classified within the *Microcerasus* subgenus or section of *Prunus*, having closer relationships with the plums.

A hybrid between P. *besseyi* and P. *watsonii*, which is called P. × *utahensis* Koehn, produces blue/black fruits with considerable bloom. Another hybrid with P. *besseyi*, or possibly with its close relative P. *pumila* as one parent, is the ornamental species P. × *cistena* (Hansen) Koehn, which is thought to have originated from a cross with P. *cerasifera* Pissardii. Its leaves are crimson in colour when young, ageing to purple; occasionally it forms small cherry-like fruits that are dark purple in colour.

The Nanking cherry: Prunus tomentosa

Originating primarily from central Asia, but also found in Tibet and China, this species is cultivated in China and Japan for its fruit. One of its merits is its extreme hardiness. Hedrick *et al.* (1915) considered the fruits tasty and the plant worthy of being brought into fruit cultivation and improved. This is one of the few esculent cherries which originate in Asia. P. *tomentosa* is usually classified within the *Microcerasus* subgenus or section.

The False cherry tree: P. pseudocerasus Lindl.

This species was introduced to Europe from China in 1819. It has white flowers, formed in March/April, and pale red fruits, which ripen in June. It is characterized also by the visible root initials which are often formed at the nodes on young stems and from which it can be readily propagated. The species, grown extensively as the Yung To in China, was, in Victorian times, 'forced' in pots under glasshouses for early fruit production in Britain.

The Dwarf cherry tree: P. pumila

This procumbent shrub, which was introduced to Great Britain in 1756, forms white flowers in May and black fruits in July. It originates in the low-lying ground of Pennsylvania and Virginia. It can make a good ornamental tree if grafted high on a suitable rootstock. Like *P. besseyi*, *P. pumila* is more closely related to the plum than to the 'true' cherry species.

Prunus humilis *Bunge*

This dwarf, shrubby cherry, a close relative of *P. glandulosa*, is cultivated in northern China for its fruits. Its downy shoots distinguish it from *P. glandulosa* and several other closely related species. The pale pink flowers are followed by brightly coloured, red fruits, 1.25 cm in diameter, which, although rather acid when grown in northern Europe, are still edible. Like *P. besseyi*, this species is a closer relative of the plums than of the 'true' cherries.

1.3 Evolution of Sweet and Sour Cherries as Cultivated Crops in Europe and North America

Cherries were first utilized by humans, possibly by nurturing trees growing in the wild, long before any written records were made of fruit culture. The fruits were definitely eaten by our prehistoric ancestors, as evidenced by the preserved cherry stones found on archaeological excavations. It is likely that, from the time humans first began to turn their attention from hunting to cultivating their food, cherries were nurtured as a supplementary food source. Kolesnikova (1975) believes that the first evidence of cherries being used as a food source dates from 5000–4000 BC.

Hedrick *et al.* (1915) contend that the Greeks were the first to domesticate cherries, as they were mentioned by Theophrastus more than 300 years before Christ. However, the Greeks may not have used the cherry for its fruits but for its fine hard wood, or its gum as a medicinal remedy for alleviating coughing. It is interesting to note how horticultural fashions move in cycles, for renewed interest has recently been shown in sweet cherry as a potential timber crop in northern Europe.

Although cultivated for more than 2000 years, the cherry has remained, at least until the last 100 years, a crop grown only around the home and for home consumption. This is attributable to the perishability of its fruits and the poor systems for transport of fruits which were in operation until recent times. This lack of large-scale production may also explain the lack of literature devoted to the culture of cherries. Not until Hedrick *et al.*'s *Cherries of New York* appeared in 1915 was a book published which was devoted entirely to the sweet and sour cherries.

Breeding of cherries by controlled hybridization has only been practised by horticulturists since the eighteenth century; prior to that, improved varieties were obtained by selection procedures from within locally grown landrace types. Many of the sour cherries still popular in northern and eastern Europe, types such as

Cigany, Pandy (Koröser), Oblácinska, Stevnsbaer and Vladimirskaya, are of this landrace type and there are many slightly differing clones of each landrace.

1.3.1 Cherry culture in Britain

Roman times – cherries for the occupying legions

Cherries are thought to have been planted and cultivated first in Kent, the most southeasterly county of England, during the Roman occupation, between AD 40 and 60. The Caesars are said to have favoured this part of the country, although for what reason is not entirely clear. It may have been its marginally warmer/drier climate compared with much of the rest of Britain or alternatively its relative freedom from marauding bands of indigenous locals.

The fifteenth and sixteenth centuries – Henry VIII's influence

Many historians contend that the varieties initially introduced were then gradually lost to cultivation during the ensuing Dark Ages, which followed the Roman withdrawal from Britain. They argue that it was not until the sixteenth century that cherry culture was revived following the reintroduction of plants from abroad by Richard Harris, principal gardener to King Henry VIII (1509–1547). This seems unlikely and indeed there is considerable evidence to suggest that it is not true. In the poem 'Lick-penny' (Lackpenny) written by Lydgate in 1415, reference is made to the hawking of cherries on the streets of London:

> Hot pescode began the cry
> Straberys ripe and cherries in the ryse.

'In the ryse' is believed to refer to the common practice of selling the fruits attached to their short branches. Also, Gerarde (1597) argued that one of Harris's introductions, the Flanders cherry, was no different from the existing 'English Cherrie Tree in stature or in forme'. Parkinson (1629), writing a little later, believed that 36 varieties were in cultivation during the reign of Henry VIII's father, i.e. during the reign of King Henry VII (1485–1509).

All of this evidence casts considerable doubt on the suggestion that Henry VIII and his gardener Harris reintroduced the cultivated cherry to Britain. A more likely explanation is that between the middle of the first century and the end of the fifteenth century little or no improvement was made to the varieties introduced by the Romans; some of these may have been lost, others undoubtedly deteriorated. It is interesting to note that the variety Montmorency was included in Parkinson's list, and is still included in a list of recommended varieties published almost 100 years later by Langley (1729). Although widely grown in the USA and previously in France, this variety never achieved great popularity in Britain.

There is little doubt that, with encouragement from Henry VIII, Harris did introduce a number of new, improved varieties to Britain. One suspects that many of the existing varieties had declined in health and performance and that an introduction of new selections would have done much to improve the productivity and quality of British cherries. One motive for these introductions may have been the high value of cherries at the time. One orchard of 32 acres produced fruit which is reported to have realized £1000 in 1540. If it is to be believed, this is a remarkable sum when one considers that land was leased at the time for about 1 shilling (£0.05) per acre per year. Harris, like the Romans, introduced his new varieties to Kent, more specifically to the area around the village of Sittingbourne; Britain's small cherry industry is still concentrated around this same village some 450 years later.

Harris and other cherry enthusiasts that followed him imported varieties mainly from Italy, but also from Flanders and Spain. One story, probably apocryphal, is that one of the oldest varieties known, Lukeward, is so named after Luke Warde, who first brought it back from Italy. The names of other varieties of the time, such as Naples and Spanish, clearly indicate their origins.

The seventeenth century – fruits for the Stuarts

A hundred years or so later cherries were still held in very high esteem at the court of King Charles I. In one garden owned by his queen and situated at Wimbledon, more than 200 cherry trees were established in 1649. The famous song 'Cherry-ripe' by Herrick is thought to date from about this time, while cherries were mentioned by Shakespeare even earlier, in *Midsummer Night's Dream*, written in about 1600.

By 1685 the following list of varieties was recommended in a book entitled *The Complete Planter and Cyderist*. They are listed in order of their ripening.

May Cherry
Duke (or May Duke)
Arch-Duke
Flanders
Red Heart
Lukeward
Spanish Black
Naples
Carnation
Amber
Purple
Bleeding Heart (so named on account of the tear-like protuberance at the fruit base)
Cluster Cherry (bearing three to five fruits on each stalk)

This same publication also mentions the 'Great Bearing Cherry of Millain' and the 'Morella', both of which were considered excellent for the making of cherry wine and preserves. This represents one of the earliest mentions in Britain of what must be the sour cherries with coloured juice, the Morello-type cherries. Importation of new varieties from the Flanders region also began at about this time.

The eighteenth century – new introductions from abroad

Improvement of cherry varieties in Britain between the seventeenth and nineteenth centuries proceeded at a slow pace, when compared with progress in several other countries. Almost all new varieties gaining popularity were introduced from abroad. Florence, a variety still occasionally grown today, was introduced in about 1700 from Florence (Firenze) in Italy. Similarly, the occasionally used name Gaskins is a corruption of Gascoigne and refers to those types originally obtained from Gascony in France. One story is that several Gaskins were introduced to Rye in Sussex by Jean of Kent at the time when her husband, the Black Prince, was commanding an army in Gascony. Gean, Guigne or Guen are all corruptions of Guienne, another area of France from which cherries were introduced. It is also suggested that the name May Duke is itself derived from Medoc, a region of France.

Richard Bradley (1718) recommended only ten varieties for planting, all of which had appeared in lists proposed by others more than 30 years earlier. Although there are reports of numerous different varieties at this time, many of these, particularly of the Heart type of sweet cherry, were thought to be synonymous with existing varieties, and their reported distinct characteristics attributable to differences in site and soil type. More recent knowledge might lead us to suspect that some of the reported differences could also have been attributable to infection by virus.

Switzer (1724) thought that at most there were only three or four Heart types available at that time, the best of the black Hearts being Orleans, the best red Gascoigne and the best white Lukeward. He also mentions the Morella or Milan cherry, which at this time was much esteemed for its ability to bear very big crops and its tolerance of cool growing conditions such as those of a northerly facing wall. Local sour cherry varieties were also in cultivation at this time; one type popular in Lincolnshire, a region of eastern Britain now thought unsuited to cherry culture, was known as Paradems (Baradems).

Accounts written slightly later (Miller, 1759) again list many of the same varieties recommended almost 60 years earlier, although several new Heart varieties are now included. The often fluctuating economics of cherry growing had, however, taken one of its downturns in Britain at this time. Even the best varieties, such as Lukeward, were thought not worth planting unless land was very inexpensive. The uncertainty of bearing, the problems of picking and, perhaps surprisingly, the small price the fruit realized in the market all made cherries a rather unprofitable crop. Many Kentish orchards were removed at about this time, although some were retained and the fruits used to distil a type of cherry wine.

The nineteenth century – the first cherries bred in Britain

Forsyth (1802) listed the principal 18 cultivars (or sortes, as he called them) of cherries growing in Britain at that time. Many of these are mentioned in the lists cited previously. Others of note were the Hertfordshire cherry, Harrison's Heart, Grassion and Ronalds' Large Black Heart. This last mentioned was said to have been introduced to Britain by a nurseryman named Mr Ronalds of Brentwood in Essex, who obtained it from what was then Circassia. Two others, Fraser's Black Tartarian and Fraser's White Tartarian, were introduced in 1796 from a garden near St Petersburg by an indefatigable plant collector of the time, Mr John Fraser of Chelsea. One variety, Lundies Gean, was even said to be suitable for cultivation in Scotland, an area now considered much too cold for successful cherry cultivation.

By 1820 cherries were again back in economic favour, for fruit orchards were considered to represent the most valuable estates in Kent, and fully productive cherry orchards, or gardens as they were called, were even more profitable than the other fruits. In this period British cherry orchards were frequently inter-cropped with strawberries.

If the list prepared by the Horticultural Society (the precursor of the Royal Horticultural Society) is to be believed, the numbers of varieties available in Britain had expanded considerably by 1826. This details eight types of British wild cherries (used mainly as rootstocks), 111 cultivated varieties of British origin (perhaps more accurately described as being not of recent introduction), eight cultivated varieties of American origin, 15 cultivated varieties of German origin and more than 100 cultivated varieties of French origin. The French varieties grown in Britain were, at this time, split into Cerisiers (49), Griottes (11), Bigarreautiers (22), Guignes (13) and Merisiers (7). Also in cultivation at this time was the 'Allsaints cherry' (*P. semperflorens* De Candolle), and three types of *P. mahaleb*, the Perfumed cherry or St Lucie, with either black or yellow fruits.

Some of the first superior varieties of sweet cherry bred in Britain were those raised by Thomas Andrew Knight in the early 1800s; some of his best known varieties were Black Eagle, Knight's Elton, Waterloo and Early Black. Until the beginning of the nineteenth century, most sweet cherries grown in Britain were of continental origin. The breeding work of Knight and later Thomas Rivers (who bred Early Rivers in 1869) redressed this imbalance.

Strangely, little mention is made of bird damage to cherries until the 1890s, when Wright (1896), in his book *Profitable Fruit Growing*, states:

> Except in the great Cherry districts, where large numbers of trees are established, cherries are the least serviceable of fruits for cottagers to grow. The reason for this: where there are thousand times more cherries than the birds can eat, those they take are not missed, but where the trees are comparatively few, the birds relatively numerous, it is most difficult to prevent their devouring the crops of the sweeter kinds especially.

Perhaps the English taste for eating various species of small birds in earlier times had a moderating effect upon bird populations.

The twentieth century – a crop of fluctuating popularity

In the early part of the twentieth century, it was traditional for Kentish cherry growers to plant large orchards of standard trees spaced very widely apart. As many as 15–20 different varieties were commonly included to provide adequate pollination and a sequence of ripening times. Problems of bird damage, erratic cropping, bacterial canker disease, rain-induced fruit splitting and the expense of managing and harvesting such large trees all contributed to the demise of the cherry as a profitable crop in Britain. From the early 1950s, the area of land devoted to cherry production has slowly but significantly declined. It is only recently, with the introduction of new improved varieties and more dwarfing rootstocks, that cherries have once again begun to make a small revival as a commercial crop in Britain.

1.3.2 Cherry culture in France

Cherries were probably first cultivated in France, as they were in Britain and other European countries, at the time of the Roman occupation. Records indicate cultivation as early as the eighth century, and by 1400 cherries were a most important crop commercially. However, not until the sixteenth century were the many different types of cherry grown classified and given distinct names. Until then they were referred to simply as Guignes, Cerises, Bigarreaux and Merises. In 1582 Charles Etienne and Jean Liebault described four varieties: Bigarreau Blanche, Cerise Heaume, Cerise Noire and Cerise Noire à Coeur. Five more varieties were added to this list by Olivier de Serres in 1600, and in 1628 Le Lectier described 13 varieties. By 1690, according to Merlet, the number of varieties had risen to 21, and approximately 50 years later the list had grown to 34 (Duhamel de Montceaux, 1768). At about this time, cherries were made into a 'paste' and preserved for use in cooking by the French royal households throughout the year.

France continued to be at the forefront of cherry selection and breeding throughout the eighteenth and nineteenth centuries, and many of the varieties introduced to Britain in this period had their origins in France. France was also renowned for 'bottling' cherries in the nineteenth century, the fruits being sterilized by heat before sealing in glass jars.

1.3.3 Cherry culture in Italy

An often quoted story concerning the early development of cherries as a cultivated crop in Europe recounts how they were, supposedly, first introduced into Italy. Between 74 and 66 BC the Roman general Lucullus was at war with Mithridates, King of Pontus. According to the historian Servius, after the capture and destruction by Lucullus of a city named Kerasoum, in what is today Armenia, he carried back to Rome various cherry trees as part of the many spoils of war. Moreover, he is said to have called the trees Kerasus (or Cerasus) after the name of the conquered city.

A more likely explanation is that the city, a maritime town ruled by the Turks at the time, took its name from the abundance of *Cerasus* trees which grew naturally in the vicinity. This latter explanation is supported by the Greek doctor Diphile, a contemporary of Alexander the Great, who was writing 250 years before the birth of Lucullus. He records three different varieties of cherry grown at the time, all of which were called Kerasias. One hundred years before Diphile, another historian, Theophrastus (*c.* 300 BC), described cherries as Kerasos. It seems certain, therefore, that the cherry was known as Kerasos long before Lucullus carried his horticultural 'triumph' back to Rome.

Also, most historians now agree that Lucullus could not have been responsible for introducing cherries to Italy. Only 100 years after his return to Rome from Kerasoum at least ten varieties were being cultivated throughout the Roman Empire; this is much too short a period for all of these to have been bred or selected from just one introduction. What is more likely is that he introduced one or more new varieties, which supplemented those already in cultivation. Although some plant historians have argued that his introduction was the

Morello, the consensus opinion now is that it was a sweet variety. One fact which lends support to this is that the modern Greek for sweet cherry types is *kerasaia*. Whatever the variety was, it obviously took the fancy of Romans both at home and abroad, for only 26 years after its introduction it had, according to Pliny (1793), been distributed throughout the Roman Empire including Britain.

When Pliny wrote his account of natural history, in approximately AD 70, he mentions ten kinds of cherry which he believed to be growing in Italy centuries before the birth of Christ. To those mentioned by Diphile he adds two others, the Macedonian and the Kerrai. He also lists those growing in Italy at the time of writing, which were:

1. Apronia – from Apronius, a Roman judge of Pliny's time, said to be the most red (perhaps the common Griotte);

2. Actia (Lutaria) – the blackest (perhaps Griotte Noire), possibly named after Lutatius Catalus, a contemporary of Pliny and the man who rebuilt Rome after its destruction by fire;

3. Caeciliana – a roundish cherry (possibly a Guigne type);

4. Juliana (Julian) – of pleasant taste but very perishable;

5. Duracine (Pliniana) – best of the firm cherries (probably a type of Bigarreau);

6. Lusitania – said to be the best grown in Picardy (possibly Griotte of Portugal);

7. Macedonian – dwarf trees which were highly esteemed by the Romans (possibly the precocious Griotte Noire);

8. Laurea – grown on rootstock of Laurel and understandably rather short-lived;

9. Tertii Colores – from the banks of the Rhine with black, red and green fruits;

10. Chamaecerasus – (possibly a type of Steppe cherry, *P. fruticosa*).

A popular Italian legend involving cherries dates back to the end of the twelfth century. According to this, St Gerard of Monza wished to spend the night praying in a local church. However, the local priest, who did not know him, told St Gerard that he would give him permission to do this only if he brought him some cherries, which Gerard duly did. However, the legend has it that this all took place in winter, suggesting prescient knowledge of fruit storage techniques beyond us even today!

Baldini (1955), in his study of the cherry varieties grown in the Florence region, cites an anonymous manuscript from the fifteenth century in which are listed several cherry varieties then in cultivation: Marchiane, Moraiole, Duracine, Amarene, Agostane, Di San Giovanni and Acquaiole were all in cultivation, as was the variety Bondi, which is still found growing in the Florence region.

Many Italian paintings, dating from as long ago as 1400, also depict cherries. Two, by Antonello da Messina (1430–1479), show the Holy Virgin offering cherries to the infant Jesus. Cherries are also featured in two paintings by Ghirlandaio (1449–1494) of the Last Supper. Crivelli (1453–1493) and Botticelli (1445–1510) similarly featured cherries in their religious paintings from this period and many other artists, such as Vincenzo Foppa, Bronzino, Mantegna, Schiavone, Pontormo and Caravaggio, have also included cherries in their works.

In 1554, the botanist Pietro Andrea Mattioli, based in Sienna, stated that at that time the cherry was the best-known tree fruit in Italy and the most popular of the 15 varieties in cultivation were Marchiane and Duracine. Mattioli also concurred with contemporary doctors in pronouncing cherries to be good for the digestion.

By the end of the seventeenth century Micheli (1679–1737) lists 37 varieties which were grown for His Royal Highness the Grand Duke of Tuscany. Amongst these, some, such as Lustrina, Maggese, Marchiana, Napoletana and Turca, are still in cultivation today.

1.3.4 Cherry culture in the Netherlands, Germany and central Europe

Trade in cherries can be traced back to some of the first fruit markets set up in the fourteenth century in what are now the Netherlands, Belgium and Germany. Records suggest that cherries were used for making wine (May Duke), their gum was used for chewing and their stones (pits) for medicinal purposes in both the fourteenth and fifteenth centuries. By the fifteenth century there are indications that cherries were used for the manufacture of jelly and morello beers in the Netherlands.

The first of the German herbals, the *Herbarius* printed in Mainz in 1491, groups cherries into sweet and sour types, indicating that both types were in cultivation at that time. By 1569 the Germans were attempting to give names to their main cherry groups, for in a popular medical herbal of the period, *Gart der Gesunheit*, they were divided into:

- the *Amarellen*, with sour juice, dark red in colour with long stems;
- the *Weichselkirschen*, red cherries with white juice and short stems;
- the *Susskirschen*, red or black sweet cherries with long stems.

Although Germans cultivated cherries more than three centuries ago, it was not until the eighteenth century that they began to give their varieties names. Undoubtedly, it was a crop more widely cultivated in Germany in these early years than in most other European countries. Some of the very oldest cultivars, whose origins are unknown, are still to be found growing in Germany. One of these, the Large Black Bigarreau (*Grosse Schwarze Knorpel*), is still widely cultivated.

Cherries are still the focal point of what is known as the Feast of Cherries in Hamburg, in which troops of children parade the streets festooned with cherries to commemorate a victory obtained in a rather strange way. In 1432, at a time when the Hussites were threatening the city of Hamburg with immediate destruction, one of the citizens, named Wolf, proposed that all of the children in the city aged from 7 to 14 should be clad in mourning dress and sent as supplicants to the enemy. Procopius Nasus, chief of the Hussites, was so touched by this act that he not only received the youngsters but regaled them with cherries and other fruits and promised to spare the city. The children returned crowned with leaves, holding cherries and crying 'Victory'. If it shows nothing else, this story indicates that cherries were probably in plentiful supply in northern Germany in the fifteenth century.

At the beginning of the eighteenth century, when cherry cultivation was spread throughout Europe, cherries were planted alongside many of the major roads in Germany. At about this time, at Guben, in central Europe, there was formed an association of arboriculturists who dedicated themselves to the selection of new improved varieties. Amongst those selected were Fromm's Herzkirsche, Bigarreau Jaune Donissen, Dönissens Gelbe, Drogens Gelbe, Noir de Guben and Germersdorf (thought to be synonymous with Schneiders Späte Knorpel); the last-named is still the premier variety cultivated in what is now Hungary.

In the same period Büttner, living in Halle, selected Büttner's Yellow, and Büttner's Late Red. It was about this time that the need for a more rational classification system for cherry varieties was recognized and in 1819 Baron von Truchess-Heim, of Battenbourg, classified 75 varieties which he later offered to the Director of the Botanic Garden in Paris. Some time later the list grew to 100 varieties and Oberdieck continued the work of von Truchess-Heim by classifying 149 varieties between 1855 and 1866 (Obeodiek and Lucas, 1875).

Germany was probably the origin of one of the most commonly planted sweet cherry varieties, Napoleon. Although its precise origin is not known, it was originally grown as Grosse Lauermanns Kirsche in Germany as early as 1791.

Further breeding and selection was obviously stimulated in the nineteenth century, for by 1877 German catalogues listed 232 varieties and by 1945 this number had increased even more, making Germany and the USA the world's most progressive postwar centres of cherry variety development. In the last century cherry culture in Germany increased rapidly, with the planting of large orchards, often near to the markets of large towns. At about this time a series of new cultivars was raised from seeds collected from cultivars growing in pomological collections. Commercially important cultivars were raised close to Werder at Potsdam, at Guben and in the central and south German production areas. Particularly influential were the Alten Landes near Hamburg, where, over approximately 100 years, cultivars were developed with characteristics very suited to the local climatic conditions. Many of these early German selections are still cultivated and are named after the breeder or the area of their development.

Distribution of cultivars was erratic and uncontrolled at this time and many were given a different name in each area of production, resulting in a nomenclatural muddle. To sort this out Heilmann (1867–1948) collected cultivars from as many of the production areas as possible and planted them in comparative trials at Diemitz in Halle. In 1928 this valuable collection was transferred to Blankenburg in Harz. This work was continued by Groh, Krümmel and Sante and has culminated in the founding of the stone fruit section of Deutsche Obstsorten (German Fruit Varieties). The pioneering initiative of this group was the establishment of selected mother trees, verified true to type. This was especially important for cultivars such as Hedelfingen and Grosse Schwarz Knorpel, where cherries of quite different characteristics had come to be grown under these names.

Even quite recently, local selections or landraces have been used in breeding programmes to produce new sweet cherry cultivars. In the lower Elbe district, near Hamburg, old local cultivars such as Allers Späte and Rule have been crossed to produce new and promising cultivars (Zahn, 1985).

1.3.5 Cherry culture in the USA and Canada

Neither *P. avium* nor *P. cerasus* is native to the USA and the only cherries available to the American Indians and early settlers from Europe were of native types, such as the racemose Choke cherries (*P. virginiana*) or species such as *P. besseyi* or *P. pumila*. None of these would today be classified within the true cherries (subgenus *Cerasus*), but would be placed within the subgenus *Microcerasus*, with closer relationships to the plums.

P. virginiana (Choke cherry) is the New World equivalent of the European *P. padus* (Bird cherry), and like the latter produces racemes of astringently flavoured and small fruits. Although now mainly cultivated for ornamental use, Choke cherry types bearing fruits of less unpleasant flavour have occasionally been found. Its fruits are usually red but sometimes amber in skin colour, as in *P. virginiana* var. *leucocarpa* Watson. The cultivated Choke cherries were reported to produce fruits up to 12 mm in diameter and were less astringent than the wild forms. Both wild and cultivated forms were usually used to produce a jelly, as most of the astringency is lost on cooking.

A very similar plant, previously called *P. demissa*, is the plains and wet lands representative of the Choke cherry group. This was at one time used for marmalade production in California and also as a rootstock for garden cherries. It differs only slightly in its leaf characteristics from *P. virginiana*.

P. serotina the wild Black or Rum cherry, is a large tree species native to America which has some similarities to the European sweet cherry *P. avium*. Its fruits are of little value but its wood was often used for cabinet making.

The Dwarf cherry group mentioned by Bailey (1894) was thought at the time of his writing to offer great potential for breeders trying to improve cultivated types of plums and cherries. They were said to be the American congenitors of the European *P. chamaecerasus* (now known as *P. fruticosa*). In 1894 Bailey separated the Western Sand cherry, *P. besseyi* (named after Professor Charles E. Bessey of the University of Nebraska) from the Eastern Sand cherry, *P. pumila*. The latter grows mainly upon rocky and sandy shores in the east of the USA; it is abundant also around the Great Lakes, where it traditionally grows on shifting beds of sand. This very variable species usually bears small and sour-flavoured fruits but occasionally larger and sweeter-fruited forms have been selected. *P. pumila* was once marketed for the medicinal value of its red roots!

P. besseyi was originally found growing from the plains of Manitoba through to Kansas and westward to the mountains of Colorado and Utah. It is bushier and more compact in habit than *P. pumila* and its fruits are larger and much more palatable. Its horticultural potential was first suggested by Fuller in 1867, and in 1888 Gipson reported that the native 'Colorado Dwarf Cherry has fruits especially valuable for pies and preserves and often pleasant to eat fresh'. Both yellow- and black-fruited sorts were described, the former being earlier to ripen and of better flavour. In 1892 a selection was introduced as the 'Improved Dwarf Rocky Mountain cherry' by Charles E. Pennock of Colorado. The breeder recommended that this selection be planted 8 feet (2.5 m) apart, when it would produce bushes only 4 feet (1.2 m) high. It ripened its fruits very uniformly, approximately one month later than the Morello. Its easy rooting from layers and ability to withstand extremely low winter temperatures made it popular in the colder areas of the USA.

Much breeding of *P. besseyi* was carried out at the turn of the century with a view to selecting improved cultivars. Although crosses with another native species, *P. americana*, were attempted, they were not successful. Crosses between *P. besseyi* and *P. hortulana* were successful and one, the Compass cherry, was introduced by Knudson. The Utah Hybrid cherry (*P. utahensis* Dieck), which is a similar hybrid, possibly between *P. besseyi* and *P. watsoni*, was bred and distributed by Johnson. If its parentage is as suggested, the Utah Hybrid cherry is of interest to breeders as one of the few hybrids between plum and cherry species. Its flowers were said to resemble those of plums and its fruits ripened simultaneously between late July and mid-August. Although attempts were made to improve these and other hybrids in breeding programmes, the results proved disappointing and these improved indigenous types were superseded by the introduction and wide distribution of sweet and sour cherry varieties from Europe.

Hedrick *et al.* in their famous book *Cherries of New York*, published in 1915, list 270 varieties of tart or sour cherry and 549 varieties of sweet cherry. This book, the first solely devoted to cherries, contains an excellent account of the origin and development of the cultivated cherries and their introduction to America. Hedrick suggests that the French were probably the first to plant cherries in Nova Scotia, Cape Breton, Prince Edward Island and in the early settlements along the St Lawrence river. These would have been brought as seeds by the early settlers from Provence or Normandy. According to Hedrick *et al.*, Francis Higgison noted that in 1629 the only cherry variety growing in Massachusetts was known as the Red Kentish, indicating some British involvement in the early introductions. There is a report in the same year suggesting that stones of all sorts of fruits were about to be sent to New England by the Massachusetts Company. These were possibly established by the late 1630s when John Josselyn records good orchards of all fruit types.

All the early cherry introductions to America were via seeds, and not for another 100–150 years were potted trees of superior varieties introduced. In 1726 Paul Dudley writes that although the apple varieties were equal to those in England and peaches better, their cherries were not so good as those in Kent and they had no Dukes or Hearts, save one or two in specific gardens.

In New York, settlement and agriculture developed more slowly than in Massachusetts. Peter Stuyvesant, an early governor of New York, was the first to import and plant fruit orchards. He spread materials from his own orchards up the Hudson river. Within a century cherries and other fruits were cultivated by the white settlers established up all the tributaries of the Hudson and also, in perhaps a less organized fashion, by the indigenous Indians. A nursery established in 1730 at Flushing, Long Island, by Robert Prince was, by the latter half of the eighteenth century, selling several named varieties of cherry. It was on the site of this nursery that the Linnaean Botanic Garden was later established. William Prince, an ancestor of the founder, prepared a list of about 20 cherry varieties available in 1804 for Willich's *Domestic Encyclopaedia*.

It is probable that both sweet and sour cherry varieties were introduced by the early settlers, although there are few written accounts describing these from the earliest times. Downing (1845) listed 54 varieties of Heart and Bigarreaux types, six Dukes and 11 Morellos then growing in the USA. Morello seedlings were suggested as possible rootstocks at this time. Certainly, accounts indicate that cherries were used to colour whisky and blend tobacco in the USA as early as the eighteenth century.

In *Hooper's Western Fruit Book*, published in 1857 in Cincinnati, 52 varieties were mentioned. Among those considered the best were Black Eagle, Carnation (a variety of Morello), Büttner's Yellow and Hildesheim (from Germany), Hortense (a Duke type from France), Large Heart-Shaped and May Duke. Several cultivars raised by Kirkland in Cleveland, Ohio, are mentioned, including Cleveland, Black Hawk, Brant, Delicate, Doctor, Early Prolific, Governor Wood, Kirkland's Mary, Kirkland's Mammoth, Osceola, Pontiac, Powhatton, Red Jacket, Rockport and Tecumseh. Although authorities at the time argued that Kirkland's varieties were better suited to USA conditions than many of the foreign imports, only Governor Wood has survived as a commercial variety, and this only in Britain up until the 1960s. Hooper also refers at this time to dwarf cherries which were raised by grafting low on to *P. mahaleb* stocks.

The distribution of cherries from the sites of early introduction on the northeast coast was first into states such as Virginia and Carolina. Cherry varieties were then moved into the Midwest and finally to the Pacific West. Franciscan monks are thought to have first introduced cherries to California, but it was not until the time of the gold rush in 1849 that their culture really expanded. However, it was the more permanent settlers in Oregon who fully established cherries in the Pacific coastal states. Henderson Lewelling is said to have brought about 300 trees planted in boxes of soil from Iowa, which he planted south of Portland at Milwaukee. The label of one of his cherries was lost and it was renamed Royal Ann; this was the variety Napoleon, named as such in Europe since 1820. Lewelling's son Seth continued the nursery business and raised new varieties himself. Many of these are still well known, for example Republican, Lincoln, Willamette Seedling and, most famous of all, Bing. His breeding and selection work was extremely successful and stimulated others to follow suit. It was from this later breeding that the popular variety Lambert emerged.

References

Bailey, L.H. (1894). The native dwarf cherries. *New York Agriculture Experimental Station Bulletin* 70, 259–265.

Baldini, E. (1955) [A contribution to the study of the cultivars of sweet cherry in the province of Florence. I. Historical notes and research on floral biology. II. Description of the crop]. *Rivista Ortoflorofrutticoltura Italiana* 39, 105–131, 233–262.

Bradley, R. (1718) *New Improvements of Planting and Gardening, Both Philosophical and Practical*. W. Mears, London.

De Candolle, A. (1884) *Origin of Cultivated Plants*. Kegan Paul, London.

Downing, A.J. (1845) *The Fruits and Fruit Trees of America*. Wiley and Putnam, New York.

Duhamel de Montceaux, H.L. (1768) *Traité des arbres fruitières*. Paris.

Forsyth, W. (1802) *A Treatise on the Culture and Management of Fruit Trees*. Nichols, London.

Gerard, J. (1597) *The Herball or General Historie of Plantes*. London.

Hayashi, Y. (1971) *Ornamental Trees and Shrubs of Japan*. Japan.

Hedrick, U.P., Howe, G.H., Taylor, O.M., Tubergen, C.B. and Wellington, R. (1915) *The Cherries of New York*. New York Agricultural Experimental Station Report for 1914. New York.

Hillig, K.W. and Iezzoni, A.F. (1988) Multivariate analysis of a sour cherry germplasm collection. *Journal of the American Society for Horticultural Science* 113, 928–934.

Hogg, R. (1860) *The Fruit Manual*. Cottage Gardener Office, London.

Hooper, E.J. (1857) *Hooper's Western Fruit Book*. Moore, Wilstach, Keys and Company, Cincinnati.

Horticultural Society of London (1826) *Catalogue of Fruits Cultivated in the Garden of the Horticultural Society of London at Chiswick*. William Nichol, London.

Ingram, C. (1948) *Ornamental Cherries*. Country Life, London.

Kawasaki, T. (1982) Classification of Japanese cherry trees. In: Honda, M. and Hayashi, Y. (eds) *Manual of Japanese Flowering Cherries*. The Flower Association of Japan, Tokyo, pp. 1–50.

Koehne, E. (1912) *Die geographische Verbreitung der Kirschen Prunus subgenus Cerasus*. Mittelungen, Germany.

Kolesnikova, A.F. (1975) *Breeding and Some Biological Characteristics of Sour Cherry in Central Russia*. USSR Priokstoc Izdatel'stvo, Orel.

Kubota, H. (1978) Notes on Japanese cherries. *Journal of Geobotany* 24(4).

Linnaeus (1753) *Specie plantarium*, 1st edn. Laurentii Salvii Holmiae.

Miller, P. (1759) *The Gardeners Dictionary*, 17th edn. John Rivington, London.

Oberdiek, J.G.C. and Lucas, E. (1875) *Illustrated Handbook of Fruit Varieties*. Verl. Ulmer, Stuttgart.

Olden, E.J. and Nybom, N. (1973) On the origin of *Prunus cerasus* L. *Hereditas* 59, 327–345.

Parkinson, J. (1629) *Paradisi in sole paradisus terrestris*. London.

Pliny (1793) *Plini historis naturalis*. Hardouins Edition, Paris.

Rehder, A. (1947) and (1974) *Cultivated Trees and Shrubs Hardy in North America*, 2nd edn. Macmillan, New York. (A revised edition was printed in 1974.)

Schmidt, H. (1973) Investigations on the breeding for dwarf rootstocks for sweet cherries. I. Flower differentiation and development in *Prunus* species and forms. *Zeitschrift für Pflanzenzucht* 70, 72–82.

Switzer, S. (1724) *The Practical Fruit Gardener*. Thomas Woodward, London.

Terpo, A. (1968) A sajmeggy (*Cerasus mahaleb* (L.) Mill.) taxonomiai problemai es a gyakorlat [Problems of *Cerasus mahaleb* (L.) Mill.]. *Szolo-es Gyumolcstermesztes, Budapest* 4, 103–131.

Theophrastus (*c.* 300 BC) *Enquiry into Plants, I and II* (Trans. 1968, L.A. Hort, London.)

Truchess-Heim, C.F. (1819) *Systematische Classification und Beschrechung der Kirschensorten*. Stuttgart.

Watkins, R. (1976) Cherry, plum, peach, apricot and almond. In: Simmonds, N.W. (ed.) *Evolution of Crop Plants*. Longman, New York, pp. 242–247.

Wright, J. (1896) *Profitable Fruit Growing*. J.S. Virtue, London.

Zahn, F.G. (1985) The cultivation of sweet cherries in Jork/FRG. *Acta Horticulturae* 169, 85–89.

Appendix

A: Classification of the *Prunus* subgenus *Cerasus* as proposed by Collingwood Ingram in 1948

Section 1 *Eucerasus*
P. *avium* Linn
P. *cerasus* Linn
P. *fruticosa* Pallas

Section 2 *Sargentiella*
P. *sargentii* Rehder
P. *serrulata* Lindley
P. *serrulata* var. *spontanea*
   (Maxim) Wilson
P. *serrulata* var. *hupehensis*
   (Ingram) Ingram
P. *serrulata* var. *pubescens*
   Wilson
P. *speciosa* (Koidz) Ingram
P. *nipponica* Matsumura
P. *kurilensis* Miyabe
P. *concinna* Koehne
P. *conradinae* Koehne
P. *campanulata* Maxim.
P. *cerasoides* var. *rubens*
   Ingram

Section 3 *Cyclaminium*
P. *cyclamina* Koehne
P. *dielsiana* Schneider

Section 4 *Confusicerasus*
P. *pseudocerasus* Lindley
P. *dawyckensis* Sealy

Section 5 *Microcalymma*
P. *subhirtella* Miquel
P. *subhirtella* var. *ascendens*
   Wilson
P. *subhirtella* var. *rosea*
   (Miyoshi) Ingram
P. *subhirtella* var. *stellata*
   Ingram
P. *subhirtella* var. *Fukubana*
   Makino

P. *subhirtella* var. *grandiflora*
   Ingram
P. *subhirtella* var. *pendula*
   Tanaka
P. *subhirtella* var. *pendula* f.
   lancelata Ingram
P. *subhirtella* var. *pendula* f.
   plena rosea (Miyoshi)
   Ingram
P. *subhirtella* var. *autumnalis*
   Makino
P. *changyangensis* (Ingram)
   Ingram

Section 6 *Magnicupula*
P. *rufa* Hooker f.
P. *mugus* Hand.-Mazz.
P. *latidentata* var. *pleuroptera*
   Ingram
P. *latidentata* var. *gracilifolia*
   Ingram
P. *setulosa* Batalin
P. *canescens* Bois.
P. *apetela* Franchet et Savatier
P. *serrula* Franchet

Section 7 *Phyllocerasus*
P. *litigiosa* Schneider
P. *pilosiuscula* Koehne

Section 8 *Phyllomahaleb*
P. *maximowiczii* Ruprecht
P. *maakii* Ruprecht
P. *pleiocerasus* Koehne
P. *macradenia* Koehne
P. *pennsylvanica* Linn.
P. *emarginata* (Hook.)
   Walpers
P. *mahaleb* Linn.

B: Classification of the subgenus *Cerasus* Pers. according to Rehder (1974)

Section *Microcerasus* Webb.

P. *besseyi* Bailey
P. *glandulosa* Thunb.
P. *humilis* Bge.
P. *incana* (Pall.)
P. *jacquemontii* Hook.
P. *japonica* Thunb.
P. *microcarpa* C.A. Mey.
P. *prostrata* Labill.
P. *pumila* L.
P. *tomentosa* Thunb.

Section *Pseudocerasus* Koehne

P. *campanulata* Maxim.
P. *cerasoides* D. Don.
P. *incisa* Thunb.
P. *kurilensis* (Miyabe) Wils.
P. *nipponica* Matsum.
P. *sargentii* Rehd.
P. *serrulata* Lindl.
P. *sieboldii* (Carr.)
P. *subhirtella* Miq.
P. *yedoensis* Matsum.
P. *canescens* Bois*

Section *Lobopatulum* Koehne

P. *cantabrigiensis* Stapf.
P. *involucrata* Koehne
P. *pseudocerasus* Lindl.
P. *dielsiana*

Section *Cerasus* Koehne

P. *avium* L.
P. *cerasus* L.
P. *fruticosa* Pall.

Section *Mahaleb* Focke

P. *emarginata* (Hook.) Walp.
P. *mahaleb* L.
P. *pennsylvanica* L.
P. *prunifolia* (Greene) Shafer

Section *Phyllocerasus* Koehne

P. *pilosiuscala* Koehne

Section *Phyllomahaleb* Koehne

P. *maximowiczii* Rupr.
P. *pleiocerasus* Koehne

---

* Schmidt (1973) has suggested that P. *canescens* is more appropriately classified within the Section *Cerasus*.

2

World Distribution of Sweet and Sour Cherry Production: National Statistics

A.D. WEBSTER[1] and N.E. LOONEY[2]

[1]Horticulture Research International, East Malling, West Malling, Kent ME19 6BJ, UK; [2]Agriculture and Agri-Food Canada, Research Centre, Summerland, BC V0H 1Z0, Canada

2.1 Introduction

Both the sweet and the sour cherry are deciduous trees originating around the Caspian and Black Seas. As such, they thrive best in areas with a temperate or Mediterranean-type climate. Although attempts have been made, often with considerable success, to cultivate sweet cherries in both hot–arid and subtropical zones of the world, more extensive horticultural inputs are needed in these areas to compensate for the climatic limitations. In particular, defoliation treatments are often necessary to impose an essential period of dormancy during the seasonal cycle of growth and cropping.

Although cherries are grown in most countries of the world offering suitable environmental conditions, consumption of cherries per capita shows huge differences between these countries. Countries such as Hungary consume almost twice the amount of sweet cherries as Germany and the former Yugoslavia, while Poland and Turkey consume only about half as much again. Consumption of sour cherries is also very different between countries, with Yugoslavs, Germans and Hungarians again consuming most. The reasons for these differences are not fully understood but are undoubtedly rooted in the culinary traditions of the respective countries and the price of cherries relative to incomes.

For those countries where it is not possible to separate consumption of sweet from that of sour cherries, Switzerland emerges as a major consumer (~ 6.0 kg), with many other nations consuming between 2 and 3 kg per capita per year. British people, in contrast, each consume only 0.2 kg cherries (both sweet and sour) per year.

The market for cherries in many western European nations and in Japan is undersupplied, unless fruits are imported from other nations. Not only do these imports supplement shortfalls in home production but they also serve to extend the short natural harvesting and marketing season for fresh sweet cherries. Moving fruit from the southern to the northern hemisphere is the most extreme example of this and is typified by the expanding exports of cherries from New Zealand to Japan at, or just after, the Christmas/New Year vacation period. Less dramatic extensions to the season are achieved by utilizing latitude and altitude differences in the areas of production within the northern hemisphere. Latitude alone greatly alters the harvesting periods within northern Europe. In Spain, southern Italy and Greece (40°N) varieties generally ripen between early May and early June, while 5° further north in southern France, northern Italy, the former Yugoslavia and Bulgaria harvesting generally occurs 2 weeks later. At 50°N, in northern France, Germany, the Czech Republic and southern Poland, mid/late June to mid-July is the usual harvesting season, while in Scandinavia the season may extend until mid-August. By transporting fruits within the European Economic Community, a $3\frac{1}{2}$-month cherry season is therefore possible.

Even within a single USA state, namely Washington State, and with a single cultivar, Bing, a 10-week marketing period is easily possible. This is achieved by combining the effects of altitude on ripening times, use of sprays of gibberellic acid to delay ripening and improve storage potential, and cool storage.

Cherries are still perceived as a luxury fruit in many countries, and affluent societies are very willing to pay large prices per kg or even per fruit (e.g. Japan) for fruit supplied out of season. These potential rewards to growers have been instrumental in stimulating research into production very early in the season in countries such as France and Belgium by growing prechilled trees under heated glass.

2.2 Environmental and Other Limitations on Cherry Production

Climatic conditions are the most important limitation on cherry production. Firstly, the sweet cherry, which is a deciduous tree, needs a period of dormancy or rest each season. This is initiated in temperate regions mainly by a fall in average temperatures in late autumn, which in turn stimulates movement of assimilates within the tree to storage organs, such as roots and wood, and the tree's subsequent defoliation. Trees only begin growth anew when the dormancy or chilling requirement has been satisfied and temperatures increase sufficiently in spring. In southern areas of Europe and in North Africa, the winter temperatures rarely cool sufficiently to satisfy the chilling requirement fully and bud break in the spring is erratic and unsatisfactory. In subtropical zones of the world, temperatures may never drop sufficiently to stimulate defoliation and dormancy; in these areas trees grow very poorly unless forced into dormancy by chemical or other dormancy-inducing treatments.

Severe cold in winter also limits the northern range of cherry production, in particular that of sweet cherries. Even in one of the most productive regions of the world for sweet cherries, Washington State, severe winter cold occasionally causes major tree losses. It is difficult to assign precise values as to how low temperatures the trees can tolerate, as much depends on the cultivar, the severity of pruning, the longevity of the freeze, the rapidity of freeze and thaw and the stage of dormancy the trees are at at the time of the freeze.

Cherries also require frequent supplies of water, from rain or irrigation, during the growing season if set fruit are to be retained and their growth sustained by the tree. Unlike other stone fruits, such as almonds, which will often withstand drought and still crop adequately, cherries will be unproductive in similar conditions. In contrast, rain occurring in the period from when the sweet cherry fruits first begin to colour through to harvest is extremely damaging. This is absorbed by the fruits, frequently causing them to split open; the whole crop may be lost by an untimely shower at harvest time. Unfortunately, many of the world's sweet cherry-producing regions suffer from rain-cracking of fruits at harvest time. When choosing new regions for cherry production in the future, this problem should be given high priority. Growers should aim to select areas where meteorological records show very low rain incidence at sweet cherry harvest time.

Cherries usually prefer a moist, retentive loam soil, of at least 0.5 m depth, overlying a subsoil with good moisture-holding capacity and reasonably free drainage. Soils with pH of 5.5–7.5 and good buffering capacity are most suitable. Light sandy or shallow soils in areas experiencing very hot summers are less suitable for production and require abundant irrigation if the trees are to size fruits adequately. Similarly, heavy clay soils with impeded drainage can create problems of root death due to anaerobic conditions and also favour damage from soil-borne fungal pathogens. Soils with high populations of damaging nematodes also limit production. Choice of the appropriate rootstock can, however, alleviate some of the problems associated with different soil types.

In many cherry-producing regions of Europe loss of fruits due to consumption by wild bird populations is severe. Such problems are generally worse close to centres of human population (large towns, cities), close to woodland and in countries where wild birds are strongly protected (e.g. Britain). Unless cherry trees can be protected from birds beneath nets, production in such areas is not usually profitable.

Many areas of the world, although environmentally (climatically and edaphically) suited to cherry production, are currently unsuited to profitable cherry production. The reason for this is the high perishability of the cherry. Even with the best 'cool chain' and storage facilities, fruits for the fresh market have a useful life of only 2–3 weeks; without such facilities their life is only a few days. This is why, traditionally, cherries (sweet cherries, in particular) were produced only by cottagers for their home consumption or sale at nearby markets. Unless markets for the crop are close by, producers and others must invest in expensive equipment to enable them to cool store or preserve the fruits; they must also have the necessary roads, railheads or airports to move the fruits rapidly to the markets.

2.3 Production Statistics for the Principal Cherry-producing Nations

Cherries are known to be produced in significant quantities in more than 40 nations of the world. Statistics of production are not easy to acquire for many of these countries and there are some notable omissions from the data presented below. Nevertheless, details of cherry production in nations in both the northern and the southern hemispheres are presented. It should be noted that the figures presented mostly relate to cherries commercially produced for sale and usually take little account of the millions of trees planted in gardens throughout the world for 'home' consumption.

On the maps accompanying the production statistics locations of sweet and sour cherry production are marked with ⬠ and ⬤ symbols, respectively.

ARGENTINA

SWEET CHERRY
Area of Production (1993):
2000 ha
Production (1993): 6000–
8000 tonnes
Trend: 30% increase in
recent years
Domestic Use:
Fresh: 80%
Processed: 20%
Exports:
Fresh: none
Processed: 20–30% to
Europe
Type of Processed Product:
canned, Maraschino or
crystallized
Varieties:
Fresh Market:
Bing
Merton Bigarreau
Van
Durone Nero 1 and 2
Early Burlat

Processing:
Graffion
Napolitana
Rainier
Rootstocks:
Prunus mahaleb seedlings
Prunus avium seedlings
Prunus cerasus seedlings

Location of Production:
Province of Mendoza
Also small area in Rio Negro
valley in south

SOUR CHERRY
None grown in Argentina

Source of Information:
Dr Enrique E. Sanchez, INTA
Alto Valle Experiment
Station, Cassila de Correo
782, 8332 General Roca, Rio
Negro, Argentina

Cordoba
Mendoza
Buenos Aires
Bahia Blanca

ARMENIA

SWEET CHERRY
Area of Production (1984):
only 200 ha in commercial
orchards and 600 ha in
home gardens
Production: average 600
tonnes per year from
commercial plantings and
1800 tonnes per year from
home gardens
Trend: information
unavailable
Domestic Use: 100%
Exports: none
Varieties:
Napoleon Black
Drogans gelbe

Vystavochnaya
Pobeda
Rootstocks:
Prunus mahaleb seedlings
Prunus avium seedlings

SOUR CHERRY
Area of Production (1984):
no commercial orchards but
approximately 1000 ha in
home gardens
Production: 3000 tonnes per
year
Trend: information
unavailable
Domestic Use: 100%
Exports: none

Varieties:
Sisian local
Kochs (Podbielski)
Rootstocks:
Mostly grown on their own
roots but also on *P. cerasus*
seedlings

Location of Production:
Sweet cherries in Ararat
valley around Erevan
Sour cherries in low
mountains near Leninakan

Source of Information:
Dr A.I. Sychov (see Ukraine)

AUSTRALIA

SWEET CHERRY
Number of Trees in Production (1992/93): 941,488
Production (1992/93): 5043 tonnes
Trend: increasing slightly
Domestic Use:
Fresh: approximately 90%
Processed: approximately 10%
Exports: 5% of production
Type of Processed Product: information unavailable
Varieties:
Empress
Supreme
Early Burlat
Merchant
Vista
Rons Seedling
Van
Lapins
Sunburst
Stella
Bing
Lambert
Celeste
Sylvia
Summerland
Kristin (small numbers)
Ulster (small numbers)
Other varieties under trial are Hartland, Oktavia, Viola, Somerset and Regina
Rootstocks:
Colt
Mazzard F.12/1
Mahaleb

Location of Production:
Victoria, New South Wales, Tasmania and South Australia

SOUR CHERRY
Sour cherries are not produced commercially in Australia

Source of Information:
Mr Ian McLachlan, Secretary, Australian Cherry Growers Federation, and Ms Liz Darmody, Executive Officer, Fleming's Nurseries and Associates Pty. Ltd, Monbulk, Victoria 3793, Australia

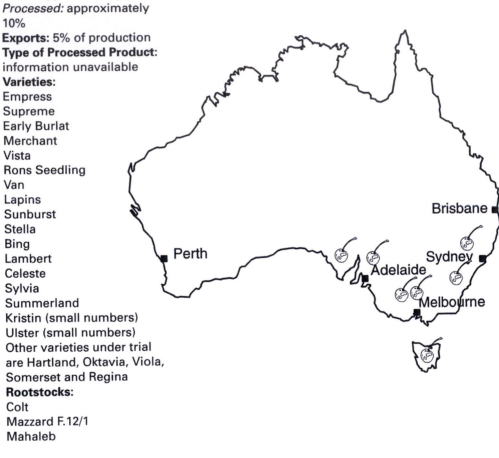

AUSTRIA

SWEET CHERRY
Area of Production (1992):
45 ha (21,000 trees)
commercial production but
910,000 trees in home
gardens
Production (1992): 20,300
tonnes total
Trend: increasing
Domestic Use:
Fresh: 90% of crop
Processed: 10%
Exports:
Fresh: none
Processed: none
Type of Processed Product:
bottled fruits, juices, jams
and cherry schnaps (cherry
brandy)
Varieties:
Grosse Germersdorfer
Grosse Prinzessinkirsche
Grosse Schwarze Knorpel
Hedelfinger
Lambert
Sam

Van
Vista
Bigarreau Burlat
Rootstocks:
P. avium seedlings
P. avium F.12/1
Colt
P. mahaleb (St Lucie 64)
P. cerasus

SOUR CHERRY
Area of Production (1992):
75.5 ha (32,700 trees)
commercial production but
237,300 trees grown in
home gardens
Production (1992): 4100
tonnes total
Trend: constant to slightly
reducing
Domestic Use:
Fresh: 80%
Processed: 20%
Exports:
Fresh: none
Processed: none

Type of Processed Product:
frozen fruits (fruit yoghurts),
juices, jams, liqueurs and
sour cherry schnaps
Varieties:
Schattenmorelle
Koröser
Ostheimer
Kelleris 14 and 16
Heimanns Rubin
Rootstocks:
P. avium seedlings
P. avium F.12/1
P. mahaleb (St Lucie 64)
P. cerasus

Location of Production: see
map

Source of Information:
Dr K. Pieber, Institut für
Obstbau, Universität für
Bodenkultur, Gregor
Mendel-Strasse 33, A-1180
Wien, Austria

AZERBAIJAN

SWEET CHERRY
Area of Production (1984):
only 600 ha in commercial
orchards; 3000 ha in home
gardens
Production: 1800 tonnes per
year from commercial
orchards; approximately
9000 tonnes from home
gardens
Trend: information
unavailable
Domestic Use: 100%
Exports: none
Varieties:
Red/Black types:
Kassins Frühe
Belle de Bianka
Ramon Oliva

White types:
Drogans gelbe
Emperor Francis
Rootstocks:
Prunus mahaleb seedlings
Prunus avium seedlings

SOUR CHERRY
Area of Production (1984):
only 400 ha of commercial
orchards but 1300 ha in
home gardens
Production: average annual
production 1200 tonnes
from commercial orchards
and 4000 tonnes from home
gardens
Trend: information
unavailable

Domestic Use: 100%
Exports: none
Varieties:
Anatolie
Kochs (Podbielski)
Early Duke
Shpanka Rannyaya
Zhukovskaya
Rootstocks:
Most of trees in home
gardens on own roots;
commercial trees on *P.
mahaleb* seedlings

Location of Production:
Northern part of the country
around Kuba

Source of Information:
Dr A.I. Sychov (see Ukraine)

BELGIUM

SWEET CHERRY
Area of Production (1988):
500 ha
Production (1988): 4000
tonnes
Trend: stable
Domestic Use:
Fresh: 100%
Processed: none
Exports: none
Varieties:
Schneiders Späte Knorpel
Hedelfingen Reisenkirsche
Early Rivers
Napoleon
Rootstocks:
P. avium F.12/1
Limburse Boskriek

SOUR CHERRY
Area of Production (1988):
1600 ha
Production (1988): 13,000
tonnes
Trend: declining
Domestic Use:
Fresh: none

Processed: 100%
Exports: none
Type of Processed Product:
fruits in their juice or tarts,
jam, beverages
Varieties:
Noodkriek (=
Schattenmorelle)
Kelleris 16
Rootstocks:
as for sweet cherries

Location of Production:
In Lumberg, Brabant and
several other areas

Source of Information:
Dr Karel Belmans,
Fruitteeltcentrum KU
Leuven, Willem de Croylaan
42, B. 3030 Leuven-Heverlee,
Belgium

BULGARIA

SWEET CHERRY
Area of Production (1982):
10,000 ha
Production (1991): 53,000
tonnes
Trend: constant since 1982
Domestic Use:
Fresh: 55–60%
Processed: 40–45%
Exports: 5000 tonnes
Type of Processed Product:
bottled (canned) fruits, jams
or jellies
Varieties:
Carna Edra
Moreau
Bing
Van
Kozorska
Grosse Germersdorfer
Hedelfinger
Rootstocks:
P. avium
P. mahaleb

SOUR CHERRY
Area of Production (1982):
5200 ha

Production (1991): 26,000
tonnes
Trend: constant since 1982
Domestic Use:
Fresh: 10%
Processed: 90%
Exports: 400 tonnes
Type of Processed Product:
frozen fruit, juice, jams and
jellies
Varieties:
Schattenmorelle
Heimanns Rubin
Fanal

Nefris
Rootstocks:
P. mahaleb

Location of Production: see
map

Source of Information:
Dr E. Karanov, M. Popov
Institute of Plant
Physiology, Bulgarian
Academy of Sciences,
Bonchev Str., Block 21, Sofia
113, Bulgaria

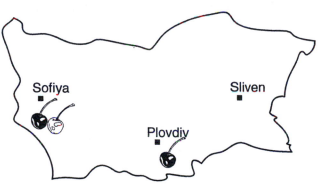

BYELORUSSIA

SWEET CHERRY
Area of Production (1984):
no commercial orchards but
1100 ha in home gardens
Production: average annual
production 3000 tonnes
Trend: information
unavailable
Domestic Use: 100%
Exports: none
Varieties:
Zolotaya loshitskaya
Krasavitsa
(both white-fruited types)
Rootstocks:
Prunus avium seedlings
Prunus cerasus seedlings

SOUR CHERRY
Area of Production (1986):
no commercial orchards but
5300 ha in home gardens
Production: average annual
production of 15,000 tonnes
Trend: information
unavailable
Domestic Use: 100%
Exports: none

Type of Processed Product:
not known
Varieties:
Lyubskaya
Novodvorskaya
Seyanetsi
Rootstocks:
Mostly *Prunus cerasus*
seedlings

Location of Production:
Sweet cherries in southwest

of country around Brest and
Grodno province, sour
cherries throughout most of
the country

Source of Information:
Dr A.S. Devjatov, The
Byelorussian Research
Institute for Fruit Growing
and Dr A.I. Sychov (see
Ukraine)

CANADA

SWEET CHERRY

Area of Production (1992)
British Columbia 680 ha
Ontario 360 ha
Production: British
Columbia (1988–1992
mean) 3000 tonnes
Ontario (1986–1990 mean)
1216 tonnes
Trend: declining slowly
Domestic Use:
Fresh: 70–80%
Processed: 20–25%
Exports:
5% of British Columbia
production exported as
fresh
Type of Processed Product:
canned cherries in British
Columbia; in Ontario >70%
brined cherries and
approximately 20% canned
cherries
Varieties:
British Columbia:
Lambert
Van
Bing
Lapins
Sweetheart

Ontario:
Hedelfingen
Vista
Bing
Viva
Venus
Van
Valera
Rootstocks:
Mazzard F.12/1 and Mazzard
seedlings
Some Mahaleb used in
Ontario

SOUR CHERRY

Area of Production: British
Columbia (1992) 200 ha
Ontario (1990) 816 ha
Production: British
Columbia (1988–1992
mean) 1082 tonnes
Ontario (1986–1990 mean)
5100 tonnes
Trend: declining rapidly in
British Columbia but more
slowly in Ontario
Domestic Use:
Fresh: 15% of the Ontario
production

Processed: 100% of British
Columbia and 85% of
Ontario production
Exports: none
Type of Processed Product:
frozen cherries and pie
fillings
Varieties:
Montmorency
Rootstocks:
Mazzard and Mahaleb
seedlings

Location of Production:
In British Columbia: the
Okanagan, Similkameen
and Creston valleys
In Ontario: Niagara (60%)
and southwestern Ontario
(40%)

Source of Information:
Dr N.E. Looney, Agriculture
and Agri-Food
Canada, Research Centre,
Summerland, BC V0H 1Z0,
Canada; John Cline,
Research Scientist,
Horticultural Research
Institute of Ontario, Simcoe,
Ont N3Y 4N5, Canada

CHILE

SWEET CHERRY
Area of Production (1993):
3002 ha
Production (1993): 14,750
tonnes
Trend: stable over the last 5
years
Domestic Use:
Fresh: 32%
Processed: 10.5%
Exports:
Fresh: 33% and increasing
Processed: 24.5%
Type of Processed Product:
Maraschino, crystallized
and canned
Varieties:
Stella
Bing
Van
Lambert
Summerland
Rootstocks:
P. avium F.12/1
Colt

SOUR CHERRY
Area of Production (1993):
70 ha
Production (1993):
information unavailable
Trend: stable
Domestic Use: 100%
Exports: none
Varieties:
Montmorency
Rootstocks:
F.12/1

Location of Production:
Mainly around Talca and
between Santiago and
Concepción

Source of Information:
Dr Y.M. Moreno, Escuela de
Agronomia, Universidad de
Talca, Chile

Antofagasta

Santiago
Talca
Concepcion

CZECH and SLOVAK REPUBLICS

SWEET CHERRY
Area of Production (1989):
12,000 ha (2.5 million trees);
possibly 1700 ha of intensive
production
Production (1989): 29,012
tonnes (5-year average
25,000 tonnes per annum);
from intensive orchards 2760
tonnes in 1989
Trend: constant: increase in
commercial production
Domestic Use:
Fresh: more than half (52%) of
the production is consumed
by the producers or sold
directly to other consumers. A
similar proportion (12,000
tonnes) is marketed through
selling organizations and 17%
is sold on the fresh market
Processed: approximately
24%, i.e. 6000 tonnes,
annually
Exports:
Fresh: 1500–2000 tonnes
Processed: small amounts
Type of Processed Product:
mainly concentrates and
canned cherries with lesser
amounts of marmalades; also
some wines and distillates
Varieties:
Karesova
Kordia
Bigarreau Napoleon
Van
Hedelfingen
Burlat
Bigarreau Moreau
Früheste der Mark
Schneider's
Rootstocks:
P. avium and *P. mahaleb*
seedlings
Colt
P-HL-A (dwarf clonal stock)

SOUR CHERRY
Area of Production (1989):
4500 ha or 1.8 million trees;

possibly 1920 ha of intensive
production
Production (1989): 13,724
tonnes (5-year average =
15,500 tonnes per annum);
8500 tonnes from intensive
orchards in 1991
Trend: increasing on account
of new plantings
Domestic Use:
Fresh: 32% are used at home
or sold directly to consumers,
68% are marketed through
selling organizations and 7%
sold on the fresh market
Processed: 52% is used for
processing, i.e. about 8000
tonnes (figure increases when
production is heavy)
Exports:
Fresh: 1200–1600 tonnes
annually
Processed: 1000–2000 tonnes
of concentrated fruit annually
Type of Processed Product:
mainly concentrates and
canned fruits but smaller
quantities used for jams and
marmalades
Varieties:
Schattenmorelle
Fanal
Morellenfeur
Favorit
Ujfehértói Fürtös
Vackova

Rootstocks:
P. avium – selected types
(seedlings)
P. mahaleb seedlings
Colt
P-HL-A (a new dwarf clonal
stock)

Location of Production:
In Czech Republic:
1. Kutná Hora, Kolín: both
sweet and sour cherries
2. Litoměřice, Louny: mostly
sour cherries
3. Jičín, Hradec Králové,
Náchod: both sweet and sour
4. Lhenice, Chelčice: both
sweet and sour
5. Znojmo, Břeclav, Hodonín:
both sweet and sour
In Slovak Republic:
1. Nové Zámky, Nitra,
Komárno: both sweet and
sour
2. Krupina, Velký Krtíš: both
sweet and sour

Source of Information:
Dr J. Zika, Vyzkumny Ustav
Ovocnarsky (Research
Institute for Fruit-Growing),
50751 Holovousy, Czech
Republic; Dr J. Blazek,
address as above

DENMARK

SWEET CHERRY
Area of Production (1987):
110 ha
Production (1987): 480
tonnes
Trend: increasing
Domestic Use:
Fresh: 90%
Processed: nil
Exports:
Fresh: 10%
Processed: nil
Varieties:
Sam
Van
Starking Hardy Giant
Rootstocks:
Prunus avium
Colt

SOUR CHERRY
Area of Production (1987):
1675 ha
Production (1987): 10,000
tonnes
Trend: increasing
Domestic Use:
Fresh: nil
Processed: 50%
Exports:
Fresh: nil
Processed: 50%

Type of Processed Product:
juice/wine and jam/jellies
Varieties:
Stevnsbär
Kelleris 16
Rootstocks:
Prunus avium
Colt

Location of Production:
Fünen
Seeland

Source of Information:
Dr J. Grauslund, Institut for
Frugt of Baer,
Kirstinebjergvej 12, 5792
Aarslev, Denmark

■ Copenhagen

■ Odense

FRANCE

SWEET CHERRY
Area of Production (1991):
14,356 ha
Production: average 85,000
tonnnes per year
Trend: after a fall between
1950 and 1985, production
is now increasing
Domestic Use:
Fresh: c. 70–80% of total
cherry production (sweet
and sour) is consumed fresh
Processed: c. 20–30%
Exports:
Fresh: 11,600–16,400 tonnes
between 1990 and 1992
Processed: not known
Imports:
2000–5000 tonnes of fresh
sweet cherries between
1990 and 1992
Type of Processed Product:
confectionery, jams and
jellies and fruits in syrup
(white cvs only)
Varieties:
The most popular varieties
in new plantings are:
Summit
Napoléon
Hatif Burlat
Camus
Stark Hardy Giant
Rainier
Van
Duroni 3
Arcina
Garnet
Noire de Meched
Sunburst
Belge
Rootstocks:
P. mahaleb, Sainte Lucie 64
P. cerasus, Edabriz Tabel(R)
Brokforest Maxma Delbard

(R) 14 plus smaller quantities
of Colt, F.12/1 and other
Mazzards

SOUR CHERRY
Area of Production: very
small – not listed separately
from sweet cherries
Production: average of
approximately 6500 tonnes
per year
Trend: reducing
Type of Processed Product:
confectionery, jams and
jellies, alcohol (kirsch)
Varieties:
Montmorency
Griotte du Nord
Ferracida
Rootstocks:
P. mahaleb, Sainte Lucie 64
P. avium, F.12/1

Location of Production:
Provence, especially
Vaucluse
Languedoc, in the Gard
Rhone valley spanning five
Departements: Drome,
Ardèche, Rhone, Loire and
Isère
Roussillon = Pyrénées
Orientales
Tarn and Garonne and the
neighbouring Aveyron
Yonne (the latest ripening
area)

Source of Information:
D. Brossard, B. Lam and I.
Jacoutet, CTIFL, 22, rue
Bégère, 75009 Paris, France;
J. Lichou, CTIFL, Centre de
Balandran, 30127
Bellegarde, France

GEORGIA

SWEET CHERRY
Area of Production (1984):
600 ha in commercial orchards; 3000 ha in home gardens
Production: average of 1800 tonnes per year from commercial plantings and 9000 tonnes from home gardens
Trend: not known
Domestic Use: 100%
Exports: none
Varieties:
Red/Black types:
Tartarian Black
Hedelfinger

Pobeditelnitsa
Priusadebnaya
White types:
Drogans gelbe
Bagration
Rootstocks:
Prunus mahaleb seedlings
Prunus avium seedlings
Prunus cerasus seedlings

SOUR CHERRY
Area of Production (1984):
no commercial orchards but 1700 ha in home gardens
Production: 5000 tonnes per year
Trend: not known

Domestic Use: 100%
Exports: none
Varieties:
Vladimir
Kochs (Podbielski)
Kartuli alubali
Rootstocks:
Mostly on own roots but also on *Prunus cerasus* seedlings

Location of Production:
All over the country

Source of Information:
Dr A.I. Sychov (see Ukraine) and Dr Devjatov (see Byelorussia)

GERMANY

SWEET CHERRY

Area of Production (1992):
5874 ha (1856 ha of this in
what was East Germany)
Number of Trees (1992): 1.5
million
Production: variable and
supplemented by large
numbers of trees in home
gardens
Trend: falling slightly
Domestic Use: 60–70% as
fresh, 30–35% after
processing
Exports: 5% fresh, small
amount after processing
Type of Processed Products:
brandy, conserves
Varieties:
West Region:
Hedelfinger Riesenkirsche
Grosse Schwarze
Knorpelkirsche
Schneiders Späte
Knorpelkirsche
Büttners Späte Rote
Knorpelkirsche
East Region:
Kassins
Knauffs
Teickners
Sam
Van
Hedelfinger Riesenkirsche
Querfurter Königskirsche
Altenburger
Melonenkirsche Grosse
Schwarze Knorpel
Rootstocks:
West Region:
Prunus avium, Limburger,
Vogelkirsche, Hüttner,
Hochzucht

F.12/1
Colt
Gisela 5
East Region:
Prunus avium Alkavo
Prunus mahaleb Alpruma

SOUR CHERRY
Area of Production (1992):
6479 ha (3014 ha of this in
what was East Germany)
Number of Trees (1992): 4
million
Production: variable and
supplemented by large
numbers of trees in home
gardens
Trend: falling
Domestic Use: fresh – very
small quantities; >90% after
processing
Exports: no fresh, 10%
variable after processing
Type of Processed Products:
canning juice, jam
concentrate, conserves
Varieties:
West Region:
Schattenmorelle
Heimanns group
East Region:
Schattenmorelle
Ujfehértói Fürtös
Röhrigs Weichsel
Leopoldskirsche
Fanal
Kelleris 16
Rootstocks:
West Region:
Prunus avium, Limburger,
Vogelkirsche,
Hüttner, Hochzucht
F.12/1

Colt
Gisela 5
East Region:
Prunus avium Alkavo
Prunus mahaleb Alpruma

Location of Production: see
map

Source of Information:
Frau Brunner, Statische
Bundesamt, Gustav-
Stresemann Ring, D-6200
Wiesebaden; Dr J. Trenkler,
Department of Economics,
Fruit Research Institute,
Genebank for Fruit,
Dresden-Pillnitz, Germany;
Dr M. Fischer, Genebank
Obst. Dresden-Pillnitz,
Dorfplatz 2, PF 69–143, 0–
8054, Dresden, Germany; Dr
R. Stehr, Obstbauver-
suchsanstalt, Betrieb
Esteburg, Moorende 53,
21635, Jork

GREECE

SWEET CHERRY
Area of Production (1991):
8450 ha
Production (1992): 40,000
tonnes
Trend: increasing
Domestic Use:
Fresh: 50–60%
Processed: 6–8%
Exports:
Fresh: 22–35%
Type of Processed Products:
juice and candied fruit
Varieties:
Bigarreau Burlat
Van
Tragana Edessis
Rootstocks:
P. avium seedlings
Colt

SOUR CHERRY
Area of Production (1992):
very small
Production (1992): 1200
tonnes
Trend: constant
Domestic Use:
Fresh: nil
Processed: 100%
Exports: nil

Varieties:
Vissino Tripoleos
Vissino Florinis
Rootstocks:
P. avium seedlings
Colt

Location of Production: see
map

Source of Information:
Mr Hatziharisis, Director of
Pomology Institute, 59200
Naoussa, Macedonia,
Greece

HUNGARY

SWEET CHERRY
Area of Production (1990–92): 700 ha in commercial orchards, but almost 2.5 million trees planted, many in home gardens
Production (1992): 30,800 tonnes
Trend: constant
Domestic Use:
Fresh: 40%
Processed: 25% (6000 tonnes) acquired for processing – either bottling or freezing – in 1990
Exports:
Fresh: 10%
Processed: 10%, mainly as bottled form
Type of Processed Product: 59–85% canned fruit and jam, 8–34% spirits, 5% deep-frozen, 2% chocolate
Varieties:
Germersdorfi clones 1, 3, 45 and 57
Bigarreau Burlat
Van
Hedelfingen
Solymári gömbölyü
Jaboulay
Popular new varieties are Margit, Linda, Katalin, and Valerij Cskalov
Rootstocks:
70% *Prunus mahaleb*
30% *Prunus avium*

Location of Production:
Pest
Bács- Kiskun
Jász-Nagykun
Szolnok and smaller quantities in other regions

SOUR CHERRY
Area of Production (1990–92): 5500 ha, mainly on cooperative and state farms

Production (1990): 78,000 tonnes
Trend: constant
Domestic Use:
Fresh: 20%
Processed: 30% or 16,000 tonnes acquired for bottling or freezing or other processing in 1990
Exports:
Fresh: 17% (80% of which to Germany)
Processed: 10% (mainly – 80% – to Germany)
Type of Processed Product: 37% canned fruit, 18% jam, 25% concentrated juice, 10% spirits, 10% frozen fruit, 8% chocolate
Varieties:
Ujfehértói Fürtös
Érdi Bötermö
Pándy clone Bb 119
Pándy clones 48 and 279
Cigány Meggy clones 3, 59, and C.404
Meteor Korai
Favorit
Popular new varieties include: Erdi Jubileum, Korai Pipacs, Csengödi,
Debreceni Bötermö, Maliga emléke and Kantorjanosi
Rootstocks:
90% *P. mahaleb*, including selected types such as CT500, CT 2753
10% *P. avium*

Location of Production:
Pest, Bács-Kiskun, Szabolcs, Szatmár, Bereg, Jász-Nagykun, Szolnok and small quantities in many other regions

Source of Information:
Dr J. Apostol and Dr Elisabeth Kállay, Enterprise for Extension and Research in Fruit Growing and Ornamentals, H-1223, Budapest, Park u. 2, Hungary; additional information from Dr T. Bubán, Station of Enterprises for Research and Extension in Fruitgrowing and Ornamentals, H.4244 Ujfeherto, PO Box 38, Hungary

INDIA

SWEET CHERRY
Area of Production (1987):
1110 ha
Production (1987): 605
tonnes
Trend: constant
Domestic Use:
90% of production sold fresh
10% used as bottled or
canned fruit
Exports: none
Varieties:

Bigarreau Noir Gross
Bigarreau Napoleon
Rootstocks:
Paja (*Prunus cerasoides* D.
Don: Syn = *Prunus puddum*
Roxb. ex Wall)
P. mahaleb Colt

Location of Production:
Jammu and Kashmir near
Srinagar

SOUR CHERRY
None produced in India

Source of Information:
Dr Suneel Sharma,
Department of Horticulture,
Harayana Agricultural
University, Hissar 125004,
India

ISRAEL

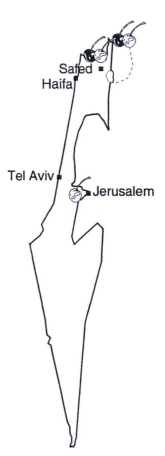

SWEET CHERRY
Area of Production: 100 ha
Production (1992): 650 tonnes
Trend: slow increase
Domestic Use: 100% of crop as fresh fruit
Exports: none
Varieties:
Bigarreau Burlat
Rainier
Bing
Chinook
Van
Hedelfingen
Rootstocks:
P. mahaleb seedlings – but these are not very successful and alternatives are currently sought

SOUR CHERRY
Area of Production: 10 ha
Production (1992): 70 tonnes
Trend: no change
Domestic Use: all of crop consumed fresh
Exports: none
Variety: Chios (this is a local self-fertile variety)
Rootstocks:
P. mahaleb seedlings

Location of Production:
Mountainous areas of the country, in order that sufficient winter chilling is achieved, e.g. Galil, Golan Heights and areas of high elevation around Jerusalem

Source of Information:
Dr Jesajahu Kovetz, Fruit Board of Israel, 20 Ha'Abra St., PO Box 7117, Tel Aviv 61070, Israel

ITALY

SWEET AND SOUR CHERRY
Area of Production (1991):
28,826 ha
Production (1991): 104,900
tonnes
Trend: sweet cherries slight
increases, sour cherries
rapidly declining
Exports:
Fresh: 4400 tonnes (1991)
Processed: 6600 tonnes
(1990)

520 tonnes of jams, jellies,
sauces and purées
Varieties:
Adriana
Bigarreau Burlat and its
clones
Durone del Monte
Durone della Marca
Della Recca
Giorgia
Durone dell'Auella
Durone Nero I

Durone Nero II
Ferrovia
Lapins
Mora di Cazzano
Bigarreau Moreau
Sunburst
Van
Vittoria
Rootstocks:
Colt
P. avium and *P. mahaleb*
types

Location of Production:
Apulia
Campania
Emilia Romagna
Veneto
Trentino

Source of Information:
Professor S. Sansavini,
Universita di Bologna, V.
Filippo Re 6, Dipartimento di
Colture Arboree, 40126,
Bologna, Italy

Imports:
Fresh: 6500 tonnes (1991)
Processed: c. 1250 tonnes
(1990)
**Type of Processed Product
(1990):**
Imported:
940 tonnes sweet cherries in
alcohol with sugar
150 tonnes canned sweet
cherries
138 tonnes jams, jellies,
sauces and purées
Exported:
4050 tonnes of canned
sweet cherries
1440 tonnes sweet cherries
in alcohol with sugar
700 tonnes canned sour
cherries

Trento
Verona
Bologna
Vignola
Forli, Cesena
Caserta
Avellino
Bari

JAPAN

Sweet Cherry
Area of Production (1991):
2650 ha (994 ha covered by
plastic against rain in 1992)
Production (1989): 15,400
tonnes
Trend: slow increase in
production
Domestic Use:
Fresh: 60–70%
Processed: 30–40%
Exports:
Fresh: trial exports to Hong
Kong and Singapore but
very small, <1.0 tonnes per
year
Processed: nil
Type of Processed Product:
bottled and canned fruits
Varieties:
Sato-Nishhiki
Napoleon
Rootstocks:
Prunus lannesiana Wils.
(Mazakura) *Prunus serrulata*
Colt

Location of Production:
65% Yamagata Prefecture
11% Hokkaido Prefecture

8% Aomori Prefecture
7% Yamanashi Prefecture
also Nagano, Iwate, Akita
and Fukushima Prefectures

Sour Cherry
None produced in Japan

Source of Information:
Mr K. Komabayashi,
Department of Agriculture
and Horticulture, Yamagata
Prefecture, 281 Matsunami,
Yamagata 990, Japan;
additional information from

Professor R. Ogata, Faculty
of Agriculture, Utsunomiya
University, Utsunomiya 321,
Japan

KAZAKHSTAN

SWEET CHERRY
Area of Production (1984):
300 ha in home gardens only
Production: average of 900
tonnes per year
Trend: not known
Domestic Use:
Fresh: 100%
Exports: none
Varieties:
Black types:
Kara Geles
Bigarreau Reverchon
White types:
Drogans Gelbe
Napoleon
Zolotaya
Rootstocks:
Prunus mahaleb seedlings
Prunus avium seedlings

SOUR CHERRY
Area of Production (1986):
400 ha only of commercial
orchards but 7000 ha in
home gardens
Production: average annual
production 1000 tonnes
from commercial orchards
and 21,000 tonnes from
home gardens
Trend: not known
Domestic Use:
Fresh: not known
Processed: not known
Exports: none
Type of Processed Product:
not known
Varieties:
Lyubskaya
Griotte d'Osteim
Kochs (Podbielski)

Vladimir
Shpanka krupnaya
Samarkand
Komsomolskaya
Rootstocks:
Home garden trees usually
on their own roots;
commercial trees on *P.
mahaleb* seedlings

Location of Production:
Sweet cherries only in
Chimkent province; sour
cherries in southeast of
country, especially around
Alma Ata, Dzhambul,
Kzylorda and Taldgkurgan,
Chimkent province

Source of Information:
Dr A.I. Sychov (see Ukraine)

KYRGYZSTAN

SWEET CHERRY
Area of Production (1984):
only 1200 ha in home
gardens
Production: 3500 tonnes per
year from home gardens
Trend: information
unavailable
Domestic Use: 100%
Exports: none
Varieties:
Red/Black types:
Leningrad Black
Mai Baumann
White types:
Drogans Gelbe

Zorka
Rootstocks:
Prunus mahaleb seedlings
Prunus avium seedlings

SOUR CHERRY
Area of Production (1984):
No commercial orchards but
3000 ha in home gardens
Production: average annual
production 9000 tonnes
Trend: information
unavailable
Domestic Use: 100%
Exports: none

Varieties:
Lyubskaya
Shpanka krupnaya
Rootstocks:
Prunus mahaleb seedlings

Location of Production:
Sweet cherries in Issykkul
and Osh provinces
Sour cherries also in
Issykkul and Osh provinces
in Chui and Talass valleys
(see map under Tadjikistan)

Source of Information:
Dr A.I. Sychov (see Ukraine)

LATVIA

SWEET CHERRY
Area of Production (1984):
no commercial orchards but
approximately 500 ha
grown in home gardens
Production: average annual
production of 1500 tonnes
Trend: information
unavailable
Domestic Use: 100%
Exports: none
Varieties:
Drogans gelbe
Vigzemes sartvidis
Rootstocks: not known

SOUR CHERRY
Area of Production (1984):
no commercial orchards but
about 1500 ha in home
gardens
Production: average annual
production 4500 tonnes
Trend: information
unavailable
Domestic Use: 100%
Exports: none
Varieties:
Latvia low
Latvia tall
Kazdangans

Daugmal Sklyanka
Rootstocks:
Most trees are grown on
their own roots but a few
are grown on *P. cerasus*
seedlings

Location of Production:
Over the whole country

Source of Information:
Dr A.I. Sychov (see Ukraine)

LITHUANIA

SWEET CHERRY
Area of Production (1984):
no commercial orchards but
approximately 500 ha in
home gardens. Severe tree
loss in cold winters
Production: average of 1500
tonnes per year
Trend: not known
Domestic Use: 100%
Exports: none
Varieties: local varieties
only, often seedlings
Rootstocks: mainly grown
on their own roots

SOUR CHERRY
Area of Production (1984):
no commercial orchards but
5800 ha in home gardens
Production: average of
15,000 tonnes per year
Trend: not known
Domestic Use: 100%
Exports: none
Type of Processed Product:
not known
Varieties:
Lyubskaya

Griotte d'Ostheim
Latvia low
Vetine rugshchoi
Rootstocks:
Mostly on own roots but
also on *P. cerasus* seedlings

Location of Production:
Over most of the country

Source of Information:
Dr A.I. Sychov (see Ukraine)

MOLDAVIA

SWEET CHERRY
Area of Production (1989):
2900 ha in commercial
orchards; 1000 ha in home
gardens
Production: average of
13,000–14,000 tonnes per
year from commercial
orchards; 5000 tonnes from
home gardens
Trend: declining slowly in
commercial orchards
Domestic Use:
Fresh: 60–70%
Processed: 30–40%
Exports: not known
Type of Processed Product:
fruits in syrup
Varieties:
Valerii Chkalov
Bigarreau Jaboulay
Kassins fruhe
Fruheste der Mark
Trushenskaya
Record
Recordnaya
Rootstocks:
Prunus mahaleb seedlings

Prunus avium seedlings
Prunus cerasus seedlings

SOUR CHERRY
Area of Production (1989):
2800 ha in commercial
orchards; 4300 ha in home
gardens
Production: average annual
production is 12,000–13,000
tonnes from commercial
orchards and 25,000 tonnes
from home gardens
Trend: slow decline since
1978
Domestic Use:
Fresh: 30–40%
Processed: 60–70%
Exports: not known
Type of Processed Product:
jams, juices, fruits in syrup,
alcohol (wine and cordial)
Varieties:
Anatolie
Kochs (Podbielski)
Rannyaya
Shpanka rannyaya
Shpanka pozdnyaya

Rootstocks:
Prunus mahaleb seedlings
Prunus cerasus seedlings
Prunus avium seedlings

Location of Production:
practically whole of country
but concentrations higher in
north and southeast

Source of Information:
Dr A.I. Sychov (see Ukraine)

■ Beltsy

Kishinev
■

THE NETHERLANDS

SWEET CHERRY
Area of Production (1992):
243 ha
Production (1991): 38
tonnes (low) but usually
between 170 and 300 tonnes
Trend: falling
Domestic Use (1991):
Fresh: most of the crop
Processed: small quantities
only
Exports (1991):
Fresh: small quantities of
sweet and sour, mostly to
Belgium and Luxemburg
Varieties:
Frühe Rote Mecken
Merton Premier
Miekers
Kordia
Castor
Lapins

Oktavia
Merchant
Viola
Rootstocks:
Prunus avium Limburgse
Boskriek and F.12/1

SOUR CHERRY
Area of Production (1992):
265 ha
Production (1991): 610
tonnes (5-year average =
1060 tonnes)
Trend: falling
Domestic Use (1991):
Fresh: none
Processed: 100%
Exports:
Fresh: see above combined
data for sweet and sour
Type of Processed Product:
jams, jellies, sauces

Varieties:
Rheinische Schattenmorelle
Kelleriis 16
Rootstocks:
Prunus avium Limburgse
Boskriek and F.12/1

Location of Production:
Mainly in Gelderland and
Limburg

Source of Information:
Mr M.L. Joosse, Information
and Knowledge Centre for
Fruit Growing, Brugstraat
51, 4475 AN
Wilhelminadorp, the
Netherlands and Ing. P.D.
Goddrie, Research Station
for Fruit Growing,
Brugstraat 51, 4475 AN
Wilhelminadorp, the
Netherlands

NEW ZEALAND

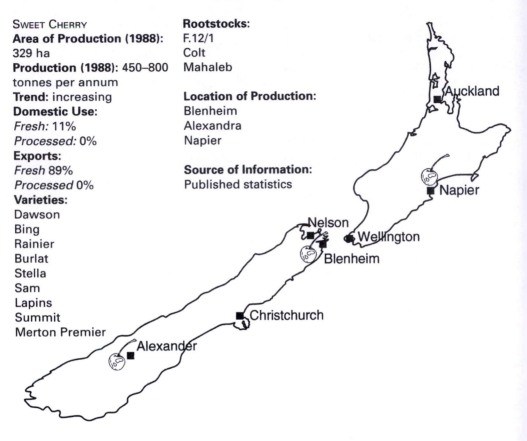

SWEET CHERRY
Area of Production (1988):
329 ha
Production (1988): 450–800
tonnes per annum
Trend: increasing
Domestic Use:
Fresh: 11%
Processed: 0%
Exports:
Fresh 89%
Processed 0%
Varieties:
Dawson
Bing
Rainier
Burlat
Stella
Sam
Lapins
Summit
Merton Premier

Rootstocks:
F.12/1
Colt
Mahaleb

Location of Production:
Blenheim
Alexandra
Napier

Source of Information:
Published statistics

NORWAY

SWEET CHERRY
Area of Production (1989):
estimate 900 ha, 242,000
trees, but commercial
production on only 173 ha
or 47,815 trees; the
remainder in private
gardens
Production (1989): 3763
tonnes (only 465 tonnes
commercially)
Trend: increasing
Domestic Use:
Fresh: almost 100% of the
crop
Processed: nil
Exports:
Fresh: very small quantities
(maximum 40 tonnes)
occasionally exported to
Sweden
Processed: nil
Varieties:
Van
Ulster
Kristin
Merton Glory
Merton Premier
Rainier
Kassin
Early Rivers
Rootstocks:
P. avium seedlings
F.12/1
Colt

Location of Production:
Near Ullensvang – 82%

SOUR CHERRY
Area of Production (1989):
682 ha (estimate); only 83 ha
of commercial production;

344,000 trees (42,526 on
commercial holdings)
Production (1989): 3873
tonnes total; 480 tonnes
commercial production
Trend: decreasing
Domestic Use:
Fresh: nil
Processed: 100% either from
home production or
commercially; 113 tonnes of
Norway-produced sour
cherries and 394 tonnes of
imported fruits processed in
1989
Exports:
Fresh: nil
Processed: nil
Type of Processed Product:
juice, wine and jams
Varieties:
Fanal
English Morello
Rootstocks:
P. avium seedlings
F.12/1

Location of Production:
Buskerud
Vestfiord
Telemark

Source of Information:
Dr A. Kvåle,
Statskonsulenten i frukt og
baerdyrking, Statens
fagtjeneste for landbruket
(SFL), Ullensvang, 5774
Lofthus, Norway

Bergen Oslo

PEOPLE'S REPUBLIC OF CHINA

SWEET CHERRY
Area of Production (1992):
3000 ha
Production (1992): 1300
tonnes
Trend: increasing rapidly
Domestic Use:
Fresh: 87%
Processed: only small
quantities
Exports:
Fresh: 13%
Processed: none
Type of Processed Product:
bottled or canned fruits;
also jellies/jams, candied
fruits, juice and wine
Varieties:
Napoleon Bigarreau (Na
Weng)
Black Tartarian (Da Zi)
Early Purple (Zao Zi)
Governor Wood (Huang Yu)
Bing (Bin Ku)
Black Eagle (Ji Xin)
Hong Deng
Hong yan
Hong mi
Zao feng
Hong feng
Wan huang

SOUR CHERRY
There are no sour cherries
produced commercially in
the People's Republic of
China

YUNG TO CHERRY
Prunus pseudocerasus
(Lindl.)
More than 60 cultivars of

Prunus pseudocerasus are
grown in China.
The principal cultivars are:
Shangxian sweet cherry
Lantain Manao cherry
Luonan Suan yingtao
Taishan yingtao
Da wo lou yie
Jian yie
Laoshan yingtao
Zhucheng yingtao
Tengxian dahong yingtao
Da ying zui
Rootstocks:
Mahaleb yingto (*P. mahaleb*)
Qing fu ying (*P. serrulata*)
Shan yingtao (*P.
sachainensis*)
Cao yingtao (*P. fruticosa*)

Area of Production (1992):
6667 ha
Production (1992): 5000
tonnes

Location of Production:
The main areas of cherry
production in China are
located in Shandong and
Liaoning provinces, with
smaller areas of production
in Shaanxi, Henan, Jiangsu,
Aahus and Gansu provinces

Source of Information:
Dr Yu-Liang Cai, PO Box 76,
Dong-Yi Road, Xian 710061,
People's Republic of China

POLAND

SWEET CHERRY
Area of Production (1993):
8000 ha; approximately 3.2
million trees
Production (1993): 32,000
tonnes; 5-year average =
20,000 tonnes annum^{-1}
Trend: increasing
Domestic Use:
Fresh: 60%
Processed: 35–40%
Exports:
Fresh: nil
Processed: nil
Type of Processed Product:
canned fruits, jams, jellies
and juice
Varieties:
Kassina
Rivan
Wczesna Rivers
Bladorózowa
Burlat
Karesova
Kunzego
Lotka Trzebnicka
Merton Premier
Buttnera Czerwona
Hedelfinska
Kordia
Schneidera Pózna
Van
Rootstocks:
Prunus avium seedlings
Prunus mahaleb seedlings
F.12/1

SOUR CHERRY
Area of Production (1993):
39,000 ha; almost 15.7
million trees
Production (1993): 147,000
tonnes; 5-year average =
100,000 tonnes annum^{-1}
Trend: increasing
Domestic Use:
Fresh: very few
Processed: more than 80%
Exports:
Fresh: nil
Processed: a few
Type of Processed Product:
canned fruits, juices, jams
and jellies
Varieties:
Ksiazeca
Northstar

Wczesna Ludwika
Groniasta z Ujfehertoi
Kelleris 16
Nefris
Pándy 103
Rootstocks:
Prunus avium
Prunus mahaleb
F.12/1

Location of Production: see
map

Source of Information:
Dr A. Holewiński, Chairman
of the Department of the
Extension Service, Research
Institute of Pomology and
Floriculture, Skierniewice,
Poland

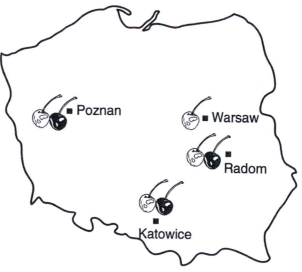

PORTUGAL

SWEET CHERRY
Area of production (1990):
2300 ha
Production (1990): 7700
tonnes
Average annual production
(1985–1990): 8350 tonnes
Fresh fruit for home
consumption estimated to
be most of the crop
Fresh exports very small –
approx. 1000 kg per year,
mainly to France, the
Netherlands and Italy
Less than 1000 kg per year
exported in processed state
Processed products are
cherry jam, cherries tinned
in syrup, cherries drained
and crystallized cherries
Trend: declining in Beira
Interior but expanding in
other areas
Varieties:
Bigarreau Napoleon (=
Napoleao, a selected clone
of the original variety)
Bigarreau Windsor
Bigarreau Burlat
De Saco
Espanhola
Francesa de Alenquer
Mirandela
Morangao
Bing
Hedelfingen

Rootstocks:
P. avium seedlings and *P.
mahaleb* St Lucie 64

SOUR CHERRY
Area of Production (1990):
471 ha
Production (1990): 576
tonnes
Average annual production
(1985–1990): 652 tonnes
Processed production 90
tonnes in 1989
Exports of processed sour
products 640 kg in 1989
Processed products are
mainly sour cherry jam
Trend: slightly down since
1985
Variety:
Garrafal
Rootstocks:
P. avium seedlings and *P.
mahaleb* St Lucie 64

Location of Production:
Beira Interior – Cova da
Beira 60%
Douro Sul e Alfandega da Fe
34%
Alenquer (others) 2%

Source of Information:
Dr Cristina M.M. Simoes de
Oliveira, DPPA
(Departamento de Producao

Agricola e Animal), Section
of Horticulture, Instituto
Superior de Agronomia,
Technical University of
Lisbon, Tapada da Ajuda
1399, Lisbon, Portugal

ROMANIA

SWEET CHERRY

Area of Production (1990):
5700 ha in commercial
production plus 500,000
trees in home gardens
Production (1990): 39,100
tonnes
Trend: constant since 1990
Domestic use:
Fresh: 74%
Processed: 24%
Exports:
Fresh: 2%
Processed: none
Type of Processed Product:
bottled and canned fruits,
jams and jellies
Varieties:
Van
Stella
Boambe de Cotnari
Grosse Germersdorfer
Rubin
Bigarreau Moreau
Negre de Bistrita
Cerna
Rootstocks:
P. avium 76–33–26 seedling
P. avium 76–25–29 seedling
P. avium VV-1 clonal
P. cerasus × *P. avium* IP-C.1
clonal

SOUR CHERRY

Area of Production (1990):
5500 ha in commercial
production, plus 3 million
trees in home gardens
Production (1990): 59,800
tonnes

Trend: constant since 1990
Domestic use:
Fresh: 38%
Processed: 60%
Exports:
Fresh: none
Processed: 2%
Type of Processed Product:
compote, jams/jellies,
juice/wine
Varieties:
Nana
Schattenmorelle
Meteor
Crisana 2
Oblacinska
Mocănesti 16
Rootstocks:
P. cerasus VG-1 seedling
P. cerasus Izvorani 78
seedling

P. avium VV-1 clonal
P. cerasus × *P. avium* IP-C
clonal

Location of Production:
Botosani
Iasi
Neamt
Vaslui
Bacau
Focsani
Galati
Buzau
Arges (Pitesti)
Dolj (Craiova)

Source of Information:
Dr Sergiu Budan and Dr
Gheorghe Mladin, Research
Institute for Pomology,
0312 – Pitesti Maracineni,
Romania

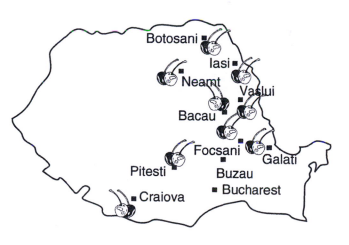

RUSSIA

SWEET CHERRY
Area of Production (1986): 4000 ha in commercial orchards; 1000 ha in home gardens
Production (1986): 10,000 tonnes from commercial orchards; 3000 tonnes from home gardens
Trend: information unavailable
Domestic Use:
Fresh: 70–80%
Processed: 20–30%
Exports: nil
Type of Processed Product: fruits in syrup
Varieties:
Red/Black cvs:
Valerii Chkalov
Badacsone Black
Bigarreau of Oratovskii
Dagestanka
Dagestankaya rannyaya
Goryanka
Izyumnaya
Iyunskaya rannyaya
Krupnoplodnaya
Melitopolskaya chernaya
Priazovskaya
Chernyavka
White cvs:
Donchanka
Cosmicheskaya
Krasa Kubani
Krasavitsa Kryma
Yantarnaya
Rootstocks:
P. mahaleb seedlings
P. avium seedlings (wild types) and also Drogans Gelbe seedlings
P. cerasus seedlings

SOUR CHERRY
Area of Production (1986): 7000 ha of commercial production and approximately 60,000 ha in home gardens
Production (1986): 20,000 tonnes from commercial orchards and 180,000 tonnes from home gardens
Trend: declining since 1970
Domestic Use:
Fresh: 30–40% consumed fresh
Processed: 60–70%
Exports: nil
Type of Processed Product: jams, juice, fruits in syrup, alcohol (cordial and wine)
Varieties:
Lyubskaya
Vladimir
Griotte d'Ostheim
Chernokorka
Turgenevka
Molodezhnaya
Pamyati Vavilova
Zhykovskaya
Anatolie
Early Duke
Rootstocks:
P. mahaleb seedlings
P. avium seedlings
P. cerasus seedlings
VP1 (*P. cerasus* × *P. maackii*) clonal

STEPPE OR GROUND CHERRY
(*Prunus fruticosa*)
Area of Production (1990): 1500 ha
Production (1990): average of 3000 tonnes each year
Trend: increasing since 1965
Domestic Use:
Fresh: approximately 30%
Processed: approximately 70%
Exports: nil
Type of Processed Product: similar to sour cherry
Varieties:
Altayskaya lastochka
Altayskaya rannjaja
Zhelannaya
Subbotinskaya
Maksimovskaya
Novoaltaiskaya
Altaiskaya Krupnaya
Rootstocks: all grown on their own roots

DOWNY OR KOREAN CHERRY
(*Prunus tomentosa*)
Area of Production (1990): 100–200 ha of commercial orchards
Production: Average of 200–300 tonnes each year
Trend: not known
Domestic Use:
Fresh: almost 100%
Exports: nil
Type of Processed Product: none
Varieties:
Amurka
Leto
Ogonek
Rannaya rosovaya
Rootstocks:
Prunus tomentosa seedlings

Location of Production:
Sweet cherry: North Caucasus, especially Krasnodar territory, Dagestan
Sour cherry: North Caucasus, especially Rostov district – 80% Central Russia 7% Volga district 7%
Steppe cherry: West Siberia, especially Altai territory, Omsk province Novosibirsk province
Downy cherry: Far east of Russia, especially Primorskii and Khabarovskii territories

Source of Information:
Dr A.I. Sychov (see Ukraine)

SOUTH AFRICA

SWEET CHERRY
Area of Production (1988):
420 ha
Production (1984): 1000 tons
Trend: decreasing
Domestic Use:
Fresh: 40%
Processed: 50%
Exports:
Fresh: 10%
Processed: none
Type of Processed Product:
candied fruits

Varieties:
Black Tartarian
Early Rivers
Bing
Giant Hedelfinger
Lambert
Napoleon
Van
Rootstocks:
Mazzard (*P. avium*)

Location of Production:
Eastern Free State
(Fieksburg)

SOUR CHERRY
No production of sour
cherries in South Africa

Source of Information: Mr
R.C. Saunders, Agricultural
Research Council,
Stellenbosch Instituut for
Fruit Technology, Private
Bag X5013, 7599
Stellenbosch, South Africa

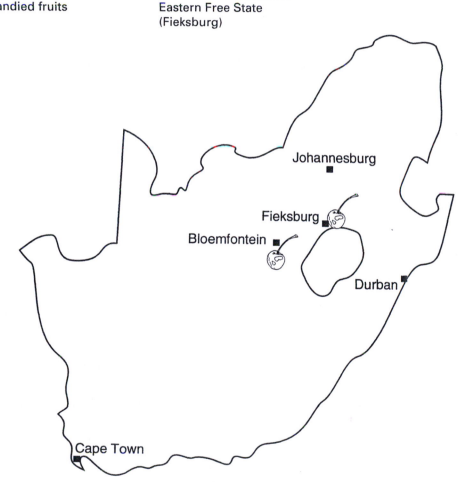

SPAIN

SWEET CHERRY
Area of Production (1990):
25,300 ha
Production (1990): 54,900
tonnes
Trend: increasing slightly
Domestic Use:
Fresh: 42,062 tonnes
Processed: 6,642 tonnes
Exports: 592 tonnes
Type of Processed Product:
canning for pies
Varieties:
New plantings:
Marvin

Burlat
Garnet
Stella
Starking Hardy Giant
Summit
Sunburst
Sweet Heart
Blanca de Provence (for
processing)
Traditional plantings:
Ambrunes
Hedelfingen
Napoleon
Bigarreau Tardif de Vignola
Tigre

Rootstocks:
P. mahaleb 70% seedlings
30% clonal
P. avium seedlings

Location of Production: see
map

SOUR CHERRY
No sour cherries are grown
in Spain

Source of Information:
Mr J.L. Espeda, Servicio de
Produccion y Sanidad
Vegetal DGA, Ministerio de
Agricultura, Pesca y
Alimentation, Madrid, Spain

SWEDEN

SWEET CHERRY
Area of Production (1987): approximately 20 ha
Production (1987): 34 tonnes but averages 50–70 tonnes in most years
Trend: information unavailable
Domestic Use:
Fresh: 100%
Processed: nil; all imported from Denmark and Poland
Exports:
Fresh: nil
Processed: nil

SOUR CHERRY
Area of Production (1987): 8 ha
Production: included in sweet cherry total

Trend: information unavailable
Domestic use:
Fresh: 100%
Processed: nil
Exports:
Fresh: nil
Processed: nil

Source of Information:
Dr V. Trajkovski, The Swedish University of Agricultural Sciences, Department of Horticulture and Plant Breeding, Balsgard, Sweden

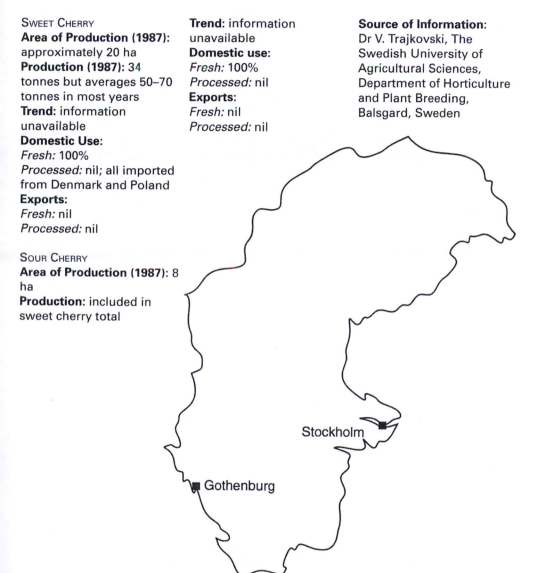

Stockholm

Gothenburg

SWITZERLAND

SWEET CHERRY
Area of Production (1993):
439 ha (802,000 trees)
Production (1993): 18,820
tonnes estimated
Trend: declining
Domestic Use:
Fresh: 20%
Processed: 20% (mainly
jam) and 60% for distillation
Exports:
*Fresh and processed
combined:* 0–300 tonnes per
annum
Imports:
Approximately 2000 tonnes
annually of sweet and sour
cherries combined
Type of Processed Product:
cherry brandy, jam, yoghurt
Varieties:
Schauenburger
Basler Adlerkirsche

Basler Langsteiler
Beta
Schumacher
Hedelfinger
Popular new varieties are:
Bigarreau Moreau, Kordia
and Star
Rootstocks:
Prunus avium F.12/1; also *P.
avium* seedlings Krakauer

SOUR CHERRY
Area of Production (1990):
less than 10 ha
Production: not known
Trend: no change
Domestic Use:
Fresh: none
Processed: 100%
Exports: see sweet cherry
exports
Imports: small quantities for
jam production from
Germany

Type of Processed Product:
confectionery
Varieties:
Aemli
Schattenmorelle
Rootstocks:
Prunus avium F.12/1; also *P.
avium* seedlings Krakauer

Location of Production:
Area of Basle, central
Switzerland, Seeland and
Waadtland

Source of Information:
T. Meli and Dr M. Kellerhals,
Swiss Federal Research
Station for Fruit-Growing,
Viticulture and Horticulture,
CH 8820 Wadenswil,
Switzerland

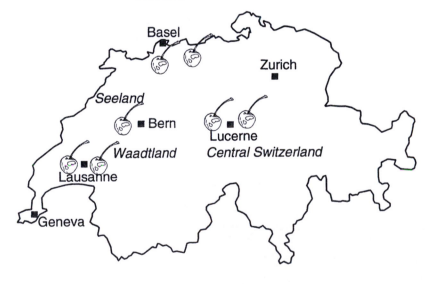

TADJIKISTAN

SWEET CHERRY
Area of Production (1984):
no commercial orchards but
1200 ha in home gardens
Production: average annual
production 3500 tonnes
from home gardens
Trend: not known
Domestic Use: 100%
Exports: none
Varieties:
Red/Black types:
Daibers Black
Negrityanka
White types:
Local pink-yellow
Drogans Gelbe
Bagration

Rootstocks:
Prunus mahaleb seedlings
Prunus avium seedlings

SOUR CHERRY
Area of Production (1984):
no large commercial
orchards; 300 ha total
commercial production; 900
ha in home gardens
Production: average annual
production 900 tonnes from
commercial plantings and
2500 tonnes from home
gardens
Trend: not known
Domestic Use: 100%
Exports: none

Varieties:
Lyubskaya 15
Griotte d'Ostheim
Kochs (Podbielski)
Samarkand
Shpanka chernaya
Anatolie
Rootstocks:
Mostly grown on their own
roots but *P. mahaleb*
seedlings also used

Location of Production:
Over most of country except
in mountainous areas

Source of Information:
Dr A.I. Sychov (see Ukraine)

TURKEY

SWEET CHERRY
Area of Production (1990):
6.3 million trees (area not known)
Production (1990): 143,000 tonnes
Trend: information unavailable
Domestic Use:
Fresh: 71%
Processed: small quantities only
Exports:
Fresh: 29%
Processed: none
Type of Processed Product:
canned fruits, jams
Varieties:
0900 Ziraat
Van
Turfanda
Bing

Lambert
Vista
Rootstocks:
Mahaleb seedlings
Mazzard seedlings

SOUR CHERRY
Area of Production (1990):
4.3 million trees (area not known)
Production (1990): 90,000 tonnes
Trend: information unavailable
Domestic Use: 5%
Fresh: 5%
Processed: 75%, of which 12% juice
Exports:
Fresh: none
Processed: 20% (18% juice)

Type of Processed Product:
juice, jams, marmalade, canning fruits, drying and some pie fillings
Varieties:
Kütakya
Montmorency
Macar
Rootstocks:
Mahaleb seedlings

Location of Production: see map

Source of Information:
Professor Dr N. Kaska, Dr Ali Küden, Dr Ayzin Küden, Department of Horticulture, Faculty of Agriculture, University of Cukurova, 01330 Adana, Turkey

UKRAINE

SWEET CHERRY
Area of Production (1993):
16,700 ha in commercial orchards; 14,100 ha in home gardens
Production (1993): 17,000 tonnes from commercial orchards; 44,300 tonnes from home gardens
Trend: declining in commercial orchards but increasing in home gardens since 1961
Domestic Use:
Fresh: 60–70%
Processed: 30–40%
Exports: not known
Type of Processed Product:
fruits in syrup
Varieties:
Red/Black types:
Valerii Chkalov
Anons
Badacson Black
Vinka
Dilemma
Dneprovka
Donetskii ugoloyk
Donetskaya krasavitsa
Izyumnaya
Iyunskaya rannyaya
Krupnoplodnaya
Melitopolskaya rannyaya
Melitopolskaya chernaya
Syurpriz
Yaroslavna
Nektarnaya
White types:
Aelita
Donchanka
Cosmicheskaya
Vystavochnaya
Rosovaya mleevskaya
Drogans gelbe
Rootstocks:
Prunus mahaleb seedlings
Prunus avium seedlings
Prunus cerasus seedlings of Griotte d'Osteim cv.

SOUR CHERRY
Area of Production (1993):
6200 ha of commercial orchards and 60,000 ha in home gardens
Production (1993): 2400 tonnes from commercial orchards and 191,000 tonnes from home gardens
Trend: significant decline since 1970 in commercial orchards but slow increase in home gardens
Domestic Use:
Fresh: 30–40%
Processed: 60–70%
Exports: not known
Type of Processed Product:
jams, juices, fruits in syrup, alcohol (cordial and wine), cherry pie (*vareniki* – a Ukrainian national dish)
Varieties:
Lyubskaya
Kochs (Podbielski)
Griotte d'Osteim
Chernokorka
Shpanka rannyaya
Ukrainian griotte
Turgenevka
Molodezhnaya
Shpanka Donetskaya
Biryulevskaya
Rootstocks:
Prunus mahaleb seedlings
Prunus avium seedlings

Location of Production:
Sweet cherry: mainly in Zaporozh'e province, especially Melitopol district Donetsk and Vinnitsa provinces
Crimea
Home gardens in south of country
Sour cherry: mainly in Donetsk, Dnepropetrowsk, Zaporozh'e, Lugansk, Vinnitsa provinces
Home gardens throughout the country

Source of Information:
Dr A.I. Sychov, Donetsk Branch of Ukrainian Horticulture Institute, 343420 Onytnoe Artemoresk, Donetsk District, Ukraine and Dr A.S. Devjatov, The Byelorussian Research Institute for Fruit Growing

UNITED KINGDOM

SWEET CHERRY
Area of Production (1993):
866 ha
Production (1993): 3400 tons
Trend: constant
Domestic Use:
Fresh: 100%
Processed: nil
Exports:
Fresh: nil
Processed: nil
Varieties
Merchant
Merton Glory
Van
Stella
Colney
Hertford
Lapins
Sunburst
Napoleon
Early Rivers
Rootstocks:
Colt
Location of Production:
Kent

SOUR CHERRY
Only a few hectares are
planted, mainly in the
southeast of the country

Source of Information:
Dr A.D. Webster, HRI East
Malling, West Malling, Kent
ME19 6BJ, UK

London

USA

Sᴡᴇᴇᴛ Cʜᴇʀʀʏ
Area of Production (1989):
19,400 ha
Production (1989): 173,800
tons
Trend: constant
Domestic Use:
Fresh: 53%
Processed: 47%
Exports:
Fresh: 18,000 tonnes
Type of Processed Product:
15% canned and 31% brined
Varieties:
Bing
Black Tartarian
Burlat
Chinook
Corum
Emperor Francis
Gold
Hedelfinger
Hudson
Lambert

Napoleon
Rainier
Republican
Schmidt
Stella
Van
Windsor
Rootstocks:
Mazzard seedlings (*P. avium*)
Mahaleb seedlings (*P. mahaleb*)

Sᴏᴜʀ Cʜᴇʀʀʏ
Area of Production (1989):
19,700 ha
Production (1989): 122,900
tons
Trend: declining slightly
Domestic use:
Processed: 100% but
includes some exports
Type of Processed Product:
frozen or canned for pie

fillings; juice
Varieties:
Montmorency
Rootstocks:
Mahaleb seedlings
Mazzard seedlings

Location of Production:
Washington State (largest
producer of sweet cherries)
Oregon
California
Michigan (largest producer
of sour cherries)
New York State

Source of Information:
Dr Don Ricks, Department of
Agricultural Economics,
Michigan State University,
East Lansing, Michigan,
USA and United States
Department of Agriculture

UZBEKISTAN

SWEET CHERRY
Area of Production (1986):
only 300 ha of commercial
orchards but 3500 ha in
home gardens
Production: average annual
production 900 tonnes from
commercial orchards and
10,000 tonnes from home
gardens
Trend: not known
Domestic Use:
Fresh: not known
Processed: not known
Exports: none
Type of Processed Product:
not known
Varieties:
Red/Black types
Bakhor
Kara-geles
Bigarreau Reverchon

White types
Drogans gelbe
Cosmicheskaya
Rootstocks:
Prunus mahaleb seedlings
Prunus avium seedlings

SOUR CHERRY
Area of Production (1986):
200 ha only in commercial
orchards but 5500 ha in
home gardens
Production: average annual
production of 500 tonnes
from commercial plantings
and 13,500 tonnes from
home gardens
Trend: not known
Domestic Use: 100%
Exports: none
Type of Processed Product:
not known

Varieties:
Lyubskaya 15
Griotte d'Ostheim
Kochs (Podbielski)
Samarkand
Shpanka krupnaya
Rootstocks:
Most trees in home gardens
are on their own roots but *P.
mahaleb* seedlings are used
as well

Location of Production:
Home gardens over whole
of country; most
commercial orchards are
near to Tashkent (see
Kazakhstan map)

Source of Information:
Dr A.I. Sychov (see Ukraine)

former YUGOSLAVIA

SWEET CHERRY
Area of Production (1990):
27,112 ha or 4,826,000 trees
Production (1990): average of
56,438 tonnes per year; yields
per ha range from 12 to 15
tonnes per ha
Trend: was increasing
Domestic Use:
Fresh: 36.6%
Processed: 56.9%
Exports:
Fresh: 1%
Processed: 4.6–5.5% exports to
many European countries
(Germany, Austria, France, UK,
the Netherlands); India,
America, Australia, Turkey and
others
Type of Processed Product:
frozen fruits and fruits in
alcohol; also compote, jam,
marmalade, Serbian slatko,
candied fruits and dried fruits
Varieties:
Primavera
Bigarreau Hativ Burlat
Souvenir des Charmes
Van
Stark Hardy Giant
Stella and its compact clone
Hedelfinger Riesenkirsche
Germersdorfer Grosse Kirsche
Bing
Lambert
Drogans gelbe
Knopelkirsche
*A number of local cultivars are
planted for local markets:*
Gomulska Rana
Vipavka
Alica
Tugarka
Dalbastija
Ohridska Crna
Kasni crveni Hrust
Donissens Gelbe Knopelkirsche
*Several new cultivars being
tested are:*
Asenova Rana
Carna
Rootstocks:
Prunus avium seedlings 97%
Prunus mahaleb seedlings 2.5%
F.12/1 a very small percentage

SOUR CHERRY
Area of Production (1990):
30,442 ha or 15,221,000 trees
Production (1990): 112,594
tonnes on average, ranging
from 57,000 to 153,000 tonnes;
yields range from 15,000 to
20,000 kg per ha (Oblacinska cv.
about 30,000 kg per ha)
Trend: small increase in
production but decrease in tree
numbers between 1985 and
1989
Domestic Use:
Fresh: 1.5%
Processed: 51.3%
Exports:
Fresh: 4.57%
Processed: 42.7%; exports to
same countries as sweet
cherries
Type of Processed Product:
frozen fruits, both with and
without stones; roll-fruits (a
special mix of fresh fruits);
concentrate juice; syrup;
compote with and without
stones; pasteurized with and
without stones; dry fruits,
especially Maraskâ cherry;
visnjevaca (fruit in alcohol)
drunk as spirit; maraschino;
Serbian slatko
Varieties:
Traditional cvs:

Oblacinska (70%)
Heimanns Konserven (Fanal)
Maraskâ and its selections
Duguljast Sokolusa
Brac No. 6
Vodica
Poljicka
Other important cvs:
Rexelle
Heimanns Konservenweichsel
Kelleriis 14 and 16
Sumadinka
Local cvs:
Meteor korai
Reine Hortensia
Cacanski rubin
Montmorency
Majurka
Double Gorsem Kriek
Rootstocks:
P. avium seedlings of wild trees
97%; the remainder *P. mahaleb*
seedlings

Location of Production:
In many regions of Serbia,
Macedonia, Croatia,
Montenegro, Slovenia and
Bosnia/Herzegovina

Source of Information:
Professor S.A. Paunovic,
Faculty of Agronomy,
Department of Horticulture,
Cara Dušana 34, 32000 Cacak,
Yugoslavia

II PLANT MATERIALS

3 Sweet Cherry Scions: Characteristics of the Principal Commercial Cultivars, Breeding Objectives and Methods

G. Bargioni

Via G. Medici 4, 37126 Verona, Italy

3.1 Introduction

Sweet cherry cultivars, which belong to the species *Prunus avium* L., are mostly diploids ($2n = 16$), although triploid and tetraploid ($2n = 24$ or 32) are also occasionally found (Fogle, 1975). The most notable biological characteristic of *P. avium* is its incompatibility; not only are most cultivars self-incompatible but the majority are also incompatible with other cultivars within the same incompatibility grouping. The inevitable limitations on fruit set imposed by these compatibility complications are discussed more fully in Chapter 8. Fortunately, there are some cultivars which, although incompatible (self-sterile), are able to pollinate any other cultivar, not just those in other incompatibility groupings; these are often referred to as universal donor cultivars. Only a few fully self-fertile cultivars have arisen naturally and these and their offspring are discussed later in this chapter.

Choice of cultivars by commercial sweet cherry producers may be strongly influenced by considerations of compatibility and by coincidence of flowering times. However good the yields and quality of its fruits, a cultivar will not be popular if it requires the planting of numerous poor and unprofitable cultivars in order to pollinate it.

Another typical characteristic of *P. avium* cultivars is their high vigour and lack of floral and/or fruiting precocity. Most other temperate tree fruits are now produced on dwarf trees, which are more productive and much easier to manage than the traditional large standard trees. The typical large size of sweet cherry trees has made them unpopular with modern orchardists. Also, many of the old, traditional cultivars grown in Europe were extremely slow to come into cropping and have become less popular with growers, who now need more precocious cropping to provide more rapid return on their orchard investments.

Sweet cherry growers in many parts of the world are also troubled by bird damage to fruits. Sprays of chemical bird repellents are in many countries environmentally unacceptable and only if the trees can be reduced in size and enclosed within nets will the problem be solved. A further problem besetting sweet cherry producers in many major production areas is fruit splitting as a result of rain at or just before harvest time. Either more dwarfed trees, enabling protection with rain shelters, or cultivars resistant to splitting are urgently needed.

Recently, the introduction of self-compatible and compact mutant cultivars, as well as the promise of more dwarfing rootstocks, has given new confidence to growers contemplating planting sweet cherries. If these advances can be sustained and built upon with the breeding of splitting-resistant cultivars, then a renaissance in cherry production can be anticipated.

3.2 The Origins of Sweet Cherry: Its Improvement and Cultivation

Although a very large number of sweet cherry cultivars have been described and some of these still form the basis of production in many countries, the great majority of cultivars resulted from natural selection carried out within the original areas of production and have spread very little outside these original areas. Many cultivars, some of which are still grown on a limited scale, were first described centuries ago (Basso, 1982; Iezzoni *et al.*, 1990), while others derived from the open pollination of these old cultivars and selection within the resultant seedlings. Production of new cultivars from controlled crosses between known parents, chosen on the basis of their desirable characteristics, is more recent, and the use of novel techniques, such as induced mutation breeding, has been used only to a very limited extent and only quite recently.

The European countries and parts of Asia (e.g. Turkey) are believed to be the most ancient areas of sweet cherry cultivation. For a very long time, little or no change occurred in the cultivars grown in each distinct area of production. This is perhaps explained by the great variation in agronomic performance which was almost certainly evident when cultivars were transferred to areas which differed in soils and climate. In addition, the new area may not have included suitable pollinating cultivars which would be vital to the productivity of the transferred cultivar. Also, sweet cherries were traditionally produced in rural, often hilly or mountainous, areas. The desire for change has always been less in such areas and this, coupled with their minimal contact with similar growers in other regions, would have greatly limited the exchange of cultivars.

Genetic improvement of the sweet cherry by controlled breeding began only in recent times, undoubtedly encouraged by the need for new genotypes able to meet the increasing and changing demands of the market for cherries, both for fresh consumption and for processing. Breeding aims have since expanded to include self-fertile cultivars, cultivars resistant to pests, diseases and inclement weather conditions, and cultivars suited to mechanical harvesting. A review of future breeding goals for cherry breeders and the germplasm available to them for use as parents is presented by Iezzoni *et al.* (1990).

Today, France, Germany, Italy, Spain, Switzerland, Turkey and the USA are the largest producers of sweet cherries (Iezzoni *et al.*, 1990), although figures for some countries, such as China, are difficult to interpret. Some cultivars are grown in almost all these countries; Burlat, Germersdorfer, Hedelfinger, Moreau, Napoleon and Van are all produced in several countries. Other cultivars are much more limited in their commercial distribution. Some are produced only in one country; examples would be Reverchon in France, Tragana of Edessa in Greece, Turfanda in Turkey, Cerna Edra in Bulgaria, Boambe de Cotnari in Romania, Ambrunés in Spain and Schauenburger in Switzerland. Occasionally the production of a cultivar may be concentrated in only one region; in Italy Mora di Cazzano is grown principally in the Veneto region and Malizia in the Campania region.

3.3 Classification of Sweet Cherry Cultivars

As discussed in Chapter 1, classification of sweet cherry cultivars is a subject about which there is only limited consensus of opinion. Most systems of classification split the cultivars into groups based upon their skin colour, colour and firmness of the fruit flesh, juice colour and time of ripening.

In Italy, sweet cherry cultivars are divided into two main groups:

1. cherries with soft flesh (tenerine);
2. cherries with firm flesh (duroni).

Each of these main groups is then further subdivided into two more groups, those with black and those with light-coloured fruits. Usually, the black subgroup have black skin and red or dark-coloured flesh and juice. The light-coloured cultivars have yellow flesh (occasionally with patches of red overcolouring), light yellow or white flesh and yellow to colourless juice.

In the UK, the USA and France, sweet cherry cultivars are distinguished simply on the basis of flesh firmness. Hearts, Geans or Guignes with soft flesh all equate with the Italian tenerines and the firm-fleshed Bigarreaux with the duroni types. Although Grubb (1949) suggested a classification based upon juice colour and subgroupings determined by time of ripening, this system is not now used, even in Britain where it had its origins.

In Romania the species is split into three different botanical varieties. *P. avium* var. *silvestris* is characterized by cultivars with small fruits with red and soft flesh, such as Früheste der Markt. *P. avium* var. *juliana* has cultivars within it with large fruits and moderately firm flesh, such as Ramon Oliva, while *P. avium* var. *duracina* equates with the Bigarreaux cultivars with hard-fleshed large fruits, like Hedelfinger or Pietroasa de Cotnari.

Classification in Germany and Belgium is quite different and is based mainly upon the ripening times of the cultivars. Cultivars are commonly listed as first-, second- through to fifth-week cultivars.

Sweet cherries are frequently referred to as small-, medium- or large-sized and several researchers have endeavoured to quantify these size divisions. Baldini (1973) based size groupings on fruit volume; large fruits were greater than $7 cm^3$, medium 4–7 cm^3 and small fruits less than 4 cm^3. Aeppli *et al.* (1982) chose to determine size on the basis of fruit diameter, which is, in fact, the basis of grade standards for sweet cherries within the European Community. Large fruits were those larger than 22 mm, medium from 18 to 22 mm, small from 16 to 18 mm and very small those less than 16 mm.

Classification of sweet cherry cultivars into groups is never simple, irrespective of the method used. Fruit firmness, the basis of many classifications, may be influenced by climate, crop load and rootstock. Time of flowering and ripening can also be greatly influenced by climatic conditions and rootstocks. These artificial classifications can, at best, give only an approximate indication of a sweet cherry's characteristics.

From the viewpoint of the sweet cherry grower the most important characteristics, in addition to fruit yield, ripening time, size and skin colour, are the cultivar's time of flowering and compatibility with other cultivars, the resistance of the fruits to rain-induced splitting, the firmness of its fruit flesh and the perishability of fruits during marketing.

3.4 Cultivars of Commercial Importance

In the following summarized descriptions of the most important commercial cultivars, the size of fruits is determined using Baldini's method (1973). Ripening dates are in relation to standards and are averaged where cultivars are grown in more than one region or country; in most instances Burlat is used as the standard. The incompatibility grouping, where mentioned, is generally that according to the classification devised by Knight (1969) and later improved by others.

3.4.1 Self-sterile cultivars

Adriana

This is an Italian cultivar which was selected by G. Bargioni at the Istituto Sperimentale di Frutticoltura of Verona, Italy, from the cross Mora di Cazzano × ISF 123 made in 1964; it was introduced in 1980 (Bargioni, 1980).

The tree has a vigorous, erect habit when young, but later develops a spreading canopy, with heavy crops on both spurs and 1-year-old shoots. The bearing tree needs heavy pruning to avoid production of blind wood. Blossoming is precocious.

The fruit is large, roundish, of very uniform size, with a thick, red or dark red skin. The flesh is firm, very juicy and sweet; the juice is pink to red in colour. The stone is large, roundish, but not free. The fruit stalk is thick and very green. The susceptibility of the fruit to cracking, to brown rot and to mould is particularly low.

The date of ripening is about 10–12 days after Burlat. Picking can be delayed without decay of the fruit.

Adriana is pollinated by Giorgia, Mora di Cazzano or Vittoria.

Ambrunés

This is an old Spanish cultivar, grown mainly in the Estremadura region of Spain.

The tree is of medium vigour; the canopy, if free-grown, shows a characteristic upside-down heart shape; it has a very low susceptibility to *Coryneum* disease. Blossoming begins early in the season but is of long duration. The tree productivity is moderate and consistent from season to season. Grafting of Ambrunés on *Prunus mahaleb* is not always successful.

The fruit, which is very suitable for export by shipping, is large, oval (a few heart-shaped), with a prominent apex. The skin is dark rose to wine-coloured. The flesh is very firm, crisp, red-wine-coloured and juicy, with a juice which is red to red-wine-coloured. The stone is large.

At maturity, an abscission layer forms between the fruit and the fruit stalk, so usually the cherries are picked and marketed without stalks.

Ambrunés ripens about 30 days after Burlat.

It is pollinated by Hedelfinger, Moreau, Napoleon, Ramon Oliva and Tigre.

Annabella

This is a German cultivar raised at the Experiment Station of Jork in northern Germany from a cross between Rube and Allers Späte.

The tree is of medium vigour to vigorous, with a spreading canopy. Blossoming time is late. Cropping is precocious and consistent.

The fruit is of medium size, roundish but some elongated slightly, with a long stalk; the skin is dark red and shiny. The flesh is firm, with a low susceptibility to cracking.

The date of ripening is 25–30 days after Burlat.

Annabella is pollinated by Castor, Corum, Early Rivers, Frühe Rote Meckenheimer, Pollux, Schneiders Späte Knorpelkirsche and Venus.

Arcina (Fercer)

This cultivar, which resulted from a cross between Starking Hardy Giant and an unknown cultivar, was bred and released by the Institut National pour la Recherche Agronomique (INRA), at Grande Ferrade in France.

Trees are very vigorous with a semierect habit. Branching is frequently average but branch angles are good. Fruiting precocity is poor but productivity moderate to good. Flowering time is early midseason, at a similar time to or slightly earlier than Burlat.

Fruit size is regular, very large (>11 g), with skin red-purple, lightly striped and red flesh. Fruits are kidney-shaped and are moderately sensitive to cracking at the fruit stalk end.

Arcina ripens 17–18 days after Burlat.

Suitable pollinators are Van, Napoleon, Hedelfinger, Rainier, Burlat and Starking Hardy Giant.

Bing

Bing is the most popular and, up until the present time, the most important cultivar of sweet cherry grown in the USA, especially in the Pacific Northwest states. Originating from a seedling of Republican in 1875, Bing was introduced by the Lewelling nursery of Milwaukee, Oregon, USA.

The tree is vigorous, with erect branches when young, later forming an upright spreading tree with rather open habit, which blossoms early to midseason.

The fruit is large or very large, roundish, with red-black-coloured skin, but very susceptible to cracking; the flesh is dark red, very firm and crisp, sweet, very flavourful, aromatic and rich in soluble solids (16–20%); the juice is dark red. Because of the firmness of the flesh, Bing is a very popular export cherry, being shipped to many parts of the world.

The date of ripening is about 20–25 days after Burlat.

Bing may be pollinated by Burlat, Larian, and Van, and is classed within incompatibility group III. In its main production area in the Pacific Northwest of USA, cultivars such as Rainier and Chinook are also used. It is intersterile with Napoleon and Lambert.

Burlat

This French cultivar, of unknown parentage, is popular throughout the world, thanks to its heavy cropping and early ripening season. It was selected originally by L. Burlat at Pierre Benite, in the Rhône valley of France. The INRA, in France, has more recently selected the Burlat clone Infel V 370.

The tree is vigorous, erect in habit when young, spreading later. Blossoming time is midseason (10–15 April in the Bordeaux–Nîmes region of France).

The fruit is large, heart-shaped, some flattened on the suture side; the skin is red or purple red, shiny and of medium thickness, but susceptible or very susceptible to cracking. The flesh is of medium firmness or soft, juicy, sweet and semifreestone.

The date of ripening in France, in the Bordeaux–Nîmes region is 26–28 May.

Burlat is pollinated by Durone Nero 1, Durone Nero 2, Durone della Marca, Giorgia, Bing, Larian, Rainier, Starking Hardy Giant, Hedelfinger, Napoleon, Garnet and Van, but is incompatible with Moreau. It is thought to be within incompatibility group VII.

The Istituto Sperimentale per la Frutticoltura of Rome, Italy, selected Burlat C 1, which was introduced in 1983. This is a compact mutant clone obtained from buds of Burlat irradiated in 1969 as part of a cooperative research programme with the Ente Nazionale per l'Energia Atomica (ENEA). The size of a Burlat C 1 tree is about 25% less than that of standard Burlat.

Büttners Späte Rote Knorpelkirsche

This is a very old German cultivar which produces trees of great vigour that are medium erect in habit, with a medium–early blossoming season and good resistance to frost. It comes into bearing early in its life and its cropping is heavy and regular.

The fruit is of medium size, heart-shaped but more broad than long and flattened on the ventral side, where the suture line is reddish. The skin ground colour is yellow, with irregular reddish areas, and it is very shiny. The flesh is very firm, crisp, yellow-coloured, of medium juiciness (juice uncoloured) very sweet and aromatic, of typical cherry taste. Its susceptibility to cracking is high.

The date of ripening is about 30–35 days after Burlat.

This cultivar is pollinated by Hedelfinger, Schneiders Späte Knorpelkirsche, and Weisse Herzkirsche.

Castor

This is a variety raised at the Instituut voor de Veredeling van Tuinbouwgewassen at Wageningen in the Netherlands (Belmans, 1986). It is thought to originate from open-pollinated Schneiders Späte Knorpelkirsche.

Tree vigour is very strong and the habit is erect and rather poor, with sparse branching. The trees flower in the midseason, 4–5 days after Early Rivers. Yield precocity is poor and productivity average. Fruit drop may be severe in some seasons.

The fruits are large and cordate to round in shape. Skin colour is dark red to black and the fruits are initially very firm but soften rapidly during ripening. Flavour is good and aromatic. They are very resistant to cracking.

The fruits ripen in the late midseason period, about 3 weeks after Early Rivers.

Castor may be pollinated by Early Rivers, Kordia, Schneiders Späte Knorpelkirsche and Venus.

Chinook

This cultivar was selected in the USA by H.W. Fogle at the Washington Agricultural Experiment Station from a cross between Bing and Gil Peck. Introduced in 1960, it is now grown particularly in Oregon and Washington states, USA, where its maturity date falls between Burlat and Bing.

The tree is vigorous, upright, spreading, high-bearing and early-blooming.

The fruit is large, and heart-shaped to round; the skin is mahogany in colour, glossy but not very attractive; the flesh is firm, red to dark red, sweet, fairly acid but less so than Bing. Like Bing, it is susceptible to cracking.

Its date of ripening is about 10 days after Burlat.

Chinook is pollinated by Bing and Van, but, classed within incompatibility group IX, it is incompatible with Rainier.

Colney

This cultivar was bred at the John Innes Institute in the UK and was originally numbered JI.12510.

Tree vigour is moderate, less than Stella, and its habit is semierect but more spreading when mature. Colney is resistant to bacterial canker. Yield precocity and productivity are average.

Fruit is large, black-purple in colour, firm and of average flavour.

Ripening season is late (end of July–early August in Britain).

Suitable pollinators are universal donors, such as Merton Glory, or self-fertile cultivars, such as Stella. Its incompatibility grouping has not yet been ascertained.

Della Recca

This is a very old cultivar from the south of Italy.

The tree is of great vigour, high-bearing and erect in habit.

The fruit is of medium size and oblate and the skin is thick and yellowish to pale red on the exposed side. The flesh is of medium firmness, white-yellowish, very juicy and very sweet. Its susceptibility to cracking is low.

The date of ripening is about 23 days after Burlat.

Suitable pollinators are not listed in the literature.

Del Monte

This is another Italian cultivar, which is cultivated mainly in the south of Italy, near Naples.

The tree is of great vigour, erect in habit and high-bearing.

The fruit, which is considered very suitable for shipping, is large and heart-shaped, with a small depression, and it has a thick yellow-pink skin, which has a shiny dark red colour on the exposed side. The flesh is very firm, yellowish, juicy, sweet and slightly acid. The susceptibility to cracking is low.

The date of ripening is about 12–15 days after Burlat.

Suitable pollinators are not recorded in the literature.

Durone Nero 1

This cultivar, typical of the Vignola region of Italy, is of unknown origin, but is considered one of the best-tasting cherries in the world by many connoisseurs.

The tree is vigorous, with long extension shoots, rather erect in habit when young; later it develops a wide 'spreading' disordered canopy with medium to high bearing capacity.

The fruit is large or very large, heart-shaped, equal in width and length, with the dorsal side more developed than the suture side, which is slightly depressed. The skin is red to dark red or nearly black, bright and of medium to light thickness. The flesh is firm to very firm, fibrous, sweet, red to dark red but not very juicy. The susceptibility to cracking is medium to high.

The date of ripening is about 12–14 days after Burlat. Ripening occurs very quickly, so it demands accuracy in choice of picking time if fruits are to be harvested in their optimum condition.

Suitable pollinators are Adriana, Burlat, Durone Nero 2. Durone Nero 1 is incompatible with Durone dell'Anella.

Durone Nero 2

Another cultivar of unknown origin which, like its near namesake, is also widely planted in the Vignola region of Italy. The tree is vigorous, spreading in habit, with late blossoming and good bearing capacity.

The fruit is large, heart-shaped, roundish, often very similar to Durone Nero 1, with moderate susceptibility to cracking. The skin is rather thick and dark red to almost black; the flesh is firm, crispy, red, very juicy, semifreestone, aromatic, and of very good taste. The date of ripening is about 25–27 days after Burlat.

Durone Nero 2 is pollinated by Mora di Vignola, Durone della Marca and Durone Nero 1.

In Italy, the Istituto Sperimentale per la Frutticoltura of Rome and the Ente Nazionale per l'Energia Atomica (ENEA) introduced a compact mutant of this cultivar in 1983, named Nero 2 C 1. In 1988, Sansavini and Lugli introduced another compact mutant, obtained by the same Institutions and named Durone compatto di Vignola. In France INRA has selected the clone Infel V 720 A, which is also named Tardive de Vignola.

Early Rivers

This is a very old English variety, still grown to a small extent in the county of Kent, where it was introduced in 1872 (Grubb, 1949).

The tree is very vigorous, with medium erect to erect canopy, and when mature it can be a very high producer. Unfortunately, it is sometimes susceptible to bacterial canker (*Pseudomonas mors prunorum*). Blossoming time in many countries is in midseason or early midseason, although in England it is one of the very earliest to blossom.

The fruit is medium-sized, roundish oblate, a little broader than long. The apical portion is flattened and the fruit stands upon this area, making an angle of 45° (or very nearly) from the vertical (Grubb, 1949). The skin is dark red, almost black. The flesh is soft, dark red, very juicy, with coloured juice, sweet and very good in quality. The susceptibility to cracking is very low.

The date of ripening is about 10 days after Burlat, usually late June in England.

Early Rivers is usually pollinated by Noir de Guben, the only cultivar of any quality to flower at the same time in England. It is intersterile with Black Tartarian and is classed in incompatibility group I.

Early Rivers is now little planted in Kent, where small-fruited early cultivars are of poor profitability.

Emperor Francis

Grown in Britain quite extensively up until 20 years ago, this variety produces rather weak trees with a compact rounded head.

Fruits have some similarities with Napoleon, i.e. roundish, heart-shaped but slightly redder in colour; if left on the tree the fruits become even darker and do not resemble a white cherry. Fruit size is medium to large and flesh firm and yellowish in colour.

It ripens approximately 18 days after Burlat.

Emperor Francis blooms in midseason and is classed as being within incompatibility group III.

Ferrovia

This is an Italian cultivar of unknown origin, similar to Bella di Pistoia; it is mainly planted in the Bari province of Italy.

The tree is vigorous and erect in habit.

The fruits are heart-shaped, occasionally pointed, and only slightly susceptible to cracking. It is considered very suitable for shipping. The skin is bright red, shiny and thick; the flesh is firm, crisp, pink, red near the stone, clinging to the stone, very sweet and slightly acid, of very good taste.

The date of ripening is about 20–22 days after Burlat.

It is pollinated by the local cultivars Forli, Bismark and Colafemmina and also by Durone dell'Anella and Durone della Marca.

Frogmore

This very old English cultivar was originally planted extensively in the county of Kent, where it proved to be a useful market cherry on account of its very regular bearing.

The tree is vigorous, with a canopy that is rather round, upright and spreading, with branches bending down in consequence of the heavy crops. Blossom occurs over an extended period relatively late in the spring and it is considered quite resistant to frost damage.

The fruit is of medium size, heart-shaped or ovoid, slightly broader than long and not very susceptible to cracking. The skin is red on a yellow ground, with evident dots and streaks. The flesh is soft to very soft, a little fibrous, very sweet, with uncoloured juice, of good quality.

The date of ripening is about 7 days after Burlat.

Frogmore, which is pollinated by Roundel and Merton Bigarreau, is not now planted in the UK.

Garnet

This cultivar, which was bred by Marvin Nies in California, USA, is gaining some popularity in France.

Trees are moderately vigorous, semierect in habit but with sparse branching. Garnet crops precociously and shows good productivity. It flowers very early, 2–9 days before Burlat.

Fruits are large, occasionally very large; the colouring of the skin is red to purple; shape is reniform to round and they have short stalks. Fruit flesh is pink in colour and of good sweetness and flavour. Sensitivity to cracking is often severe.

Ripening is quite slow but fruits remain firm on the tree. Time of ripening is 10–12 days after Burlat.

Suitable pollinators are Van and Starking Hardy Giant.

Germersdorfer

This very interesting cultivar which is often confused with Schneiders Späte Knorpelkirsche, is mainly grown in Hungary but was first discovered in Guben. There is much variability in the varieties known as Germersdorfer in different parts of Europe.

The tree is vigorous, with a canopy shape similar to a wide pyramid; it blossoms late in the season and has high bearing capacity. The blossoms are, however, quite susceptible to frost damage.

The fruit is large and rather susceptible to cracking with skin red to dark red in colour, shiny, with small light spots. The flesh is firm, crisp, light red in colour with pink juice, and has a very aromatic balanced taste, sweet to slightly acid.

The date of ripening is about 30–35 days after Burlat.

Germersdorfer is pollinated by Hedelfinger and Techlo 2 but is incompatible with Napoleon and Bing.

Improved clones of Germersdorfer have recently been selected in Hungary.

Giorgia

This is an Italian cultivar, introduced in 1985; it was selected by G. Bargioni at the Istituto Sperimentale di Frutticoltura of Verona, Italy, from the cross ISF 123 × Caccianese made in 1964.

The tree is vigorous, rather erect in habit but comes into bearing early in its life and is a very high cropper and requires heavy pruning. Blooming is in early midseason and of long duration.

The fruit is large, with a large cavity, heart-shaped and more wide than long. The susceptibility to cracking is moderate. The skin is red, very shiny and rather thick. The flesh is firm, sweet, rather juicy and pale pink.

The date of ripening is about 7 days after Burlat.

Giorgia is pollinated by Adriana, Moreau, Durone Nero 1 and Van.

Gold

This is a small-fruited variety, grown in New York State and elsewhere in the USA for brining and processing. It is a particularly winter-hardy cultivar. It has no red pigment and is easily bleached for brining.

Gold ripens in mid-July in the USA.

It is in a unique pollination group.

Grosse Schwarze Knorpelkirsche

This is one of the very oldest cultivars from Germany, and is considered very suitable for shipping.

The tree is vigorous, high-bearing, erect in habit, with long limbs and very little lateral branching. Blossoming occurs medium to late and is quite resistant to frost damage.

The fruit is large, heart-shaped, roundish, but rather susceptible to cracking. The skin is dark red to brown-black; the flesh is very firm, crisp and of very good quality.

The date of ripening is about 25–30 days after Burlat.

A suitable pollinator is Büttners Späte Rote Knorpelkirsche.

Hartland

This is a new variety raised at New York State Agricultural Experiment Station, Geneva, New York, USA.

The trees are large, vigorous and with a spreading tree form, and produce many lateral branches. This variety is particularly winter-hardy. Blossoming is midseason, and blossoms remain receptive for long periods. Cropping productivity is very good, with trees bearing well on both spur and 1-year-old wood.

Fruits are medium large but may be smaller if the trees are allowed to overset. Fruit shape is round with a slight oblong tendency; the skin colour is deep purple and glossy. Flesh colour is similar to that of the skin. Sugar concentrations, at 14.5–17.0% are lower than many other commercial cultivars, but acceptable to taste panels. Fruits are medium firm and moderately resistant to cracking.

Ripening date is early midseason, about 15–20 days after Burlat.

Hartland is self-sterile and incompatible with other cultivars in group VI but compatible with most commercial cultivars flowering early midseason.

Hedelfinger

Hedelfinger, like Bing, is one of the most widely grown cultivars in the world, particularly in central Europe, but is popular also in Ontario, Canada, where it is considered precocious in bearing. Bradbourne Black, a cultivar very similar to Hedelfinger, is popular in England.

The tree is of medium vigour to vigorous, with upright growth when young; later it develops wide and spreading branches, assuming a typical wide pyramidal shape. Blooming is medium late to late, and quite resistant to frost.

The fruit is medium to large, ovoid to heart-shaped; the susceptibility to cracking is medium to low. The skin is thick, brown-red in colour, with light red speckling and striping. The flesh is firm, red to dark-violet-coloured, juicy and clingstone, with a slight flavour of almonds.

The date of ripening is about 25 days after Burlat.

Hedelfinger is pollinated by Basler Langstieler, Heidegger, Kordia, Schauenburger, Germersdorfer, Star and Vittoria. It is possibly within incompatibility group VII, although clones held at Vineland in Canada were classed as universal donors.

Hertford

This is a variety raised by the John Innes Institute in the UK which shows some resistance to bacterial canker.

Tree vigour is moderate to vigorous and habit is good, with spreading branches.

Hertford fruits are firm, large, dark red/purple in colour, with a firm flesh and good flavour. Fruit stalks are short.

Ripening season is midlate, similar to Van.

Suitable pollinators are universal donors, such as Merton Glory, or self-fertile cultivars, such as Stella. Its incompatibility grouping has not yet been ascertained.

Hudson

This is another cherry cultivar bred at the Geneva, New York State, Research Station, USA.

It produces very large trees with an open habit and is a late-blooming cultivar. Productivity is only moderate.

Fruits have a good sweet flavour and are firm, dark in colour and medium to large size.

It is classed as the latest-ripening sweet cherry grown in New York State. Hudson is classed within incompatibility group IX.

Karabodur

This is a cultivar extensively grown in Turkey.

The tree is vigorous, erect in habit, with late blooming and regular productivity.

The fruit is large, heart-shaped, rather broad; the skin is yellow, covered by a pink to reddish blush on the sunny side. The flesh is firm and sweet, with a very good taste.

The date of ripening is about 25 days after Burlat.

Karabodur is pollinated by Van.

Katalin (Katalin 261)

This is a variety bred in Hungary by S. Brozik from a cross between Germersdorfer and Podjebrad.

The tree is only moderately vigorous, weak on some sites, and of upright habit when young, drooping later in life. Flowering time is medium late and the variety is reported to be extremely productive. It is suitable for dessert use as well as for freezing and canning. It is sometimes machine-harvested in Hungary.

The fruit is large (24–27 mm diameter) and cordate in shape. The flesh colour is red and the skin colour vermilion to mahogany. The fruit is firm with a good acid : sweet flavour balance.

Katalin ripens late in the season, i.e. the seventh week in relation to Burlat.

It is pollinated by Van, Linda and Germersdorfer 3.

Kordia

This variety, which is of unknown parentage, was discovered by scientists in the Czech Republic growing as a chance seedling in Techlovice, a village near Hradec Kralove in eastern Bohemia. Discovered in 1963 and first selected with the name Techlovicka II (Aeppli *et al.*, 1982) the variety was registered in the national variety list of Czechoslovakia in 1981.

Tree vigour is very strong in young trees, later moderating to form trees of rounded habit. The variety flowers late in the spring and is very productive.

Fruits are heart-shaped and slightly elongate in shape. The skin is carmine-red to dark violet and the flesh is red with pinkish streaks. The fruits are firm and have an excellent, slightly spicy taste with good balance of sweetness and acidity.

Fruits ripen about 26–28 days after Burlat.

Kordia is self-incompatible and may be pollinated by many other varieties, such as Regina, Hedelfinger, Star, Schauenburger.

Kristin

Kristin originated from the cross Emperor Francis × Gil Peck made in 1936 at the New York State Agricultural Experiment Station, Geneva, USA, and was introduced to commerce in 1982.

The tree is vigorous and very winter-hardy; in Norway it showed no damage to fruit buds or trunks even after several days at $-18°C$. Blooming is midseason in comparison with other cultivars.

The fruit is large, with dark red or purplish black skin and very attractive; the flesh is red or dark red, firm, juicy, sweet and of very good eating quality. The pit is small. Kristin is only slightly susceptible to cracking and is suggested as a replacement for the cultivar Schmidt.

It ripens about 16–18 days after Burlat.

Good pollinators for Kristin are not listed, but any variety which is not in incompatibility group III and flowers midseason should suffice.

Lambert

Lambert was obtained from a seedling found in Oregon, USA, by J.H. Lambert.

The variety forms vigorous trees, erect in habit, with medium to late blossom time.

The fruit is medium to large, heart-shaped, some flattening on the ventral side, with a pointed apex and susceptible to cracking. The skin is purple-red with dots and some streaks, moderately shiny. The flesh is firm, dark red, a little fibrous, juicy and of good taste.

Lambert ripens approximately 20 days after Burlat.

Lambert is pollinated by Sam, Summit and Van but is incompatible with Bing, Emperor Francis, Napoleon, Star and Vernon; it is classed within incompatibility group III.

A mutant with compact habit (Compact Lambert) was produced by K.O. Lapins at the Summerland Research Station, British Columbia, Canada, in 1963.

Larian

This is a cultivar selected by R.M. Brooks at the California Agriculture Experiment Station from the cross Lambert × UCD 50 (Bing × Black Tartarian) made by G.L. Philp in 1946. Larian was introduced in 1964.

The tree is moderately vigorous, upright but also spreading, and productive, with a midseason blooming time.

The fruit is large and round to cordate, with low susceptibility to cracking. The skin is dark red and thin and the flesh is moderately firm to firm, red to dark red, with red juice, which is sweet to slightly acid and of good taste.

The ripening date is about 10 days after Burlat.

Larian is pollinated by Black Tartarian, Burlat, Van, Bing, Bada, Napoleon and Starking Hardy Giant.

Linda (Linda 156)

This variety originated in Hungary and was selected by S. Brozik from a cross between Hedelfinger and Germersdorfer.

The tree is of medium vigour with a drooping habit. It flowers late in the spring and is reported to be heavy-yielding and very easy to harvest.

The fruit is of medium size (22–26 mm diameter) and slightly elongated in shape. The flesh is purple in colour, while the skin is deep mahogany. It is particularly juicy, firm and of excellent flavour.

Linda ripens midseason to late, i.e. in the fifth week of harvest in Hungary. It is pollinated by Germersdorfer 3, Katalin and Van.

Malizia

This is a cultivar grown in the Campania region of Italy, of unknown origin.

The tree is of great vigour, with a spreading canopy and good productivity.

The fruit is large, heart-shaped, very suitable for shipping; but susceptibility to cracking is high. The skin is dark red; the flesh, which is very firm and crisp, is red, juicy and of very good taste.

The ripening date is about 20 days after Burlat.

Suitable pollinators are not recorded in the literature.

Margit (Margit 66)

This is a variety which was selected in Hungary by S. Brozik from seed collected from the cross Germersdorf Orias cl. 3 × Békéscsaba and introduced in 1991.

The tree is of medium vigour with an elongate to round habit. It flowers medium late and has good productivity.

The fruit is medium in size (23–26 mm diameter), flat-round in shape and very firm and has very good flavour. It is suitable for mechanical harvesting. Skin colour is mahogany and flesh colour purple.

The ripening date is medium early, about 6–7 days, after Burlat in Hungary.

Margit is pollinated by Van, Hedelfinger, Linda and Germersdorfer 3.

Merchant

This cultivar was raised by the John Innes Institute in the UK from open-pollinated Merton Glory.

The tree is moderately vigorous with a spreading habit when mature and good branch angles. Cropping is precocious and highly productive. Merchant flowers early midseason at a similar time to Merton Glory and before Van. Trees show good resistance to bacterial canker.

Fruit is of moderate size with black skin colour and good flavour. Merchant has a tendency to overcrop on some sites, reducing fruit size. It is only moderately susceptible to splitting.

Merchant ripens early midseason, about 15 days after Burlat.

Merchant is a universal donor which may pollinate or be pollinated by most other cultivars that have similar flowering dates.

Merton Bigarreau

This is a cultivar selected from the cross Knight's Early Black × Napoleon made by M.B. Crane at the John Innes Institute in the UK.

The tree is vigorous and very spreading, with regular and very high cropping.

The fruit is large or very large, roundish to almost spherical. Its susceptibility to cracking is moderate. The skin, of medium thickness, is deep mahogany-coloured. The flesh is very firm, dark red and extremely rich in flavour.

The ripening date is about 18–22 days after Burlat.

Merton Bigarreau is pollinated by Merton Glory.

Merton Glory

This cultivar was produced by the John Innes Institute in the UK.

Trees are of moderate vigour with an excellent habit, with many wide-angled branches. Fruiting precocity and productivity are both very good. The trees are particularly resistant to bacterial canker. Flowering is midseason, slightly earlier than Van.

Fruit is large, with a creamy yellow background colour flushed with red on the exposed surfaces. Flavour and sweetness are good but the fruits have a tendency to be rather soft and are difficult to market unless handled very carefully.

Merton Glory ripens semiearly, approximately 10 days after Burlat.

It is a universal donor cultivar and can pollinate or be pollinated by most other cultivars that flower at the same time.

Merton Heart

This is an English cultivar, selected from the cross Schrecken × Elton made by M.B. Crane at the John Innes Institute in the UK.

The tree is vigorous to very vigorous, very upright, but irregular in cropping. Blossom time is midseason or a little later; blossoms are very susceptible to *Monilinia laxa*.

The fruit is large, rather long heart-shaped, regular, with a slight furrow on the dorsal side. The susceptibility to cracking is low. The skin is moderately shiny, deep purplish crimson, becoming almost black when fully ripe. The flesh is moderately firm or slightly soft, dark red, juicy, of very good, slightly acid taste.

The ripening date is about 12 days after Burlat.

Merton Heart is pollinated by Merton Glory, Merton Bigarreau and Roundel Heart.

Mora di Cazzano

Of unknown origin, this cultivar is planted extensively in the Verona province of Italy. It is considered very suitable for shipping.

The tree is vigorous, rather erect in habit when young and later wide-spreading.

The fruit is large, roundish heart-shaped, with a small depression on the ventral side. The susceptibility to cracking is very low, almost as good as that of Adriana. The skin is thick and uniformly red or dark red and shiny; the flesh is very firm, crisp, pink, juicy, sweet and slightly acid.

The date of ripening is about 15–18 days after Burlat.

It is pollinated by Adriana, Giorgia, Ferrovia, Caccianese, Moreau, Starking Hardy Giant and Van, but it is incompatible with Mora dalla Punta.

Moreau (Bigarreau Moreau)

This is a French cultivar of unknown origin found by M. Sandrin at Hameau des Charmes, France. It was introduced by the nurseryman E. Moreau at the beginning of the twentieth century.

The tree is of medium vigour with fairly erect habit and of medium productivity.

The fruit is large, slightly larger than Burlat, roundish and some are flattened on the ventral side, with a furrow on the dorsal side, of greater width than length. Its early to midseason blossoms are susceptible to frost damage. The susceptibility to cracking is low to medium. The skin, which is thinnish, is dark red to red-black and moderately shiny. The flesh is quite firm, red, sweet and juicy.

The ripening date is similar to Burlat or sometimes 1–2 days earlier.

Moreau is pollinated by Bing, Durone Nero 1, Durone Nero 2, Adriana, Giorgia, Mora di Cazzano, Napoleon and Van but is incompatible with Burlat. It is classed in incompatibility group VII.

In France INRA has selected the clone Infel 1439, while in Italy F. Faccioli has selected the clone B, which is said to be more productive than the standard clone.

Nalina

This is a cultivar bred in 1962 at the Institut für Obstforschung Dresden-Pillnitz in Germany from open pollination of the cv. Braunauer and obtained using *in vitro* culture of embryos; it was introduced in 1986.

The tree is of medium to high vigour, with a wide roundish canopy and very green leaves; the blossom time is early season; it is precocious in bearing and the cropping is good.

The fruit is large, heart-shaped, with red-coloured skin and with medium susceptibility to cracking. The flesh is of medium firmness, juicy, sweet, slightly acid and slightly aromatic.

The ripening date is similar to that of Burlat.

It is pollinated by Namosa and Early Rivers, but it is incompatible with Burlat and Van.

Namosa

This German cultivar was obtained in 1961 at the Institut für Obstforschung Dresden-Pillnitz in Germany from open pollination of Farnstadter Schwarze.

The tree has a medium to vigorous habit and a round canopy when young, later developing a flattened canopy. It blooms early in the season, has a good resistance to frost and has good cropping productivity.

The fruit is large, roundish, with a brown-coloured skin. The retention force between fruit and its stalk is low, so Namosa is suitable for mechanical harvesting. The flesh is of medium firmness, sweet and aromatic. The susceptibility to cracking is very low.

The date of ripening is about 15–18 days after Burlat.

Namosa may be pollinated by Burlat, Early Rivers, Napoleon and Van.

Napoleon (Royal Ann)

Napoleon is one of the older European cultivars which originated in Germany in the second half of the eighteenth century, originally with the name of Lauerman. It was later named Napoleon in honour of Napoleon Bonaparte.

Subsequently, it was distributed in the USA and Canada with the name of Royal Ann. This cultivar, of unknown parentage, is one of the most important cultivars in the world on account of its widespread use for processing.

The tree is vigorous and rather erect when young, medium erect to wide spreading when 12–15 years old. Blossoming time is medium late.

The fruit is of medium or large size, heart-shaped, nearly as long as broad, regular, a few flattened on the suture side. The skin is pale yellow, covered by a shining red blush, especially on the exposed side. Napoleon has moderate susceptibility to cracking. The flesh has a pale yellow colour, is of medium firmness, a little fibrous, very juicy and of very good quality.

The date of ripening is about 18–22 days after Burlat.

Napoleon is pollinated by Burlat, Durone della Marca, Larian, Mora di Cazzano, Rainier, Van and Vittoria. It is incompatible with Bing, Lambert and Star and is classed within incompatibility group III.

In the south of Italy they grow a very similar cultivar, named Imperiale. Two interesting clones of Napoleon have been selected in France by INRA: the clone Infel 1884 with large fruit and the clone Infel 1007 with smaller fruit.

Noble

This is a cultivar of English origin that demands a very good site, particularly with respect to soil depth and fertility.

The tree is vigorous with a large canopy, and is late in blossoming and with good bearing productivity.

The fruit is large to very large, regular, heart-shaped, a little broader than long, often slightly flattened on the suture side. The stem cavity is deep, considerably shallower at the suture side. The susceptibility to cracking is medium to high. The skin is dark red, with some dots and streaks. The suture line is darker. The flesh is firm or very firm, dark red-coloured and of good quality.

The ripening time falls about 30 days after Burlat.

Noble is pollinated by Bing, Lambert and Napoleon and is classed within incompatibility group XII.

Noir de Meched

This cultivar, which originates in Iran, is quite extensively grown in France.

Trees are of moderate vigour, semierect in habit, with good sub-branch angles. Precocity is average and cropping productivity very good. Flowering time is late, 2–8 days after Burlat.

Fruit size is only moderate and skin colour dark purple. Fruit shape is round to oval and flesh is red in colour and firm. Eating quality is good and the fruit transports well. It is not very susceptible to fruit splitting and is very easy to pick.

Ripening is 23–25 days later than Burlat.

Suitable pollinators are Ulster, Tragana of Edessa, Hedelfinger and Duroni 3.

Rainier

This cultivar was bred at the Washington Experiment Station in Prosser, USA, from the cross Bing × Van made by H.W. Fogel, and was introduced in 1960.

The tree is vigorous, with canopy semispreading to spreading, and is early-blooming and high-bearing.

The fruit is large, slightly obovate, similar in shape to that of Bing, with moderate susceptibility to cracking; the skin is yellow and blushed pink and a few are completely covered by red colour on the exposed fruit surface. The flesh is firm, white-yellowish, with colourless juice, of very good taste. The stone is fairly free and small to medium.

The date of ripening is about 18–20 days after Burlat.

Rainier is pollinated by Bing, Burlat, Napoleon, Starking Hardy Giant, Ulster and Van, it is classed within incompatibility group IX.

Three clonal variants of Rainier, differing in the degree of red skin colouring, have been selected.

Ramon Oliva

This is an old Spanish cultivar of unknown origin.

The tree is vigorous, wide-spreading, with a flattish round head canopy, considerably drooping. It is early-blossoming and of high productivity.

The fruit is large, roundish, short truncate or heart-shaped and is rather susceptible to cracking. The skin is dark red or nearly black, and shining; the flesh, of medium firmness, is pale yellow, juicy, sweet and of good taste.

Ripening time is similar to that of Burlat.

Ramon Oliva is pollinated by Ambrunés, Marmotte and Moreau.

Regina

This variety was produced by the Jork Institute in Germany in 1957 from a cross between Schneiders Späte Knorpelkirsche and Rube. It was introduced to commerce in 1981.

Tree vigour is strong and trees are pyramidal in habit, with spreading, drooping branches. The trees flower 6–8 days later than Early Rivers and yield productivity is excellent.

The fruits are flat-round to round in shape and largish and are reported to have a low propensity to crack. Skin colour is dark red to black and the fruits are firmish, with a good, juicy, aromatic, sweet flavour.

The ripening period is late to very late, more than a month after Early Rivers.

Regina is self-incompatible, being pollinated by Annabella, Schneiders Späte Knorpelkirsche, Summit and Sam.

Reverchon

This ancient cultivar, of unknown parentage, probably originated in France.

The tree is very vigorous, erect in habit, with few branches, and late-blossoming; it is late coming into bearing and of medium productivity.

The fruit is large, heart-shaped, roundish, very susceptible to cracking but very suitable for shipping. The skin is carmine to dark-purple-coloured, and shining; the flesh is very firm, pink to pale red in colour, juicy and of good quality.

The date of ripening is about 20–25 days after Burlat.

Reverchon is pollinated by Hedelfinger.

In France INRA has selected the clone Infel V 1814 A.

Ron's Seedling

Ron's Seedling is an Australian cultivar, very widely planted in New South Wales. It originates from the cross Eagle Seedling × Noir de Guben made by S.A. Thornell at Young, New South Wales, Australia, in about 1928.

The tree is of medium vigour, blossoms medium early and is high-bearing, especially in dry weather conditions.

The fruit is medium to large, about 15% longer in length than in breadth, flattened on the suture side, with small humps on the suture line on top of the fruit. The skin is shiny, dark red to purple, rather susceptible to cracking. The flesh is firm, slightly fibrous, juicy, with a very good taste. The juice is dark-coloured. The stone is rather small and semifree.

Ripening time of Ron's Seedling is about 20 days after Burlat.

Early Lyons and Supreme are the preferred pollinators, but Bing, Black Bigarreau, Black Eagle, Burgsdorf, Chapman, Early Rivers, Lambert and Napoleon are also compatible with Ron's Seedling.

Roundel Heart

This is probably an old renamed variety which was for many years widely planted in Kent, UK.

The tree is very vigorous, rather erect when young, later the canopy is vase-formed and the basal branches are almost in the horizontal position. Cropping is high and regular. Bloom is in midseason.

The fruit is large, considerably broader than long, roundish, short heart-shaped, flattened on the suture side. The skin is very shiny, bright red, nearly black if fully ripened. The suture line is dark in a stripe of pale colour. The flesh is soft, dark red-coloured, juicy, with coloured juice; the taste is very good.

The ripening time is about 20 days after Burlat.

Roundel Heart is pollinated by Merchant, Merton Bigarreau and Van.

Royalton

This cultivar, which was released in 1991, was bred at the New York State Agricultural Experiment Station, Geneva, New York State, USA, from an open-pollinated selection of NY 1725 (Giant × Emperor Francis).

Tree habit is extremely upright and vigorous, with few lateral branches, and it needs careful pruning and training in the formative years to achieve good orientation of scaffold limbs. Blossoming is earlier than the majority of cultivars. Flowering and fruiting precocity is poor, the trees taking more years than many other cultivars to come into full bearing. Consistency of cropping is not yet determined.

Fruit size is very large, and quality, especially flavour, is exceptional. Fruit shape is round oblate with deep greyish purple skin colour. Flesh colour is slightly lighter in shade than the skin colour. Firmness is similar or slightly less than that of Bing. It is quite resistant to cracking, similar in this respect to Ulster and Kristin.

Ripening time is late midseason, or approximately 60 days from full bloom in New York conditions.

Royalton is incompatible with Schmidt and is thought to be in group VII or VIII of the incompatibility groupings. Further tests will be needed to find ideal pollinating cultivars which are both compatible and flower sufficiently early in the season.

St Margaret

This cultivar is very popular in New South Wales, Australia, particularly in the Orange District.

The tree is vigorous, with a large canopy, late blossoming and good cropping, but very susceptible to bacterial canker. A considerable variation is found in the habit and cropping of the trees. Some consider St Margaret to be the same as Noble, some as Black Republican, and some as Tradescant's Heart.

The fruit is large and heart-shaped and resembles the fruit of Noble, with a dark red skin and firm flesh, of very good taste and excellent preserving qualities.

Ripening time is late, about 30–35 days after Burlat.

St Margaret is pollinated by Bing, Florence, Lambert, Merton Heart, Napoleon, but is not compatible with Black Bigarreau, Black Republican and Noble.

Sam

This is an open-pollinated seedling of the Vineland Research Station selection V-160140, which is itself a seedling of open-pollinated Windsor. It was bred and introduced by Summerland Research Station, BC, Canada in 1953.

Trees are vigorous with initially upright habit but later spreading.

Sam blooms 4 to 5 days later than many other cultivars but is still frequently used as a pollinator in orchards. It is slow to come into cropping but when mature yield productivity is good.

Fruits are of medium size, fully black, medium firm with average flavour. many studies have shown that Sam fruits are very resistant to rain-induced cracking.

Sam matures approximately 9 days after Burlat.

Its incompatibility group is not known.

Sandra

This Italian cultivar, of unknown origin, is planted mostly in the Venetia region and is mainly of interest on account of its very early time of ripening.

The tree is vigorous to very vigorous, very erect in habit and comes late into bearing; the blossoming time is medium early.

The fruit is medium to large, heart-shaped, a little sharpened at the apex, very susceptible to cracking. The skin is thick, red-violet in colour. The flesh is of medium firmness to firm, pale red, of medium juiciness and of very good taste. The stone is semifree.

The date of ripening is 1–3 days before Burlat.

Sandra is pollinated by Visciola, Romana and Roana Precoce.

Schauenburger

This cultivar, of unknown parentage, originated from a tree introduced into Switzerland from Lebanon by the hotel-keeper B. Flury and was subsequently distributed by Mr Bad Schauenburg from Liestal.

The tree is vigorous, erect in habit, with late blooming. This high-bearing cultivar is adapted for growing in most types of locations.

The fruit is of medium size, heart-shaped and regular, and very few are susceptible to cracking; the skin has a medium thickness and is black-brown in colour. The flesh is of medium firmness to firm, red-coloured, sweet and acid, but at the same time aromatic.

The date of ripening is very late, about 30–35 days after Burlat.

Schauenburger is intersterile with Basler Adler, Beta, Heidegger and Schumacher; it is usually pollinated by Basler Langstieler, Hedelfinger and Star.

Schmidt

Of unknown origin, Schmidt was, and to some extent still is, one of the most important sweet cherry cultivars grown in the northeastern states of the USA. However, it is currently being replaced by Venus and Valera because of their superior bearing.

The tree is vigorous, late-bearing and moderately productive.

The fruit is large, very attractive, with a purplish black skin of medium thickness. The flesh is firm and is of good quality, both for dessert and for canning.

The date of ripening is about 15–18 days after Burlat.

Schmidt, which is classed within incompatibility group VIII, is usually pollinated by Napoleon.

Solymari Gömbölyu

This cultivar, of unknown origin, is widely planted in Hungary.

The tree is vigorous and high-bearing, with a wide conical canopy, and it blossoms late in the season.

The fruit is large, heart-shaped to spherical, with a shining claret-red skin colour. The flesh is firm and of very good taste.

The date of ripening is about 14–16 days after Burlat.

Solymari Gömbölyu is pollinated by Munchebergi Korai and Van.

Star

This is an open-pollinated seedling of Deacon, introduced by Summerland Research Station, BC, Canada in 1949.

Trees are vigorous with initially upright habit but later spreading.

Star produces large attractive black cherries, similar in type to Lambert. Unfortunately, it can overset and produce soft fruits of much poorer quality.

Fruits are moderately firm, of good flavour and show some resistance to rain-induced cracking.

Star ripens approximately 5 days after Burlat.

It is classed in incompatibility group III.

Starking Hardy Giant

This US cultivar originated from a tree of unknown origin found in Cedarburg, Wisconsin, USA, in the orchard of Mrs Ottilie R. Mayer. It was originally named Mayer but subsequently renamed and commercialized by Stark Brothers Nurseries in 1949.

The tree, which is very susceptible to virus diseases, has medium vigour and a spreading growth habit; the shoots develop with wide branch angles. Blossoming is early or medium early.

Starking Hardy Giant is very high-bearing and demands heavy pruning for best results. The fruit is large, round heart-shaped, firm, juicy and of very good taste. Unfortunately, it is very susceptible to cracking.

The ripening date is about 12–15 days after Burlat.

Starking Hardy Giant is pollinated by Burlat, Garnet, Hedelfinger, Mora di Cazzano, Napoleon, Rainier, Tragana of Edessa, Ulster and Van.

Summit

This cultivar was produced at the Summerland Research Station, Summerland, British Columbia, Canada, from a cross between Van and Sam (Lopin, 1974).

The tree, vigorous to very vigorous, is erect in habit when young, few-branched, with a midlate blossoming; it comes moderately late into bearing and has a medium productivity.

The fruit is very large and heart-shaped, with medium susceptibility to cracking. The skin is bright red, shining and very attractive; the flesh is of medium firmness, pale pink, very sweet and aromatic even before full maturity.

Summit ripens about 16–18 days after Burlat.

It is pollinated by Hedelfinger, Tragana of Edessa and Van and other varieties in incompatibility groupings II and III.

Sylvia (4 C-17–31)

This new cultivar, which was bred by the Agricultural Research Station, Summerland, Canada, resulted from a cross between Compact Lambert and Van.

The tree is semicompact in habit and bears moderate to heavy crops consistently. It warrants more extensive testing in medium-density planting systems.

Fruits are large to very large, with dark red skin colour, firm texture and good flavour. Fruits are resistant to splitting and have excellent short-term storage potential. Its dense protective leaf canopy makes fruits less liable to sun scald.

Ripening time is approximately 16–20 days after Burlat.

Sylvia is self-sterile but its pollination grouping has not been determined.

Szomolyai Fekete

This is an ancient Hungarian cultivar of unknown origin.

The tree is of medium vigour, with conical globular habit, and is a heavy producer. The blossoming time is very early.

The fruit is spherical to heart-shaped, of medium or small size, with black skin and semifirm flesh. Flesh and juice are dark red to black, very aromatic and of delightful flavour.

The date of ripening is about 15 days after Burlat.

Szomolyai Fekete is pollinated by Burlat and Markikorai.

Techlovan

This variety was produced at the Research and Breeding Institute at Holovousy in the Czech Republic from a cross between Van and Techlovicka II. It was registered in 1991.

It produces trees of medium vigour with a spreading habit.

Fruits are globose to heart-shaped and have a skin colour of carmine-red with small lighter-coloured flecks. The flesh is firm and the fruits have an excellent sweet aromatic taste.

The variety ripens about 6 days before Napoleon.

The variety is self-incompatible.

Tragana of Edessa

This very old cultivar, of unknown origin, is very widely planted in Macedonia (Greece) in the region of Edessa and is the most important sweet cherry cultivar in Greece.

The tree is of medium vigour to vigorous, with erect habit and good yielding; its blossoming time is medium late in the season.

The fruit is large, roundish heart-shaped, a few flattened on the suture side, with a medium susceptibility to cracking. The skin is of medium thickness, red-purple to black in colour; the flesh is firm, red to dark red, sweet and lightly acid, with a small stone.

Tragana of Edessa ripens about 25 days after Burlat.

It is pollinated by Hedelfinger, Summit and Ulster.

Turfanda

This cultivar, of unknown origin, is planted mainly in Turkey.

The tree, very vigorous and heavy-bearing, is erect in habit, with an early or very early blossoming time.

The fruit is of medium size or small, heart-shaped, with a skin that is red-violet or violet-coloured. The flesh is soft, juicy and fairly pleasant.

Turfanda ripens with Burlat or a little earlier.

Suitable pollinators are Burlat, Merton Premier and Noir de Guben.

Ulster

This cultivar was bred at the New York Experiment Station, Geneva, New York State, USA, from a cross between Schmidt and Lambert.

Trees are quite vigorous and semierect in habit. Ulster flowers are reported to have good resistance to frost.

Fruits are of moderate size and firm, with purplish skin colour and cordiform shape, somewhat similar to Schmidt. Fruit flavour is good when fully ripe. The cultivar is of particular interest on account of its resistance to cracking on some sites.

Ulster ripens 19–21 days after Burlat.

Ulster, which is in incompatibility grouping XIII, is pollinated by Burlat, Napoleon, Hedelfinger and Starking Hardy Giant.

Valera

This cultivar, which resulted from a cross between Hedelfinger and Windsor, was introduced in 1967 by the Horticultural Research Institute of Ontario, Canada.

The trees are vigorous, but come into cropping precociously and crop consistently thereafter.

Fruits are dark, almost black, in colour and of good flavour. Sensitivity to brown rot fungus is reportedly less on this cultivar.

Valera belongs to Group 'O'.

Van

This cultivar originated in 1936 at the Research Station in Summerland, British Columbia, Canada, from open-pollinated Empress Eugenie. Today it is distributed and planted extensively throughout the world.

The tree is of medium vigour to vigorous, upright in habit, a very early producer and also heavy-bearing. On account of this very heavy bearing, the final fruit size may not be as large as may be desired. Blooming is midseason.

The fruit is large, with a bright red skin which is not too thick; the flesh is firm (slightly firmer than Bing), very good, not very juicy and pale red-coloured. The fruit stalk is very short so picking is not very easy. The susceptibility to cracking, especially at the stylar point, is medium to high but less than that of Bing.

Van ripens about 20 days after Burlat.

Van, which is in compatibility grouping II, may be pollinated by Adriana, Bing, Burlat, Giorgia, Hedelfinger, Lambert, Mora di Cazzano, Napoleon, Rainier and Vittoria. Van is intersterile with Abundance, Gil Peck, Jubilee, Merton Beauty, Schrecken, Sodus, Venus and Windsor.

An early compact clone of Van was bred at the Summerland Research Station, British Columbia, Canada.

Viscount

This new Canadian cultivar, which was introduced (1984) by G. Tehrani at the Horticultural Research Institute of Ontario, Vineland Station, Canada, was obtained from a cross between the Vineland Selections V.35024 and V.35029.

The tree is vigorous, with a spreading canopy, and highly productive; blossoming time is midseason. Viscount has the same degree of resistance to brown rot and bacterial canker as Bing.

The fruit is large, kidney-shaped (wider than long), with a low susceptibility to cracking. The skin is dark red, glossy and very attractive. The flesh is red and firm, with dark juice, and of very good quality.

Viscount ripens about 20–25 days after Burlat, the same ripening date as Bing in Ontario, and it may be planted as a substitute for Bing in some areas.

Viscount, which is in incompatibility grouping IV (or IX according to Tehrani and Lay, 1991), is pollinated by Bing, Hedelfinger, Lambert, Moreau and Star.

Vittoria

This Italian cultivar was selected by G. Bargioni at the Istituto Sperimentale di Frutticoltura in Verona, Italy, from the cross Moretta di Cazzano × Durona di Padova, made in 1958. It was introduced in 1970 as the first sweet cherry especially bred for mechanical harvesting.

The tree is very vigorous, and erect in habit. It has a tendency to bear on 1-year shoots as well as spurs, so when adult it requires severe pruning to avoid formation of blind wood. It shows good resistance to bacterial canker.

The fruit is round kidney-shaped, medium in size; the skin is rather thick, red or dark red, very shiny and attractive. At maturity, an abscission layer forms between the fruit stalk and the fruit so the retention force is very low and mechanical harvesting is made easier. The flesh is very firm, crisp, sweet and slightly acid, aromatic, with red or dark red juice, very suitable for processing purposes, particularly canned fruits. The stone is small and semifree. The susceptibility of the fruit to cracking and of the tree to bacterial canker is very low.

Ripening time is about 35 days after Burlat.

Vittoria is pollinated by Adriana, Burlat, Corinna, Durone Nero 1, Flamengo SRIM, Francesca, Giorgia, Moreau and Van.

Viva

This is another cultivar bred by the Horticultural Institute at Vineland, Ontario, Canada and introduced by G. Tehrani and G.H. Dickson in 1973.

The tree is vigorous and produces medium-sized, semifirm, dark red fruits. One of its main merits is its reported resistance to rain-cracking when grown in Ontario.

Viva flowers midseason, matures with Black Tartarian and is classed in incompatibility grouping IV.

Windsor

This is a very old cultivar, originating at Windsor, Ontario, in Canada from unknown parentage.

The tree is vigorous, upright-spreading, with a large roundish head; it blossoms early in the season.

The fruit is large or very large, roundish, heart-shaped, broader than long and slightly flattened on the suture side. The susceptibility to cracking is high. The skin is dark red, with dots and conspicuous streaks, and moderately shining.

The flesh is red and of medium firmness, becoming fairly soft at maturity, juicy and of good quality. The juice is pale red.

Windsor ripens about 30 days after Burlat.

It is pollinated by Bing, and is classed within incompatibility group II.

3.4.2 Self-fertile cultivars

The recent development of self-fertile cultivars of the normally self-sterile sweet cherry represents a significant advance in breeding. Effective pollination of normal self-sterile cultivars is dependent upon the transfer to them of compatible pollen from other suitable cultivars planted close by. Choice of these pollinating cultivars is often difficult, as they must flower at approximately the same time as the main cultivar and they must also be compatible with it. Effective pollen transfer is also dependent upon suitable weather conditions during blossoming, conditions that encourage the activities of bees and other pollen vectors. In situations experiencing inconsistent spring weather conditions, fruit set is frequently poor. With the development of self-fertile cultivars, some of these constraints on fruit set are alleviated.

Some naturally self-fertile cultivars have been noted. Bou Argoub in Tunisia, Cristobalina and Llosetina in Spain and Kronio in Italy all fall into this category. However, none of these have been considered worthy of large-scale planting.

As well as being self-fertile, all the cultivars listed below will also successfully pollinate many other cultivars.

Celeste

This new self-fertile cultivar was selected at the Summerland Research Station in British Columbia, Canada, by W.D. Lane from a cross between Van and New Star. Originally numbered 13S 24 28, Celeste was named and released in 1993.

The tree is vigorous, but with compact habit. It is not very precocious in bearing but is very high-cropping in maturity.

The fruit is large, with short stems, dark flesh, medium firmness and a very good flavour; it is rich in soluble solids.

Celeste ripens about 6 days after Burlat.

Isabella

This Italian cultivar was selected by G. Bargioni, F. Cossio and C. Madinelli at the Istituto Sperimentale di Frutticoltura of Verona, Italy, from the cross Starking Hardy Giant × Stella made by G. Bargioni in 1975; it was introduced in 1993.

The tree is of medium vigour, erect in habit, early-bearing and high-cropping. Blossom is in midseason.

The fruit is large, heart-shaped and with medium susceptibility to cracking. The skin, which resembles that of the female parent, is shiny, red-coloured and very attractive. The flesh is medium firm to firm, pale red-coloured and sweet.

Isabella ripens about 6 days after Burlat.

Lapins

This cultivar, which originated from the cross Van × Stella made by K.O. Lapins in 1965, was selected by Lapins and named by W.D. Lane at the Summerland Research Station, British Columbia, Canada and was introduced in 1983.

The tree is vigorous, very erect in habit and thought by some to be suitable to hedgerow training; it is a high producer, with early blooming.

The fruit is large to very large, roundish to heart-shaped and some are flattened on the ventral side. It is of low susceptibility to cracking. The skin is purple-red to dark red and the flesh is quite firm, sweet, juicy and very slightly acid.

The ripening date is about 25–28 days after Burlat.

New Star

This self-fertile cultivar was selected by W.D. Lane at the Summerland Research Station, British Columbia, Canada, and by S. Sansavini at the Centro per il Miglioramento Varietale in Frutticoltura of the Bologna University, Italy, from the cross Van × Stella made by K.O. Lapins in 1965. New Star, introduced in 1987, is not related to Star.

The tree is of medium vigour, wide-spreading and rather compact in habit and it blooms early in the season.

The fruit is large, rather spherical in shape, similar to Van, and rather susceptible to cracking. The skin is shining, dark red to red-black; the flesh is quite firm, dark red, very sweet and aromatic. The stone is semifree.

The ripening date is about 13–15 days after Burlat.

Starkrimson

A self-fertile cultivar derived from genetic improvement, Starkrimson was bred by Floyd Zeiger at Modesto, California, USA, from the cross Garden Bing × Stella. It has the merit of compact habit. Introduced in 1980, it is primarily distributed to home gardeners.

Stella

Stella was the first self-fertile cultivar of commercial merit to be derived from work on the genetic improvement of sweet cherries and was bred by K.O. Lapins at the Summerland Research Station, British Columbia, Canada, from the cross he made between Lambert and the self-fertile John Innes seedling 2420. It was introduced to commerce in 1970.

The tree is vigorous, rather erect in habit and very high-bearing.

The fruit is large, heart-shaped and regular but rather susceptible to *Monilinia fructigena*. The skin is thin, dark red-coloured, shiny and very attractive. The flesh is medium firm to firm, and the stone is semifree.

The ripening time of Stella is about 15 days after Burlat.

Sunburst

Sunburst is another self-fertile cultivar selected by W.D. Lane at the Summerland Research Station, British Columbia, Canada (Lane and Schmid, 1984) from the cross Van × Stella made by K.O. Lapins in 1965. It was introduced to commerce in 1983 (Sansavini and Lane, 1983).

The tree is of medium high vigour, wide-spreading in habit; it has a medium to late blooming date and comes into bearing early in its life.

The fruit is large, spheroidal and a little heart-shaped, and is of medium susceptibility to cracking. The skin is shiny, red-coloured and of medium to light thickness. The flesh is semifirm to firm, pale red-coloured, juicy, of very good taste and aromatic. The stone is rather small and adheres to the flesh.

The date of maturity is about 18–20 days after Burlat. Ripening occurs very quickly so this cultivar demands rapid harvesting and accuracy in choice of the time of picking.

Sweetheart

Sweetheart is a self-fertile cultivar selected by W.D. Lane at the Summerland Research Station in British Columbia, Canada, from a cross between Van and New Star. Originally it was distributed for trialling under the number 13S 22 8, before it was named and released in 1993.

The tree is vigorous, with a fairly spreading growth habit; it crops very early in its life and is an abundant cropper.

The fruit is medium to large, with red skin and red flesh, firm and of good flavour, with a low susceptibility to cracking.

Sweetheart is particularly interesting for its late maturity; it ripens approximately 30 days after Burlat (10 days later than Lambert at Summerland).

3.5 Genetic Improvement of Sweet Cherries: The Objectives

The main problems experienced by sweet cherry growers are:

- very large tree size, which makes management difficult and expensive;
- poor precocity, which delays cropping and return on investment;
- self-incompatibility, which limits fruit set in many locations and may oblige growers to plant unprofitable pollinating cultivars;
- susceptibility of the fruits to rain-induced cracking, which frequently results in the loss of a large proportion of the marketable yield;
- susceptibility to bacterial canker or other pathogens, which may kill many trees in orchards;
- damage to the fruits and yield loss in regions with high populations of starlings and other birds that eat cherries;
- poor tolerance of trees to severe winter cold;
- small or soft fruits, which are unpopular with markets and consumers;
- short harvesting season and an inability to store fruits for extended periods.

Many of these problems may, in principle, be overcome by breeding new improved cultivars. The production of smaller or compact cultivars (possibly spur types) should, by reducing tree stature, make tree spraying, harvesting and management much less expensive, while also making the enclosure of the trees within nets or polythene structures (against bird or rain damage) much more feasible. Cultivars which flower early in their life, following planting in the orchard (precocious cultivars) will provide the rapid return on investment demanded by today's growers. If these cultivars are also self-fertile then fruit set should be more consistent from season to season and only the most profitable cultivars need be planted. Cultivars resistant to bacterial canker (*Pseudomonas mors prunorum* and *P. syringae*) are essential in the more northern and wetter cherry-growing areas, such as Germany and Britain, where the effects of this disease can be devastating to sweet cherry growers. Also, cultivars which ripen very early or late in the season should spread the labour requirements at harvesting time and extend the cherry marketing season.

3.5.1 Tree vigour and size

Excessive tree vigour is still the principal problem besetting most sweet cherry growers throughout the world; the sweet cherry tree remains the largest of the temperate fruit species produced commercially. In favourable soils and environmental conditions a sweet cherry tree may easily attain a height of 15–20 metres. The vigour of all other temperate fruit species is now controlled very effectively and much of the equipment used by growers, such as sprayers and harvesting aids, are largely tailored to the needs of these smaller trees. The necessity for specialized spraying and pruning equipment and the problems of harvesting very large trees have prompted many previous growers of sweet cherries to abandon the crop. Picking cherry fruits, which are relatively small but easily bruised, from extremely large trees has become prohibitively expensive and large reductions in the size of sweet cherry trees, or efficient methods of mechanical harvesting, will be essential if the crop is to become as important commercially as apples, pears and peaches. A method of dwarfing the sweet cherry would allow increased planting densities, reduce costs of pest and disease control, pruning and harvesting. It would also facilitate the erection of protective structures against birds, frost and rain.

Many different methods of dwarfing sweet cherry trees have been tried. Pruning and training techniques are, except on the poorest soils, in themselves insufficient to control vigour. Although sprays of chemical plant growth retardants, such as paclobutrazol or daminozide, have been shown to control tree growth very effectively, regular and often expensive treatments are necessary to sustain growth control and such treatments are not popular with those who wish to see a reduction in chemical use on fruit.

Probably the two best methods of controlling tree size are to use either dwarfing rootstocks or compact (spur-type) scion cultivars. The preferred solution of these two would be to use dwarfing rootstocks, as the development of just one rootstock could, in theory, enable all the existing scion cultivars to be dwarfed. However, rootstock:scion:environment interactions are not uncommon in studies evaluating rootstocks, and to evaluate fully all the interesting rootstock/scion/site combinations would take huge resources and many years. For these reasons we should not discount the potential value of breeding compact scion cultivars and some promising results have already been achieved using this strategy. Sweet cherry cultivars with compact habit have been raised by conventional hybridization techniques (Matthews, 1970; Trajkovski, 1984) or as selections from natural mutations (Cireasa, 1972; Frecon, 1980); Starkrimson is a new compact cultivar raised from conventional hybridization which has the additional advantage of being self-fertile. A third breeding method employed involves subjecting scions to ionizing radiation and then selecting the promising mutants induced (Lapins, 1973, 1975; Saunier *et al.*, 1987). These dwarf types may have very different growth patterns (Lane, 1978); some have strong apical dominance, as in Compact Stella, while others have weak apical dominance, as in Compact Lambert. Full genetic dwarfs, which are double recessives, have rugose leaves and crop very poorly; they are occasionally utilized as rootstocks.

An alternative to dwarfing trees is to develop methods of mechanically harvesting large trees. Although this does not solve the problems of spraying and pruning large trees, it can reduce the largest cost, that of harvesting. Most of the existing cultivars cannot be mechanically harvested without use of fruit looseners (ethylene-liberating chemicals). The efficacy of these sprays may vary greatly depending upon environmental conditions at the time of application. In some conditions the sprays may produce phytotoxic symptoms. One method of overcoming this problem is to breed cultivars of sweet cherry which at maturity form, naturally, an abscission layer between the stalk and the fruit (Bargioni, 1970, 1985). Fortunately, although such cherries lack stalks, they are perfectly acceptable for the fresh market as, after detachment, there is no loss of juice at the abscission layer (Baldini *et al.*, 1979, 1981). Cultivars amenable to mechanical harvesting may, therefore, offer an attractive alternative to dwarf trees for growers seeking to reduce costs of production.

3.5.2 Yield precocity and productivity

Most of the old traditional sweet cherry cultivars were extremely slow to come into bearing following planting. Today, fruit growers require a rapid return on the capital invested in orchards if their business is to remain profitable. This is almost impossible to achieve using the older cultivars budded on vigorous Mazzard or Mahaleb rootstocks.

Improved precocity may be achieved by use of certain dwarfing rootstocks and this solution, which is the only one possible with many of the older existing cultivars, is discussed elsewhere in this book. A better long-term solution is to breed cultivars that crop precociously, and considerable progress has already been made towards this objective with the release of cultivars such as Van. Breeding for a short juvenile period following orchard planting would also aid vigour control; young trees that crop precociously also invariably make less shoot growth, due to fruit: shoot competition for the trees' vital resources. Blazek (1985) has recommended using cultivars such as Van, Kordia and Techlovicka as parents, as these appear to pass on their precocity to their offspring.

Two factors that are particularly important in determining successful fruit set of self-incompatible cultivars are the longevity of the ovule sac viability and the speed of pollen tube growth (see Chapter 8). Differences between cultivars are known to exist for these two factors (Guerrero-Prieto *et al.*, 1985) and, if bad weather temporarily delays pollen transfer or slows pollen tube growth, then only those cultivars with extended longevity of ovule viability can produce a good crop. Greater attention should be given to this criterion in selecting new sweet cherry cultivars.

Late flowering of a cultivar may reduce the risk of frost damage in the spring and increase the probability of good fruit set. It has been suggested that late-flowering cultivars, such as Lambert, Jubilee and Sam, should be used more in breeding programmes with this objective in mind (Way, 1966).

3.5.3 Self-fertility (compatibility)

Although several self-compatible cultivars are now available to commercial growers (Lapins, 1971; Sansavini and Lane, 1983; Lane and Sansavini, 1988; Lane, 1992; Bargioni et al., 1993; Lugli et al., 1993), the majority of cultivars still express complex compatibility relationships. Way (1966) recognized 13 incompatibility groupings to which sweet cherry cultivars were assigned; studies by Knight (1969) and others suggest that 13 groupings plus a universal donor group may be more accurate. The reason for the confusion is probably due to the inconsistency in naming of many of the older cultivars.

The necessity for cross-pollination demands that the grower organizes very carefully the relative positioning of cultivars within the orchard. It also limits his/her choice of cultivars and may necessitate the planting of unprofitable ones. Self-sterile (incompatible) cultivars must be planted no further than 20–25 metres from compatible pollinating cultivars, which must also flower at approximately the same time. Ideally, cultivars planted adjacent to one another in the orchard should also ripen at much the same time. If they do not, it can mean that considerable time is wasted moving picking ladders long distances within the orchard. Even in the best-designed orchards poor weather conditions at blossom time can severely restrict pollen transfer between cultivars and result in poor yields.

Most of the above constraints on production are overcome by planting self-compatible cultivars and it is hoped that sweet cherry breeders will produce more of these to add to the current very limited range. Apart from the well-known breeding programme at Summerland, British Columbia, Canada, which in recent years has focused on producing superior self-fertile sweet cherry cultivars, several other cherry breeders have now identified self-fertility as a goal. The breeding programme at Vineland, Ontario, Canada, has also produced a number of promising self-fertile cultivars, mainly from crosses made between Van and Stella. None of these are yet named but they are available for field trialling under the codes V.69061, V.69062, V.69068, V.690616, V.690618 and V.690620 (Tehrani, 1988). Breeders in Hungary and several other European countries also have promising self-fertile cultivars undergoing selection.

Most of the promising self-fertile cultivars now in commercial cultivation originate from three self-fruitful seedlings (JI.2420, JI.2434 and JI.2538), which were produced and introduced from the breeding programme at the John Innes Institute in Britain (Lewis and Crowe, 1954; Matthews, 1970). Two of these were produced as a result of X-ray treatment of Napoleon pollen and Emperor

Francis as the maternal parent; the third was simply a self from a cultivar originally raised from a cross between Bigarreau Schrecken and Governor Wood. JI.2420, which has a mutation at the S4 locus which affects pollen activity, was crossed with Lambert to produce the first commercially acceptable self-fertile cultivar, Stella. A detailed review of breeding for self-fertility in sweet cherries is given by Tehrani and Brown (1992).

Despite their obvious merits, in favourable conditions for fruit set some self-fertile cultivars have a tendency to overset and to produce small fruit size (Tehrani, 1988). Indeed, among the self-fertile cultivars currently available only Sunburst is reported to consistently attain large fruit size.

3.5.4 Resistance to cracking

Susceptibility of fruits to cracking, following wetting by rain, affects the majority of the existing commercial cultivars and is a particular problem in regions that experience frequent rain at, or close to, harvest time. The problem of sweet cherry cracking and possible remedies are discussed at length in Chapter 12 of this book. Although the factors that determine why cultivars differ in their susceptibility to cracking are not fully understood, there is little doubt that several cultivars, such as Adriana, Ambrunés, Castor, Kordia, Kristin, Merton Late, Sam, Ulster, Viscount and Vittoria, show considerable resistance, in some if not all situations. Amongst the self-fertile cultivars Lapins also reported to show some resistance. Future sweet cherry breeding programmes should perhaps use these cultivars as parents, in attempts to extend the available range of cultivars resistant to rain-induced cracking.

3.5.5 Resistance to diseases

Sweet cherries, like most other tree fruit species, are sensitive to a number of pathogens. Perhaps the most serious, in the cooler and wetter areas of production, is bacterial canker (*Pseudomonas mors prunorum* and *P. syringae*). The organism lives harmlessly on the leaves of most cultivars throughout the summer and only causes problems if it gains entry to the wood of the tree. This usually occurs through leaf scars or branch wounds in the autumn or winter. The problem was considered so severe in Britain that a programme of breeding for resistance was carried out by the John Innes Institute. Unfortunately, although the programme was initially very successful, some of the cultivars released have since shown some susceptibility to canker. The problem is that the bacteria causing canker are capable of mutating to new and often more virulent forms to which the 'resistant' cultivars are not always fully resistant (Freigoon and Crosse, 1974). While it may not be possible to achieve full resistance to bacterial canker in new cultivars, breeders should at least select for field tolerance. Highly sensitive cultivars, such as Van, despite their other considerable merits, are often excluded from new orchard plantings in Britain on account of the high risk of tree loss due to canker.

Most sweet cherry cultivars are highly susceptible to the brown rot fungus *Monilia fructicola* (Wint) Honey, a damaging pathogen in damp humid conditions close to harvest time. Several of the self-fertile cultivars recently released for testing from the Vineland, Canada, breeding programme are reported to show some tolerance to this organism.

3.5.6 Resistance to pests

One of the most troublesome pests attacking cherries in northern Europe is the black cherry aphid (*Myzus cerasi*). Breeding research at East Malling in the UK has focused on selecting for resistance to this pest, and *Prunus canescens*, *P. incisa*, *P. kurilensis* and *P. nipponica* clones have all shown resistance to colonization. These species and their hybrids, raised by crossing these clones with the sweet cherry cultivar Napoleon, have undergone further glasshouse inoculation tests in which the species clones and some of the hybrids proved resistant but not immune to colonization by black cherry aphid. It is hoped to improve fruit size on the resistant hybrids by further back-crosssing them to sweet cherry cultivars.

3.5.7 Resistance to winter cold injury and the chilling requirement

This is difficult to achieve as it involves first specifying the problem in a particular cherry production area. The problem may be due to slow acclimatization in the autumn leading to damage in early winter, to insufficient cold tolerance during deep dormancy, or to low chill requirement, early dormancy breaking and damage in the spring.

Resistance during deep dormancy to severe winter cold damage, as experienced in the Ukraine and Russia, is probably best achieved by crosses using resistant cultivars of *Prunus fruticosa*, the Steppe or Ground cherry. This will involve much back-crossing and take considerable time to achieve. Where winter temperatures are less severe some benefit may be gained by use of cultivars such as Windsor, Black Eagle, Vic, Kristin and Hudson, all of which are reported to be more resistant to winter cold than average.

Some resistance to blossom damage from spring frosts is exhibited by the cultivars Ambrunés, Burlat, Frogmore, Sam and Grosse Schwarze Knorpelkirsche. The cultivars Lambert, Sekunda and Trajana Edemis are reported to show resistance to midwinter cold damage and to spring frost damage to the blossoms.

Attempts have also been made to select cherry cultivars with a low winter chilling requirement which could be of value in subtropical and other regions experiencing very mild winters. The cultivars Black Tartarian, Chapman and Republican are all reported to need less winter chill units than other cultivars such as Bing, Lambert and Napoleon. Most cultivars need between 1050 and 1900 hours at temperatures below 7°C to satisfy their dormancy chilling requirement. Attempts to breed new scions with much lower chilling requirements, involving crosses between sweet cherry cultivars and *Prunus campanulata* or *P. pleiocerasus*, have not proved successful to date (Sherman, 1977).

3.6 Breeding Methods

To summarize, breeders of sweet cherry scions should in the future work towards the following objectives:

- trees of reduced stature;
- improved yield precocity;
- reduced susceptibility of fruits to cracking;
- self-compatibility;
- heavy and consistent yields;
- large, attractive and good-flavoured fruits;
- small freestone fruits with colourless juice for the processing market;
- improved resistance or tolerance to diseases;
- extended harvesting season with earlier- and later-ripening cultivars;
- improved adaptation to climate (frost tolerance, chilling requirement).

There are three principal methods of achieving the above objectives.

- Clonal selection, involving the screening and selection from within natural populations of a cultivar for improved clones or types.
- Conventional breeding, by crossing parents with known desirable characteristics and then selecting from within the siblings.
- Use of novel breeding methods. These may involve inducing mutations using gamma irradiation or chemical mutagens, either *in vitro* or on scion graft or budwood. Future techniques may permit the direct introduction of useful genes to existing cultivars, using one or more methods of 'genetic engineering'.

3.6.1 Clonal selection

Methods of clonal selection applied to the populations of existing cultivars of sweet cherry do not differ from the methods employed for other fruit species. They are not, therefore, described in detail in this chapter. Improved clones of cultivars, such as Germersdorfer, Reverchon, Burlat, Durone Nero 2 and Napoleon, have been selected and should always be chosen in preference to the original cultivar sources in their country of selection.

3.6.2 Conventional breeding by hybridization

Conventional breeding, i.e. by crosses made between two cultivars, is still today the most important method of obtaining new sweet cherry cultivars.

In comparison with other fruit species, *P. avium* possesses a high variability (heterogeneity) that has not yet been well explored and exploited. Therefore, crosses made between existing varieties can still have very interesting results. On the other hand, the various sources of genes that are associated with traits of importance in sweet cherry scion cultivars are already known; they were recently reviewed by Iezzoni *et al.* (1990). Also, cultivars with similar useful characteristics are known. For instance, Adriana and Viscount can be grouped for cracking resistance.

Traditional breeding concerns the controlled pollination of the pistil of a cultivar (maternal parent) with the pollen of another cultivar (paternal parent), collecting the seeds obtained from the fruit, stratifying and germinating them and carefully selecting superior seedlings from among the many siblings.

If the maternal parent is a self-incompatible cultivar, emasculation of the flowers is not always essential, particularly if open-pollination is the aim, but it can be convenient when the pollen of a particular cultivar must be used to pollinate different cultivars in controlled crosses. Emasculation is essential if the mother plant is self-compatible and must be made before the flowers reach the balloon stage, removing the calyx cup after excising it at the rim near its base.

Emasculated flowers, because they are without petals and because their nectar evaporates rapidly, are very rarely visited by insects, so it is not usually necessary to isolate flower limbs using paper bags or other bee-proof material. However, flowers that are not emasculated must be isolated to avoid visits by insects. The isolation of the flower clusters or branches of the maternal parent, by bagging or other means, is done a little before the floral balloon stage.

Pollination can be carried out just once or on more than one occasion. If only one pollination is to be done, the optimal time is at full bloom of the tree; repeated pollinations during blooming are beneficial in that they often produce a higher percentage of fruit setting.

For collecting pollen, one can collect branches with flowers mostly at the balloon stage from the paternal parent (removing flowers that have already opened). These limbs are preserved in water in a greenhouse (carefully avoiding visits by insects) and the pollen is collected when the anthers are dehiscing. If flowering limbs are preserved in buckets of water until the flowers are open, water may become polluted by those chemicals used for pest control that are on the bark. This may result in absorption of toxins and, to avoid this, maintenance of limbs in running water is to be preferred. Alternatively, flowering limbs are isolated on the tree before the flowers open; then they are collected at full bloom. In the laboratory the single flowers are removed from the limbs and the anthers removed from their filaments and subsequently the pollen grains are extracted from the anthers. According to Fogle (1975), the best results are obtained by rubbing the opened flowers across a close-mesh wire net placed over a Petri dish. By giving care to detail the collection of pollen is very easy. If some stamen filaments fall through the mesh, they must be removed with tweezers.

Pollen can be transferred to the maternal stigma by means of a small artist's paint brush or, more simply, with the forefinger. Another method of pollination, of use when the blossoming time of the parents is at the same time, is as follows. Isolated flowering limbs of the paternal parent are collected at full bloom. Then, using three to five flowers each time, the blossoms are rubbed over the opened maternal flowers, so that their pollen is transferred to the stigmas.

After pollination, the unemasculated flowers must again be isolated, using bags until blossoming of the mother tree is completely finished. Immediately after petal fall, the bags are opened and the fruits left to grow freely.

A very useful bagging material for isolation of limbs is Viledon, a non-woven polyester tissue, permeable to air but not to water or insects.

If the ripening time of the maternal plant is medium late or late, fruits are collected at maturity. The seeds of these midseason and late cultivars are generally well developed in a high proportion of the fruits. After stratification (100–120 days) they are sown to produce seedlings. The germination percentage is usually not high and various treatments (e.g. cracking of the stone pits) have been suggested to improve germination percentage. Practically, good results are obtained by simply sowing seeds (or entire cherry fruits) immediately after picking and maintaining sufficient moisture in the soil until autumn.

The seeds of early-ripening cultivars are very rarely viable; they usually abort at the end of the second growth phase. *In vitro* culture of the embryos is necessary to obtain seedlings. The optimal time of fruit picking for embryo excision is before the fruit is ripe and can differ depending on the characteristics of the cultivar and the agronomic conditions experienced in the period between 20–23 (Enikeev *et al.*, 1984) and 30–40 (Ivanicka and Pretova, 1986) days after pollination. To achieve the best results Bassi *et al.* (1984) suggest: (i) peeling off the seed integuments in order to obtain a surface easy to sterilize, and (ii) applying a chilling treatment of + 5°C for 2–4 months to the embryos *in vitro*.

Within the same variety, larger embryos are usually more suitable and germinate better than smaller embryos in *in vitro* culture.

3.6.3 Production of new dwarf sweet cherry cultivars using ionizing radiation techniques

The research aimed at producing dwarf scion cultivars by use of ionizing radiations originated from the experiments of Zwintzscher (1955), in which he irradiated sweet and sour cherry budsticks with X-rays. Lapins (1963), at the Summerland Research Station, British Columbia, Canada, obtained the first mutant of a sweet cherry cultivar with compact habit (cv. Lambert Compact).

The methods used in irradiation breeding, including the plant material used, irradiation techniques, postirradiation treatments and selection techniques, were fully reviewed by Lapins (1973).

Usually, the buds on 1-year-old scion wood of a cherry cultivar are treated with X-rays or gamma rays after they have finished their rest period. X-rays are used at an intensity of 1000 R h^{-1} or slightly more and receive a total exposure of 3–4 K R. Gamma rays (from a cobalt-60 source) can be used at 530–610 R h^{-1} and a total exposure of 3–4 K R is recommended. The part of the graft stick to be grafted must be protected with a lead sheet against the effects of the radiation.

Ionizing radiation can, inevitably, stimulate mutants that are not only dwarf or changed in habit, but also changed in other characteristics. In fact, mutagenic treatments, when applied to normal leaf buds, involve treatment of multicellular organs or relatively large parts of the plant and often result in diplontic selection and the formation of chimeras. The possibility of applying ionizing radiation to *in vitro*-cultured material, so-called '*in vitro* mutagenesis', can provide an opportunity of avoiding, to a high extent, these disadvantages, for, by treating single cells of small groups of cells *in vitro*, the chances of forming chimeras is diminished (Walther and Sauer, 1985). Nevertheless, similar results can be obtained by using, in the selection phase, shoots derived from accessory buds (Lapins, 1973).

Walther and Sauer (1985) have studied the radiosensitivity to X-rays of *in vitro*-derived apices and also the basal segments of microshoots of Mazzard clones. They found that the median lethal dose (LD_{50}) was 22 Gy for apices and 29 Gy for basal segments (12 mA, 150 RV, 1.7 mm Al filter resulting in a dose rate of 0.9 Gy min^{-1} at a distance of 55 cm).

A particular and careful selection of the treated material must be made in order to eliminate the inferior individuals.

The procedure involves a series of agamic propagations (grafting) after the ionizing treatment, usually applied to budsticks of two buds. The treated scion is grafted on to a rootstock; the shoots (V1) originating from the irradiated buds are then cut back up to the fourth or fifth bud. Donini *et al.* (1972) suggest using the fifth or sixth to tenth buds for new grafts, the highest frequency of deviants being generally obtained at the fifth to tenth buds. In fact, they noted that there were about ten primordia in each irradiated main bud. Saamin (1988) recommended the use of buds 11–30 on V1 shoots derived from the irradiated main buds for the efficient recovery of mutants with reduced growth.

In the second year of vegetative growth (V2), both the shoots developed from the buds of the cut-back scion wood of the V1 grafts and the shoots obtained from the grafted buds are observed and the interesting deviants selected. A similar process is adopted for another year to make sure of identifying all the possible mutants.

This procedure is rather laborious and takes some considerable time because it is necessary to exclude any possible chimeral selections. It is even more difficult and time-consuming to use the procedure in order to obtain mutants involving new fruit characteristics. In this case, the time required is generally much longer than that required for traditional breeding methods.

Mutants may also be obtained using colchicine or other mutagenic agents. However, no sweet cherry mutant scions derived from this process have yet been introduced.

Acknowledgements

The author would like to express his sincere thanks to K. Belmans, Fruitteeltcentrum, Leuven, Belgium; G. O'Connor, Institute of Plant Sciences, Knoxfield, Victoria, Australia; E.E. Domeney, Domeney Brothers, Flowerpot, Tasmania, Australia; V. Georgiev, Research Institute for Fruit Growing, Kustendil, Bulgaria; D. Lane, Agricultural Research Station, Summerland, British Columbia, Canada; A. Roversi, University Sacro Cuore, Piacenza, Italy; R. Saunier, INRA, Bordeaux, France; B. Timon, University of Agriculture, Budapest, Hungary; and J. Ystaas, Ullensvang Agricultural Research Station, Lofthus, Norway, for information received in relation to several of the cultivars described.

In particular, the author expresses heartfelt thanks to Dr A.D. Webster, Horticulture Research International, East Malling, Kent, UK, for the indispensable assistance afforded in the preparation of the English version of this chapter.

References

Aeppli, A., Gremminger, U., Nyfeler, A. and Zbinden, W. (1982) *Kirschensorten*. Stutz & Co., Wädenswil.

Baldini, E. (1973) Scheda pomologica. In: *Indagine sulle Cultivar di Ciliegio Diffuse in Italia*. Consiglio Nazionale le delle Ricerche, Bologna, p. 6.

Baldini, E., Bargioni, G. and Costa, G. (1979) Giudizio dei consumatori sulle ciliegie raccolte a macchina. *L'Informatore Agrario* 21, 6057–6066.

Baldini, E., Bargioni, G. and Costa, G. (1981) Giudizio dei consumatori sulle ciliegie raccolte a macchina: il mercato di Milano. *L'Informatore Agrario* 26, 16275–16278.

Bargioni, G. (1979) 'Vittoria', nuova cultivar di ciliegio dolce. *Rivista Orto-florofrutticoltura Italiana* 6, 3–12.

Bargioni, G. (1980) Una nuova ciliegia per il Veronese: 'Adriana'. In: *Rinnovamento e Sviluppo della Coltura del Ciliegio*. Ente Sviluppo Agricolo Veneto, pp. 175–181.

Bargioni, G. (1985) Due nuove cultivar di ciliegio dolce per la raccolta integralmente meccanizzata: 'Corinna' e 'Francesca'. In: *Indirizzi nel Miglioramento Genetico e nella Coltura del Ciliegio*. Amministrazione Provinciale di Verona, pp. 117–122.

Bargioni, G., Cossio, F. and Madinelli, C. (1993) Isabella. *Rivista Frutticoltura* 3, 75.

Bassi, D., Gaggioli, G. and Montalti, P. (1984) Chilling effect on development of immature peach and sweet cherry embryos. In: *Efficiency in Plant Breeding. Proceedings of the 10th Congress of European Research on Plant Breeding:* Eucarpia. Pudoc, Wageningen, p. 293.

Basso, M. (1982) Ciliegie. In: *Agrumi, Frutta e Uve nella Firenze di Bartolomeo Bimbi Pittore Mediceo*. Centro di Studio per la Tecnica Frutticola del Consiglio Nazionale le delle Ricerche, Bologna, pp. 57–70.

Batikov, S.G. (1983) Self-fertility in sweet cherry varieties. *Nauchno-teknicheskii Byulleten Vsesoyznogo Ordena Lenina* 134, 63–65.

Belmans, K. (1986) Op zoek naar nieuwe rassen: Castor (zoete kers) en Reine Claude Souffrian (pruin). *Boer en Tuinder* 41, 19.

Blazek, J. (1985) Precocity and productivity in some sweet cherry crosses. *Acta Horticulturae* 169, 105–113.

Cireasa, V. (1972) Una cultivar di ciliegio dolce romena di scarso sviluppo utilizzata come portinnesto nanificante. In: *Atti 2° Convegno del Ciliegio*. Verona Camera di Commercia IAA, pp. 133–138.

Donini, B., Fideghelli, C. and Rosati, P. (1972) Mutanti compatti indotti con radiazioni in varietà di ciliegio. In: *Atti 2° Convegno del Ciliegio*. Camera di Commercio IAA, Verona, pp. 33–51.

Enikeev Kh.K., Vysotskii, V.A. and Plotnikova, G.A. (1984) Development of sour and sweet cherry embryos in *in vitro* culture after isolation at early stages of embryogenesis. *Breeding Abstracts* OPO 560 1322, OCO 560 1600.

Fogle, H.W. (1975) Cherries. In: Janick, J. and Moore, J.N. (eds) *Advances in Fruit Breeding*. Purdue University Press, West Lafayette, Indiana, pp. 348–366.

Frecon, J. (1980) Starkrimson cherry (Zaiger 1G–200), introduced by Stark Brothers' Nurseries and Orchards Co. *Fruit Varieties Journal* 34, 18.

Freigoon, S.O. and Crosse, J. E. (1974) Host relations and distribution of a physiological and pathological variant of *Pseudomonas morsprunorum*. *Annals of Applied Biology* 81, 317–330.

Grubb, N.H. (1949) *Cherries*. Crosby Lockwood and Son, London.

Guerrero-Prieto, V.M., Vasilakakis, M.D. and Lombard, P.B. (1985) Factors controlling fruit set of 'Napoleon' sweet cherry in western Oregon. *HortScience* 20, 913–914.

Iezzoni, A., Schmidt, H. and Albertini, A. (1990) Genetic resources of temperate fruit and nut crops. 3: Cherries. *Acta Horticulturae* 190, 111–173.

Ivanicka, J. and Pretova, A. (1986) Cherry (*Prunus avium* L.). In: Bajai, Y.P.S. (ed.) *Biotechnology in Agriculture and Forestry. Vol. I, Trees I.* Springer Verlag, Berlin, pp. 154–169.

Knight, R.L. (1969) *Abstract Bibliography of Fruit Breeding and Genetics to 1965,* Prunus. Eastern Press Limited, London.

Lane, W.D. (1978) Compact sweet cherries. *Fruit Varieties Journal* 32, 37–38.

Lane, W.D. (1992) Le nuove varietà di ciliegio canadesi. *Rivista Frutticoltura* 1, 19–24.

Lane, W.D. and Sansavini, S. (1988) New Star. *Rivista Frutticoltura* 9, 60.

Lane, W.D. and Schmid, H. (1984) Lapins and Sunburst sweet cherry. *Canadian Journal of Plant Science* 64, 211–214.

Lapins, K.O. (1963) Note on compact mutants of Lambert cherry produced by ionizing radiation. *Canadian Journal of Plant Science* 43, 424–425, 524–525.

Lapins, K.O. (1971) Stella, a self-fruitful sweet cherry. *Canadian Journal of Plant Science* 51, 252–253.

Lapins, K.O. (1973) *Induced Mutations in Fruit Trees.* International Atomic Energy Agency, Vienna.

Lapins, K.O. (1974) 'Summit' sweet cherry. *Canadian Journal of Plant Science* 54, 851.

Lapins, K.O. (1975) Compact Stella cherry. *Fruit Varieties Journal* 29, 20.

Lewis, D. and Crowe, L.K. (1954) The induction of self-fertility in tree fruits. *Journal of Horticultural Science* 29, 220–225.

Lugli, S., Sansavini, S. and Baldassari, M.T. (1993) Valutazione di nuove varietà e selezioni canadesi di ciliegio dolce. *Riv. Frutticoltura* 2, 27–35.

Matthews, P. (1970) The genetics and exploitation of self-fertility in the sweet cherry. In: *Proceedings of Eucarpia Fruit Breeding Symposium.* Angers, pp. 307–316.

Saamin, S. (1988) Radiation-induced mutations in sweet cherry (*Prunus avium* L.) cvs. Napoleon and Bing. *Dissertation Abstr. Intern., B (Sciences and Engineering)* 8, 2204.

Sansavini, S. and Lane, W.D. (1983) 'Sunburst' e 'Lapins' ciliegie autofertili durone-simili. *Rivista Frutticoltura* 9/10, 55–57.

Saunier, R., Fos, E., Tauzin, Y., Edin, M., Tronel, C., Chartier, A. and Labergère, M. (1987) Les variétés de cerisier. *Arboriculture Fruitière* 397, 55–58; 398, 29–36 and 400, 52–55.

Sherman, W.B. (1977) Attempts at cherry breeding in Florida. *Fruit Varieties Journal* 31, 60–61.

Tehrani, G. (1988) Self-fertility: A new dimension in sweet cherry breeding. *Highlights of Agricultural Research, Ontario* 11, 1–4; 12, 29.

Tehrani, G. and Brown, S.K. (1992) Pollen-incompatibility and self-fertility in sweet cherry. *Plant Breeding Reviews* 9, 367–388.

Tehrani, G. and Lay, W. (1991) Verification through pollen incompatibility studies of pedigrees of sweet cherry cultivars from Vineland. *HortScience* 26, 190–191.

Trajkovski, V. (1984) Summary of activities during the biennium 1982–83. *Växtförädling av Frukt och bär, Balsgard* 151–159.

Walther, F. and Sauer, A. (1985) Analysis of radiosensitivity, a basis requirement for *in vitro* somatic mutagenesis. I. *Prunus avium* L. *Acta Horticulturae* 169, 97–104.

Way, R.D. (1966) Identification of sterility genes in sweet cherry cultivars. In: Marshall, R.E. (ed.) *Proceedings of the XVII International Horticultural Congress* I. Paper 145. Michigan State University. East Lansing, Michigan.

Zwintzscher, M. (1955) Die auslösung von Mutationen als Methode de Obstzuchtung. I. Die Isolierung von Mutanten in Anlehnung an primäre Veränderungen. *Züchter* 25, 290–302.

4 Sour Cherry Cultivars: Objectives and Methods of Fruit Breeding and Characteristics of Principal Commercial Cultivars

A.F. Iezzoni

Department of Horticulture, Michigan State University, East Lansing, MI 48824-1325, USA

4.1 Introduction

Sour cherry is an allotetraploid ($2n = 32$) with sweet cherry (*Prunus avium*) and Ground cherry (*P. fruticosa*) as the presumed progenitor species (Olden and Nybom, 1968). In Europe and Asia, where the habitat of sour cherry overlaps with that of sweet and ground cherry, hybridization between sour cherry and its two progenitor species is common. Due to the prevalence of interspecific hybridization, some of the sour cherry cultivars listed for countries rich in cherry germplasm may in fact be hybrids of *P. cerasus* with either sweet or ground cherry.

As a result of its polyploid nature and continued interspecific gene flow, sour cherry is a polymorphic species with variation spanning the morphologies characteristic of the two progenitor species (Hillig and Iezzoni, 1988). The variability in sour cherry presents a wealth of diversity for the breeder and has resulted in many cultivars that differ dramatically in fruit type and tree structure (Iezzoni *et al.*, 1989). The number of cultivars grown in a particular area is an indication of the local genetic diversity of sour cherry. In the centre of diversity in eastern Europe, the cultivars grown span a ripening period of approximately 40 days. In contrast, the USA and Canada, which have no natural diversity for sour cherry, have remained monocultures of a 400-year-old cultivar, Montmorency.

4.2 Breeding Objectives

Consistent yields and superior fruit quality are the two main objectives of sour cherry breeding programmes. Yield is a complex trait influenced by various aspects of tree and fruiting structure, flower characteristics and abiotic and biotic

113

stress tolerance. Fruit quality varies dramatically and selection criteria for quality will be based on the desired use.

4.2.1 Tree and fruiting structure

Tree size and fruiting structure vary widely within sour cherry. Tree size ranges from vigorous cultivars, of comparable size to sweet cherry, to dwarfs which reach a mature height of only 1 m and which would lend themselves to over-the-row machine harvesting. Within this range of tree size are cultivars considered of low vigour, which reach an approximate height of 2–3 m. In Romania, such low-vigour cultivars are planted at spacings of 4 × 2.5 m and trained to a palmette system on three wires (Gozob *et al.*, 1978).

Most sour cherry cultivars fruit both on 1-year-old wood and on spurs formed on older wood, while certain cultivars fruit exclusively on one or the other. Since sour cherry buds produce either floral or vegetative growth, never both, 1-year-old wood that bears only flower buds is barren the following year. It follows, therefore, that trees which fruit primarily on 1-year-old wood require more pruning than spur-type trees if significant proportions of the branch framework are not to become barren (bare wood).

Ultimately, the tree size and fruiting structure desired will be largely determined by the chosen harvesting technique (i.e. trunk shakers, over-the-row harvesters, or hand harvesting) and the amount of pruning which the growers are willing to do. Within these practical limitations, the plant breeder strives to optimize yields by maximizing whole-tree yield efficiencies among different selections. For example, the cultivar Meteor has a significantly higher limb yield efficiency (yield/limb cross-sectional area) than the cultivar Montmorency because Meteor bears 70% of its fruit on spurs compared with Montmorency's 32% (Chang *et al.*, 1987). However, the Montmorency trees sampled had approximately four times as many 4-year-old limbs as did Meteor and therefore a considerably higher tree yield. Therefore, attempts to increase yield per hectare must take into account not only floral bud position, density and setting efficiency but also branch density in relation to tree size.

4.2.2 Flower characteristics

Sour cherry breeders can increase yields by selecting trees with high efficiency of fruit set. Flore (1985) reported that a minimum of two leaves per fruit was necessary for optimum fruit size and development in Montmorency. However, as Montmorency leaf-to-fruit ratios are generally above this threshold, with fruit set efficiency rarely higher than 30%, there are likely to be factors other than this usually limiting yields.

Sour cherry is frequently considered to be self-compatible, although self-incompatible and partially self-incompatible cultivars do exist. Self-incompatibility in sour cherry resembles gametophytic self-incompatibility, with complete inhibition of self-pollen tube growth in the style (Lansari and Iezzoni, 1990). In partially self-incompatible cultivars, only a percentage of self-pollen reaches the ovules. The inheritance of self-incompatibility in sour cherry is not understood. In breeding programmes, self-incompatibility is often encountered and can be a major problem. For example, self-incompatible progeny frequently occur in crosses where both parents are self-compatible. Low fruit set frequently limits yield in self-incompatible cultivars, and, as a result, breeders attempt to select those cultivars which are completely self-compatible.

4.2.3 Abiotic stress

Cold-hardiness of sour cherry wood and flower buds is one of the most important breeding objectives for the colder production areas. In Russia, sufficient midwinter hardiness is extremely important and Kolesnikova (1975) has described two ecotypes of sour cherry which differ in their midwinter cold-hardiness. When fully acclimatized, the cultivars of the middle Russian group have flower buds that can be hardy to $-38°C$ while cultivars in the European group have wood and flower buds that can survive $-34°C$ and $-24°C$, respectively. *P. fruticosa*, which is more cold-hardy than sweet cherry and sour cherry, is frequently in the pedigree of cultivars in the hardier middle Russian group.

Failure to acclimatize sufficiently in the autumn and early deacclimatization in the spring can also result in significant cold damage, because the expanding pistil is less hardy than a fully dormant flower (Callan, 1990). For example, temperatures of $-9.4°C$ cause damage to flower primordia in Montmorency when the buds begin to swell, whereas $-3.3°C$ causes injury during anthesis (Dennis and Howell, 1974). Additionally, drought stress and disease or insect pressure, which result in early defoliation and weak trees, dramatically reduce both wood and flower bud hardiness of Montmorency, predisposing the tissues of this cultivar to cold injury (Howell and Stackhouse, 1973; Flore *et al.*, 1983).

A delay in the onset of spring floral bud development would decrease the probability of crop loss from a spring freeze. Consequently, sour cherry breeders frequently select for late blooming (Simoviki, 1959–1960; Apostol and Iezzoni, 1992). For example, the cultivar Fruchtbare von Michurin blooms 6–9 days later than Montmorency in Michigan. This represents a difference of approximately 100 heat units accumulation with a base temperature of 4.5°C (Iezzoni and Hamilton, 1985). Fruchtbare von Michurin, which originated in Russia, is reported to be extremely winter-hardy and frost-resistant, and is suspected to be a natural hybrid between *P. cerasus* and *P. fruticosa* (Papapov and Dutova, 1973).

Although sour cherry cultivars grafted on Mahaleb rootstock are generally considered to be drought-tolerant, no published reports are known of sour cherry cultivars showing drought tolerance independent of the rootstock effect.

4.2.4 Biotic stress

Resistance or tolerance to diseases

Breeding for disease resistance in sour cherry has concentrated on resistance to cherry leaf spot caused by *Blumeriella jaapii* (syn. *Coccomyces hiemalis*). This fungal disease is prevalent in all the major cherry-growing areas of the world. Cherry cultivars exhibit a range of tolerance to this disease, with the cultivars Csengodi and North Star being the most tolerant.

Breeders have sought to incorporate resistance to cherry leaf spot from other cherry species into the sour cherry. Sweet cherry is generally more resistant than sour cherry (Sjulin *et al.*, 1989) and, in Romania, several *B. jaapii*-tolerant sour cherry cultivars, which have sweet cherry parents, have been selected. Hybridizations made in Russia between sour cherry and *P. maackii* have produced selections that are immune to *B. jaapii*. Unfortunately, maintaining this disease immunity while attempting to increase fruit size in back-crossing programmes is proving difficult (E. Jigadlo, personal communication).

Brown rot, caused by *Monilinia* spp., is a prevalent fungal disease of sour cherry which reduces yields by infecting and decaying blossoms, twigs and fruit. *M. laxa* occurs primarily on sour cherry and generally causes blossom and spur blight, while *M. fruticola*, which occurs on 11 different *Prunus* spp., causes fruit rot. Although cultivars are known to differ in tolerance to *Monilinia*, data on this are currently limited. As a large number of sprays are required to control fruit brown rot, especially near harvest, the expense of such sprays and environmental pressure to reduce them may make breeding for tolerance to brown rot a more important objective in the future.

Prunus necrotic ringspot virus (PNRSV) and prune dwarf virus (PDV) are the most important virus diseases affecting sour cherry. Differences in virus tolerance have been reported among a few sour cherry cultivars, but no programme has focused directly on breeding for PNRSV and/or PDV resistance. Other breeding objectives have to date been considered of higher priority. Nevertheless, planting of only virus-free nursery trees is strongly recommended.

Although sour cherry is generally not as susceptible to bacterial canker, caused by *Pseudomonas* spp., as is sweet cherry, some sour cherry selections are susceptible to the disease and it is recommended that superior selections be tested for tolerance to *Pseudomonas* before release.

Resistance or tolerance to pests

Some 40 species of insects are known to cause damage to cherries by attacking the foliage, fruit or woody portions of the tree. Some, like the cherry fruit fly (*Rhagoletis cingulata*) and black cherry aphid (*Myzus cerasi*), are specific to cherry, while others, such as the curculio (*Conotrachelus nenupha*) may infest a number of species of tree fruits. Tolerance to cherry fruit fly would be most beneficial, because of the numerous sprays required to control this insect and the rejection of fruit by the processor if cherry fruit fly larvae are found. Unfortunately, no source of resistance has yet been identified among sour cherries and, although early ripening cultivars may avoid the larval stage of the cherry fruit fly, this breeding possibility has not been explored. Within cherry breeding populations, differences in tolerance to the two-spotted mite (*Tetranychus urticae*) are apparent, but these have yet to be verified in systematic studies. Resistance of the cultivar Meteor Korai to the San Jose scale (*Quadraspidiotus perniciosus*) (Jenser and Sheta, 1969) is, in fact, the only published report of insect resistance within the sour cherry species.

4.2.5 Fruit quality and maturity period

Sour cherry cultivars are generally classified as Morellos or Amarelles, which refer to red or clear flesh and juice colour, respectively. In the USA, the clear-fleshed Amarelle cultivar, Montmorency, is used almost exclusively for the production of cherry pies. In Europe, it is the red-fleshed Morello cultivars that are preferred for use in a wide range of processed products, including juice, jam, consumer glass pack and canned products for bakery use in pies, tortes, cakes and other pastries. Within the Morello types, regional preferences have resulted in an array of cultivars whose flesh and juice colour range from light red to purple. When breeding for a highly pigmented cherry cultivar, intended for juice or liqueurs, anthocyanin content and sugar/acid ratio of the fruits are critical. Frequently, the highly pigmented Danish cultivar Stevnsbär is used as a standard (Christensen, 1976).

A third group of sour cherry cultivars, called Marasca, are grown predominantly in the former Yugoslavia and adjacent countries. Marasca cultivars are characterized by small, very dark-coloured fruit which are of the best quality for making cherry wine and liqueurs. Marasca cultivars are sufficiently distinct to have been classified by early botanists as a subspecies of *P. cerasus*.

Because the majority of sour cherries grown in the USA are harvested and pitted mechanically, new cultivars should be adapted to these technologies. Uniform formation of abscission layers between the fruit and its stalk and a low fruit removal force (i.e. the force needed to separate the fruit from its stalk) would reduce the energy required to shake the fruit from the tree. Also, uniformity of ripening among the fruits on a tree is necessary if mechanically harvested cherries are to be of consistent quality. Increased fruit firmness would also be very beneficial, thereby reducing bruising caused by wind damage and mechanical handling. Similarly, cherry cultivars with fruits that are mechanically pitted (stoned) must have round or only slightly oval pits (the stone or endocarp) to minimize pit chipping. Processing losses could be greatly diminished by an improvement in the percentage pitted yield (i.e. the percentage of the harvested weight usable after pitting). Hand-picked Montmorency cherries normally lose 12–14% of their total weight when pitted, and approximately half of this loss is attributed to pit weight (Burton *et al.*, 1979). However, mechanically harvested Montmorency fruit lose 14–24% of their total weight; much of the additional weight loss is due to juice lost through the ruptured skin of bruised cherries and torn stem scars. New cultivars are needed, therefore, with stones that are easily freed from the flesh and which have a greater flesh-to-pit ratio and a dry stem scar.

Although most sour cherries are processed, a small portion of the Morello fruit in Europe is sold for a premium on the fresh market. Fresh fruit is picked with the stems attached and an individual fruit weight of 6–8 g is desired.

Since the majority of the sour cherry production is processed, the ripening date is not as critical as in sweet cherry. However, an extended ripening season provides a longer supply of fresh market fruit plus more efficient use of harvesting and processing labour and equipment. Breeding programmes in several countries, such as Serbia, Hungary and Romania, have successfully identified or bred an assortment of sour cherry cultivars which ripen over a 40-day period.

4.3 Breeding Techniques

For centuries, peasants and gardeners in Europe and western Asia have selected the most productive sour cherry cultivars with the highest-quality fruit. The best clones, propagated either by root suckers or grafting, were then planted in garden plots and along public roads. In eastern Europe and Russia, where sour cherries flourished in great abundance, the following landraces arose: Cigany, Crisana, Mocanesti, Oblacinska and Vladimskaia. Each landrace can be characterized by a specific description under which many closely related clones can be grouped. In the countries where these landraces arose, clones selected from within these groups still represent a significant part of the production acreage. When the breeding programmes in these areas began, the breeders first collected the best selections from the villages. Some of the best of these selections were then named as specific clones. Open-pollinated seed was collected of others, while yet more were used as parents in hybridizations. In most cases, the cultivars grown today are no more than one generation removed from these local selections.

Currently, breeding programmes focus on generating progeny from controlled hybridizations or open pollination. For controlled crosses, pollen is collected from newly opened or opening flowers just before the petals separate. Prior to pollen dehiscence, anthers are removed from the flowers by rubbing the inverted blossoms over a wire mesh and these are then collected and allowed to dry at room temperature (22°C). Usually the anther wall will dry and the pollen will be released within 24 h. Unless needed for immediate use, the pollen is usually stored for a short period in glass vials in a refrigerator in the presence of a desiccant (anhydrous $CaSO_4$); for long-term storage, pollen must be frozen. Pollen viability can be checked by germinating it at room temperature in a liquid medium containing 10 ppm boron and 15–20% sucrose.

Flowers from maternal parents in the intended crosses must be emasculated if they are self-compatible. To achieve this, the perianth tube is cut below the area where the stamens are attached and then gently pulled off the flower, thus removing the calyx, corolla and stamens. With the removal of the petals, bee visitation is unlikely and therefore the possibility of contamination with foreign pollen is reduced. Following emasculation, the intended paternal pollen is applied to the pistils using a finger, glass rod or brush. In general, the temperature must remain above 12°C for pollen germination and tube growth to be effective.

The cherry seeds produced, either by controlled crossing or open pollination, are removed from ripe fruit and then stratified. The first step in this procedure is to sterilize the seeds by soaking in a dilute solution of chlorine and bleach for 2–5 min and/or a fungicide solution for 1 h. The seeds are then placed in bags of moist, sterile vermiculite, sand or sphagnum moss and stored at 2–5°C. After 4 months, the seeds are periodically checked for germination and seedlings are transplanted to the greenhouse when the radicle is 1 cm in length. In early-ripening cultivars, the embryos often abort in the later stages of fruit development and may need to be excised and cultured on nutrient agar to increase the percentage of viable seedlings (Havis and Gilkeson, 1949).

In those breeding programmes where resistance to *Blumeriella* (leaf spot) is a high priority, the 1-year-old seedlings are screened, either by natural infection in a nursery or by artificial inoculation in the greenhouse. Only the most tolerant seedlings, which can be identified by their minimal defoliation, are planted in the field for fruiting and further evaluation.

Sour cherry seedlings begin to fruit 3–7 years after germination. When a seedling fruits for the first time, any determination of optimum maturity date is inevitably subjective. However, since a majority of the new cultivars are intended for mechanical harvesting, fruit is generally considered to be mature when the force required to pull the fruit from its stalk is lower than 300 g. In general, at least five to ten fruit must be sampled per seedling to make an adequate assessment of the genetic potential for fruit size (Iezzoni, 1986). Only five pits need be measured for assessment of their length and width, as the environmental influence on pit size is minimal. Since soluble solids content is influenced by ripening date, year and sampling error, accurate measurements can only be made when fruit from more than one tree is available. The records frequently taken on fruits of new selections include: fruit weight, fruit length and diameter, pit length and width, pit as a percentage of total weight, soluble solids, juice acidity and colour, stem scar damage at abscission and pedicel length.

Yield potential of the seedlings can be evaluated by dividing yield into its component parts. Tree yield in cherry is defined as the product of the primary yield components, fruit number and individual fruit weight (Chang *et al.*, 1987). The secondary yield components, whose product equals fruit number, are the number of spurs and lateral buds per limb, the number of flower clusters per spur or lateral bud, the number of flowers per cluster and percentage fruit set. By evaluating these characters, those seedlings which produce their fruits primarily on spurs or lateral buds can be identified. Also, those seedlings which may have a particularly low fruit set or high fruit weight may be identified. Measuring branch angle with the trunk or major scaffold limbs and internode length may also give some indication of the potential growth habit and vigour of the seedling (Krahl *et al.*, 1991).

Cold-hardiness of selections is frequently measured by subjecting twigs and flower buds to freezing temperatures in growth chambers and evaluating xylem and pistil injury at various temperatures. Frequently, the data are presented as LT_{50} values, or the lowest temperature required to kill 50% of the xylem tissue or flower primordia. Because cold-hardiness is a complex character, including autumn acclimatization, then midwinter hardiness and finally deacclimatization evaluations must span the autumn, winter and spring months.

So-called 'marker-assisted' selection would undoubtedly increase the efficiency of the breeding effort. In such a scheme, markers are identified which cosegregate with traits of importance. Therefore, instead of selecting for the trait directly, seedlings are selected for the cosegregating marker(s). This selection tool would be particularly useful for: (i) traits such as flower and fruit characteristics which are expressed only in mature plants, so enabling undesirable seedlings to be discarded before field planting; and (ii) major genes which are responsible for most of the variation in a quantitative trait and/or traits with low heritabilities.

In sweet cherry, polymorphisms for 34 putative loci have been described and seven additional loci were confirmed by inheritance studies (Santi and Lemoine, 1990). The inheritance of the seven isozyme loci involved two alleles each for the following loci: acid phosphatase, Acp1; leucine aminopeptidase, Lap1; shikimate dehydrogenase, Sdh1; malate dehydrogenase, Mdh1; glutamate oxaloacetate transaminase, Got1; isocitrate dehydrogenase, Idh1; and malic enzyme, Me1. The following two loci pairs were linked: Lap1 and Got1 (r = 0.03 ± 0.02), and Lap1 and Me1 (r = 0.05 ± 0.07). Chloroplast restriction fragment length polymorphisms (RFLPs) are able to distinguish between sweet cherry (P. avium) and ground cherry (P. fruticosa) chloroplasts (Iezzoni et al., 1989). Although a majority of sour cherry selections have patterns characteristic of the ground cherry chloroplasts, a few cultivars have the patterns characteristic of sweet cherry chloroplasts. As a result, it may be important for the sour cherry breeder to consider which cytoplasm, sweet cherry or ground cherry, would be preferred.

Various laboratory techniques provide useful tools for the sour cherry breeder. Micropropagation can be used to generate own-rooted plants. Normally the cultures are initiated using unexpanded vegetative buds in the spring because of the ease of sterilizing such buds as opposed to elongated meristems. Frequently, three media are used sequentially: a bud establishment medium, a shoot proliferation medium and finally a rooting medium (Snir, 1982; Ivanicka and Pretova, 1986). As previously mentioned, embryo excision and rescue can increase the percentage of viable seedlings recovered from those crosses where the maternal parent aborts a large number of embryos. The use of embryo culture is more prevalent in sweet cherry breeding programmes, however, where the development of early-ripening cultivars is a higher priority.

In vitro culture of anthers or microspores from the allotetraploid sour cherry should, if successful, give rise to haploid plants ($2x = n = 16$), which could then be crossed directly with the diploid sweet cherry. No confirmed reports of haploid microspore-derived plants have yet been published for cherry; however, it is possible that the anther culture procedures of Jordan (1974) and Seirlis et al. (1979) would be successful. Selection among single isolated somatic cells in protoplast culture could result in the identification of useful somaclonal variants. Such useful variants could possibly be selected following the imposition of stresses such as those induced by toxins, salt (Ochatt and Power, 1989) or cold. Protoplast fusion would permit the development of interspecific hybrids between those species of Prunus that are not sexually compatible with sour cherry, allowing for the transfer of desirable genes to the sour cherry (Ochatt et al., 1989). Once protoplasts have been selected or fused, plants need to be regenerated from these protoplasts. In cherry, protoplast regeneration has only been successful to date with the rootstock cultivar Colt, a triploid interspecific hybrid between P. avium and P. pseudocerasus (Ochatt et al., 1987), and with the sour cherry rootstock clones CAB 4D and CAB 5H (Ochatt and Power, 1988).

As single genes are identified which confer economically important traits, a method must be found to transfer these single genes into sour cherry. Gene transfer via transformation of cherry with the Agrobacterium rhizogenes vector shows some promise in this respect. Sour cherry is susceptible to infection by Agrobacterium (James et al., 1985; Dandekar and Martin, 1986), suggesting that gene transfer via *Agrobacterium* is indeed possible. However, if the *Agrobacterium* gene transfer system is to be successful, plants must be regenerated from the transformed cells. Unfortunately, regeneration in cherry from leaf or other somatic tissue has not yet been reported and, although there are promising reports of regeneration from immature cotyledons or embryo-derived callus (Lane and Cossio, 1986; Mante *et al.*, 1989) this tissue would be genetically different from that of the maternal parent.

4.4 Characteristics of the Principal Commercial Sour Cherry Cultivars

Érdi Bôtermô

Origin: Pándy 38 × Nagy Angol, bred in Hungary.
Fruit: Ripens midearly, approximately 5.5 g, dark red skin and juice.
Tree: initial fruiting is on spurs, however, older trees fruit on 1-year-old wood.
Flowers: self-compatible, early bloom.

Kelleriis 14

Origin: (Ostheimer × Früheste der Mark) × open-pollinated, bred in Denmark.
Fruit: red juice, ripens midearly, 4.7 g.
Tree: low vigour.
Flowers: self-compatible, late bloom.

Kelleriis 16

Synonym: Morellenfeuer.
Origin: (Ostheimer × Früheste der Mark) × open-pollinated, bred in Denmark.
Fruit: 5 g, red juice, ripens midearly.
Tree: vigorous, early-bearing.
Flowers: self-compatible, early bloom.

Meteor

Origin: Montmorency × Vladimir, bred at the University of Minnesota.
Fruit: middle–late-ripening, 4 g, red skin and light red juice, Amarelle.
Tree: medium vigour, 70% of the fruit on spurs.
Flowers: self-compatible, late bloom.

Meteor Korai

Synonym: Hungarian Meteor.
Origin: Pándy × Nagy Angol, bred in Hungary.
Fruit: ripens early, approximately 5 g, dark red skin and juice.
Tree: vigorous, bearing mainly on 2-year-old spurs.
Flowers: self-compatible, early to midbloom.

Montmorency

Origin: a 400-year-old variety from France.
Fruit: middle ripening period, 4 g, bright red skin, clear juice.
Tree: 30% of the fruit on spurs, medium vigour.
Flowers: self-compatible, midbloom period.

Nana

Origin: Crisana open-pollinated, bred in Romania.
Fruit: 5 g, red juice.
Tree: low-vigour tree, the majority of the fruit is on 1-year-old wood.
Flowers: self-compatible.

Nefris

Origin: variety from Poland.
Fruit: 5 g, dark red juice.
Tree: low-vigour tree, fruits on 1-year-old wood.
Flowers: self-compatible.

North Star

Origin: English Morello × Serbian Pie No. 1 (a seedling of *P. cerasus* from Serbia), University of Minnesota.
Fruit: late-ripening, 4 g, dark red skin and juice.
Tree: low vigour.
Flowers: self-compatible, midbloom period.

Oblačinska

Origin: local variety in the former Yugoslavia.
Fruit: 2.8–3.6 g, dark red juice.
Tree: low vigour, fruits on spurs.
Flowers: the majority of clones are self-compatible, midbloom period.

Pándy

Synonyms: Crisana, Körös, Köröser Weichsel.
Origin: local selection in Hungary (Pándy) and Romania (Crisana).
Fruit: midseason ripening, 6 g, light red skin and juice.
Tree: medium vigorous to vigorous.
Flowers: self-incompatible, varies depending on the clone, Pándy 148 blooms early while Pándy 279 blooms in the midseason.

Schattenmorelle

Synonyms: English Morello (USA and England), Griotte du Nord (France and Belgium), Dubbelte Morelkers, Morel (Holland), Lutowka (Poland).
Origin: an ancient cultivar, most of the clones were selected in Germany.

Fruit: 4.8 g.
Tree: the majority of the fruit is on 1-year-old wood.
Flowers: self-compatible, midlate bloom.

Stevnsbär

Synonyms: Lörskal, Heeringbär.
Origin: clones of wild *P. cerasus* which have been grown under different local names in Denmark.
Fruit: late-ripening, 3 g, very dark.

Tschernokorka

Origin: former USSR.
Fruit: 5 g, dark red to purple.
Tree: most of the fruit is on 1-year-old wood.
Flowers: self-incompatible.

Turgenevka

Origin: seedling of Zukovskaya open-pollinated, bred in Orel, former USSR.
Fruit: 7 g, red juice.
Tree: most of the fruit is on 1-year-old wood.
Flowers: self-compatible.

Ujfehértōi Fürtös

Origin: local Hungarian selection.
Fruit: late-ripening, 5 g, dark red skin and red juice.
Tree: vigorous.
Flowers: late bloom, self-compatible.

Zukovskaya

Origin: selected from Michurin's seedlings, former USSR.
Fruit: 3.4 g, almost black fruit with dark red juice.
Tree: most of the fruit is on 1-year-old wood.
Flowers: self-compatible.

General Reading

Iezzoni, A.F., Schmidt, H. and Albertini, A. (1990) Cherry spp. Genetic resources of temperate fruit and nut crops. *Acta Horticulturae* 190, 110–173.

References

Apostol, J. and Iezzoni, A. (1992) Sour cherry breeding and production in Hungary. *Fruit Variety Journal* 16, 11–15.

Burton, C.L., Tennes, B.R., Brown, G.K. and Hazebrigg, A. (1979) *The Effect of Bruising on the Percent Pitted Yield of Tart Cherries*. Michigan State University Agricultural Experimental Station Research Report, No. 377. East Lansing, Michigan.

Callan, N.W. (1990) Dormancy effects on supercooling in deacclimated 'Meteor' tart cherry flower buds. *Journal of the American Society for Horticultural Science* 115, 982–986.

Chang, L.S., Iezzoni, A. and Flore, J. (1987) Yield components in 'Montmorency' and 'Meteor' sour cherry. *Journal of the American Society for Horticultural Science* 112, 247–251.

Christensen, J.V. (1976) Description of the sour cherry variety Stevnsbär. *Tidsskrift for Planteavl* 80, 911–914.

Dandekar, A.M. and Martin, L.A. (1986) Genetic transformation of some California fruit and nut tree species with *Agrobacterium tumefaciens*. In: *Communications Abstracts. Conference on Fruit Tree Biotechnology*. Moet-Hennessy, Paris, p. 32.

Dennis, F.G. Jr and Howell, G.S. (1974) *Cold Hardiness of Tart Cherry Bark and Flower Buds*. Michigan State University Agricultural Experimental Station Research Report, No. 220. East Lansing, Michigan.

Flore, J.A. (1985) The effect of carbohydrate supply on sour cherry fruit size and maturity. *HortScience* 20, 90.

Flore, J.A., Howell, G.S. and Sams, C.E. (1983) The effect of artificial shading on cold hardiness of peach and sour cherry. *HortScience* 18, 321–322.

Gozob, T., Bodi, I. and Ivan, I. (1978) Varieties of sour cherry adapted to intensive cultivation. *Productia Vegetala Horticultura* 27, 22–26.

Havis, A.L. and Gilkeson, A.L. (1949) Starting seedlings of Montmorency cherry. *Proceedings of the American Society for Horticultural Science* 53, 216–218.

Hillig, K.W. and Iezzoni, A.F. (1988) Multivariate analysis of a sour cherry germplasm collection. *Journal of the American Society for Horticultural Science* 113, 928–934.

Howell, G.S. and Stackhouse, S.S. (1973) The effect of defoliation on acclimation and dehardening in tart cherry (*Prunus cerasus* L.). *Journal of the American Society for Horticultural Science* 98, 132–136.

Iezzoni, A.F. (1986) Variance components and sampling procedures for fruit size and quality in sour cherry. *HortScience* 21, 1040–1042.

Iezzoni, A.F. and Hamilton, R.L. (1985) Differences in spring floral bud development among sour cherry cultivars. *HortScience* 20, 915–916.

Ivanicka, J. and Pretova, A. (1986) Cherry (*Prunus avium* L.). In: Baja, Y.P.S. (ed.) *Biotechnology in Agriculture and Forestry*, vol. 1. Trees. I. Springer-Verlag, Berlin, pp. 154–169.

James, D.J., Murphy, H. and Passey, A.J. (1985) Transformation of apple, cherry and plum rootstocks. *Annual Report of East Malling Research Station for 1984*, p. 118.

Jenser, G. and Sheta, I.B. (1969) Investigation of the resistance of a few Hungarian sour cherry hybrids against the San Jose leaf scale (*Quadraspidiotus perniciosus*). *Acta Phytopathology* 4, 313–315.

Jordan, M. (1974) Multicellular pollen in *Prunus avium* anther culture *in vitro*. *Zeitschrift für Pflanzenzuchtung* 71, 350–363.

Kolesnikova, A.F. (1975) *Breeding and Some Biological Characteristics of Sour Cherry in Central Russia*. Priokstoc Izdatel'stvo, Orel, USSR.

Krahl, K.H., Lansari, A. and Iezzoni, A.F. (1991) Morphological variation within a sour cherry collection. *Euphytica* 52, 47–55.

Lane, W.D. and Cossio, F. (1986) Adventitious shoots from cotyledons of immature cherry and apricot embryos. *Canadian Journal of Plant Science* 66, 953–959.

Lansari, A. and Iezzoni, A. (1990) A preliminary analysis of self-incompatibility in sour cherry. *HortScience* 25, 1636–1638.

Mante, S., Scorza, R. and Cordts, J.M. (1989) Plant regeneration from cotyledons of *Prunus persica*, *Prunus domestica* and *Prunus cerasus*. *Plant Cell Tissue Organ Culture* 19, 1–11.

Ochatt, S.J., and Power, J.B. (1988) An alternative approach to plant regeneration from protoplasts of sour cherry (*Prunus cerasus* L.). *Plant Science* 56, 75–79.

Ochatt, S.J., and Power, J.B. (1989) Selection for salt and drought tolerance in protoplast- and explant-derived tissue cultures of Colt cherry (*Prunus avium* × *pseudocerasus*). *Tree Physiology* 5, 259–266.

Ochatt, S.J., Cocking, E.C. and Power, J.B. (1987) Isolation, culture and plant regeneration of Colt cherry (*Prunus avium* × *P. pseudocerasus*) protoplasts. *Plant Science* 50, 139–143.

Ochatt, S.J., Patat-Ochatt, E.M., Rech, E.L., Davey, M.R. and Power, J.B. (1989) Somatic hybridization of sexually incompatible top-fruit tree rootstocks (*Pyrus communis* var. *pyraster* L.) and Colt cherry (*Prunus avium* × *pseudocerasus*). *Theoretical and Applied Genetics* 78, 35–41.

Olden, E.J. and Nybom, N. (1968) On the origin of *Prunus cerasus* L. *Hereditas* 59, 327–345.

Papapov, S.P. and Dutova, L.I. (1973) The results of a comparative study of the sour cherry in different ecological conditions [in Russian]. *Izvestiya Timiryazevskoi Selskokhozyaistvennoi Akademii* 1, 138–150 (*Plant Breeding Abstracts* 44, 7162, 1974).

Santi, F. and Lemoine, M. (1990) Genetic markers for *Prunus avium* L.: inheritance and linkage of isozyme loci. *Annales des Sciences Forestières* 47, 131–139.

Seirlis, G., Mouras, A. and Salesses, G. (1979). *In vitro* culture of anthers and organ fragments of *Prunus*. *Annales Amélior. Plantes* 29, 145–161.

Simoviki, K. (1959–1960) Relation between flowering time of fruit trees and frost period in the Macedonian People's Republic. *Godisen Zb. Zemjod-sum. Fak. Univ. Skopie* 13, 5–57.

Sjulin, T.M., Jones, A.L. and Andersen, R.L. (1989) Expression of partial resistance to cherry leaf spot in cultivars of sweet, sour, duke, and European ground cherry. *Plant Disease* 73, 56–61.

Snir, I. (1982) *In vitro* propagation of sweet cherry cultivars. *HortScience* 17, 192–193.

5 Rootstocks for Sweet and Sour Cherries

A.D. WEBSTER[1] and H. SCHMIDT[2]

[1]*Horticulture Research International, East Malling, West Malling, Kent ME19 6BJ, UK;* [2]*Bundesforschungsanstalt für Gartenbauliche Pflanzenzüchtung, Bornkampsweg 31, 2070 Ahrensburg/Holst, Germany*

5.1 Introduction

Neither sweet nor sour (tart) cherries reproduce true to type if propagated from seeds; propagation of selected cultivars must be by asexual methods. The most common method of asexual (vegetative) propagation is by budding or grafting the scion on to a compatible rootstock. Sour and, more particularly, sweet cherry scion cultivars are usually difficult to propagate using conventional summer or winter cutting techniques, whereas budding on to a rootstock is much easier and more reliable. Originally, all cherry rootstocks were raised from seed and were mainly Mazzards (Geans or *Prunus avium*). These were in plentiful supply growing wild in Europe and were relatively easy to propagate. Using only a shield bud or small scion graft, many new trees could be raised each year from just a single scion 'mother' tree.

For many years rootstocks were used only as a means of propagating selected scion cultivars. However, observant horticulturists soon realized that some rootstocks conferred additional benefits to the compound tree (or stion) by modifying its growth and/or cropping, or giving it tolerance to unfavourable or harmful edaphic or climatic conditions. Although ease of propagation still remains an important criterion in rootstock selection, other factors, such as the rootstock's effect upon scion vigour, precocity and abundance of cropping and resistance to unfavourable soil conditions or soil-borne diseases, have become of equal or often greater importance.

It is unlikely that any single rootstock will ever be completely suited to all the environmental conditions in which sweet and sour cherries are grown throughout the world. Hence, it is of great importance that growers select the rootstocks most appropriate for their local environmental conditions and also for their chosen system of management.

Unfortunately, very little is understood of how rootstocks bring about their many and varied effects upon scion growth and cropping. Several theories have been advanced to explain this, most, in recent years, seeking to link balances of the various endogenous plant hormones with the observed effects upon scion growth. None of these theories has yet fully explained rootstock effects on scion growth and cropping.

5.2 History of Rootstock Use

It is thought that Mazzard (*Prunus avium* L.) seedlings were used as rootstocks for cherries more than 2400 years ago. Greek and Roman horticulturists used them in cherry tree production and this practice spread to most other cherry-producing countries, where it has remained little changed up to the present day. The principal advantage offered by Mazzards were, and still are, their good graft compatibility with both sweet and sour cherry cultivars and their ready availability as seeds collected from trees growing in the wild.

The use of Mahaleb (*Prunus mahaleb* L.) as a cherry rootstock is, according to horticultural literature, a more recent practice and the first known use of Mahaleb rootstocks was recorded in France in 1768. Thomas Rivers writing in his book *The Miniature Fruit Garden*, published in 1865, recommends using the 'Mahaleb' or 'Perfumed cherry' – named on account of the most agreeable perfume given off by its wood when burnt – as a stock for dwarfed sweet cherries. The stock was at that time known in France as the Bois de St Lucie and had been used there to dwarf cherries for many years previously. Thomas Rivers noted that he had used Mahaleb rootstocks in Britain since the early 1850s and that they enabled cherries to be grown in strong white clay or soils with chalky subsoil, i.e. those not traditionally recommended for the crop. These recommendations hardly correspond with what we know today of Mahalebs' dislike of clay soils. However, Rivers did note that Bigarreaux and Heart cherries were too short-lived in many kinds of soils when grafted on this rootstock. He recommended double grafting with the Morello cherry as rootstock and Mahaleb as interstock.

Thomas Rivers further reported that trees on Mahaleb grew quite vigorously, particularly in the first few years following planting. Later in their lives, especially if root-pruned, they were made to form dwarf trees, which cropped prolifically and which could be easily covered with muslin or tiffany, so protecting them from frosts in the spring and damage from birds later in the season. He suggested planting trees on Mahaleb spaced 5 or 6 feet apart and training them to produce seven or nine branches arising from the centre of the plant.

Mahaleb began to gain popularity in the USA in the mid-1800s and by the early twentieth century had superseded Mazzard as the most popular rootstock for cherries. This change was undoubtedly prompted more by nurserymen than by fruit growers. Mahalebs are easier to germinate from seed and resistant to leaf spot disease (*Coccomyces hiemalis* Higg. syn. *Blumeriella jaapii* (Rehm) v. Arx.) to which Mazzards are very susceptible. Mazzards remained popular in some parts of the USA, however, and by the mid-1920s research indicating short life of trees on Mahaleb stocks had swung the balance of popularity back towards Mazzards. Mahaleb has continued, nevertheless, to be the main rootstock used for sour cherry culture.

Sour cherry (*Prunus cerasus*) clones became popular as rootstocks for sweet cherry grown on heavy clay soils with poor drainage. In California, the sour cherry rootstock selection, Stockton Morello, at one time accounted for 15% of the planted acreage, but this figure has since declined to less than 5%. Sour cherry stocks have always proved popular in central Europe (parts of Russia), where their greater tolerance of low winter temperatures, compared with Mazzard and Mahaleb, made them more suitable.

5.3 Factors to Consider when Choosing a Cherry Rootstock

5.3.1 Graft compatibility

When great difficulty is found making a tree, whether fructiferous, or ornamental, of any species, or variety, produce blossoms, or in making its blossoms set when produced, success will probably be obtained in almost all cases, by budding or grafting upon a stock which is nearly enough allied to the graft to preserve it alive for a few years, but not permanently.

So wrote the famous horticulturist Thomas Andrew Knight, when discussing rootstocks in 1809. While horticulturists today would aim for more than a few years' life for their cherry trees, the principles suggested by Knight remain valid. Wolfram (1979) has argued that, to achieve a reasonable longevity for sweet cherry trees, the growth reduction brought about by the rootstock should not be more than 30% of that on Mazzard rootstocks. Although trees much more dwarfed than this can be produced, they are likely to be too short-lived.

With a few exceptions, budding sweet cherry scions on to Mazzard, or sour cherry scions on to sour cherry rootstocks, has little effect upon the scion's intrinsic vigour or cropping precocity. In contrast, budding on to some other, albeit closely related, *Prunus* species or hybrid can often greatly reduce vigour and improve scion precocity and productivity. The difficulty is in choosing a rootstock which will confer these benefits upon the scion but at the same time sustain its growth over a reasonable orchard life. A type of beneficial 'congeniality' or 'partial incompatibility' is to be aimed for.

Severe incompatibility between scion and rootstock may be indicated by poor bud take in the nursery or death of the tree in the maiden year of growth. Alternatively, trees may grow quite successfully for as long as 8–10 years, only for the union to snap off in a wind or for the scion to suddenly decline and die. This delayed incompatibility is often foreshadowed by one or more typical symptoms developing prior to tree mortality. Trees that develop small and/or yellow leaves and stunted growth should always be suspected, as should those that bear smallish fruits. Early reddening and fall of leaves in the autumn or excessive rootstock suckering is also a common symptom of incompatibility. Many reports suggest that scion overgrowth of the rootstock at the union is another indication of developing incompatibility. This is not necessarily symptomatic, however, as there are many cherry trees showing considerable scion overgrowth which are perfectly healthy and have robust and efficient graft unions after 20 or 30 years in the orchard. It is also important to be aware that incompatibility may be exacerbated by virus infection and, wherever possible, only virus-tested rootstocks and scions should be used. Recent research suggests that some reported 'delayed incompatibility' may be attributable to hyper-sensitivity to certain races of *Prunus* necrotic ringspot virus (PNRSV) or prune dwarf virus (PDV).

The causes of graft incompatibility in cherry are not fully understood. Some authorities have suggested a relationship with the presence of specific phenolics and their infusion into tissues close to the union (Treutter, 1985). Other reports have shown reduced amounts of auxins and enzymes in the cambium of incompatible combinations (Feucht and Schmid, 1984). Further research will be needed if graft incompatibility is to be fully understood.

All Mazzard rootstocks are fully compatible with sweet and sour cherries, whereas problems of compatibility are often experienced with Mahalebs. Prema-ture tree death of sweet cherries on Mahaleb rootstocks is frequently ascribed to delayed incompatibility. This may, indeed, often be the cause of tree death, but a more likely explanation in many situations is the Mahaleb's sensitivity to temporary root anaerobis in loam and clay soils. Mahaleb clones certainly differ in their compatibility with sweet cherry scions; the broad-leaved Mahalebs found in southeastern Europe are usually less compatible than the small-leaved forms found and used in France.

Some clones of sour cherry rootstock are also not fully compatible with many sweet cherry scions and care should be taken when choosing sour cherry clones as rootstocks. German research work has suggested that, although incompatibility of sweet cherry cultivars on *P. cerasus* rootstocks is genetically controlled, its expression is greatly affected by internal stresses and environmental conditions (Feucht and Schmid, 1984). It was also shown that the amount and integrity of phloem formed in the tissues above and below unions was greatly influenced by compatibility (Feucht *et al.*, 1984). Sweet cherry scions showing incompatibility with the Weiroot 13 (see below) clone of *P. cerasus* formed less phloem, the sieve plates had narrower pores and a high proportion of these pores were blocked with callose when compared with trees on the fully compatible Mazzard, F.12/1.

Unfortunately, compatibility between a specific sweet cherry scion and a clonal rootstock does not guarantee similar compatibility if the scion cultivar is changed. The hybrid rootstock Colt is compatible with most sweet cherry scions but in some situations is reported to show incompatibility with the cultivars Van and Sam. When evaluating new rootstocks selections it is therefore essential to test their compatibility with as many scion cultivars as possible.

5.3.2 Cold-hardiness

Resistance or tolerance to winter cold injury is a complex attribute and the results from tests in different regions of the world are rarely consistent. The problem is that damage may be the result of slow acclimatization of the trees in the autumn, poor tolerance to low temperatures while fully dormant or spring damage due to trees breaking dormancy too early in the year. Rootstocks may influence all of these processes.

Sour cherry selections are considered the most cold-hardy of the commonly used rootstock types. The sour cherry Weiroot rootstock selections proved more cold-resistant than the Mazzard F.12/1, Colt or *P. fruticosa* in German tests and even increased the hardiness of Hedelfingen scions (Schmid *et al.*, 1982).

Mahalebs are more hardy than Mazzards, which are in turn more hardy than the hybrid rootstock Colt. The roots of Mazzards will usually tolerate only $-10°C$, whereas Mahalebs will tolerate an extra 5°C drop in temperature before significant damage is observed. Scions are also believed to acclimatize earlier in the autumn/winter when worked on Mahaleb and this may be one explanation for the greater cold tolerance of scion cultivars often recorded on these rootstocks.

Some seedling-raised selections of Mazzard, such as HZ 170 or HZ 170 × 53, are reported to show better cold tolerance than other Mazzards and Alkavo or Caucasian Mazzards are more cold-resistant than the Limburger strain.

Work in Michigan has shown that the sweet cherry cultivars Gold and Hedelfingen were hardiest when worked on M×M 39 (a hybrid between Mazzard and Mahaleb); trees on seedling Mazzard, although less hardy, were hardier than trees on Colt (Howell and Perry, 1990). In similar tests Napoleon was hardiest on Mahaleb seedlings but, perhaps surprisingly, trees on Colt were more hardy than those on Mazzard.

In German trials interspecific hybrids of *P. cerasus* with either *P. fruticosa* or *P. subhirtella* showed the least frost damage (Strauch and Gruppe, 1985). Quite good frost resistance with little response to dehardening was shown by the standard clone *P. mahaleb* St Lucie 64 and by the Giessen clones 150/6 and 149/1 selected from crosses between *P. cerasus* and *P. subhirtella*. Among *P. avium* types tested in the Giessen research, selections Gi81, Gi82 and Gi84 were resistant to cold injury. Rootstock selections from open-pollinated *P. concinna* or *P. wadai* or hybrids with these species showed greatest sensitivity to winter cold injury.

Colt was shown to be sensitive to winter cold injury when grown in the Franken region of Germany (Schmid *et al.*, 1982); most damage occurred to Colt planted as liners in the nursery.

5.3.3 Soil type, pH and nutritional status

Mahalebs are very sensitive to wet soils, and even temporary anaerobic root conditions during the winter dormant period may cause tree decline and subsequent death. Mahaleb is best suited to light-textured, gravelly or sandy soils with very free drainage, where trees on it produce sparsely branched, deep, almost vertical roots (Larsen, 1972). These deep roots may partly explain the greater drought tolerance of trees on Mahaleb rootstocks, although this attribute is not entirely due to roots exploring the lower and possibly moister levels of the soil profile.

Loams or light clay loams are better suited to Mazzard rootstocks, which form a high proportion of their roots in a dense mat in the top 20 cm of the soil profile, while at the same time producing sufficient deep roots to anchor the tree.

Sour cherry stocks are best able to tolerate wet, imperfectly drained soils. However, they produce fewer deep scaffold roots, and anchorage on sour cherry rootstocks is often poor. Trees grown on sour cherry stocks planted on sandy soils will often be excessively dwarfed and unproductive.

Little is yet known of the adaptability of newer rootstock selections to soil conditions. The hybrid Colt appears to tolerate soils with some impeded drainage (Parnia *et al.*, 1985), but does not thrive on droughty soils. Similarly, some of the M×M series are rather more tolerant than their Mahaleb parents of temporary anaerobic soil conditions. German research suggests that *P. canescens* and all hybrids with this species are very sensitive to anaerobic conditions and the pathogens associated with wet soils (Roth, 1986). In contrast, Gisela 1 (*P. fruticosa* × *P. avium*) and Gisela 10 (*P. fruticosa* × *P. cerasus*) show good tolerance of such soils. German tests of 95 Giessen clones, grown in containers and subjected to flooding for 3–5 weeks in June–July, showed promising material which was tolerant of poor drainage within siblings of *P. cerasus* × *P. fruticosa* and *P. fruticosa* × *P. avium* crosses (Roth, 1986).

Soils with a high content of free lime, such as those commonly found in parts of the cherry-growing regions of Spain and France, pose particular problems to growers. Symptoms of lime-induced chlorosis develop on these soils, possibly as a result of the plants' inability to take up sufficient iron and manganese. The Belgian selection GM 8 ((*P. subhirtella* × *P. yedoensis*) × *P. subhirtella*) is particularly sensitive to high-pH soils, and other new rootstocks such as Edabriz, Damil and Inmil also show some sensitivity. In contrast, all Mahaleb types are well suited to calcareous soils and are usually the preferred choice on sites with high soil pH.

Rootstocks influence the mineral content of both the leaves and the fruits of the scion. Fortunately, these effects are usually small and are not likely to result in symptoms of either deficiency or toxicity in the scion, unless the soils themselves are markedly deficient in a particular element or of unsuitable pH. Montmorency is one example where rootstocks have been shown to influence mineral status. Research in Michigan revealed that trees on Mazzard rootstocks were generally higher in leaf potassium, calcium, boron, nitrogen and manganese concentrations but lower in leaf magnesium than trees on Mahaleb rootstocks (Hanson and Perry, 1989). These rootstock differences did not, however, appear to be related in any way to crop load or tree vigour. Mahalebs are thought to be more efficient than Mazzards in taking up zinc from soils and are therefore better suited to sites where zinc deficiency may prove a problem.

Other research has shown that the leaf potassium content of young trees of the sour cherry variety North Star on Mazzard seedling rootstocks was generally higher when compared with those on Mahaleb seedling rootstocks. In response to potassium fertilizer, a relatively greater increase in leaf potassium content was found in the rootstocks and maiden trees which absorbed more potassium, even from an unfertilized soil. This suggests that a genetically determined capacity for absorbing potassium cannot be easily overcome by application of high fertilizer doses. German research on the Giessen rootstock clones showed that leaf-rolling symptoms on sweet cherry scions were associated with low potassium uptake, while premature yellowing was associated with low nitrogen content in leaves.

Magnesium deficiency is frequently recorded in leaf samples taken from sweet cherries grown on Mazzard rootstocks in Britain. Work at East Malling has demonstrated that trees on Mazzard rootstock pick up much less magnesium from soils slightly deficient in this element than do similar trees on Colt.

5.3.4 Damage by nematodes

Several lesion nematodes attack cherry rootstock roots and may seriously affect growth. Species of *Pratylenchus* and *Xiphenema* are both damaging to cherry growth in the USA. Mazzard and sour cherry rootstocks are both sensitive to these pests, while Mahaleb clones offer some degree of tolerance. Root knot nematodes (*Meloidogyne inconita* and other similar species) also cause damage, but here Mazzard and sour cherry rootstocks are more tolerant than Mahalebs. Sensitivity is not always as straightforward as this, however: French research has shown that, although Mazzard seedlings and F.12/1 are very sensitive to *Pratylenchus vulnus* and Mahalebs are not, the reverse is true when considering sensitivity to another related species, *Pratylenchus penetrans* (Zepp and Szczygiel, 1985).

5.3.5 Damage by fungal pathogens

Several species of *Phytophthora* attack the roots of cherries and may on some sites cause severe problems (Wicks *et al.*, 1984). Fortunately, Mazzard, sour cherry and Colt rootstocks all show some tolerance to these root pathogens. In contrast, Mahaleb stocks are susceptible to *Phytophthora* damage and, on soils with impeded drainage, the problem may be most severe. Tests in Australia demonstrated that Colt, Mahaleb, Mazzard and even Stockton Morello rootstocks were to some degree sensitive to *Phytophthora cambivora*, with Mahaleb again being the most susceptible. The inherent susceptibility of the rootstocks was not changed when they were grafted with a more resistant scion cultivar (Wicks *et al.*, 1984). However, tests by Cummins *et al.* (1986) showed Stockton Morello, Vladimir and many hybrids with *P. cerasus* as one parent to be resistant to the problem. Research in New York State has shown that *Phytophthora megasperma* and *P. dreschsleri* were present in local soils and were probably responsible for the death of sweet cherry trees from root rot, although *P. cambivora* and *Pythium* spp. have also been implicated in this problem. Research in Germany has shown that the Belgian rootstocks Inmil and Camil and the German hybrid clones 196/4 and 196/13 (both *P. canescens* × *P. avium*) are all very susceptible to *Phytophthora* damage.

Occasionally the honey fungus (*Armillaria*) may attack cherries, and trees are more susceptible on Mahaleb and sour cherry stocks than on Mazzards. One of the Giessen selections, 195/1, and the Mahaleb × Mazzard hybrid M×M 60 also showed less sensitivity: Colt, Mahaleb and Inmil were all more sensitive than average (Proffer *et al.*, 1988).

Unfortunately, all of the common cherry rootstocks are susceptible to damage by the fungi causing wilt diseases, *Verticillium* spp.

Leaf spot disease (*Coccomyces hiemalis* or *Blumeriella jaapii*) is often very severe in parts of the USA and Europe unless protective fungicides are sprayed. Mahalebs are tolerant of leaf spot while sour cherries are mostly very susceptible; Mazzard cherries and Colt are only moderately susceptible. The hybrid rootstock VP-1 (*P. cerasus* × *P. maackii*) is also reported to be resistant to leaf spot (Kolesnikova *et al.*, 1985).

No rootstock resistance has been reported to silver leaf disease, caused by *Chondrostereum purpureum*, but the Mahaleb clone SL.64 is known to be particularly sensitive. Research in Britain showed that 'replant disease' of sweet cherries on Mazzard rootstocks was attributable to *Thielaviopsis basicola*; most Mazzards and *P. avium* × *P. incisa* hybrid clones tested proved sensitive to this organism, whereas hybrids between *P. avium* and *P. pseudocerasus*, such as Colt, were resistant (Pepin *et al.*, 1975).

5.3.6 Damage by bacterial pathogens

Crown gall (*Agrobacterium tumiefaciens*) may reduce growth of cherry trees planted on droughty soils, particularly where trees are grown in areas experiencing very hot summers. The Mazzard F.12/1 and Colt are particularly sensitive to crown gall infections of the roots, while Mazzard seedlings are less susceptible and sour cherries and Mahalebs only moderately sensitive (Breton, 1980). Research by Schmidt (1989) showed that hybrids between *P. avium* and *P. canescens* were very susceptible to crown gall, whereas those between *P. cerasus* and *P. fruticosa* were not.

Bacterial canker and blight (*Pseudomonas mors prunorum* and *P. syringae*) are particularly damaging pathogens on sweet cherries in many temperate areas. Both F.12/1 and Mahaleb are tolerant of bacterial canker, as is Colt. Mazzard seedlings, in contrast, are usually much more susceptible. Scions differ in their sensitivity to the damaging species and races of this pathogen and this sensitivity may be influenced by rootstock. Unfortunately, the relationship is a complex one, with some cultivars showing more sensitivity on one rootstock than another, while with other cultivars the reverse effect may be recorded (Garrett, 1986). The problem is complicated by the emergence of distinct races of the pathogen. Research by Garrett (1977) demonstrated that although resistant to race 1 of *Pseudomonas mors prunorum*, F.12/1 was sensitive to race 2. The Mazzard rootstock Charger and many *P. avium* × *P. pseudocerasus* or *P. avium* × *P. incisa* hybrid rootstock clones were shown to be resistant to both races of the pathogen. Research at the John Innes Institute in Britain (Matthews, 1975) indicated that some *Prunus* species, such as *P. conradinae* and *P. fruticosa*, were very resistant to bacterial canker, whereas others, such as *P. cerasus*, *P. avium*, *P. serrula* and *P. yedoensis*, were all very sensitive.

5.3.7 Damage by viruses and mycoplasmas

Originally, it was believed that there were no great differences between rootstocks in their tolerance to most of the viruses infecting cherries. However, recent studies conducted by scientists in Washington State, USA, indicate that many of the hybrid rootstock clones developed at Giessen in Germany differ greatly in their sensitivity to PDV or PNRSV. These viruses, all of which are pollen-transmitted, may prove very damaging and it is essential, when collecting seed for raising rootstocks, to use only virus-free mother trees.

Hybrid rootstocks between *P. fruticosa* and either *P. cerasus* or *P. avium* showed hypersensitivity when infected with PNRSV and/or PDV; shoots and in some cases whole plants died in the 3 years following infection (Uphoff *et al.*, 1988). Some clones of *P. cerasus* were particularly sensitive to PDV infection, whereas hybrids with *P. canescens* as parent were more sensitive to PNRSV. In contrast, tolerance to this combination of viruses was shown by F.12/1, Colt and hybrids between *P. cerasus* Schattenmorelle and either *P. avium* or *P. canescens*. The promising 148 series showed good tolerance to these two viruses in the German tests but some of this series (e.g. 148/1) have proved less promising in virus tests conducted in Washington State.

Mycoplasmas also infect cherries and one causing a disorder known as Molière's decline has proved a problem in France (Carles, 1986). The disease was first noted in the Tarn-et-Garonne department of France in 1952 and susceptible sweet cherry cultivars include Burlat, Hedelfingen, Stark Hardy Giant, Napoleon, Reverchon and Précoce de la Marche. Trees on St Lucie (*Prunus mahaleb*) rootstocks appear to be more susceptible than those on Mazzard cherry rootstocks. The disease seems to be caused by a mycoplasma which is transmitted by the leafhopper *Fieberiella florii*; this same leafhopper also transmits western X disease in the USA. The type of soil, the site and water stress may also be implicated in the development of the disease.

Western X disease is also caused by a mycoplasma and tests have shown that both Colt and Mazzard rootstocks are badly affected by the disease. Mahalebs are hypersensitive to western X and this can be a useful attribute in limiting its spread.

5.3.8 Control of scion vigour

Rootstocks offer the most effective and permanent method for control of sweet and sour cherry tree vigour. The sweet cherry is naturally a vigorous forest tree and, when grown on the same species (Mazzard) as rootstock, trees are similarly very vigorous. Although 'genetic dwarf' selections from sweet cherry seedlings will, when used as rootstocks, dwarf sweet cherry trees, such rootstocks cannot be propagated clonally and trees worked upon them are usually very unproductive. Mazzard rootstocks offer very little opportunity, therefore, for growth control of sweet cherries.

Some trial results in the USA suggest that trees on the clonal Mazzard F.12/1 are more vigorous than similar trees on Mazzard seedlings. However, this is not always true and much will depend upon the provenance of the seed chosen. In the USA, trees on Mahaleb initially grow more vigorously than those on Mazzard but are eventually overtaken in size by the latter. Many trees grown on Mahaleb in the Bordeaux area of France, where soils are light, sandy or gravelly and summers very hot, are quite small in stature. In contrast, trees grown on the same rootstocks but planted on loam soils in Britain are much larger and similar in size to those on Mazzard rootstocks.

Trees on many *Prunus cerasus* (sour cherry) rootstocks are usually smaller than similar trees on Mazzards. Several clones of *P. cerasus* selected in Germany and France (Weiroot clones and Edabriz) show useful control of scion vigour.

Trials in Britain show many sweet cherry cultivars worked on the hybrid rootstock Colt to be only two-thirds the size of trees on F.12/1 (Pennell *et al.*, 1983). Unfortunately, the vigour of many sweet cherry cultivars is not reduced by Colt and this lack of response is exacerbated in certain cherry-growing areas, such as California and Washington State, USA. The reasons for these soil, climate and scion interactions with rootstock are not understood.

German research demonstrated that trees which were severely dwarfed by rootstocks also exhibited early senescence, decreasing shoot elongation each year as they aged and death of the flowering spurs on older wood. Fruitfulness also declined with age on such trees. This would indicate that, if such rootstocks are to be used in future cherry plantings and reasonable shoot growth sustained, there will be a need for much more regular and severe pruning than is currently practised.

Influence of budding height

It is well known that vigour control by apple rootstocks may be increased by budding the scion higher on the rootstock stem and similar effects have been recorded for European pears worked on quince (*Cydonia oblonga*) clonal rootstocks. Attempts have also been made to improve vigour control of sweet cherries by high budding. As might be expected, high budding on to vigorous Mazzard rootstocks usually has little effect upon scion vigour. However, in German trials 28% vigour reduction was achieved by budding sweet and sour cultivars at more than 23 cm above ground level, compared with at 7 cm (Schimmelpfeng and Vogel, 1985). Work at East Malling comparing four different budding heights (15, 30, 60 and 75 cm above ground) of two sweet cherry cultivars on Colt rootstock showed no difference in the final weight (i.e. size) of the trees at the time of grubbing. Similarly, German research showed no significant reduction in scion crown volume when the sour cherry rootstock Weiroot 13 was budded at heights ranging from 15 to 90 cm. However, studies in France showed that the variety Vittoria budded on SL.64 was 20% smaller, as judged by trunk girth, if budded at 50 cm rather than 10 cm above ground level.

5.3.9 Modification of scion habit

Mahaleb, when not worked with a scion and when left unpruned, forms a spreading tree with a round-shaped crown, whereas Mazzard is naturally more erectly branched and upright in habit. Despite these differences, they are not often reflected in the growth habit of scions worked upon them and the natural habit of the scion seems to be dominant. Work in Britain has shown, however, that sweet cherries budded on some *P. avium* × *P. pseudocerasus* rootstock clones develop much wider branch angles than similar trees on Mazzard rootstocks.

5.3.10 Precocity, consistency and abundance of scion yields

Sweet cherry trees budded on most of the vigorous Mazzard or Mahaleb rootstocks exhibit poor cropping precocity, trees often taking up to 12 years to come into full production on F.12/1. Trees on Colt and some of the newer size-controlling rootstocks are, in comparison, much more precocious. This increased precocity is not, however, necessarily linked with the dwarfing effect of a rootstock; the Belgian selections Inmil and Damil, both dwarfing, often induce very poor cropping precocity in the scions worked upon them.

German studies have shown that many of the Giessen hybrid rootstocks influence branching and habit as well as vigour. Also, although the stocks tested had little influence on the number of flower buds per spur, they had a big effect on the numbers of flowers per bud. Flowers per unit branch size were often much more numerous on the more dwarfing rootstocks, such as 172/9. Although several Giessen-bred stocks (195/1, 195/2, 196/4 and 196/13) induced good scion flowering efficiency, the trees had low fruiting efficiency due to increased fruitlet abscission. When numbers of spurs, flowers and fruits per metre length of 2-year-old wood were determined and the ratio of spurs : flowers : fruit calculated, the most productive rootstock clone produced a ratio of 1 : 27 : 10, while for trees on F.12/1 rootstock the ratio was only 1 : 9 : 0.3.

Dwarfing rootstock clones which stimulated precocious cropping in sweet cherry scions were more frequently selected from among triploids and tetraploids derived from crosses within the *Eucerasus* section of *Prunus* than from among diploid hybrids between the *Eucerasus* and *Pseudocerasus* sections.

Cropping precocity may also be improved by use of an interstem, such as Montmorency, grafted between the rootstock and the scion. Whether this response is attributable to the interstock clone *per se* or to indirect effects associated with the different method of tree propagation employed in raising interstock trees remains a cause for debate.

5.3.11 Drought tolerance

Mahaleb rootstocks are generally much more tolerant of drought conditions than Mazzards; sour cherry and Colt rootstocks are particularly sensitive to such conditions. Several of the M×M hybrid rootstocks (*P. mahaleb* × *P. avium*) also show good tolerance to drought conditions. Many of the shallow-rooted dwarfing rootstocks, both pure species and hybrids, also exhibit poor tolerance of drought conditions.

5.4 Mazzard (*Prunus avium*) Rootstocks

Mazzard rootstocks are still the most popular rootstocks for sweet cherries grown throughout the world, and they are also widely used for raising sour cherry trees. They are easy to raise from seed and invariably show excellent compatibility with scion cultivars. With the exception of 'genetic dwarf' types (small stunted plants with crinkled leaves which often occur among batches of seedlings), Mazzards are very vigorous rootstocks which produce sweet cherry trees with poor yield precocity. Seedlings and also some clones of Mazzards frequently delay fruit ripening by several days when compared with Mahaleb and other species and hybrid rootstocks. The French seedling stocks, Pontavium and Pontaris, delay ripening by 3–5 days compared with F.12/1. Difficulties in propagating Mazzards vegetatively has meant that many nurseries offer for sale only seedling-raised rootstocks.

5.4.1 Seedling selections of Mazzard

It was recognized many years ago that rootstocks raised from Mazzard seed collected from fruiting orchards usually produced trees which cropped poorly, were variable in growth and were inconsistent in their tolerance to winter cold. This prompted researchers in several countries to select improved lines of seedling Mazzards for use as rootstocks.

US seedling selections

Seedlings were originally imported into the USA from Europe, as *Prunus avium* originates in the Old not the New World. Nowadays, however, it is widely planted in the USA and Canada where it crops abundantly. Selections of seed sources were made over many years by researchers in Washington and New York States; the main objectives were cold-hardiness and resistance to leaf spot disease (*Blumeriella jaapii*). Eventually, this led to the selection of Mazzard No. 570 (which produces small, dark, red fruits) and Sayler (another clone with lighter red fruits). No. 570, which originates from Mazzards collected from the Harz mountains in Germany, is the more cold tolerant and, therefore, generally the more popular of the two. These stocks are now available in virus-free forms. OCR 1 is another Mazzard seedling line which was selected and is still used in Oregon.

French seedling selections

Pontavium (Fercahun) and Pontaris (Fercadeu), which were selected from commercial seedling lots by Institut National de la Recherche Agronomique (INRA) at Bordeaux, in France, are cross-compatible and form the basis of the French virus-free seed orchards (Edin and Claverie, 1987). Both give 60–70% germination if stratified for 120–130 days at temperatures between 0 and 2°C. Seedling rootstocks from reciprocal crosses between the two selections are quite homogeneous and fairly resistant to crown gall (*Agrobacterium tumefaciens*) in the nursery. Fruiting and yield of sweet cherries grown on these rootstocks are similar to those on the clonal F.12/1. Tree vigour on rootstocks with Pontaris as the maternal parent is similar to that on F.12/1, but growth on rootstocks where Pontavium is the maternal parent is more vigorous. Trees on these seedling rootstocks frequently exhibit delays in fruit maturity of approximately 3 days in comparison with F.12/1.

Merisier Commun is a Mazzard rootstock which originated from a collection of stones (pits) from the forests of the Massif Central in France. Poor germination (only 20–30%) and irregular performance in the nursery have prompted a decline in the use of this rootstock.

German seedling selections

Mazzards collected from the Harz mountains are renowned for their hardiness, and have formed the basis of selections made in the USA, Belgium and Germany. The selections made by Küppers (Gi81, 82 and 84) and their derivatives (Gi90 and 94) are very resistant to winter cold injury. These selections are still used in parts of Germany. Hüttner clones 170 and 53 originated as selected seedlings of Limburger Vogelkirsche in Germany; these two clones are crossed to provide seedling Mazzards for many parts of Germany.

Belgian seedling selections

The Limburger Vogelkirsche strain of Mazzard seedling rootstocks originated from selections of Harz mountain Mazzards made at Limburg in Belgium.

Mazzard seedlings used in Bulgaria, Romania, Ukraine and Moldavia

Two Mazzard seedling selections used in Bulgaria are IK, selected at Plodiv, and 123, which was selected at Dryanova in Bulgaria. Seedlings from open-pollinated trees of Napoleon or Drogans Gelbe are also used in Bulgaria. In Romania seedlings from crosses between F.12/1 and Dönissens Dwarf are used as rootstocks; two selected seedling lines are 76–33–26 and 76–25–29. In the Ukraine three Mazzard lines, numbered 3, 4 and 5, are utilized. Seed from the open-pollinated sweet cherry cultivars Susleny and Napoleon are used in Moldavia.

5.4.2 Vegetatively propagated selections of Mazzard

F.12/1 and Charger

Selected at East Malling from a batch of rootstocks received from an English nursery in the 1920s, F.12/1 is more vigorous than Mazzard seedlings in many situations and shows very good compatibility with both sweet and sour cherry cultivars. A virus-free (EMLA) clone of this rootstock is available. F.12/1 is propagated, usually, by trench layering, but can also be propagated from softwood cuttings, taken under mist or fogging environments; propagation in Australia is by root cuttings. It exhibits useful resistance to bacterial canker but is sensitive to crown gall and it is this sensitivity which has made it less popular with nurserymen and growers in recent years. Suckering may also be a problem of sweet cherry trees grown on F.12/1, although the severity of this problem varies with site and scion cultivar. Trees on F.12/1 are reported to make growth late in the season and may therefore acclimatize poorly in some locations, leading to winter cold damage.

Charger is a clonal Mazzard rootstock, which was also raised at East Malling. It is easier to propagate than F.12/1, more resistant to bacterial canker and forms trees slightly smaller than those on F.12/1. A virus-free clone is now available.

Cristimar IAI

By seeding, selecting, reseeding and reselecting among superior sweet cherry forms from the local Pletoase Mazzards, Romanian researchers are reported to have selected a dwarfing form of Mazzard rootstock, which they have named

Cristimar IAI. Cristimar is said to be weaker in vigour than Mahaleb, with reductions of between 7 and 27% reported. Many researchers now believe this to be a hybrid between *P. avium* and *P. cerasus*.

A clonal Mazzard selection, VV-1, is also used in Romania.

5.5 Mahaleb (*Prunus mahaleb*) Rootstocks

The principal merits of Mahalebs are their better adaptability to calcareous or droughty soils than Mazzards.

Trees on Mahaleb also ripen their wood earlier in the autumn than Mazzards (an aid to winter cold tolerance) and show better resistance to zinc deficiency and lime-induced chlorosis. However, animals such as gophers prefer them and may cause more damage to Mahaleb than to Mazzard roots. Mahaleb is a very heterogeneous species with types differing quite considerably in leaf characteristics, habit and vigour; they exhibit much more natural variation than Mazzards.

5.5.1 Seedling selections of Mahaleb

US seedling selections
The USA produces most of its Mahaleb seed from sources introduced in 1944 (Nos 902, 904, 908 and 916), which were originally selected in Michigan. These cross-pollinate well, setting seeds which, after harvesting and blending, are made available as Mahaleb 900. Seed-source orchards, which are well isolated from other cherry plantings, and thus from sources of infection from viruses and mycoplasmas, are essential for maintaining healthy stocks of Mahaleb 900.

French seedling selections
The Mahalebs most frequently used in France in the past were raised from seed collected randomly, with no attempts made to select improved lines. Although such seed germinated well (80–90% after 3 months at 0–2°C), it gave rise to a significant percentage of trees which developed symptoms of delayed incompatibility and were inconsistent in their growth. These Mahalebs were also very sensitive to *Phytophthora cambivora* and *P. megasperma*. They are not now recommended for new plantings.

St Lucie 405 (SL.405). The French are now testing a new line of seedling Mahaleb which is self-fertile and numbered SL.405. Its advantages are reported to be its slightly reduced sensitivity to root asphyxiation compared with SL.64, good compatibility with sweet cherry scions and induction of improved fruit size on the scion. Trees on SL.405 and SL.64 are similar in vigour, i.e. slightly less than for trees grown on F.12/1 in French conditions.

German seedling selections
Alpruma is a seedling-raised Mahaleb previously grown in East Germany and selected originally from a large seedling population. It exhibits good winter

hardiness. Hüttners Heimann 10 is another seedling type of Mahaleb used in the western parts of Germany; Heimann is self-fertile.

Hungarian seedling selections

Hungarian researchers have selected several seedling races of Mahaleb which are still undergoing comparison and evaluation. Korponay, which they classify within the subspecies *simonkai* with large leaves with no hairs, is allowed to self to produce seedlings very suitable as rootstocks for sour cherry cultivars. Magyar, which is classified within the subspecies *cupaniana*, with leather-like leaves and no hairs, is usually pollinated by either Soroksar or Egervar (both in the subspecies *mahaleb*, with small leaves and hairs) to produce seedlings for use as rootstocks for either sweet or sour cherry cultivars. Two other selections, CT500 and CT2753 are also utilized in Hungary.

Other seedling selections of Mahaleb

Mahaleb No. 24 is a selection commonly used in the Ukraine, while Rozovaya Prodolgovataya, Chernaya Kruglaya iz Bykovtsa, and No. 1 iz Solonchem are selections used in Moldavia.

5.5.2 Vegetatively propagated selections of Mahaleb

St Lucie 64 (SL.64)

Selected at Bordeaux in France in 1954, SL.64 propagates easily from soft or semihardwood cuttings, but can prove difficult by micropropagation. Budding of SL.64 should be carried out later in the summer than for Mazzard rootstocks. It shows good compatibility with sweet cherry cultivars, particularly the Bigarreaux types. Trees on SL.64 thrive best on well-drained soils but are more adaptable to different soil types than many other Mahaleb types. The rootstock is often recommended for use on sites suffering from cherry replant syndrome. Scions on SL.64 are compact in vigour when grown on the sandy and gravelly soils around Bordeaux, but more vigorous than Colt and similar in vigour to Charger when grown on the deep loam soils in the UK. Trees on SL.64, irrespective of vigour, usually exhibit good productivity and precocity. SL.64 is moderately sensitive to the nematodes *Meloidogyne incognita* and *Pratylenchus penetrans* but resistant to *Pratylenchus vulnus*.

Dunabogdány and SM 11/4

These are clones of Mahaleb selected and occasionally used in Hungary. They are propagated from leafy cuttings and show good compatibility with both sweet and sour cherries.

5.6 Sour Cherry (*Prunus cerasus*) Rootstocks

Sour cherry rootstocks have the advantage of being more cold-hardy than Mazzard or Mahaleb stocks and they also tolerate wet soils better. They perform poorly, however, on droughty or highly calcareous soils. Some types exhibit inconsistent compatibility with several sweet cherry scions, such as Sam and Van. Sour cherry rootstocks may also be more poorly anchored than Mazzard or Mahaleb types.

5.6.1 Seedling selections of sour cherry

Seedlings of *P. cerasus* are usually very variable as rootstocks and are not commonly used for this purpose.

VG.1 (VT.1)

This Romanian selection, which is self-fertile, was released in 1985. Yield of seed, its germination and bud take are reported to be good. Trees on VG.1 are less vigorous than trees on Mazzard and are very productive (Kolesnikova *et al.*, 1985).

Other P. cerasus *seedling selections*

Seedlings of the sour cherry cultivars Ilva, Meteor and Mocanesti are sometimes used as rootstocks in Romania, as is the seedling selection Izvorani 78. Open-pollinated seedlings from the cultivar Trevnenska are utilized in Bulgaria.

5.6.2 Vegetatively propagated selections of sour cherry

Tabel/Edabriz

This clone, which was raised from sources collected in Iran, was selected and released by INRA in France. It has, as yet, undergone only limited testing in countries other than France. Edabriz is best propagated from semihardwood cuttings, although micropropagation is also possible. Unlike many other sour cherry rootstocks, Edabriz has so far exhibited good compatibility with all sweet cherry scions tested in France; compatibility is also good with the sour cultivar Montmorency. Trees on Edabriz are dwarfed, on some sites attaining only 15–20% the size of similar trees on F.12/1 (Edin, 1989). However, this dwarfing effect is greatly influenced by both soil type and environmental conditions and in some situations trees on Edabriz are much larger (up to 60% the size of those on F.12/1). Nevertheless, trees on Edabriz have in all situations proved smaller than trees on the hybrid rootstocks Colt or Maxma (M×M) 14. Sweet cherry trees on Edabriz crop precociously and abundantly, are well anchored and produce few suckers.

Edabriz is likely to be best suited to loam or clay soils and may thrive less well on droughty soils, particularly those with a high pH. Its sensitivities to *Phytophthora* species, nematodes or crown gall are not yet determined.

Weiroot series

These sour cherry rootstocks were selected at Weihenstephan near Munich in Germany from wild material growing in Bavaria, and originally three clones were released for testing, numbers 10, 13 and 14 (Schimmelpfeng and Liebster, 1979). Propagation of these clones is best by soft or semihardwood cuttings. Although compatibility with Van, Merton Glory and Merpet sweet cherry cultivars was good in UK trials, reports from Germany and Switzerland indicate problems of incompatibility with some sweet cherry cultivars. This was particularly severe in Swiss trials where 20–40% of the trees on Weiroot selections died. Heavy clay, wet soils, tending towards temporary anaerobic conditions, are thought to exacerbate this problem of incompatibility. In German trials the cultivar Sam shows incompatibility with Weiroot selections. This is not altogether surprising, however, as Sam exhibits poor compatibility with many of the new rootstocks for sweet cherry.

Trials at East Malling showed sweet cherry trees on Weiroot clones 10 and 13 to be extremely productive and of vigour similar to or slightly less than that of trees on Colt, i.e. less than that of trees on Mazzard rootstocks (Webster, 1993).

Several new Weiroot selections (clone numbers 53, 72 and 158) have recently been tested by the breeders in Germany and results indicate that these may be more compatible with sweet cherry scions. Trees on clones 53 and 72 are reported to crop very well and be very weak in growth but they do require a stake for support. Clone 158 is also reported to be slightly more dwarfing than clone 10 and to induce precocious and heavy crops on the scion.

Vladimir

Vladimir, which originated in Russia, was introduced into the USA in 1900 as part of a group of Morello sour cherries; its final selection was carried out in the USA. It is a semidwarfing rootstock which produces poorly anchored sweet cherry trees that sucker profusely. Cropping of trees on Vladimir is very precocious and abundant. Its resistance to *Phytophthora* spp. and tolerance to cold wet soils are its only real merits. The cultivar Vladimirskaya is grown on its own roots in Russia or its seed is used as rootstocks.

Stockton Morello

This rootstock is said to have originated in Illinois, USA, where it was first known as American Morello. However, it became very popular in the Stockton area of California, where it acquired its now accepted name. It propagates from root suckers and by softwood cuttings, but trench layering is not successful. It became popular on account of its tolerance of the heavy, clay soils common in the Stockton area of California and also the reduced vigour of trees worked upon it. This latter influence was probably attributable to virus infection, however, as virus-free clones exhibit no significant control of vigour. This inadequate vigour control and the poor anchorage of trees on Stockton Morello have contributed to its recent lack of popularity in the USA.

CAB selections

Clones CAB-6P and CAB-11E are possibly the best of those selected from wild *P. cerasus* material collected in the Emilia Romagna region of Italy. These clones may be propagated by softwood cuttings or micropropagated. They are reported to give 20–30% reduction in scion vigour compared with Mazzard rootstocks. Clone CAB-6P is considered the best of this series for the cultivar Bigarreau Moreau grown in Italy, although both 6P and 11E have given good results with another Italian sweet cherry cultivar, Durone della Mora. Other clones tested in the series are CAB-8H and CAB-4D. Several of the CAB series rootstocks have recently performed well in trials in New Zealand.

5.6.3 Sour cherry scion cultivars used as rootstocks

Montmorency

When used as a rootstock, the Amarelle type Montmorency produces trees with good anchorage and suckering of the trees is not too severe. The Morello cultivar Schattenmorelle produces vigorous and highly productive trees on Montmorency rootstocks when grown in Belgium. Montmorency grown on its own roots is said to be a smaller and more productive tree than when worked on F.12/1, but in this instance it does sucker very badly.

Oblăcinska

Yugoslav research showed that this cultivar performed quite well as a rootstock for sweet cherries, in comparison with Mazzard seedlings. Tree vigour was reduced and the trees cropped precociously and abundantly.

5.7 The Steppe Cherry (*P. fruticosa*) Rootstocks

Although *P. fruticosa* is easily raised from seed, sweet cherry compatibility on seedlings of the species is very variable and clonal rootstocks are generally preferred. Unfortunately, the Steppe cherry is difficult to propagate vegetatively, other than from the abundant suckers it produces. Sweet cherries are dwarfed on most clones of *Prunus fruticosa*. Many of the clones tested as rootstocks in Europe are believed to be hybrids between *P. fruticosa* and *P. avium*.

5.7.1 Oppenheim and its derivatives

This selection of *P. fruticosa* was made at Oppenheim in Germany and hence its name (Plock, 1972). One of its principal merits, compared with most other selections from this species, is its relative freedom from suckering. Propagation is generally by softwood cuttings or by micropropagation, although propagation by root cuttings is also a possibility. Compatibility in most trials has been good, although both Sam and Van are incompatible on Oppenheim. Twelve-year-old Hedelfingen grown on Oppenheim in Germany were two-thirds the size of trees on F.12/1, well anchored and productive. Suckering was only moderate and resistance to *Phytophthora* good. In contrast to the German results, other reports suggest anchorage is poor on Oppenheim and suckering much worse than originally reported. For instance, trials in Italy showed it to have poor compatibility with sweet cherry scions and trees worked on it were unproductive and bore poor-quality fruits. Some researchers now believe that this clone may be a hybrid between *P. fruticosa* and *P. cerasus*.

Other selections of *P. fruticosa* which were derived from the Oppenheim material and have been tested in Germany are Bisamberg, Himmelhof, D1 and D16. Problems of incompatibility with sweet cherry scions have been experienced with all of these and they cannot be recommended.

5.8 Other *Prunus* Species Occasionally Used as Cherry Rootstocks

5.8.1 *Prunus canescens*

Trials at East Malling showed that sweet cherries were compatible on this species and produced trees of similar vigour to those on Colt. However, fruit size was reduced in comparison with trees on Colt. German research shows that clones of *P. canescens* and many hybrids with this species as one parent are very sensitive to anaerobic conditions and to pathogens, such as *Phytophthora*, associated with wet soil conditions. Researchers in Belgium have selected a superior clone of *P. canescens*, which has been named Camil (see below).

5.8.2 *Prunus × dawyckensis*

A clone of this species, which is perhaps more accurately classified as a natural hybrid between *P. canescens* and *P. dielsiana*, has been selected in Belgium and named Damil (see below). Germplasm of this species held in the British Botanic Gardens is, however, quite distinct morphologically from the Belgian Damil clone.

5.8.3 *Prunus incisa*

A selection of this species tested at East Malling produced trees which were less than 50% the size of trees on Colt. Several clones of this species raised at Wageningen in the Netherlands (Nos 33, 123 and 128) produced very dwarf trees in trials at Giessen in Germany; often the trees were more dwarfed than many of the Giessen (Gisela) selections. Yield precocity and fruit size was poor in the Giessen trials of this species. Unfortunately, anchorage is usually very poor on *P. incisa* rootstocks.

5.8.4 *Prunus concinna*

Tests of this species in Germany show that it may dwarf sweet cherry scions by approximately 40%. Its effects on cropping and its long-term compatibility with sweet cherry scions are still being investigated. Clones of this species usually show poor tolerance to winter cold.

5.8.5 *Prunus serrulata*

Researchers at Gembloux, in Belgium, found that more than 20 cultivars of this species had very poor compatibility when used as rootstocks for sweet cherries. However, one cultivar, Daikoku, which was tested at Gembloux as GM 156, appeared quite promising as a dwarfing rootstock in preliminary trials. Unfortunately, the virus status of the *P. serrulata* used in these Belgium experiments was not always known.

5.8.6 *Prunus subhirtella*

Clones of this species tested in Germany dwarfed sweet cherry scions significantly. However, yield precocity, productivity and fruit size were poor. Several clones, including the cultivar Pendula Autumnalis, are currently under test in Britain.

5.8.7 *Prunus mugus*

Sweet cherry cultivars grown on a clone of *Prunus mugus* at East Malling have formed very small trees, less than 50% the size of trees on Colt. Unfortunately, the clone is extremely difficult to propagate vegetatively and this has limited its more extensive use. In trials in Germany the clone showed poor winter-hardiness.

5.9 *Prunus* Hybrids as Cherry Rootstocks

5.9.1 Colt

Colt is a hybrid between *P. avium* L. and *P. pseudocerasus* Lind. which was bred at East Malling in 1958 and released to nurseries in the 1970s. It is extremely easy to propagate by all conventional methods of vegetative propagation and this has made it a particularly popular rootstock with nurserymen. When grown in favourable nursery conditions, hard-pruned hedges form extension (1-year-old) shoots that have rudimentary roots already visible on their bases at the time of collection in early winter (Fig. 5.1). Compatibility is generally very good on

Fig. 5.1. Root initials formed on the base of 1-year-old shoots of hard-pruned Colt hedges growing in Britain.

Colt, although occasionally problems are experienced with the cultivars Sam and Van. Bud take in the nursery is particularly good on Colt and trees develop with abundant wide-angled feathers (lateral branches).

Vigour of trees on Colt is inconsistent, depending upon the scion cultivar chosen and the site. Van, Merton Glory and many of the newer UK cultivars bred at the John Innes Institute usually form trees approximately two-thirds the size of trees on F.12/1, when grown in UK conditions (Pennell *et al.*, 1983). However, many of the more traditional cultivars may show only minimal vigour reduction on Colt.

Trees on Colt crop precociously and on favourable sites very productively. Fruit size is often larger from trees grown on Colt compared with F.12/1 rootstock. The sour cherry cultivar Montmorency is also extremely productive when grown on Colt.

One problem with Colt is its sensitivity to cold damage when grown in areas experiencing low winter temperatures. Colt is usually damaged when unworked or recently budded in the nursery; semimature and mature trees suffer much less winter damage. Recent trials in Spain indicate that Colt requires fertile, well-drained and adequately irrigated soils if it is to perform well. Poor growth and chlorotic symptoms were observed on trees grown in shallow, dry and highly calcareous soils.

Mutants of Colt

Colt is a sterile triploid and therefore of only limited use for further rootstock breeding. Recent research at East Malling, using *in vitro* techniques, has succeeded in doubling the chromosome numbers and producing hexaploid clones of Colt (James *et al.*, 1987). Although primarily produced for potential breeding purposes, these hexaploids are also being tested as rootstocks, and preliminary results suggest that they may be slightly more dwarfing than normal triploid Colt (Webster, 1993).

Other East Malling research has sought to mutate Colt by irradiation of *in vitro* cultures. Several clones produced show some promise as dwarfing rootstocks.

5.9.2 The M×M series

This series of rootstocks were selected by an Oregon, USA, nurseryman from 3000 open-pollinated seedlings of *P. mahaleb*. On the basis of their growth habit and leaf size most are thought to be hybrids with *P. avium*. The six most promising of the series were included in cherry rootstock trials planted at several locations in the USA. All of them sucker extensively when planted on US sites but suckering has been minimal in some French experiments. Clone numbers 2, 39, 60 and 97 show some resistance to *Phytophthora* spp., possibly associated with their greater tolerance of clay soil conditions compared with many pure Mahaleb types. Clones 14, 39, 60 and 97 are believed to be more resistant to winter cold injury and canker than either Mazzards or Colt.

Trees on this series of rootstocks have usually cropped less precociously than trees on Colt in US trials.

M×M 14 (Brokforest or Maxma Delbard 14)

This clone, undoubtedly the most dwarfing in the series, has gained greater popularity in France than in its country of origin, the USA. Although the vigour of sweet cultivars and Montmorency was reduced and cropping productivity was good on M×M 14 in Michigan trials, fruit size, particularly of Montmorency, was poor (Perry, 1985). The foliage of trees on M×M 14 also appeared stressed in midsummer, possibly indicating delayed incompatibility in these growing conditions. In French trials, trees on Maxma Delbard 14 (= M×M 14) showed good resistance to lime-induced chlorosis and fruit weight was similar to that from trees on other rootstocks (Edin *et al.*, 1989). Fruit production began 2 years earlier than with SL.64 rootstock. M×M 14 is classed as semivigorous, forming trees 40–60% the size of trees on F.12/1, or 60–80% the size of trees on SL.64. In trials at Valence in France, tree size of the sweet cultivar Burlat was similar on Colt and M×M 14, whereas Montmorency formed more dwarfed trees on M×M 14 than on Colt in Michigan trials. Very little suckering was noted on the trees on M×M 14, and the stock is also thought to have some resistance to two species of *Phytophthora* and to be more tolerant to bacterial canker than is sweet cherry.

M×M 97 (Brokgrove or Maxma Delbard 97)

This clone is usually considered semidwarfing and when high-worked with Napoleon in Oregon its dwarfing influence was increased, producing trees only one-third the size of trees on F.12/1. Montmorency forms more dwarfed trees on M×M 97 than on Colt in Michigan trials. Unfortunately, symptoms of delayed incompatibility have frequently been noted when sweet cherry cultivars were worked on this rootstock in France and the UK. Although yield precocity is often good on M×M 97, subsequent yield productivity is poor. Unlike many other Mahalebs it shows some resistance to *Phytophthora cambivora* and *P. megasperma*. M×M 97 is very variable in performance and on present evidence cannot be recommended for use on many loam or clay soil types.

M×M 60 and other M×M series clones

The main merits of M×M 60 are its resistance to *Phytophthora cambivora* and *P. megasperma,* its good compatibility and very good yields per tree once trees reach maturity. Sweet cherry trees are, however, extremely vigorous when grown on M×M 60 and their cropping precocity is poor.

Trials, mainly in the USA, have also tested two other M×M clones, numbers 2 and 39. Clone 2 is of strong vigour, whereas on some sites clone 39 has proved less invigorating than most others in this series. However, in trials at East Malling, trees on M×M 39 showed delayed incompatibility.

5.9.3 The Gisela (Giessen) series

Cherry rootstock breeding, which began at Giessen in Germany in 1965, has produced several dwarfing rootstocks worthy of further appraisal. Studies on the compatibility and growth of pollen *in vitro*, together with the manipulation of flowering times by growing parent plants partially under glass, enabled many interesting *Prunus* hybrids to be produced. These were then screened for their ability to propagate from layers, softwood, hardwood or root cuttings. In parallel with these tests, representative clones from all the successful hybridizations were budded with sweet cherry cultivars Hedelfingen and Büttners Rote Knorpel, and preliminary screening was carried out for their effects on tree vigour and cropping (Gruppe, 1985b, c). After extensive studies at Giessen, 13 selections were released for further orchard testing in the rest of Europe and the USA; further selections have recently been made available. Table 5.1 lists most of the more promising Giessen selections.

Table 5.1. Promising clonal rootstocks bred at Giessen in Germany.

Clone no.	Species or hybrid	Vigour
107/1	*P. cerasus* × *P. avium*	Moderate–vigorous
148/1	*P. cerasus* × *P. canescens*	Moderate–vigorous
148/2	*P. cerasus* × *P. canescens*	Semidwarfing
148/8	*P. cerasus* × *P. canescens*	Dwarfing, spreading habit
148/9	*P. cerasus* × *P. canescens*	Dwarfing, spreading habit
148/13	*P. cerasus* × *P. canescens*	Moderate–vigorous
154/4	*P. cerasus* × *P. fruticosa*	Moderate vigour
154/5	*P. cerasus* × *P. fruticosa*	Dwarfing
154/7	*P. cerasus* × *P. fruticosa*	Moderate vigour
169/15	*P. cerasus* × *P. avium*	Moderate vigour
172/7	*P. fruticosa* × *P. avium*	Dwarfing
172/9	*P. fruticosa* × *P. avium*	Dwarfing
173/9	*P. fruticosa* × *P. cerasus*	Dwarfing
195/1	*P. canescens* × *P. cerasus*	Moderate vigour
195/2	*P. canescens* × *P. cerasus*	Moderate vigour
195/4	*P. canescens* × *P. cerasus*	Moderate vigour
195/20	*P. canescens* × *P. cerasus*	Moderate vigour
196/4	*P. canescens* × *P. avium*	Moderate vigour
497/8	*P. cerasus* × *P. avium*	Moderate vigour

The most promising selections, in terms of compatibility, reduction of scion vigour and good cropping, have proved to be those originating from: *P. fruticosa* × *P. cerasus*, *P. fruticosa* × *P. avium*, and *P. cerasus* × *P. canescens and P. cancescens* × *P. cerasus* crosses. The best of these reduced vigour by 30–50% compared with F.12/1. To date only three of these preliminary selections have been named, Gisela 1, Gisela 5 and Gisela 10, of which Gisela 5 appears to be the most promising.

Gisela 1 (172/9)

This *P. fruticosa* × *P. avium* hybrid is considered the most dwarfing of the first series of Giessen selection. Trees in a German trial were only 17% the size of trees on F.12/1 after 5 years in the orchard, and trees in an unirrigated trial at East Malling made extremely poor growth. Trees planted in Norway appear to be dying due to delayed incompatibility. This stock will undoubtedly prove too weak unless grown on very fertile soils with adequate irrigation. The clone shows some resistance to wet, anaerobic conditions and the associated damage from *Phytophthora* spp., but is hypersensitive to infections by PNRSV and PDV.

The clone is, however, unlikely to gain widespread acceptance.

Gisela 5 (148/2)

Also classed as a semidwarfing rootstock, this *P. cerasus* × *P. canescens* hybrid is reported to produce sweet cherry trees with only 50% of the canopy volume of trees on F.12/1 after 5 years in German orchard trials. Tests in Germany indicate that it may not be suited to anaerobic conditions on heavy clay soils, where *Phytophthora* may prove damaging, but it does show some tolerance to infections by PNRSV and PDV. It is a most promising dwarfing clone which merits much more extensive testing.

Gisela 10 (173/9)

This clone, which was raised from crosses between *P. fruticosa* and *P. cerasus*, is considered semidwarfing and produces trees with canopy volumes 60–80% of those on F.12/1, approximately 5 years after orchard establishment. In trials in the USA, results suggest that the stock is much more dwarfing than the German results would suggest. It is similar to others of the same parentage in showing some resistance to winter cold injury, but it is hypersensitive to infections by PNRSV and PDV. The clone also shows some resistance to wet, anaerobic conditions and the associated damage from *Phytophthora* spp. and may also prove slightly resistant to crown gall. Although yield efficiency of trees on Gisela 10 has been good in some USA trials and fruit size quite good in Norwegian trials, it is unlikely to prove successful on many sites.

Other promising Giessen rootstocks

Although not a dwarfing rootstock, clone 148/1, a *P. cerasus* × *P. canescens* hybrid, has performed particularly well in trials in Britain and the USA. Sweet cherry trees on this rootstock have excellent yield precocity and productivity and on most sites are intermediate in vigour. Other hybrids in this same 148 subseries are also showing considerable promise. In trials in the USA, 148/8, which is more dwarfing than 148/1, also shows potential; 148/9 (dwarfing) and 148/13 (intermediate vigour) may also be worthy of further testing. Several hybrids between *P. canescens* and *P. cerasus* (the 195 subseries) are also showing promise in rootstock trials in several countries. They are rootstocks of intermediate vigour which induce good yield productivity in scions budded on to them.

More recent tests in Germany suggest that other hybrids selected at Giessen, Ahrensburg and Witzenhausen in Germany, clone numbers 107/1, 173/1, 209/1, 318/17, 473/10 and 497/8, may also warrant further testing as dwarfing or intermediate vigour rootstocks for sweet cherry.

5.9.4 The Gembloux series

In research which began in 1963 at Gembloux in Belgium, more than 220 species and hybrids of ornamental cherry were collected; most of these had their origins in Japan. These clones were then evaluated for their ease of vegetative propagation, compatibility with sweet or sour cherry cultivars, influence on scion vigour and cropping, resistance to bacterial canker and other diseases, propensity to sucker and suitability to different soil and climatic conditions. After many years' trials, four clones (GM 8, GM 9, GM 61/1, and GM 79) were selected for more widespread testing in Europe and America (Trefois, 1985a, b). Although other Gembloux clones, such as GM 1, GM 11, GM 19, GM 26 (*P. subhirtella* 'Autumnalis'), GM 54 (*P. yedoensis* Moerh.) and GM 60, also controlled scion vigour, their variable compatibility with sweet cherry scions precluded their further distribution.

All the selected GM clones, numbers 9, 61/1 and 79, can be propagated by softwood cuttings; GM 79 is best if propagated at the beginning of June, the other two slightly later in the growing season.

GM 9 (Inmil)

This is a selection of the hybrid *P. incisa* Thunb. × *P. serrula* Franch. which, when used as a rootstock for sweet cherries, is the most dwarfing of the Belgian selections. Tree size may be reduced by two-thirds to three-quarters, depending upon the site conditions and the scion cultivar. Belgian scientists recommend tree densities of up to 740 ha^{-1} on this rootstock. The sour cherry Montmorency has shown variable vigour on Inmil, from trees approximately two-thirds to three-quarters the size of trees on F.12/1, when grown on deep fertile soils, to trees only 30% the size of F.12/1 on poorer soils. When growth is weak the trees must be regularly pruned and irrigated to sustain adequate new growth.

One major problem with trees on Inmil is their sparse branching and erect growth habit.

Compatibility is generally good with the virus-tested clone of Inmil, but poor compatibility with Early Rivers was noted in one of the early Belgian trials and Hedelfinger Froschmaul failed on Inmil in German experiments.

Although the results of Belgian trials indicated average cropping precocity and productivity, trials in the USA have shown trees on Inmil to be particularly slow to come into cropping in comparison with trees on many other rootstocks.

Trees on Inmil are moderately sensitive to lime-induced chlorosis and have less tolerance to winter cold than trees on Damil or Camil. Inmil does not thrive on wet soils and is very sensitive to damage by *Phytophthora* species. Trees on Inmil need staking and will require regular pruning if adequate renewal shoot growth is to be sustained.

GM 61/1 (Damil)

Damil is a selection of the hybrid *P.* × *dawyckensis*, which as a rootstock for sweet cherries produces trees of moderate vigour, one-half to two-thirds the size of trees on F.12/1 based on trunk girth measurements but only one-third to one-half based on canopy volume. Planting densities of 370 to 570 trees ha^{-1} are suggested for trees on this rootstock. Damil propagates easily from softwood cuttings taken under mist. French results indicate that precocity on this rootstock is good, in that it is similar to that of trees on SL.64. However, recent trials in the USA and several other countries show relatively poor cropping precocity and yield productivity for sweet cherries grown on Damil in comparison with other dwarfing rootstock clones.

Trees on Damil have adequate anchorage but may need staking in the first few years following establishment. The stock produces few suckers, is moderately sensitive to soils with high pH, but shows moderate resistance to two damaging *Phytophthora* spp. Trials in New York show Damil to be extremely sensitive to crown gall on sandy soils. Damil also exhibits good tolerance to winter cold.

GM 79 (Camil)

This clone was selected from a *P. canescens* population containing types exhibiting non-fastigiate growth. In trials in Belgium, trees on Camil have one-third to one-half the trunk girth and one-half to two-thirds the canopy volume of trees on F.12/1. Trees on Camil were reported to require six to eight times less pruning than similar trees on F.12/1. Trials in the USA show sweet cherry cultivars on Camil forming trees 70% the size of trees on F.12/1 after 10–12 years of growth. Belgian researchers have suggested planting densities of 300–450 trees ha^{-1}.

Compatibility with sweet cherry scions is generally good, although incompatibility between Camil and the Canadian variety Summit has been reported in France. Yield precocity and productivity are very good on this rootstock. Unfortunately, Camil is sensitive to wet soils and to damage or death from infection by *Phytophthora* species. Camil shows good tolerance of severe winter cold, but suckering is a problem, similar to that noted on F.12/1. Unlike the two other Belgian selections, trees on Camil do not need staking.

GM 8

This Belgian selection, which is a hybrid between *P. subhirtella* × *P. yedoensis* and *P. subhirtella*, has given variable results in orchard trials and has not been named and released by Gembloux. It shows variable compatibility with sweet cherry scion cultivars and often poor anchorage. Trees on GM 8 are intermediate in vigour, often quite similar to trees on Colt, exhibit good yield precocity and produce few suckers. Unfortunately, GM 8 is extremely sensitive to highly calcareous soils and also to winter cold injury.

5.9.5 Hybrids raised at Dresden-Pillnitz, Germany

An extensive and long-standing breeding programme at Dresden-Pillnitz in Germany has produced a number of hybrids showing some promise as rootstocks for sweet cherries (Wolfram, 1979). In early work species and hybrid selections involving *P. incisa*, *P. nipponica kurilensis*, *P. canescens*, *P. tomentosa* and *P. avium* were not promising. In later crosses, in which *P. cerasus*, *P. pseudocerasus* and the Prunus hybrids Okame (*P. campanulata* × *P. incisa*), Kursar (*P. campanulata* × *P. nipponica kurilensis*) and Ivensii were introduced into the breeding programme, the rootstock potential improved.

Pi-Ku 4.20, which is a hybrid between *P. avium* and (*P. canescens* × *P. tomentosa*), is one of the intermediate vigour selections and trees of Kassins, Van and Hedelfingen have fruited very heavily on this rootstock (Wolfram, 1993). There is some concern, however, that tree longevity may be insufficient on this rootstock when grown on light sandy soils. Another sibling clone showing some promise is Pi-Ku 4.17, which is slightly more dwarfing. Three clones raised from a cross between *P. avium* and a hybrid between *P. canescens* and *P. kurilensis*, namely clone numbers 4.11, 4.13 and 4.15, also show some promise, being slightly more dwarfing than those in the previous mentioned series. Pi-Ku 4.83, a hybrid between *P. pseudocerasus* and (*P. canescens* × *P. incisa*), produces sweet cherry trees of intermediate vigour, similar to Colt, and with good yields.

5.9.6 Other *Prunus* hybrids tested as rootstocks

Several clones selected at Holovousy in Czechoslovakia (the P-HL series) have shown some promise as rootstocks for sweet cherries. One of these, P-HL-6, produced trees smaller and more productive than those on F.12/1 and the stock is readily propagated by softwood cuttings. Several P-HL numbered selections, 4, 6, 50 and 84, are listed, of which number 84 is used commercially. However, trials in Poland indicate poor compatibility of the cultivar Burlat on P-HL-84. These clones are thought to be natural hybrids between *P. avium* and *P. cerasus*.

Another hybrid, selected in Hungary is Prob, which was found as a vigorous seedling among a seedbed of *Prunus fruticosa* (Hrotko, 1994). It is thought to be a hybrid between this species and *Prunus mahaleb*. Graft compatibility between sweet cherry scions and Prob rootstock appears good and it reduces tree vigour by about 50% compared with trees on seedling Mahaleb rootstocks. Tree habit on Prob is not entirely satisfactory with sparse branching; suckering is minimal. As either a rootstock or an interstock, Prob induces precocious cropping. It is easily propagated from softwood cuttings.

Romanian researchers have selected a hybrid between the Mocanesti sour cherry and a clone of *P. avium*, which they have named IP-C1 (Parnia *et al.*, 1985). This rootstock is propagated by softwood cuttings, by micropropagation or from layers. Vigour of sour or sweet cherry trees on IP-C1 is similar to or slightly less than that on Colt. IP-C1 has better tolerance to wet soils than F.12/1, suckers less and induces better yield precocity.

5.10 Interstocks

Interstocks are very often used in apple or pear tree raising, but much less frequently used for sweet or sour cherry tree production. An interstock (interstem) is a rootstock (scion) clone which is budded or grafted between the rootstock and the fruiting scion, to form part of the lower trunk. The reasons for doing this are various. Their principal use for pears is to overcome problems of incompatibility between scion cultivars, such as Williams Bon Chretien, and quince (*Cydonia oblonga*) rootstocks. However, interstocks may also be used to reduce apple or pear scion vigour or improve yield precocity and/or productivity. Use of a dwarfing clone as an interstock, rather than as a rootstock, would be the norm where the desired rootstock clone is very difficult to root or where it is sensitive to soil-borne diseases, pests or other edaphic conditions.

Most dwarfing clones of apple and pear rootstocks usually also dwarf scions when used as interstocks, although their effect as interstocks may be less and is influenced by the length of interstock used. Unfortunately, many cherry rootstock clones that dwarf scions when used as a rootstock produce no or very little dwarfing effect when used as interstocks.

An interstem (another scion cultivar which is grafted between the rootstock and the main scion cultivar) is frequently used in raising apple trees in Holland. Trees raised with interstems are reputed to show improved yield precocity. Whether this effect is attributable to the interstem clone *per se* remains subject to debate. It has been suggested that when raising interstem trees by budding twice in successive years, their root systems and their root : shoot ratios, at the time of planting in the orchard, are much greater than on single worked trees and that it is this that influences yield precocity.

5.10.1 Sweet cherry clones as interstocks and interstems

Mazzards as rootstocks generally have little effect upon sweet cherry tree vigour and cropping and it would therefore be surprising if they had effects when used as interstocks. Even the Genetic Dwarf types, which as rootstocks do reduce tree vigour, had no effect on tree vigour when used at East Malling as 30-cm long interstocks between F.12/1 rootstocks and sweet cherry scions.

Occasionally sweet cherry clones used as interstems have been shown to have some effect on scion performance. For instance, Czech research showed that interstems of either P-HL-4 or P-HL-84 produced less growth in the crown of the sweet cherry cultivar Querfurter Königskirsche on Mazzard rootstocks. The trees with interstocks also produced more flowers and fruits, and there were no deleterious effects upon either fruit size or quality.

5.10.2 Sour cherry clones as interstems

The sour cherry cultivar Montmorency is occasionally used as an interstem between Mazzard or Mahaleb rootstocks and sweet cherry scions; USA results show reduced tree vigour (20–30%) but also reduced fruit size, with these combinations. However, trials in Britain showed no reduction in tree vigour when virus-free Montmorency was used as an interstem between F.12/1 and several sweet cherry cultivars.

Research in the USA has tested several sour cherry cultivars as interstems between Mazzard or Mahaleb rootstocks and the scion cultivars Bing and Chinook (Larsen *et al.*, 1987). After 20 years in the orchard, trees with interstems had cropped better than trees without and cumulative yields were higher for trees on Mahaleb than for trees on Mazzard rootstocks. The smallest Bing and Chinook trees were produced with North Star interstems. Yield efficiency of Bing/Northstar/Mahaleb trees was the highest among the combinations tested. Italian trials comparing effects on the vigour of five sweet cherry cultivars also showed that use of a sour cherry interstem of low vigour, such as North Star, gave good results (sometimes a 50% reduction in canopy volume).

Yugoslav research shows that the sour cherry cultivar Oblăcinska, used as an interstem between Mazzard seedling rootstocks and three sweet cherry cultivars, reduced vigour and greatly improved early yield productivity. Similarly, in German trials in which the effects of nine different sour cherry interstems were compared with two sweet cherry cultivars, all on F.12/1 rootstocks, the interstem Reine Hortense increased the yield per unit crown volume but reduced the crown volume to two-thirds and the ground area occupied to three-quarters that of control trees without interstems. Interstems of the Morello cherry cultivar Schattenmorelle had no effect on tree size but also increased the yields by 50%. Interstems of the variety Vladimir have produced similar effects on yield precocity, whereas Stockton Morello and the CAB rootstock clones have had poor effects.

Interstems of the cultivar Köröser Weichsel have given the most consistent results in German trials, reducing sweet cherry tree vigour by up to 30% in several trials and sometimes increasing yield productivity.

5.10.3 Mahaleb clones as interstocks

A dwarf selection of Mahaleb, selected at the University of California, Davis, has shown some promise as an interstock between sweet cherry scions and Mazzard rootstocks. Where bacterial canker is a problem but soils are unsuited to Mahalebs, this combination of rootstock and interstock may be appropriate.

5.10.4 *Prunus fruticosa* clones as interstocks

Work in Poland testing five clones of *P. fruticosa* as interstocks showed that all the clones reduced the vigour of the sweet cherry cultivar Büttners Rote Knorpel and increased yield productivity, but there were no interstock effects on cropping precocity. In other research, conducted in Germany, it has been suggested that the selection Oppenheim is better used as an interstock over F.12/1 or SL.64 rootstocks than as a rootstock itself.

5.10.5 Other *Prunus* species or hybrids tested as interstocks

Trials at East Malling have shown that *Prunus mugus,* which dwarfs sweet cherries very effectively when used as a rootstock, has almost no dwarfing effect if used as an interstock between Colt rootstocks and Merchant scions.

Earlier studies, also at East Malling, with the hybrid cherry rootstock Fb.2/58/15 (*Prunus avium* × *P. pseudocerasus*), a sibling of Colt, indicated that grown on its own roots and not budded with a scion it is very dwarfing, but when used as a rootstock or interstock it is very invigorating. However, if interstock trunks of Fb.2/58/15 were allowed to develop just a few small shoots and leaves, this produced 20–30% dwarfing of the scion (Jones and Quinlan, 1981). The reasons for this effect are not understood and warrant further investigation.

5.11 Rootstock Breeding

5.11.1 Breeding objectives

Although sweet cherries are of much less importance than apples, pears or peaches, in terms of the numbers of trees planted worldwide, cherry rootstock breeding has attracted a disproportionate amount of research effort in recent years. The high value of the crop is undoubtedly one reason for this as is the relatively 'undeveloped' nature of the crop. Until recently, the scion cultivars grown were mostly of ancient origin, were variable in cropping and had small fruit size. The absence of any dwarfing rootstocks meant that all trees were extremely large and therefore difficult and expensive to manage.

Objectives in breeding and selection of rootstocks for sour and sweet cherries have, in recent years, been:

1. tree size control (not so important for sour cherry);
2. increased precocity of cropping;
3. increased yield productivity;
4. uniformity of growth and cropping;
5. cold-hardiness;
6. adaptation to:
   (a) different soil types;
   (b) droughty soils;
   (c) poorly drained soils;
   (d) highly calcareous soils;
7. tolerance or resistance to:
   (a) soil pests;
   (b) soil-borne diseases;
   (c) aerial pests and diseases;
8. ease of propagation;
9. freedom from suckering;
10. good tree longevity (productive life of 12–15 years).

Moreover, any new rootstock must be compatible with the majority of commercial scion cultivars, if it is to be considered suitable for recommendation and release.

The importance of each of these criteria, and hence the priority they assume in breeding programmes, should largely be influenced by the environmental and managerial constraints which limit cherry production in any particular area. For instance, where labour is inexpensive and plentiful, dwarfing of trees will be of less importance than perhaps resistance to some particularly troublesome pest or disease. Unfortunately, not all breeders adhere to these simple principles, often putting easy-to-measure rootstock attributes, such as ease of propagation, higher than they really warrant in their ranking of breeding priorities.

5.11.2 Breeding strategies

Selection within Prunus *species*

Much of the early rootstock research focused on crossing and selecting superior clones of Mazzard, Mahaleb, sour and Steppe cherries. Although Mazzards were produced with improved tolerance to bacterial canker and with better propagation, none of these gave effective control of tree vigour or improved cropping productivity. Superior Mahalebs, in terms of ease of propagation, graft compatibility and induction of cropping, were also produced but, as with the

Mazzards, none were fully dwarfing and most were unsuited to growing on loam or clay soils.

Of more promise as dwarfing rootstocks are selections of the sour and Steppe cherries, which also induce excellent precocity in scions (Plock, 1972). However, this rootstock selection strategy does have its own problems. Incompatibility is often observed between sour or Steppe cherry rootstock clones and sweet cherry scions, and poor anchorage and abundant suckering are also problems with both these species. Nevertheless, some extremely promising rootstock selections of the sour cherry (*Prunus cerasus*) have been made by researchers at the Weihenstephan Institute near Munich, in Germany.

More recently other species from within the *Pseudocerasus* section of the *Prunus* genus, such as *P. incisa*, *P. concinna*, *P.* × *dawyckensis*, *P. mugus* and others, have also shown promise as dwarfing rootstocks. Although many of the selections made from these and other closely related species exhibit good compatibility, control of scion growth and minimal suckering, scion cropping precocity and productivity on these clones is often very poor. However, clones of species such as *P. canescens*, *P. incisa*, *P. kurilensis* and *P. nipponica* have shown significant benefits, in, for instance, their resistance to the damaging blackfly pest (*Myzus cerasi*).

Although breeding and selection of rootstock clones from within related *Prunus* species still continue, increased effort in recent years has turned towards breeding hybrids between two or more of these species.

Breeding and selection of Prunus hybrids

Most sweet cherry rootstock breeding effort now focuses on producing hybrids between two or more cherry species. Crosses made between *P. avium* and *P. pseudocerasus* produced the popular rootstock Colt, and open-pollinated hybrids between *P. mahaleb* and *P. avium* produced the M×M series, one of which, M×M 14, has gained considerable popularity in France. More recently, selection in Belgium among *Prunus* hybrids usually grown as ornamental trees or shrubs has produced Inmil and Damil, both selections of natural *Prunus* hybrids.

The most extensive and coordinated efforts to produce new cherry hybrid rootstocks have undoubtedly come from the breeding programme initiated at Giessen in Germany (Gruppe, 1985a). Work there began in 1965 when initial crosses were made between what were considered appropriate parent species, all of which were growing outside (i.e. without any glasshouse protection). This was of limited success with only 45 seedlings produced from 24,000 pollinated flowers. Later, by moving parental plants into glasshouses, by synchronizing flowering times of different species and by testing pollen style compatibility *in vitro*, breeding success was greatly increased.

The first task facing the German breeders, after generation of several thousand different hybrids at Giessen, was to test their ability to propagate and at the same time generate plants for further screening trials. This was achieved using stooling methods supplemented by girdling the 1-year-old shoots with wire to encourage root formation, i.e. marcotting. The hybrids which initially proved easiest to propagate were those between *P. cerasus* or *P. fruticosa* and *P. wadai*; the latter is itself a hybrid between *P. pseudocerasus* and *P. subhirtella*. These all developed aerial roots at the base of 1-year-old wood, similar to those formed on Colt. The tetraploid hybrid clones generally rooted very poorly, even after girdling treatments. In some cases root cuttings or propagation from softwood cuttings under mist environments were used to effect propagation.

Attempts to preselect the clones for their vigour potential on the basis of estimates of root : bark ratios were not entirely successful; similar problems using this technique on cherry hybrid rootstocks have been noted at East Malling.

The German work also showed poor relationships between the vigour of the rootstocks when unworked and their vigour when worked with sweet cherry scions. Nevertheless, annual shoot length was reasonably correlated with time of shoot growth cessation and there was some correlation between worked and unworked rootstocks in this respect (Franken Bembenek and Gruppe, 1985a, b). The most reliable, albeit the most time-consuming, method of estimating root-stock effects on scion vigour and cropping continues to be by budding with scions and testing in the orchard, and the German researchers wasted no time in establishing many field screening trials for their new rootstock clones (Gruppe, 1985b).

It was soon recognized that trees budded on many of the dwarfing rootstock clones exhibited early leaf senescence in the autumn, decreasing shoot elongation each year and death of the flowering spurs on older wood. Fruitfulness also declined with age. This indicated that, if growing sweet cherries on these dwarfing rootstocks was to be commercially successful, regular pruning of trees, to stimulate the growth of new wood, would be essential.

Many *Prunus* hybrids, other than those currently selected or in multisite tests (see above), have been evaluated as part of the Giessen breeding programme (Gruppe, 1985c; Schmidt, 1985). *P. cerasus* × *P. concinna*, *P. cerasus* × *P. incisa* and *P. cerasus* × *P. subhirtella* hybrids all reduced tree vigour by 40–50% when used as rootstocks, while *P. fruticosa* × *P. incisa* hybrids produced trees with more than 50% size reduction. However, yields of trees on these hybrids were only moderate and all suckered, some very badly. These hybrids were not, therefore, selected for further evaluation. Other hybrids also screened for their rootstock potential included *P. Pandora* × *P. nipponica* hybrids, which reduced scion vigour by over 40%, and *P. nipponica* × *P. incisa* and *P. incisa* × *P. speciosa* hybrids, which dwarfed trees even more. However, yield precocity and productivity were very poor and fruit size was small on these hybrid rootstocks. Clones of *P. pseudocerasus* × (*P. incisa* × *speciosa*), *P. hillieri*, *P. incisa* or *P. incisa* × *speciosa*, as well as crosses between these last two and *P. subhirtella* or *P. concinna*, all greatly dwarfed trees but floral precocity and flower bud quality were poor. Other crosses between stocks within the *Pseudocerasus* group, involving *P. Pandora*, *P. pseudocerasus*, *P. incisa*, *P. concinna*, *P. nipponica*, *P. subhirtella* and *P. hillieri*, produced very dwarfed but short-lived trees. Hybrids in which *P. fruticosa* was one parent produced clones which suckered, often quite badly, when tested as rootstocks.

It must be concluded that, while many *Prunus* hybrids have been developed which effectively dwarf sweet and sour cherry trees when used as rootstocks, many of these have severe agronomic shortcomings. With hybrid cherry root-stocks, the induction of high floral precocity and efficiency of fruit set is not closely linked with dwarfing, as it is, for the most part, with dwarfing apple rootstocks. Careful tree management will also be required when using many of the dwarfing cherry rootstocks if tree longevity is to be adequate and good fruit size sustained.

Novel methods of rootstock breeding

In the future, fruit breeders may choose to adopt novel methods, other than conventional hybridization, for the production of new cherry rootstocks. Some of these methods, such as irradiation techniques, have already been used to a limited extent. Others, involving the production of transgenic plants using techniques of molecular biology, have yet to be exploited. Use of *in vitro* techniques for the production of somaclones, polyploids, irradiation mutants or transgenic clones is currently the focus of efforts at several research centres. Other studies focused on gene mapping in *Prunus* will, when complete, greatly expedite progress with these novel techniques.

Mutants from irradiation breeding. Gamma irradiation techniques have been used to produce dwarf mutants of many fruit species, including *Prunus avium* (Walther and Sauer, 1985). Much of the early research on mutant induction using irradiation involved treatment of dormant graftwood of the cultivar using a cobalt-60 source. Although this approach was in some cases successful, a large proportion of the mutants produced were unstable chimeras.

In more recent research in Switzerland, Germany and Britain, irradiation of *in vitro* cultures has been tried on the assumption that by treating smaller propagules and fewer cells the chances of producing unstable chimeras are diminished. Theiler-Hedtrich (1990) treated the apical and axillary buds in *in vitro* cultures of the Mazzard rootstock clone F.12/1 with 30 Gy for 30 min, using a cobalt-60 source (1 Gy min^{-1}). The surviving plants were subcultured several times before rooting and weaning. In subsequent growth in the nursery several dwarf clones were apparent and these have yet to be fully evaluated as rootstocks.

In similar work at East Malling, James and colleagues irradiated cultures of the rootstock Colt. Several mutant clones originating from this research have since been budded with the variety Stella and planted in orchard trials. Preliminary results suggest one clone may be more dwarfing than conventional Colt. If this early promise is sustained, it will be necessary to check the stability of this mutant clone.

Induction of polyploid clones using in vitro *techniques.* Polyploid clones of citrus species are generally more dwarfing as rootstocks than the equivalent clones of lower ploidy, and there is research with the apple rootstock M.13 that indicates increased scion dwarfing when using the tetraploid rather than the conventional diploid clone. James *et al.* (1987) produced several hexaploid clones of the triploid rootstock Colt, by treating callus cultures with the chemical mutagen colchicine. Although the prime object of the research was to try to produce a fertile allopolyploid from the sterile triploid Colt and use this in future breeding programmes, the hexaploid clones have proved interesting in other respects.

Orchard trials, in which these hexaploid clones are compared with conventional triploid Colt as rootstocks for the sweet cherry varieties Van and Merchant, show slightly reduced scion vigour on the hexaploid clones.

Theiler-Hedtrich (1990) has used similar techniques to produce polyploid clones of F.12/1; these have yet to be evaluated as rootstocks.

Transgenic mutants. Finally, it is appropriate to consider the future possibilities of using molecular biology techniques for the introduction of useful genes into existing rootstock clones, i.e. the production of transgenics. Before this can be achieved, 'useful' genes must first be identified, sequenced and transcripts made; suitable marker genes must also be attached to the transcript so that production of the transgenic can be easily confirmed. Secondly, suitable promoters must be used such that the introduced gene is expressed efficiently in the target tissues.

Before any of this technology can be put into practice, regeneration systems must be developed for the recipient rootstock. There are several ways of achieving this but the most frequently used, currently, involves introduction of genes using leaf disc cultures and clones of *Agrobacterium rhizogenes* as vectors. Use of biolistics, where the gene is incorporated into a form of bullet and propelled into the receptor tissues using a gun, is also being evaluated. All of these techniques are expensive and require special skills. To date they have not been used for producing transgenic cherry rootstocks. However, research progress over the last decade has been very rapid and further developments in the future may provide great opportunities for the successful use of novel breeding techniques.

The techniques may also permit the production of bigeneric hybrids, which may be of great value in introducing resistance to damaging pests and diseases. Work in Britain has already suggested that mesophyll protoplasts of wild pear can be chemically fused with cell suspension protoplasts of the cherry rootstock Colt, following an electroporation treatment of the separate parental protoplast systems (Ochatt *et al.*, 1987, 1989).

General Reading

Gruppe, W. (ed.) (1985) International Workshop on Improvement of Sweet and Sour Cherry Varieties and Rootstocks – New developments and methods. *Acta Horticulturae* 169, 380 pp.

Iezzoni, A., Schmidt, H. and Albertini, A. (1990) Cherries. In: Moore, J.N. and Ballington, J.R. Jr (eds) Genetic resources of temperate fruit and nut crops. *Acta Horticulturae* 190, 109–173.

Rom, R.C. and Carlson, R.F. (eds) (1987) *Rootstocks for Fruit Crops.* John Wiley & Sons, New York, USA, 494 pp.

References

Breton, S. (1980) *Le Cerisier*. Centre Technique Interprofessionel des Fruits et Legumes, Paris, pp. 21–31.

Carles, L. (1986) Le dépérissement de Molières. [Molières' decline.] *Arboriculture-Fruitière* 33 (383), 23–24.

Cummins, J.N., Wilcox, W.F. and Forsline, P.L. (1986) Tolerance of some new cherry rootstocks to December freezing and to *Phytophthora* root rots. *Compact Fruit Tree* 19, 90–93.

Edin, M. (1989) Tabel Edabriz, porte-greffe nanisant du cerisier. [Tabel Edabriz, a dwarfing rootstock for cherry trees.] *Infos Paris* 55, 41–45.

Edin, M. and Claverie, J. (1987) Porte-greffes du cerisier: Fercahun-Pontavium, Fercadeu-Pontaris, deux nouvelles sélections de merisier. [Cherry rootstocks: Fercahun-Pontavium, Fercadeu-Pontaris, two new wild cherry selections.] *Infos, Centre Technique Interprofessionnel des Fruits et Legumes, France* 28, 11–16.

Edin, M., Masseron, A., Tronel, C., and Claverie, J. (1989) Porte-greffe nanisants pour cerisiers: premiers resultats d'experimentation en France. [Dwarfing and semi-dwarfing rootstocks for cherry trees: preliminary results of experiments in France.] *Fruit Belge* 57 (427), 235–241.

Feucht, W. and Schmid, P.P.S. (1984) Zur Veredlungskombination Susskirsche auf *Prunus cerasus* [The graft combination of sweet cherry on *Prunus cerasus*.] *Erwerbsobstbau* 26, 195–199.

Feucht, W., Schmid, P.P.S. and Christ, E. (1984) Verzogerte Unvertraglichkeit der Susskirschensorte Sam auf einigen Klonen von *P. cerasus* L. II. Anatomische Aspekte zur Degeneration des aktiven Phloems im Veredlungsbereich im Juni und Juli. [Delayed incompatibility of Sam sweet cherry on certain *Prunus cerasus* L. clones. II. Anatomical aspects of active phloem degeneration in the zone of the graft union in June and July.] *Mitteilungen Klosterneuburg Rebe und Wein, Obstbau und Fruchteverwertung* 34, 132–137.

Franken Bembenek, S. and Gruppe, W. (1985a) Variability in vegetative growth of different cherry hybrids (*Prunus* X spp.). *Acta Horticulturae* 169, 257–262.

Franken Bembenek, S. and Gruppe, W. (1985b) Growth relationships of ungrafted and grafted hybrid cherry rootstocks. *Acta Horticulturae* 169, 245–250.

Garrett, C.M.E. (1977) Selection of resistant rootstocks. *Annual Report of East Malling Research Station for 1976*, 128.

Garrett, C.M.E. (1986) Influence of rootstock on the susceptibility of sweet cherry scions to bacterial canker, caused by *Pseudomonas syringae* pvs *morsprunorum* and *syringae*. *Plant Pathology* 35, 114–119.

Gruppe, W. (1985a) An overview of the cherry rootstock breeding program at Giessen 1965–1984. *Acta Horticulturae* 169, 189–198.

Gruppe, W. (1985b) Evaluating orchard behaviour of cherry rootstocks. *Acta Horticulturae* 169, 199–207.

Gruppe, W. (1985c) Size control in sweet cherry cultivars (*Prunus avium*) induced by rootstocks from interspecific crosses and open pollinated *Prunus* species. *Acta Horticulturae* 169, 209–217.

Hanson, E.J. and Perry, R.L. (1989) Rootstocks influence mineral nutrition of 'Montmorency' sour cherry. *HortScience* 24, 916–918.

Howell, G.S. and Perry, R.L. (1990) Influence of cherry rootstock on the cold hardiness of twigs of the sweet cherry scion cultivar. *Scientia Horticulturae* 43, 103–108.

Hrotko, K. (1994) Unterlagen und Schnittmassnahmen an Spindelbaumen von Susskirschen. *Proceedings of the 6th International Symposium for Fruitgrowing (Poldi),* Lednice, Moravia 26–27 October 1994 (in press).

James, D.J., MacKenzie, K.A.D. and Malhotra, S.B. (1987) The induction of hexaploidy in cherry rootstocks using *in vitro* regeneration techniques. *Theoretical and Applied Genetics* 73 (4), 589–594.

Jones, O.P. and Quinlan, J.D. (1981) Effect of cherry rootstock clone 15 (FB2/58, *Prunus avium* × *P. pseudocerasus*). *Journal of Horticultural Science* 56, 237–238.

Knight. T.A. (1801) *A Treatise on the Culture of the Apple and Pear and on the Manufacture of Cyder and Perry.* 2nd edn. H. Proctor, London.

Kolesnikova, A.F., Ossipov, Yu.V., Kolesnikov, A.I. and Osipov, Yu.V. (1985) New hybrid rootstock for cherries. *Acta Horticulturae* 169, 159–162.

Larsen, F.E. (1972) Characteristics of available sweet cherry rootstocks. *Good Fruit Grower* 22, 14–15.

Larsen, F.E., Higgins, S.S. and Fritts, R. Jr (1987) Scion/interstock/rootstock effect on sweet cherry yield, tree size and yield efficiency. *Scientia Horticulturae* 33, 237–247.

Matthews, P. (1975) In: Brown, A.G., Watkins, R. and Aston, F. (eds) *Proceedings of Eucarpia Fruit Section Symposium V. Top Fruit Breeding Symposium.* University of East Anglia, UK, pp. 84–107.

Ochatt, S.J., Cocking, E.C. and Power, J.B. (1987) Isolation, culture and plant regeneration of colt cherry (*Prunus avium* × *pseudocerasus*) protoplasts. *Plant Science, Irish Republic* 50, 139–143.

Ochatt, S.J., Patat Ochatt, E.M., Rech, E.L., Davey, M.R. and Power, J.B. (1989) Somatic hybridization of sexually incompatible top-fruit tree rootstocks, wild pear (*Pyrus communis* var. *pyraster* L.) and Colt cherry (*Prunus avium* × *pseudocerasus*). *Theoretical and Applied Genetics* 78, 35–41.

Parnia, P., Mladin, G. and Popescu, M. (1985) A new autochthonous vegetative rootstock for sweet and sour cherry. *Acta Horticulturae* 169, 169–176.

Pennell, D., Dodd, P.B., Webster, A.D. and Matthews, P. (1983) The effects of species and hybrid rootstocks on the growth and cropping of Merton Glory and Merton Bigarreau sweet cherries (*Prunus avium* L.). *Journal of Horticultural Science* 58, 51–61.

Pepin, H.S., Sewell, G.W.F. and Wilson, J.F. (1975) Soil populations of *Thielaviopsis basicola* associated with cherry rootstocks in relation to the effects of the pathogen on their growth. *Annals of Applied Biology* 179, 171–176.

Perry, R.L. (1985) Progress with cherry rootstocks. *Compact Fruit Tree* 18, 107–108.

Plock, H. (1972) [The importance of the Ground Cherry, *Prunus fruticosa* Pall., as a dwarfing rootstock for sweet and sour cherries.] In: *Atti II. Convegno del Ciliegio, Verona-Vignola.* Industria Grafica Moderna SpA, Verona, pp. 119–132.

Proffer, T.J., Jones, A.L. and Perry, R.L. (1988) Testing of cherry rootstocks for resistance to infection by species of *Armillaria. Plant Disease* 72, 488–490.

Rivers, T. (1870) *The Miniature Fruit Garden,* 6th edn. Longmans, London.

Roth, M. (1986) Bodenvernassungstoleranz von Unterlagenklonen verschiedener Kirscharten (*Prunus* spp.) und interspezifischer Kirschhybriden (*Prunus* X spp.). [Tolerance to soil flooding in clonal rootstocks of various cherry species (*Prunus* spp.) and interspecific cherry hybrids (*Prunus* X spp.).]. Thesis, Justus Liebif University, Giessen, Germany, 183 pp.

Schimmelpfeng, H. and Liebster, G. (1979) *Prunus cerasus* als Unterlagen Selectionsarbeiten Vermehrung eignung für Sauerkirschen. *Gartenbauwissenschaft* 44, 55–59.

Schimmelpfeng, H. and Vogel, T. (1985) Einfluss unterschiedlicher Veredlungshohen auf Wachstum und Rtragsverhalten von Susskirschen auf Schwachwuchsinduzierenden Unterlagen in den ersten 8 Tandjahren. [Effect of different grafting heights on growth and yield of sweet cherries on dwarfing rootstocks in the first 8 years.]. *Obstbau* 10 (3), 104–107.

Schmid, P.P.S., Wagner, C. and Schimmelpfeng, H. (1982) Relative hardiness of sweet cherry graft combinations in relation to the rootstocks used. *Gartenbauwissenschaft* 47, 116–123.

Schmidt, H. (1985) First results from a trial with new cherry hybrid rootstock candidates at Ahrensburg. *Acta Horticulturae* 169, 235–243.

Schmidt, H. (1989) Beobachtungen zum Befall von Kirschhybridunterlagen mit Wurzelkropf. [Observations on the incidence of crown gall on cherry hybrid rootstocks.] *Erwerbsobstbau* 31, 42–43.

Strauch, H. and Gruppe, W. (1985) Results of laboratory tests for winter hardiness of *P. avium* cultivars and interspecific cherry hybrids (*Prunus* X spp.). *Acta Horticulturae* 169, 281–287.

Theiler-Hedtrich, R. (1990) Induction of dwarf F.12/1 cherry rootstocks by *in vitro* mutagenesis. *Acta Horticulturae* 280, 367–371.

Trefois, R. (1985a) Dwarfing rootstocks for sweet cherries. *Acta Horticulturae* 169, 147–155.

Trefois, R. (1985b) Two dwarfing rootstock selections for sweet cherries. *Acta Horticulturae* 169, 157–158.

Treutter, D. (1985) Polyphenol patterns in the phloem of the union with respect to early selection for compatible graft combinations in cherries. Thesis, Technical University of Munich, Freising-Weihenstephan.

Uphoff, H., Eppler, A. and Gruppe, W. (1988) Reaction patterns of some cherry hybrid rootstock clones towards infection with PNRSV and a PDV isolate. *Mededelingen van de Faculteit Landbouwwetenschappen, Rijksuniversiteit Gent, Belgium* 53 (2a), 491–498.

Walther, F. and Sauer, A. (1985) Analysis of radiosensitivity – a basic requirement for *in vitro* somatic mutagenesis. I. *Prunus avium* L. *Acta Horticulturae* 169, 97–104.

Webster, A.D. (1995) Cherry rootstock evaluation at East Malling. *Acta Horticulturae* 410 (in press).

Wicks, T.J., Bumbieris, M., Warcup, J.H. and Wallace, H.R. (1984) Phytophthora in fruit orchards in South Australia. *Biennial Report of the Waite Agricultural Research Institute, 1982–1983.* 1984, 147.

Wolfram, B. (1979) [Results from many years of testing *Prunus* hybrid progenies as rootstocks for sweet cherries.] *Tangungsbericht Akademic der Landwirtschafts Wissenchaften der DDR, Berlin*, 174, 257–262.

Wolfram, B. (1995) Advantages and problems of some selected cherry rootstocks in Dresden-Pillnitz. *Acta Horticulturae* 410 (in press).

Zepp, L. and Szczygiel, A. (1985) Pathogenicity of *Pratylenchus crenatus* and *Pratylenchus neglectus* to three fruit tree seedling rootstocks. *Fruit Science Reports* 12 (3), 109–117.

III CROP PHYSIOLOGY AND HUSBANDRY

6 Propagation of Sweet and Sour Cherries

A.D. WEBSTER

Horticulture Research International, East Malling, West Malling, Kent ME19 6BJ, UK

6.1 Introduction

Sweet and sour (tart) cherry cultivars are genetically heterozygous and, in the characteristics of their fruits, trees raised from seed often show little similarity with the parent tree. Consequently, if selected horticultural scion cultivars are to be reproduced true to type, propagation must be by some asexual (vegetative) method. Direct propagation of the scion on its own roots is difficult to achieve, particularly with cultivars of the sweet cherry, and very few cherry trees are propagated in this way.

Propagation is usually achieved by budding or grafting a scion of the desired cultivar on to a graft-compatible rootstock. Traditionally, the rootstocks used were all raised from seed. As their only purpose was to provide a root system and a means of propagating the desired scion, some seedling variability was tolerable and effects on vigour of scion growth and cropping were not considered of great importance. Nowadays, many rootstocks are propagated vegetatively, particularly those which impart additional desired characteristics to the scion tree, such as disease resistance or vigour control.

With the development of *in vitro* techniques of propagation (micropropagation), it is now much easier to propagate sweet and sour cherry varieties on their own roots. Nevertheless, the performance of such trees is still largely untested and raising trees by budding or grafting on to rootstocks still remains the principal method of sweet or sour cherry propagation.

The costs of young cherry trees purchased from a nursery are relatively high compared with the other costs associated with establishing an orchard; trees and their supporting stakes may amount to approximately half the total cost of establishing a semi-intensive orchard. It is increasingly important, therefore, that only healthy, true-to-type trees are planted and that they are brought into cropping as soon as possible.

6.2 Choice of Materials

6.2.1 The rootstock

Traditionally, rootstocks were used solely as a means of propagating those scion cultivars which were very difficult to propagate vegetatively on their own roots. Nowadays, use of the appropriate rootstock may offer many more benefits than this, both to the nurseryman and to the orchardist. Choice of rootstock is, however, often a compromise between the needs and demands of the fruit grower and those of the nurseryman. The attributes of an ideal rootstock will always be prioritized very differently by these two groups. The nurseryman will seek rootstocks that are easy to propagate, take the bud well and produce large, abundantly feathered maiden trees. The fruit grower, while also appreciating the last of these attributes, will emphasize the importance of the rootstock's effects on the orchard performance of the tree. Rootstock effects on tree vigour, cropping, fruit size and resistance to unfavourable soil conditions, pests or diseases are of much greater importance to the fruit grower than to the nurseryman.

The majority of sweet and sour cherry trees are raised on either Mazzard (*Prunus avium* L.) or Mahaleb (*Prunus mahaleb* L.) rootstocks; the former are chosen for deep loam soils while the latter are superior on freely draining sandy or gravelly soils. Where soils are heavy clays and more poorly drained, sweet cherries are often raised on sour cherry (*P. cerasus* L.) rootstocks and this is also sometimes the favoured rootstock for sour cherry scion cultivars. The respective merits of these three and of other *Prunus* species and hybrid rootstocks for sweet and sour cherries are described more fully in Chapter 5.

Traditionally, all rootstocks for cherries were raised from seed, although occasionally suckers dug up from the base of fruiting trees provided an alternative source. In many parts of the world the majority of cherry trees are still raised from seed. Nurserymen, if given a choice, usually favour the production of Mahaleb rather than Mazzard rootstocks, as the former are easier to germinate, less susceptible to leaf spot disease and more resistant to both winter injury and drought than Mazzard stocks. Unfortunately, trees on Mahaleb may prove short-lived if planted on the wrong soil types and for this reason this rootstock is less popular than in the past.

Vegetatively propagated Mazzard, Mahaleb and sour cherry rootstock clones have been selected over the last 50–70 years and several of these, such as F.12/1 and St Lucie 64 (SL.64), have gained considerable popularity. Clonal Mazzards are not easy to propagate, however, and many producers have sought to avoid these problems by offering only seedling-raised stocks. Clonal Mahaleb stocks, such as SL.64, have gained popularity in France and may be propagated quite easily from summer softwood cuttings. Many sour cherry rootstock clones are very easy to propagate from softwood cuttings and this is the preferred method of raising clonal stocks of this species.

Two other species of *Prunus*, *P. fruticosa* and *P. canescens*, are also occasionally recommended as rootstocks for sweet cherries. The former, which, when used as a rootstock, greatly dwarfs sweet cherry trees, suckers profusely; it is best propagated by some modification of the layering technique, as it is most difficult to propagate from cuttings. *P. canescens*, which as a rootstock produces trees of intermediate stature, will root quite easily from softwood cuttings.

Recently, several *Prunus* hybrids have shown promise as rootstocks for sweet cherries. Colt (*P. avium* × *P. pseudocerasus*), the most popular of these, is extremely easy to propagate by almost any of the techniques described below. Trees on Colt perform well in the nursery, forming trees with many wide-angled branches. Other hybrid rootstocks, bred and selected in Germany or Belgium, may provide a range of vigour control for sweet cherry growers. Most of these new hybrids can be raised using softwood cutting techniques, or by micropropagation.

6.2.2 The interstock

Interstocks are quite widely used in raising temperate fruit species other than cherry. Interstocks are used to overcome incompatibility problems between pear scions and their quince rootstocks, or as a means of dwarfing apple trees. They may be particularly useful where the interstock clone is itself difficult to propagate vegetatively or is poorly anchored when used directly as the rootstock.

Interstems, which differ from interstocks in that they are intermediate grafts of other scions rather than rootstock cultivars, are also occasionally of value. An interstem of a winter-hardy cultivar can provide useful cold tolerance to the susceptible trunk of some apple cultivars, while interstems are also reported to improve apple scion precocity of cropping.

Although there have been very many research trials evaluating the use of interstocks in cherry tree production, the reported effects are inconsistent and neither interstocks nor interstems are widely used in sweet cherry tree raising. Unfortunately, few of the trials testing cherry interstocks have taken adequate account of the virus status of the materials used and it is therefore difficult to assess whether any of the observed effects were due to the interstock *per se* or indirectly attributable to its health status. Most researchers have tested clones (cultivars) of sour cherry or Mahaleb as dwarfing interstocks for sweet cherries. Some reports suggest reduced scion vigour associated with use of an interstock, others show increased cropping. More work is needed to prove or refute these claims.

There is little doubt that, where traditional trees of large stature are grown and where bacterial canker (*Pseudomonas mors prunorum* and/or *P. syringae*) is a severe problem, there is some merit in using a resistant rootstock clone as a 'stem builder' and frameworking the sensitive scion on to the resistant scaffold branches. If this 'stem builder' clone is difficult to propagate on its own roots, then its use as an interstock should be considered.

Use of interstocks does add to the cost of raising trees, irrespective of the grafting or budding method employed. The most popular and possibly the cheapest method is to graft the scion to the interstock and then graft this interstock plus scion to the rootstock, both graft unions being achieved at the same time. However, this does limit the length of interstock which can be used and trees raised in this way require very careful weaning if they are to establish successfully and grow uniformly. Usually, this double grafting is conducted using rootstocks lined out in the nursery, although occasionally the whole operation may be completed as a double-bench graft using bare-rooted rootstocks. A more lengthy procedure is to bud the rootstock with the interstock and then bud the interstock with the scion in the subsequent summer. This method, although more expensive, is usually more reliable and may also give benefits of improved tree precocity.

6.2.3 The scion

A fruit grower's choice of scion cultivar may be based upon objective study of trials results comparing new cultivars or upon previous success with a particular cultivar. The choice may be quite specific, some growers demanding a particular clonal selection of a cultivar. Superior clones of sweet and sour cherry cultivars showing improved cropping, fruit size, colour and quality have been selected in several countries and should be chosen wherever available.

The nurseryman's inputs into scion choice will focus on the health status of the scion material and its suitability for budding or grafting. Wherever possible only trees of prove clean health status should be used as 'mother trees', i.e. as a source of cuttings, buds or graftwood. Also, the wood must be of the right size and in the right physiological condition to ensure success in the propagation process. This is discussed further in later sections of this chapter.

6.2.4 Quality and health of plant materials

All rootstocks, interstocks and scions used for raising cherry trees should be of the best health status available. Rootstocks should be free from soil-borne diseases such as crown gall (*Agrobacterium tumefaciens*), and pests, such as nematodes. Both scion and rootstock should be free of insect pests, bacterial and fungal diseases and all known viruses and mycoplasmas.

Many virus diseases of cherry are transmitted by bees or other insects carrying infected pollen from a diseased to a healthy tree. For this reason, virus-free mother trees of both scion cultivars and seedling-raised rootstock types must be grown sufficiently distant from other *Prunus* trees, such that virus spread into the site by insects carrying infected pollen is extremely unlikely. Scion mother trees are severely pruned to encourage the production of abundant supplies of budwood and graftwood and also to minimize flowering and potential virus spread. Regular virus indexing of mother trees is essential to facilitate the identification and removal of infected trees. Mother trees have a limited useful life and should be replaced by newly tested virus-free trees after every 10 or 15 years.

Further details of the pests and diseases affecting sweet and sour cherries may be referred to in Chapters 14 and 15 of this book.

6.3 The Fruit Tree Nursery

6.3.1 The site

The ideal site for a fruit tree nursery is on level rather than undulating land, as this facilitates the operation of the undercutting and lifting machinery used. In areas experiencing cold springs, a slightly south-facing slope may be preferred to help warm the soil early in the year and speed bud break and growth. Also, the site should be sheltered from strong winds, as these not only make support of young trees difficult, but may also blow out newly emerging scion buds. Windbreaks of species such as *Alnus*, *Populus* and *Salix* are commonly used to filter the strong winds and their beneficial effects will usually extend for a distance equivalent to ten times their height. Care should be taken in choice of windbreak species, however, as many are alternative hosts for damaging diseases such as silver leaf (*Chondrostereum purpureum*). In areas where radiation frosts are a problem in the spring, sites lying in hollows should be avoided, as frost damage to young shoots may be considerable; dense windbreaks across the line of a sloping site may give similar problems by impeding the natural flow of cold air.

6.3.2 The soils

Deep, fertile and well-drained soils of medium texture with a good organic matter content and a balanced supply of nutrients are most appropriate for nursery production of fruit trees. Ideally, the soil should not have grown fruit trees previously. Shallow and light-textured soils are rapidly depleted of moisture and nutrients and, although very easy to cultivate in almost all weather conditions, they need careful management and irrigation if they are to produce quality trees. Heavy clay soils are difficult to cultivate when wet and often slow to warm up in the spring. Nursery soils should have a good nutrient buffer capacity and not be subject to either surface 'capping' or erosion after heavy rain.

Nursery soils should be ploughed in late autumn or winter and then cultivated until a fine tilth is produced. Never cultivate when the soil is too wet or soil texture may be irreversibly damaged. Soils with a hard pan are best avoided as this will impede drainage and root growth. However, where this is not possible the pan must be broken up with a suitable subsoiling implement.

Replant disease may prove a problem on sites previously planted with cherries or other stone fruits. The problem can be greatly reduced by good crop rotation and long-duration grass leys. Green cropping with nitrogen-enriching legumes will also reduce the effects of some causes of replant problems and improve soil fertility prior to its use for nursery production. Where soils with replant symptoms must be used and there is insufficient time for rotations, fumigation is the only alternative. The choice of chemical fumigant will be determined by the nature of the replant problem and by legal constraints on chemical use. Where nematode damage is anticipated incorporation of dichloropropane-dichloropropene (DD) may prove to be the best remedy. However, growers favouring a more 'organic' approach to the problem may prefer to try to reduce nematode populations by planting a crop of marigolds prior to use of the land as a nursery. Where fungi are perceived to be the main cause of replant problems, chemicals such as methyl bromide, formalin or chloropicrin have all proved effective. None of these products is pleasant to use, however, and local environmental constraints may prevent their use on nurseries situated close to urban developments. Soils contaminated with crown gall (*Agrobacterium tumefaciens*) should not be used for nursery production.

Prior to planting it is important to ascertain the nutritional status of the soil and its pH. Deficiencies of all major and many minor elements may be corrected by application of an appropriate base fertilizer dressing. Where available, incorporation of well-rotted manures may supplement nutrition as well as improve soil structure. Low pH (< 5.0) can lead to severe physiological problems on young trees, such as 'measles', which is associated with manganese toxicity, whereas very high pH (> 8.0) may result in problems of iron deficiency and leaf chlorosis. Liming the soil will increase pH, while use of acidifying fertilizers may, if used consistently, lower it.

6.4 Rootstock Propagation

Although cherry rootstocks are still raised from seed in many countries, including the USA, nurseries in northern Europe began using clonal Mazzard (*P. avium*) rootstocks more than 50 years ago. The greater variability in supply and quality of seed sources in Europe may partly explain this change of practice. Cropping and thus seed supplies of wild cherries has always proved variable from season to season and the wild populations in northern Europe are much more heterogeneous than those derived from introductions to countries such as the USA, where the Mazzard is not indigenous. Another factor promoting the change was the desire among fruit growers for a rootstock which conferred additional benefits to the scion tree, as well as aiding its propagation.

The clonal *Prunus avium* rootstock F.12/1 was selected for its beneficial effects on scion cropping and also its ability to propagate well on the layer bed. Another strong reason for using F.12/1 was its partial resistance to the damaging effects of bacterial canker (*Pseudomonas mors prunorum*). Since the release of F.12/1 in the 1920s, many other clonal rootstocks have become available, each selected for particular attributes, such as ability to adapt to soil/site conditions or to control tree growth and cropping (see Chapter 5).

Vegetative methods of rootstock propagation can be divided into two main categories, techniques based on division and techniques based on cuttings (see flow chart).

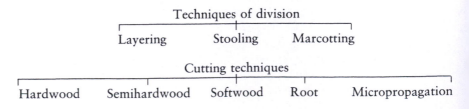

The main distinction between the two categories is that, with all techniques of division, the propagule remains wholly or partially attached to the mother stock plant during rooting, while with cutting techniques the propagule is completely severed from the mother plant prior to rooting.

While attachment to the stock plant during rooting, as practised in division techniques, offers many advantages to successful rooting and establishment, it also has its problems. All division techniques, including the allied practice of lifting suckers from beneath orchard trees, run the risk of transferring soil-borne diseases (such as *Agrobacterium tumiefaciens* – crown gall) or nematodes from the stock bed to the budding nursery. The one clear advantage of cutting techniques is that they allow this cycle to be broken. Aerial shoots cut from stock hedges and planted into disease-free (possibly fumigated) soils should not carry these pests and diseases with them.

6.4.1 Sexual propagation by seed

As noted previously, sweet and sour cherries are extremely heterozygous species and produce seeds sexually, not apomictically. Consequently, cherry seedling rootstocks are genetically very variable and might be expected to be similarly variable in growth, appearance and their effects upon scions. It is surprising, therefore, that most Mazzard seedlings, irrespective of provenance (seed origin), are quite similar in morphology and potential vigour and, when worked with sweet cherry scions, form trees which are similar in their mature habit and vigour. Seedlings of Mahaleb of different provenance are more variable in appearance than are Mazzards, although, like Mazzards, trees grown upon them are often quite uniform in vigour and habit.

Source of seed

The provenance or origin of cherry seed is of considerable importance where seedling rootstocks are to be raised in areas experiencing climatic extremes. For instance, a source of seeds collected from trees indigenous to an area of mild or warm climatic conditions may prove slow to acclimatize and may suffer if grown in areas experiencing severe winter cold. Wherever possible, seed should be used which is collected locally or from an area of similar climatic conditions; the important climatic parameters are latitude, length of growing season, mean temperatures during the season, frequency of summer droughts and severity or timing of winter frosts.

Rather than collecting seed from trees growing in the wild, which will have some intrinsic variability and will certainly be of unknown health status, it is preferable to obtain seed grown on virus-free trees planted in an isolated block, i.e. mother trees. Where seed cannot be obtained from such a source, the nurseryman should try to obtain seed from known parent or mother trees and test the seeds' performance, using a progeny test (Hartmann and Kester, 1983). Unfortunately, in many countries seeds of cherries are not obtained in this way, a more common practice being to obtain seed extracted by commercial processors of cherries. While this is acceptable where the seed is predominantly from a single cultivar and where previous tests have shown this to produce good rootstocks, it is less desirable where seed batches are of several cultivars from trees of unknown health status and provenance.

It should be noted that early-ripening cultivars of sweet cherry usually abort most of their seeds prior to maturity and pits (stones) of these cultivars are of no value as sources of cherry seeds.

Once cherry fruits are collected for seed it is important to separate the seed and its protecting pit (stone or endocarp) from the flesh as soon as possible. This separation may be achieved using special equipment for stone removal, high-pressure water or maceration procedures.

If it is necessary to store seeds of *Prunus avium*, they should first be cleaned, as described above, and then kept in polythene bags at 1°C and 8–12% moisture content for 2–8 weeks before transfer into sealed containers held at temperatures below freezing ($-1°C$ to $-18°C$). So treated, seeds will retain an acceptable viability for approximately 3 years.

Seed dormancy

The seeds of cherries, like those of most other stone fruits, are enclosed within a hard endocarp (pit or stone) which slows down embryo development by delaying water imbibition by the seed and the subsequent expansion of the embryo. The delay in water ingress also slows the leaching of germination inhibitors from the seed. Chemical inhibitors which prolong dormancy, such as abscisic acid (ABA), are thought to be important in preventing the germination of some types of stone fruit seeds. In contrast, the seed's content of growth-promoting substances, such as gibberellins, increases as dormancy is broken and the 'after-ripening' requirement is fulfilled. Most of the germination inhibitors are leached out by water once the hard endocarp is cracked or removed.

Cherry seeds, which exhibit what is known as embryo dormancy, require a period of 2–6 months' moist chilling in order for germination to occur naturally. The traditional method of speeding and aiding this process is to place seeds between moist layers of sand or some other similar medium, either in containers within a refrigerator or, in temperate areas, in the open ground during winter where the cold temperatures will successfully provide the necessary chilling. This process, which is known as 'stratification', helps overcome embryo dormancy by providing conditions which allow imbibition of water by the seed, expose the seed to chilling temperatures (ideally between 2 and 7°C) and provide the seed with sufficient oxygen.

Dormancy breaking of both Mahaleb and Mazzard seeds is best achieved at temperatures of 5°C. However, the species differ in the length of time required to complete the process: 90–100 days for Mahaleb and 120–140 days for Mazzard. The time required for dormancy breaking and after ripening of Mazzard seed may be reduced to approximately 60 days if the seeds are soaked in gibberellins after a shortened period of stratification (Fogel, 1958).

Failure to fully stratify cherry seeds may result in seedlings with short root (radical) growth but no epicotyl development. Alternatively, some seedlings growing from seeds which have been insufficiently chilled will develop normal roots but dwarfed and rosetted epicotyls; the latter are often referred to as 'physiological dwarfs'. Gibberellin sprays may help force shoot extension on these rosetted plants.

Seed quality assessments and germination

Three factors need to be taken into account when assessing seed quality and condition. The first consideration is seed viability, which is expressed by the percentage of seeds which successfully germinate given favourable soil conditions. Secondly, the speed of germination is measured, as this is an important indicator of the success of 'after-ripening'. Finally, a seedling's vigour is of critical importance to its subsequent value as a rootstock. Poor seed provenance, harvesting when immature, improper storage and poor seed health will all reduce seed viability, speed of germination and seedling vigour. Embryo excision or a tetrazolium test may be used to determine seed viability. Details of these and other tests to assess seed quality are to be found in standard textbooks on propagation (Hartmann and Kester, 1983).

Seeds need to imbibe large amounts of water before they will germinate. Once the seed's impervious testa or endocarp is breached, water imbibition is initially rapid but then declines for a short period before increasing again as the seed radicle emerges. The rate of imbibition is influenced by the availability of water in the growing medium, the colloidal properties of the seed and the degree of imperviousness of the seed coat; it is also positively correlated with increasing temperatures up to 20°C.

The ideal soil medium for germination is one which has a good moisture-holding capacity, low salinity and does not surface-'cap' after heavy rainfall. If the soil is prone to 'capping', it may be alleviated by covering the seed with a thin layer of moist peat or softwood sawdust before finally covering with a thin layer of soil.

Temperature of the seed-bed soil is important as, like several other stone fruit species, germination of cherry seeds may be inhibited at temperatures greater than 20°C. Exposure to high temperatures induces a secondary dormancy and temperatures of between 10 and 17°C are optimum. In areas experiencing very cold winters, it may be necessary to protect the seeds by mounding soil over the seed rows during the winter months; these mounds should be pulled down again in the spring.

In those temperate zones which experience mild winters, no pre-stratification is practised, nurserymen preferring to plant seeds directly into the nursery seed-bed in the autumn. The warm soil temperatures at that time of year favour the initial stages of stratification and after-ripening; thereafter the winter soil conditions provide the necessary chilling and completion of stratification. Where winter soil temperatures are consistently below freezing, it is preferable to stratify the seeds before sowing and plant only when temperatures begin to rise in the spring.

For successful germination, seeds require a consistent supply of oxygen within the soil medium. Too much water, soils of too fine structure or soil 'capping' may all reduce the oxygen supply and inhibit germination.

Seeds are often broadcast in seed-beds of 1–1.2 m width which are raised several centimetres above the height of the intervening paths as an aid to water drainage. An alternative to broadcast sowing is to drill the seeds into closely spaced rows on the beds, using specialized seed planters. Density of sowing will depend upon whether it is planned to retain the seedlings in the beds until they are of a size suited for lining out and budding, in which case density should not be too high, or whether one or more transplanting stages are planned. It is common practice to undercut the roots of seedlings at the end of 1 year's growth and transplant for a further 1–2 years prior to use as rootstocks. Alternatively, seeds may be drilled directly into nursery rows and the seedlings grown on and budded *in situ* when of sufficient basal stem callibre. In this case the rows are spaced approximately 1.2 m apart (depending upon machinery used) and the seeds planted 10 cm apart in the rows. Seed spacing in the rows will, however, depend greatly upon the expected percentage germination of the particular seed batch. Depth of seed planting is usually equivalent to two to three times the seed diameter.

6.4.2 Propagation by division

Layering (trench layering)

The principles of layering (shown in diagrammatic form in Fig. 6.1) and the related technique of stooling (see below) are similar. Firstly, the severe winter pruning, which is an essential component of the technique, encourages the growth of vigorous 'juvenile' shoots from the basal stubs of shoots left after harvesting the layers. Such shoots are generally much easier to root than the more 'mature' shoots formed on plants only lightly pruned. Secondly, partial covering of the young shoots as they develop with soil, sawdust or some other medium (referred to as 'earthing up') creates the ideal edaphic environment for the initiation and development of roots on the developing shoots. Exclusion of light (etiolation) is the critical prerequisite which, if accompanied by adequate

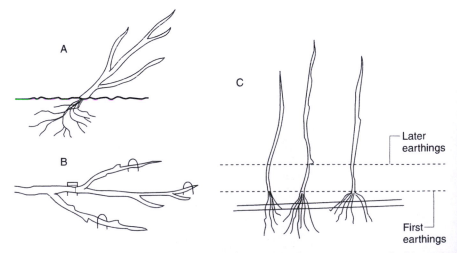

Fig. 6.1. The technique of trench layering showing (a) planting liner at 45° angle, (b) pegging the liner down horizontally into the trench and (c) earthing up the shoots arising from lateral buds on the original liner with earth or other suitable medium.

available soil moisture and air, favours the initiation of rooting. Etiolation is known to increase starch concentrations in shoots while at the same time decreasing phloem and sclerenchyma production. These effects, and possibly others not yet understood, in some way predispose the etiolated shoots to form roots.

Mazzard cherry rootstocks generally root better from layered than from stooled shoots; this is probably because layered shoots are etiolated by covering them with soil before they emerge while stooled shoots are only blanched, i.e. covered with soil soon after emergence. The more uniform covering of the shoots achieved using layering, which is generally not mechanized, also improves rooting success.

Layer beds of sweet cherry rootstocks, such as F.12/1, are established by planting rootstocks 45–75 cm apart in rows 1.5 m apart. The stocks are planted at an angle of 45° to the vertical, all pointing down the row. The soil between adjacent layer beds is maintained in a good tilth during the first season following planting and all weed growth suppressed. In the subsequent winter, the shoots produced during the summer, i.e. the parent layers, are pegged down to the horizontal along the row. Layer beds need regular checks following frosty weather, which may cause some of the securing pegs to ease out of the ground. All weak lateral growths on the stocks should be cut back to within 1 cm of the main stem and strong laterals lightly tipped.

The layer is pegged down into a 5-cm-deep narrow trench made along the length of the layer bed; this trench should be wide enough (usually approximately 20 cm) to accommodate the trimmed layer. When pegging down, it is essential to ensure that the layer is held horizontal and that its tip remains very slightly higher than its base. This can be facilitated by removing a little of the soil from around the base of the layer just above the roots. Traditionally, the pegs are made from sharpened pieces of wood, each approximately 30 cm long, into which a 5-cm wire nail is driven about 3 cm from the upper end. Care should be taken to make pegs only from those types of timber which will not root when inserted into the ground. Sweet chestnut (*Castanea sativa*) is the preferred choice in the UK. The thinner lateral shoots are usually pegged down using 50-cm-long pieces of 10-gauge wire bent to form U-shaped pieces.

The entire layer is then covered to a depth of about 3 cm with fine soil or some other suitable medium; this must be done before the buds begin to swell. As the young curled shoot tips push through the soil covering in the early spring, it is essential to cover them again with another 3 cm of soil before their leaves begin to expand or their tips uncurl. Further soil must not be added, however, to any sections of the layer bed where shoots have not yet emerged. If etiolation of the shoot bases is to be successful, this operation, known as earthing up, must be repeated several times in the first few weeks of the growing season. Failure to do this will invariably result in poor rooting of layers.

After the initial earthing up and when the young shoots have grown to about 10 cm above the soil surface, the layer beds are managed much the same as stool beds. Earth from between the beds or artificial medium is pushed up and around the shoots, so as to cover half of the length of each exposed shoot, taking care that no young shoots are completely covered. Further earthing up is usually carried out when the shoots are approximately 20 cm and 45 cm high; the final earthing up is usually completed by midsummer. Soil or other medium used for earthing up must be moist, fine and friable and contain nothing phytotoxic to the cherry (e.g. certain herbicides). Earthing up may be done by hand, although special ridging plough equipment is now available for the task.

In the winter following earthing up, the covering ridge of soil is forked away from the base of the layers. The rooted layers are then removed using sharp secateurs (pruning shears), but the unrooted shoots left intact. These will later be pegged down to form the basis for the next season's crop of layers. It is essential to leave one shoot intact for each 30 cm linear length of layer bed. As the layer bed ages, it will be necessary to cut or break out old unproductive pegged-down shoots to make space for younger more productive shoots.

Infection of trench layer beds with crown gall can be a serious problem. Production of F.12/1 layers may be reduced by almost half in infected parts of the bed and the initial growth of maiden trees on galled rootstocks may also be reduced. Most galls develop in cracks formed where the shoot is bent over or, alternatively, near to the previous season's cut. Surprisingly, although shoot numbers produced by the bed are greatly reduced by infection, shoot height on the layer bed is not. Some rootstocks, such as Colt, are particularly sensitive to crown gall and, where the disease is a problem, techniques of propagation based upon aerial cuttings should be used. Chemical treatments against crown gall have proved fairly ineffective to date and cannot be recommended.

The fungus *Thielaviopsis basicola*, thought to be the causal organism of cherry replant disease in Britain, may also reduce growth and productivity of Mazzard layer beds. The regular seasonal use of the same soil for earthing up may favour the development of replant effects attributable to this fungus and with it the production of many short, unrooted shoots. Drenching the beds with a systemic fungicide such as benomyl in May and July can alleviate this problem.

Stooling

Mound or stool layering, a technique widely used for the propagation of apple rootstocks, is also occasionally used for cherry rootstock propagation. Most comparisons of the two techniques have indicated, however, that cherries root better from trench layering than from stooling. The principles and many of the practices of the two techniques are similar, the object being to etiolate or blanch shoot bases early in the season and create a favourable edaphic microclimate at their bases to encourage rooting.

Rootstocks are usually established 30 cm apart in rows wide enough to allow tractors to pass between. Immediately after planting, the stocks are cut back to 45 cm in height. The shoots are allowed to develop unchecked during the first season to enable the plant to establish well and achieve a favourable ratio of shoot to root size. After one season's growth, the shoots are cut back hard to within 3 cm of the ground level. As the new shoots grow in the subsequent spring, they are earthed up two or three times, each time to half their height. These operations, which are nowadays usually mechanized, are carried out when shoots are 7.5 cm, 25 cm and 45 cm high. For best results it may be necessary to supplement the mechanized earthing up by working soil or other media between the clusters of stool shoots by hand. As with trench layering, part of the success of stooling is in maintaining the soil in excellent tilth to facilitate the earthing-up operation.

After leaf fall the mounds or ridges are ploughed or forked away from the stooled shoots and the rooted and unrooted shoots removed with sharp secateurs (pruning shears) as close to their bases as possible. The stool bed is then left exposed until new shoots begin to sprout the following spring, when the cycle of operations begins again. Well-managed stool beds should remain productive for 10–15 years.

Marcotting

In some climatic or edaphic conditions, layers of sweet cherry rootstocks may be difficult to root; this can be a particular problem in hot dry soils. Work in India has shown that layering success may be increased in these conditions by supplementary treatments to aid rooting. Ring-barking (girdling) the emerging layer shoots in May and at the same time treating the girdled stem with indolyl-3-butyric acid (IBA) (2500–10,000 mg l^{-1}) prior to earthing up increased rooting considerably.

A similar effect of marcotting was achieved in German work, where nearly 250 different hybrid rootstock clones were successfully rooted on stool beds by wrapping sharp, strong wire around the bases of the young shoots before earthing up. This partial constriction of the stem base improved rooting significantly.

6.4.3 Propagation by cuttings

Cutting techniques all involve removal of a vegetative propagule from the mother plant prior to rooting. The propagule may be a young actively growing shoot or its tip, usually referred to as a summer or softwood cutting or a micropropagule. Alternatively, it may be a shoot which has stopped elongating and begun to lignify slightly (known as a semihardwood or greenwood cutting) or a longer shoot harvested after leaf fall (a hardwood or winter cutting). Some cherry clones may also be regenerated from segments of shoots comprising just two internodes, i.e. a stem node and its leaf and axillary bud; these are leaf bud cuttings. Finally, a few rootstock clones are propagated from small root pieces, which are known as root cuttings.

All cuttings are encouraged to root and establish by placing them in favourable environments; often root initiation is further stimulated by localized application of rooting hormones. The principles involved in micropropagation are very similar to those for softwood cutting techniques, differing only in the size of the propagules used and the propagation environment used to root them (see below).

The cutting methods have several advantages compared with division techniques. The principal one is that the propagator has much more flexibility in organizing nursery management operations; the management of stock hedges for cuttings is simple and relatively cheap irrespective of whether cuttings are harvested or not. In contrast, layer and stool beds are much more expensive to manage and the tasks involved cannot be neglected in seasons when demand for the product is low. A second advantage is that propagation by cuttings can help prevent transmission of soil-borne diseases, such as crown gall, from the site of the stock plant to the liner nursery.

Cutting methods do, however, have one major disadvantage. Whereas the rooted layer is planted directly into the liner nursery and may be budded the following summer, rooted cuttings usually require one, or sometimes two, extra seasons of growth before reaching a suitable size for budding. Cutting techniques do, therefore, limit the nurseryman's ability to respond rapidly to changes in demand for specific rootstocks.

Summer (softwood) cuttings

Problems of plant establishment are often the major constraint encountered when raising cherry rootstocks from softwood cuttings. Many species and hybrid clones root slowly and then make very little growth in the season of rooting. Such plants are often particularly sensitive to overwatering in the first winter after rooting and many fail to survive. Research has shown that when propagating these recalcitrant *Prunus* clones it is better if the cuttings are rooted into deep boxes or directly into pots and then carefully weaned under fogging or misting systems; transplanting should not be undertaken until the subsequent spring. Foliar feeding can help speed the growth of cuttings, once rooted.

Prunus cuttings transpire more rapidly than similar cuttings of most other temperate fruit species and often produce mucilaginous substances at their cut surfaces, which block xylem vessels and exacerbate water stress. For this reason, if desiccation is to be prevented, particular care must be taken to maintain high humidities around the cutting during the rooting process. It is likely that the more recently developed 'fogging' systems may prove more successful than the conventional misting systems for rooting recalcitrant cherry rootstock clones. Important considerations in the softwood propagation environment are represented in Fig. 6.2.

Rooting of clones which traditionally prove difficult to propagate from softwood techniques may be improved if basal cuttings are selected from severely pruned stock hedges. It is possible that further improvement in rooting may be achieved if the bases of the cuttings are blanched by wrapping them with black polythene tape. Such techniques have proved beneficial when propagating *Malus* and several other woody ornamental species. The black tape must be applied close to the apex of the young shoot as it begins its extension in the spring; this taped section later becomes the base of the cutting. An alternative strategy, also successful with cuttings of *Malus*, is to erect black polythene tunnels over stock hedges in the spring to etiolate, temporarily, the developing shoots which will later be used as cuttings (Fig. 6.3). Research indicates that total exclusion of light is not desirable; approximately 1% light gives the best results. The structures are removed when the cuttings reach the desired length and the cuttings are taken immediately, as the beneficial effects on rooting are often rapidly lost once the tissues are exposed again to light. Although the benefits of etiolation or blanching are well documented for propagation of apple rootstocks and some ornamental shrub species (Howard *et al.*, 1988), little work has yet been done on stone fruit species.

Softwood cuttings of most *Prunus* species benefit from treatment of their

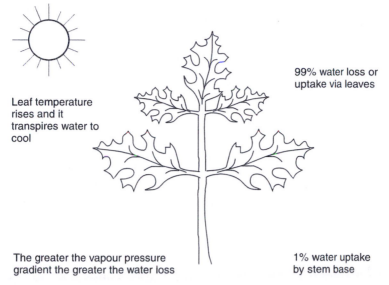

99% water loss or
uptake via leaves

Leaf temperature
rises and it
transpires water to
cool

The greater the vapour pressure
gradient the greater the water loss

1% water uptake
by stem base

Fig. 6.2. Important considerations when designing suitable environments for softwood cutting propagation.

Fig. 6.3. Tunnels of black polythene temporarily erected over hedges of *Malus* rootstocks in order to etiolate the cuttings and aid their propagation (photograph from B.H. Howard, HRI).

bases with auxins. IBA is the most popular treatment, although naphthalene acetic acid (NAA) is also occasionally recommended. Where hormone treatments are not used, it is essential to retain the apical bud of the cutting; removal of the apical buds of cuttings of *P. mahaleb* rootstocks greatly reduced rooting unless IBA dips were used.

Growth of rooted softwood cuttings is checked by transplanting, and the plants often form a rosette of closely spaced leaves and no extension growth. Shoot extension on rosetted plants of some rootstock clones can be stimulated by foliar sprays of gibberelic acid (GA$_3$) at 200 mg l^{-1}. If transplanting stress causes the cutting leaves to abscind and the plants to go into dormancy, then a chilling treatment (approximately 1500· h at temperatures less than 7°C but greater than 0°C) may be necessary before the plants can be forced back into active shoot growth.

One problem with propagating many of the weaker (dwarfing) cherry rootstocks by softwood cuttings is that they invariably require 2 years in the nursery before they grow to a size sufficient to bud. Several researchers have sought to shorten the length of time needed to raise a tree by carrying out some of the propagation procedures in parallel, rather than in series, as is usually practised. Most of these short cuts involve grafting the scion on to either non-rooted or recently rooted softwood cuttings of the rootstock and then rooting and/or growing on the rootstock plus scion (stion) under glass. The time taken to produce trees is reduced by up to 18 months using these techniques, but tree quality and uniformity are invariably poor.

The majority of species and hybrids used as rootstocks for the sweet or sour cherry can be propagated from softwood or summer cuttings. Only a few experimental rootstocks, such as clones of *P. mugus*, *P. fruticosa*, and 'genetic dwarf' types of *P. avium*, are particularly difficult to root and establish using this technique.

Mazzard (*P. avium*) rootstock clones do propagate from softwood cuttings, although considerable care and skill are needed if the technique is to prove successful. Mazzard clones have been successfully propagated by taking 10-cm-long cuttings of the current season's growth, treating them with a hormone dip and inserting them under mist. In one series of experiments, cuttings inserted in June or November rooted best and a medium of pure gravel was much better than peat and sand. Research in Poland has shown that the Mazzard rootstock F.12/1 may be propagated from softwood cuttings under mist. Cuttings 'quick-dipped' in 2500 mg l^{-1} IBA and given 21°C bottom heat rooted well (69%) and 80% of these subsequently established in the field. Recommendations concerning the optimum season for taking Mazzard cuttings differ and it must be concluded that it is the physiological condition of the cutting rather than a chronological time which determines success. F.12/1 softwood cuttings are extremely sensitive to rotting, particularly if taken early in their growth, when they are very soft; fungicide dips prior to insertion can help alleviate this problem.

Many sour cherry (*P. cerasus*) rootstock clones may also be rooted from softwood cuttings. The Stockton Morello clone is easily propagated from terminal cuttings of actively growing shoots inserted under mist. IBA dips are essential for maximum rooting success and best rooting occurs between April and June. The cultivar Vladimirskaya, which is used as a rootstock for other sour cherry cultivars in Russia, may be rooted from softwood cuttings if these are first dipped in 50% ethyl alcohol solutions of IBA (2–5 g l^{-1}) for 5 s before insertion under mist. Russian research demonstrated that treatment of the cuttings with indolyl acetic acid (IAA) or IBA improved the numbers of roots per cutting but not the rooting percentage. Rooting was best if conducted under plastic covers enclosing mist irrigation.

Clones of the Steppe cherry (*P. fruticosa*) and its hybrids with *P. avium* are often most difficult to root from softwood cuttings. The cuttings callus profusely but are very difficult and slow to root. Success was achieved, however, in Austria, where 82% rooting was achieved after treating cuttings of three clones with 0.8% IBA and inserting them into a medium containing perlite beneath a polythene tunnel. The optimum physiological condition is vital to success in rooting this species and it has been suggested that cuttings should only be taken when 15–25% of the young shoot radius comprises xylem; less than this and the cuttings rot, more than this and rooting is very poor.

Other research has shown that the clonal Mahaleb (*P. mahaleb*) rootstock SL.64 rooted best from softwood cuttings taken in May under mist and within plastic tunnels or glasshouses. IBA powder (0.2–2.0%) improved rooting greatly. Other research on the same clone has shown that 90% of SL.64 cuttings root at all timings between June and early September. Cuttings, which were 15 cm long, were dipped in hormone before insertion into a perlite medium under mist. Rooting was achieved after only 3–4 weeks. Clones of Mahaleb do, however, differ greatly in their rooting response to treatments with IBA, and cuttings which contain high concentrations of reducing sugars but lower concentrations of sucrose generally root best.

Softwood cuttings of the popular M×M (Mahaleb × Mazzard) series of hybrid rootstocks were reported to root best in Oregon if taken between July and mid-August. Cuttings 30–35 cm long were selected and collected from the stock plants early in the morning to minimize stress and desiccation. The cuttings were soaked in an insecticide and a fungicide, such as benomyl, for approximately 15 min and finally dipped in rooting hormone before insertion into pots of perlite under mist irrigation over bottom heat. Rooting took approximately 1 month, after which time the cuttings were regularly fed with balanced foliar feeds. Misting continued for 2 weeks after rooting but bottom heat was maintained until October, when the pots were moved outside under lath shade houses.

Many of the interspecific hybrids of *Prunus* bred and tested as sweet cherry rootstocks at Giessen in Germany have been successfully rooted from softwood cuttings inserted under outdoor mist beds. Experiments testing several rooting media showed the advantage of inclusion of some partially sterilized potting compost beneath a shallow layer of a 2:1 peat:sand mixture. The beds were irrigated with mist nozzles and enclosed within a low polythene tunnel, which could be shaded when necessary.

Semihardwood cuttings

Several cherry rootstocks may be propagated from semihardwood (greenwood) cuttings taken from current season's shoots before leaf fall, but after termination of active extension growth. Research in the south of France suggests that the base of the cuttings should be first dipped in 1000 mg l^{-1} IBA before insertion under mist irrigation. The cuttings are usually inserted outside but covered with a light shelter (Bernhard and Claverie, 1986). Once rooting is confirmed, usually after approximately 3 weeks, the misting is stopped and the shelter removed. Several Mahaleb clones, including SL.64, rooted well (95%) using this technique, as did a sour cherry clone (70%). However, for reasons not explained, Colt rooted very poorly (10%) using this technique.

Research in Italy has demonstrated that rooting of semihardwood cuttings of Mahaleb rootstocks was improved by foliar sprays of boron, manganese and iron. The sprays, which were shown to reduce defoliation and to act synergistically with IBA dips (1000 mg l^{-1}), had only minimal effects if applied without the hormone. In contrast, sprays of nitrogen, phosphorus and potassium reduced rooting.

Winter (hardwood) cuttings

Most Mazzard and Mahaleb rootstocks root very poorly from hardwood cuttings and the technique is rarely used for propagating these two species. Hardwood cuttings are, however, an extremely effective way of rooting the hybrid rootstock Colt and much of the production of this rootstock in northern Europe is by this method. Colt roots so easily and establishes so quickly using this technique that often the cuttings are suitable for budding only 8–10 months after collection from the hedges. Occasionally, sour cherry rootstocks have been propagated from hardwood cuttings, although scions such as Montmorency have responded poorly to the treatment. Attempts to propagate P. fruticosa clones from hardwood cuttings have proved unsuccessful.

Hardwood cuttings propagate best if grown on stock hedges raised from rootstocks planted 30–100 cm apart in the row and with 2–3 m between rows. It is vital that the hedges are severely pruned each winter if the best cuttings are to be produced. Each current season's shoot should be cut back to within two or three buds of its base during the winter.

Usually, rooting is best if hardwood cuttings include the basal nodes of the current season's growth. Shoots emerging from the main stems of hard-pruned hedges or even 'water shoots' from the trunks of trees always root better than those collected from more distal positions on the branch framework. This can sometimes be explained by the adventitious nature of the buds which are encouraged to develop and emerge following severe pruning; shoots developing from deeply dormant buds, root suckers and sphaeroblasts root particularly well.

There is no optimal length of hardwood cuttings, although they should be of sufficient size to support and sustain first the initiation and development of roots and later bud burst and the development of leaves. Small cuttings may have insufficient reserves of carbohydrates to support these processes. Research in Britain has shown that best results are achieved with cuttings of the current season's growth which are 60 cm in length. The cuttings should also be as straight as possible to facilitate the insertion of scion buds at a later stage. Wounding or splitting the base of the cutting up to 2 cm deep may also increase rooting, possibly by exposing to auxin uptake more of the cambium and phloem from which new adventitious roots emerge.

Some form of auxin treatment is also usually beneficial as a supplement to the endogenous auxins present within the cutting (Howard, 1986). IBA is usually the preferred auxin; IAA which is the endogenous auxin is very unstable and therefore not suitable for use as a cutting dip. NAA is occasionally used in powder mixtures with IBA, but is easily translocated within the cutting and may inhibit bud growth.

IBA is commonly used dissolved in ethanol or acetone, which is then diluted 50:50 with distilled water and applied as a 'quick dip' treatment to the base of the cutting. The solution containing the hormone is taken up by the exposed vascular elements at the cutting base. The auxin then moves laterally from the xylem into the secondary phloem and cambium, where the roots are initiated. Uptake through the undamaged epidermis of the cutting is believed to be of only minor importance. The amount of IBA taken up by the cutting may be influenced by the depth of dipping, the length of time it is dipped and the extent of cutting desiccation. Research in Britain has shown that 24 h after collection cherry cuttings should be dipped for 5 s to a depth of approximately 1 cm. Alternatives are to use the potassium salt of IBA which may be dissolved in water and used similarly, or IBA in powder formulations. Prewetting the cutting bases with ethanol or acetone increases the amount of the IBA powder formulation which adheres to the cutting and aids IBA uptake and subsequent rooting. Powder formulations are taken up more by the epidermis of the cutting than are liquid formulations.

Species differ in their requirements for IBA treatment. Colt, which, if grown on hard-pruned hedges in a favourable humid nursery environment, forms cuttings with preformed root initials, may require no IBA treatment and only minimal basal heat to survive and establish. However, Colt cuttings lacking root initials or cuttings of other rootstock clones benefit from powder or liquid treatments with IBA. A quick dip for 5 s in a 50:50 acetone:aqueous solution of 1000 mg l^{-1} IBA is ideal for Colt cuttings, whereas higher concentrations (2500 mg l^{-1}) may give better results with other clones.

Traditionally, hardwood cuttings were planted directly into rows in the nursery, although in the coldest areas they were often protected from the severest weather by surrounding them with straw bales and a thatched roof cover. However, for optimal root initiation, the best environment is a base temperature of 20–30°C in a well-drained and aerated compost. This is best provided by specially constructed heated bins or beds (Howard, 1971; Fig. 6.4). Usually, hardwood cuttings should remain in the heated bins for only 2–3 weeks before being moved out to the nursery or to specially constructed raised beds outside.

The compost used for propagation in heated bins or beds must be sufficiently well aerated to support respiration at relatively high temperatures (Harrison-Murray and McNeil, 1984). The commonest cause of failure with hardwood cuttings is rotting of the cutting base, and insufficient air spaces in the rooting medium exacerbate this problem. Composts containing a high proportion of sphagnum moss peat are particularly bad in this respect, and granulated pine bark is preferable. A 7.5-cm layer of pine bark spread over heating cables, which are themselves laid over a 20-cm layer of fine sand, has given good results in Britain. The bundles of cuttings are twisted into the bark so that their bases are all 3 cm above the heating cables. The aerial environment must be cool but not frosty and with a humidity greater than 90%. Wrapping or partially wrapping

Fig. 6.4. Propagation from hardwood (winter) cuttings. (a) Bundles of cuttings inserted in a well-drained medium within a propagation bin with basal heat; the wrapping prevents undesirable desiccation. (b) Cuttings lifted from the heated bin which ideally should have their bases callused and roots initiated but sometimes show early root development.

the cutting bundles in polythene can help maintain humidity where the ambient air is too dry.

In favourable conditions, callus initially forms on the cut surfaces of the cutting base, and roots subsequently push through this as cambial outgrowths. Where heated bins are used to promote rooting, it is preferable to merely callus and initiate roots before moving them outside for the roots to develop. Bottom heat of too high a temperature or applied for too long may result in rapid root development within the heated bin but may also deplete the cuttings' reserves and reduce establishment success.

Hardwood cuttings are usually taken just after leaf fall in the autumn or in the late winter or early spring. Differences in rooting success of hardwood cuttings taken at different times during the dormant season are thought to be attributable to changes in the activity of phenolic rooting cofactors which synergize with auxins to induce rooting. Sensitivity to the water content in the rooting medium also changes between autumn, winter and spring, and this may contribute to observed seasonal differences in rooting and establishment success. Several difficult-to-root *Prunus* species (e.g. *Prunus mugus*) have rooted much better if taken as early as possible in the autumn, even if leaf fall is accelerated by hand removal. In contrast, cuttings of the easy-to-root Colt may be taken at any time during the winter with success, provided that they exhibit root initials.

Leaf-bud cuttings

Cuttings comprising one or two stem nodes and the surrounding internodes have been used for cherry propagation only occasionally and offer no clear advantages over other techniques. Work conducted in Norway has shown, however, that both sweet and sour cherry clones may be propagated using this technique. The hybrid rootstock Cob and several *P. fruticosa* × *P. avium* hybrid rootstocks have also been propagated this way.

Root cuttings

The Mazzard rootstock F.12/1 is largely propagated from root cuttings by nurseries in Australia. Roots of suitable length and basal calibre are harvested from rooted plants and planted vertically with their tops level with the soil surface. As shoots begin to emerge they are earthed up to aid rooting. Where several shoots emerge from the same root, they are singled to provide one strong shoot. At the end of the year, they are lifted and the rooted stocks detached from the old root pieces; some of the latter may be replanted and used for generating more shoots in a second season. *Prunus fruticosa* rootstocks, which sucker profusely in the orchard, but which are extremely difficult to propagate from either softwood or hardwood cuttings, may also be propagated quite easily using root cuttings.

Root cuttings are best collected from 1- or 2-year-old plants by lifting in late autumn and keeping the plants in humidified cold storage until the roots are removed. Roots 1–2 cm in diameter are selected and cut into 10-cm lengths. Slanting cuts are made at the distal and level cuts at the proximal ends to help cutting orientation at insertion. The root cuttings may be stored in moist sawdust or sand until soil conditions in the nursery are suitable for planting in early spring. The cuttings are planted vertically 5–7.5 cm apart in shallow trenches.

6.4.4 Micropropagation

In micropropagation, axillary buds in the meristems of shoot tips are encouraged to produce small lateral shoots under sterile laboratory conditions. These shoots are then separated from the original shoot and either rooted or used to produce more shoots. Micropropagation has become a fashionable area for research in recent years, with numerous publications describing the successful *in vitro* propagation of many cultivars of temperate fruit scions and rootstocks, including cherries. It is perhaps too easy to assume that because the technique is at the forefront of modern technology it must be superior to the other techniques of propagation that preceded it. This is by no means true. Indeed, in many instances micropropagation may lead to problems previously not experienced with other more conventional methods of propagation. It is important, therefore, to consider the benefits and disadvantages of propagation by micropropagation before making the decision to invest in this technique.

The advantages of micropropagation in comparison with most other techniques may be summarized as follows.

1. It facilitates the rapid multiplication of materials which are in very short supply.
2. It facilitates the multiplication of many clones which are extremely difficult to propagate using other methods.
3. It enables propagation to continue all year round, irrespective of weather conditions.
4. It raises material in an environment which can more easily be kept free of

pests and diseases than can that of the conventional techniques.

5. It often induces epigenetic changes in the material, which may subsequently improve its conventional propagation.

In contrast, micropropagation also has a number of disadvantages.

1. The equipment needed to establish a micropropagation laboratory is much more expensive than that needed for nursery or simple glasshouse production.

2. The facilities needed to wean plants successfully from micropropagation are sophisticated and expensive. Micropropagated plants are generally much more difficult to wean and establish than conventionally propagated plants.

3. Plants which are 'rejuvenated' during micropropagation may sucker or burr-knot profusely (rootstocks) or be delayed in cropping (scions).

4. Regeneration from adventitious rather than axillary buds, which sometimes occurs during micropropagation, may lead to undesirable somatic mutation.

The propagator should carefully weigh the above advantages and disadvantages before choosing to micropropagate rootstocks or scions. The technique is probably best used for bulking up materials which are in very short supply or which are slow to multiply using conventional techniques, or for propagating subjects which cannot be propagated using other techniques. One distinct advantage of micropropagated plants is that they are usually more acceptable to plant health authorities operating at the borders of countries, making them easier to export than plants raised by more conventional methods. For a full description of the techniques of micropropagation, Ahuja (1993) should be consulted.

Establishment of micropropagation cultures

Micropropagation cultures are initiated from explants, the distal tips of newly and preferably rapidly growing shoots. Explants are obtained from shoot tips harvested from healthy, true-to-type stock plants; the best time for initiating cultures is in early spring. Containerized stock plants may be forced into growth early in the year, if first chilled in a cold store. Explants from glasshouse-grown stock plants often show less bacterial contamination than those collected from plants growing outside. Care should be taken to avoid water touching the leaves and shoots of stock plants during watering, as this can increase the populations of contaminating microflora on the explants.

Recent research on *Malus* (Webster and Jones, 1989) has indicated that tips from shoots growing at different positions, or at different orientations, on the stock plant may ultimately give rise to cultures which differ in their productivity and ease of rooting. The speed of shoot proliferation and the success of rooting, both in culture and subsequently when weaned, may differ between cultures derived from shoots originating from different locations on the same stock plant. Although these effects, often referred to as topophytic effects, have not yet been noted in micropropagation of *Prunus* species, they may yet prove to be a contributory reason for the often recorded variability in culture productivity.

Shoot tips 3 cm long are first cut from stock plant shoots 5–10 cm long. As many as possible of the leaves and leaf primordia are removed and the tips then washed in running tap water for approximately 50 min before surface-sterilizing by immersion in a solution of sodium hypochlorite for 25 min; ethanol is occasionally substituted for sodium hypochlorite. Finally, the tips are washed a further four times in sterilized distilled water. The next step in the procedure is to dissect the most distal 3–5 mm of shoot tip, including the apical meristem. This is then treated with 10% calcium hypochlorite for a period of 10–30 min before again rinsing in distilled water. The dissected tip is then placed within some sterilized vessel, such as a 15 × 2.5-cm test-tube, on a suitable culture medium.

Dissection of the sterilized tips and all the subsequent subculturing operations are conducted within a stream of filtered air beneath a laminar-flow cabinet to avoid contamination with fungal or bacterial spores. Instruments used for dissection and subsequent subculture transfers and the necks of all vessels should be sterilized by flaming.

Subculturing and proliferating cultures

Once cultures have established, after approximately 8 weeks' growth, they are usually transferred to a larger vessel, such as a 100-ml conical flask, containing suitable proliferation (shoot multiplication) medium. This encourages axillary buds to develop into small shoots, which, when large enough, are cut off, teased apart and used to start more cultures. This division of cultures and the starting of fresh cultures from the dissected shoots is called subculturing. Portions of the parent culture of two to three buds each are split off every 2–4 weeks and transferred into fresh flasks of the proliferation medium. Proliferation rates are generally rather poor at the first few subcultures but after 4–5 months the proliferation rates usually increase considerably. All tubes are closed with metal caps and all flasks with double layers of aluminium foil.

Rooting and weaning the micropropagules

When the time comes for rooting micropropagules, the propagator has two choices: rooting *in vitro* or alternatively rooting under fogging or misting installations, often referred to as direct sticking. Rooting *in vitro* involves excising shoots from proliferating cultures and then transferring them for a brief period into vessels containing a different medium; one which favours root induction (see below). They remain on this medium for only a short time (4–7 days) before they are moved on to yet another medium, this time one suited to root development. Roots usually take about 3 weeks to form. Plants raised by this method must then be weaned on to peat, sand or perlite media and acclimatized to a glasshouse environment.

The alternative procedure is to transfer the micropropagules, after root initiation, directly into a conventional medium (peat/sand/perlite) and allow the roots to develop and the acclimatization to occur simultaneously under mist or fogging environments. Occasionally, the process has been shortened even more by inserting shoots removed from the proliferating cultures directly under mist or fog after first treating them with a rooting hormone. This considerably shortens and cheapens the procedures, by facilitating acclimatization (weaning) simultaneously with the initiation and development of roots. The processes of acclimatization of micropropagules is covered more fully by Hutchinson and Zimmerman (1987).

Media for micropropagation

Both liquid and solid media can be used in the micropropagation process. Solid types of media, based on agar, are the most popular, and inorganic salts (see Table 6.1) as well as growth hormones, such as cytokinins, auxins and gibberellins, and other substances, such as vitamins, are added to this. Media for encouraging root initiation differ from proliferating media in that auxin concentrations are increased and cytokinins omitted (see Table 6.1).

All media used are usually sterilized using an autoclave but a process of fine filtration may be used if the media contain growth regulators unstable in heat. A special medium containing bacterial peptone is occasionally used to check for contamination of cultures with microorganisms.

Table 6.1. Inorganic salt content in two popular media.

Salts	Medium (g l^{-1})	
	Murahige and Skoog	Quoirin and Lepoivre
KNO_3	1.9	1.8
$MgSO_4\ 7H_2O$	0.37	3.6
KH_2PO_4	0.17	2.7
NH_4NO_3	1.65	0.4
$Ca(NO_3)_2\ 4H_2O$	None	2.0
$CaCl_2\ 2H_2O$	4.4	None

Problems experienced in micropropagation

One problem commonly experienced in micropropagation is the vitrification of cultures, often evident where high cytokinin (e.g. benzyladenine (BA)) concentrations are incorporated into the medium. The reasons for vitrification are ably discussed by Kevers *et al.* (1984).

Perhaps the greatest problem concerning micropropagation of cherries, however, concerns the acclimatization and weaning of the young rooted micropropagules once they are transferred from the culture tubes or flasks and transplanted into conventional media. Cherries transpire rapidly and, unless high humidities are maintained during this critical weaning stage, severe desiccation may occur, leading to plant mortality. The reasons for a micropropagule's sensitivity to desiccation are not fully understood. What is known is that *in vitro* the development of epicuticular waxes may be reduced, so increasing the possibilities for cuticular transpiration at low humidities. Also stomatal function is impaired, the stomata remaining open irrespective of plant–water relations. This leads to even greater transpiration and potential stress.

Occasionally, sprays of antitranspirants have been tried as alternatives to the maintenance of high humidities under mist or fogging systems. Unfortunately, many of the products tried have proved phytotoxic to the young, sensitive leaves, although sprays of hydrocarbon waxes have helped weaning of apple micropropagules. Other studies have indicated that roots formed in culture may not be functional once transferred from micropropagation to conventional media (potting composts), where they are replaced by more functional roots.

Inconsistency in the rate of axillary branching is recognized as a major constraint on lowering the costs of *Prunus* micropropagation. Researchers have shown that these rates vary greatly throughout the season, even though the microclimatic conditions in which the cultures are grown remain constant. Despite maintaining cultures of *Prunus* for 16 h each day in cool white fluorescent light, significant differences in the numbers of shoots formed by proliferation were noted at different times of the year. Most shoots were formed in either May and June or September. For rooting, a 1-week dark period followed by 3 weeks' light has proved of benefit with some clones.

The simplest method of increasing axillary shoot proliferation is to increase cytokinin concentrations in the medium. Unfortunately, this has the negative effect of increasing vitrification of the cultures, promoting callus formation and possible mutation. Research has shown that, depending upon the *Prunus* species under test, it is likely that inclusion in the medium of methionine and vitamin D_2 (at rates of between 50 and 100 mg l^{-1}) can increase the rate of axillary branching in cultures. Also, rates of vitrification may be reduced by inclusion of galactose in the subculture medium, if and when vitrification becomes a problem.

Micropropagation of cherry clones

Workers in Belgium micropropagating a range of *Prunus* species and hybrids used a medium for proliferation containing Lepoivre mineral nutrients (Quoirin and Lepoivre, 1977), thiamine (0.4 g l^{-1}), *meso*-inositol (100 mg l^{-1}), gibberellic acid (0.1 mg l^{-1}) and Merk agar (5 g l^{-1}). Where long-term micropropagation was contemplated, the Belgian researchers found that the above medium should be supplemented with L-methionine (100 mg l^{-1}) and L-tyrosine (100 mg l^{-1}) and the gibberellic acid (GA_3) replaced by an ammoniacal solution of IBA (0.1 mg l^{-1}). Also, preautoclaving (at 120°C) the sucrose (3%) in the presence of activated charcoal (150 mg) and adjusting the pH to 5.5 before adding the agar gave the best results. The micropropagation techniques used in Belgium (Druart, 1985) have proved extremely successful for propagating many cherry rootstocks and scion varieties.

German research demonstrated that the Mazzard rootstock F.12/1 and the sour cherry rootstock Weiroot 10 were easily raised *in vitro*, starting from explants with two to ten primordia. F.12/1 explants derived from root suckers had higher multiplication rates in the first 6 months *in vitro* than explants from the outer tree canopy. Interestingly, for the sour cherry the situation was different, with cultures establishing much more successfully from dormant buds. Also, F.12/1 explants collected from the spring flush of growth had a higher multiplication rate during the first 8 months of culture than those taken from dormant buds. Nevertheless, between 80 and 90% of the explants established and grew, irrespective of the season collected. Starting with explants 0.2–2.0 mm long, shoot proliferation rates usually increased after the third subculture and then remained at this higher level. Rates of 0.5–1.0 BA in Murashige and Skoog medium were recommended for long-term subculturing.

F.12/1 micropropagated at East Malling (Jones and Hopgood, 1979) multiplied four- to fivefold every 3 weeks and, when transferred to a rooting medium, 90% rooting was achieved within 4 weeks over a 5-month test period. Approximately 80% of the rooted micropropagules established successfully, with no rosetting. Addition of phloroglucinol to the medium gave no benefits. The Mazzard rootstock Charger also micropropagates quite easily, using media similar to those recommended for F.12/1. However, 'genetic dwarf' selections of *P. avium* (dwarf seedling selections with curled atypical leaves and stunted growth) proved very difficult to micropropagate in earlier trials at East Malling, producing very few shoots, all of which were extremely difficult to root and establish.

Research in Germany showed that the Weiroot clones of *P. cerasus* rootstocks were best cultured on a modified Murashige and Skoog medium with half strength macronutrients and BA concentrations between 0.1 and 1.0 mg l^{-1}. The survival of stored cultures of *P. cerasus* rootstocks has been shown to depend upon interactions between storage temperature, light and the age of the subculture. For short-term storage (up to 170 days), temperatures of $+4°C$ and 16-h photoperiods were adequate. For longer storage (up to 300 days), complete darkness and cooler temperatures of -3 or $-4°C$ were optimal.

6.5 Scion Propagation

Most cherry trees are formed by budding or grafting the desired scion cultivar on to a seedling or clonally raised rootstock. The tree thus formed is sometimes referred to as the stion. Occasionally a third component, an interstock or interstem, is grafted between the rootstock and scion.

For success with budding or grafting, several principles must be understood.

1. The scion and rootstock must be graft-compatible. The combination must produce trees of the desired vigour which have strong graft unions and an economic longevity.
2. Cambium tissues of the scion and rootstock (and interstock) must be brought into close contact at a time of active cambial growth.
3. The young unions must initially be held together tightly and protected against desiccation and against attacks by pests and diseases.

In the northern hemisphere cherry rootstocks planted in the nursery in early spring are either budded in July/August or grafted in March/April of the following year. In temperate areas the buds do not begin to grow out until the spring after budding and trees are usually lifted as 'maidens' in the subsequent winter.

Occasionally nurserymen and/or growers have tried to shorten the time taken to raise trees. One method is to bud rootstock shoots before their removal from the layer bed. This is not to be recommended, as the careless use of buds infected with virus may result in the whole layer bed becoming infected. Also, there is no guarantee that the shoot budded will prove to be one of the rooted ones on the layer bed.

Another technique, which is sometimes recommended, is to transfer the trees to the orchard in the first winter after budding, i.e. when the bud is still dormant ('sleepy'). This method may have marginal benefits in speeding the establishment of the orchard, but should not be used with dwarfing or weak-growing rootstocks.

A third technique used to speed up tree production is to chip-bud the rootstock early in the year (June), using budwood cold-stored from the previous season. Scion growth commences in the same year as budding using this technique but, unfortunately, maiden tree quality is much poorer with this technique, unless the trees are left for a further year (i.e. the standard length of time) in the nursery.

6.5.1 Budding

Budding involves the insertion of a vegetative (not floral) scion bud into the stem of the rootstock. It is generally preferred to grafting as it is easier and quicker to perform and stock plants yield many more buds than grafts, so reducing the number of stock plants needed to raise the same number of young trees. For successful budding, the rootstock and scion must be graft-compatible and be in the correct physiological condition; the bark should lift easily from the wood and the buds should be 'ripe'. Rootstock liners of suitable size (preferably 7.0–9.0-mm stem calibre at the height of budding) are trimmed of all side branches and cut back to approximately 40 cm in height before planting in the early spring, spaced 30–45 cm apart in rows at least 1 m apart. Rootstocks may be planted by hand or by machine and should be planted 15 cm deep.

Rootstocks should be planted in nurseries of virgin or fumigated soil and balanced nutrition ensured by a base dressing of fertilizer. The fertilizer requirement should be based upon the results of soil nutrient analyses. Adequate irrigation should be supplied during spring and summer to ensure vigorous growth of the rootstocks. The part of the rootstock stem to be used for budding should be kept free of sprouts by rubbing these out during the spring and summer.

The optimum time for budding in the northern hemisphere is between July and September, when cambial growth is active and sap flows are favourable. By autumn the bud should have formed a good union with the rootstock. The shoots on the rootstock above the inserted scion bud are cut back in the winter following budding. If budding is carried out earlier in the season, using budwood collected in the previous season and cold stored, the scion bud will grow out into a short 'whip' in the same season as budding. Similar effects are noted when trees are budded at the conventional time but in Mediterranean or subtropical areas.

Cherries are usually budded 1–2 weeks before apples and pears and the rootstocks differ in when they are best suited to budding. Mazzard stocks are usually in a suitable condition from mid-July to mid-August in temperate zones of the northern hemisphere; Mahalebs are usually better budded up to a month later than this. Research in Germany showed that budding sweet cherries on F.12/1 during July was not always very satisfactory, owing to excessive callus production and thickening of the rootstock. Similarly, budding too late in early September resulted in insufficient callus production and the formation of a necrotic layer at the union during the ensuing winter. This response was worse in cold winters and frequently led to bud failure.

Budwood, which is harvested from the current season's extension shoots, should only be collected from healthy, true-to-type stock plants (mother trees). All leaf laminae and stipules are removed immediately after collection to prevent undue desiccation. The budsticks, as they are referred to, should be kept moist by wrapping in damp cloth, or alternatively the basal ends should be put into containers of water. This will facilitate storage for several days prior to budding. The unripe buds on the tip and the small buds at the base of the budstick are discarded and not used for budding.

The bud is inserted into the rootstock stem at least 10 cm above ground level; clonal stocks are usually budded higher than seedling stocks. Height of budding is largely determined by the calibre of the rootstock stem; budding too high into thin, spindly stems is unlikely to be successful. Higher budding is occasionally practised if rootstocks of sufficient age and/or size are available. High budding, 1.5–2.0 m above ground level, on to rootstocks resistant to bacterial canker was a common practice some years ago but is now rarely practised. Budding at approximately 30–50 cm above ground level can be of benefit in preventing the growth of scion feathers which are too low to be of value in branch training. Budding at this height may also favour deep planting in the orchard; this can be of particular value for trees on poorly anchored rootstocks, such as clones and hybrids of *Prunus incisa* or *P. cerasus*. There is, however, conflicting evidence as to whether high-working sweet cherries on dwarfing rootstocks further increases the dwarfing effect of the rootstock upon the scion, as has been noted with apples and pears.

Buds should be inserted all on the same side and facing the prevailing wind. Occasionally, buds which appeared to be vegetative at the time of insertion prove to be floral. Although this is to be avoided if possible, all is not lost, as eventually a vegetative bud will emerge from the centre of the flower cluster to form a maiden tree of less than average stature.

T or shield budding

The method of T or shield budding is fully described by Garner (1979) and is similar for cherries and most other temperate fruit species. Good bud take is usually indicated by the abscission of the leaf petiole. Should this shrivel and remain, it portends poor bud take. The rootstock is cut back to immediately above the inserted scion in the winter or early spring following budding. Some authorities have recommended delaying the cutting back of budded cherry stocks until new growth has just begun. It is argued that earlier cutting back occasionally results in delayed growth or even bud failure. Shoots developing on the rootstock stem are usually removed as they appear; this may need to be done several times during the season.

Anatomical studies have shown that cambial contact is frequently poor in T-budded unions and this can result in rosetting of the bud and delayed scion extension growth. The problem is due to the fact that, when the T flaps are eased back on the rootstock stems, the majority of the cambium remains on the flap and not attached to the xylem. Only cambium remaining on the xylem can effect a union with the cambium of the inserted bud.

A modification of the T-budding technique involves inverting the T cut on the rootstock stem. Some propagators argue that this technique is better than ordinary T budding, particularly for ornamental cherries such as *P. sargentii* and *P. × hillieri* Spire.

Chip budding

This technique, which has largely replaced T budding in British and European nurseries in recent years, is not new, being mentioned in grafting textbooks written many years ago. However, prior to the introduction of modern tying materials, which help prevent bud desiccation during healing, the technique was usually less successful than T budding. Flexibility of timing is one of the major advantages of chip budding. Sudden checks to growth such as temporary drought in the nursery, which may render materials unsuited to T budding, cause no delay to chip budding. Bud take has also proved better than that for T budding in many experiments. The technique is fully described by Howard *et al.* (1974).

Failure of stock and scion buds to unite following chip budding may be attributable to poor rootstock growth after budding. In some situations, root or shoot pruning of the stock may reduce bud take, as have other factors which result in checks to growth, such as herbicide damage or replant effects. Polythene ties may sometimes overcome these problems by remaining in place longer than degradable rubber ties, allowing more time for a successful union to be achieved. It is recommended that rubber ties are best used with vigorous stocks growing in ideal nursery conditions; in all other circumstances polythene ties will be preferable.

Summer chip budding is now preferred to T (shield) budding in Britain and many other countries. It leads to better and quicker formation of the union, thereby minimizing the risk of cold damage and favouring rapid and uniform growth in the subsequent spring. These advantages are even greater in areas experiencing cool summers, where bud union with the stock is even slower. Another advantage of chip budding compared with T budding is that it can be carried out in the spring before active cambium growth makes tissues amenable to T budding. However, work at East Malling comparing spring chip budding with spring grafting or budding in the previous summer showed that the largest sweet cherry trees produced on F.12/1 were those grafted and the smallest those spring chip-budded.

6.5.2 Grafting

Field or nursery grafting

The objective of all field grafts is to join together, in a strong, effective and long-lasting union, a rootstock and the desired interstock or scion. Adequate contact between the cambia of the two components is essential for success. Nursery grafting of cherries, once the only method of propagation employed, is now used much less. Occasionally, rootstocks which were budded unsuccessfully in the summer are grafted in the following spring. Unless the bud failure was due to poor technique or to climatic or nursery factors checking rootstock growth, this practice is not recommended; unions that fail due to incompatibility or to virus infection of one of the components should not be attempted again.

Grafting is also employed when rootstocks are worked *in situ*. Unworked rootstocks are planted in the orchard and, when well established, grafted with the desired scion cultivar. Scion growth is often variable following grafting *in situ*, possibly due to the less favourable environment within the orchard compared with the nursery.

Another reason for grafting is to produce standard trees on a rootstock stempiece. An example would be the production of sweet cherry trees on bacterial-canker-resistant rootstock stems. Strong rootstocks are lined out and within one season should have reached the desired stem height. The scion is then grafted on to the top of the stem, using a whip and tongue or similar graft.

An even better protection against tree loss due to bacterial canker is to raise trees on staddles. These are trees in which the stem or trunk and the initial branch framework are formed from disease-resistant rootstocks or interstocks. The theory is that, while infection of a disease-susceptible scion trunk usually kills the tree, infection of an individual branch has less severe consequences. Normally a 1- or 2-year-old branch framework is grafted with scions of the desired variety, using whip and tongue grafts. Another alternative is to build a substantial crown or scaffold canopy with the resistant rootstock and then framework the crown, using stub grafts (Garner, 1979).

Although such techniques are undoubtedly of value in avoiding the worst effects of severe infections with bacterial canker, staddle and high-worked trees are expensive to produce and are now only occasionally used. The demand for smaller trees with a bush or spindle-shaped crown is not compatible with the practice and the high costs of production are also prohibitive.

German comparisons of budded trees with high-grafted sweet cherry trees on the F.12/1 rootstock showed the budded to be superior on account of their better cropping precocity.

A cherry graft comprises one or more buds on a short section of 1-year-old wood. The shoots to be used as grafts should be collected during the winter but not in periods of frosty weather. The graftwood shoots are tied into small bundles, labelled and then stored inserted 15–20 cm into soil, sawdust or peat in a cool (4°C) position. Alternatively, they can be wrapped in polythene and stored in a jacketed (humidified) cold store. The most commonly used graft is the whip-and-tongue type, which is very fully described by Garner (1979).

New growth from both scion and rootstock usually appears in the spring 1–2 months after grafting. All shoots emerging from the rootstock stem are rubbed out when quite small, whereas all scion growth is left intact until the longest is approximately 15–20 cm in length. The strongest upright shoot is then selected to form the tree leader and the other shoots pinched back to five or six leaves. At about the same time (May or early June in Britain), the graft ties are cut by running a sharp knife up the back side of the rootstock stems, level with the union.

Bench grafting

When inexpensive trees are required or nursery conditions are unsuited to field budding or grafting, bench grafting may be used. Good-quality uniform root-stocks should be established (preferably in pots) in the season before they are required for grafting. For a short period before grafting the rootstocks should be kept dry, watering only occasionally. The rootstocks either are grafted while still in their pots or may be removed and bare-root-grafted. Scions of four to six buds and a whip-and-tongue graft are generally used, although saddle, veneer, spliced-side, wedge or similar grafts are also occasionally used (Garner, 1979). The grafts are tied with elastic strip and sealed with heated paraffin wax. Weaning of bench grafts is particularly critical and often difficult.

The technique is laborious and requires considerable skills on the part of the propagator if success is to be achieved. Also, maiden tree quality is generally inferior and more variable than with trees raised from field budding or grafting. It is occasionally popular with fruit growers wishing to raise large numbers of trees cheaply; growers who cannot afford to purchase nursery-raised trees often resort to this technique.

Micrografting

Researchers trying to speed up the processes of propagation have evaluated a technique called micrografting, which is conducted *in vitro*. In Russia the stem apices of the sour cherry cultivar Lyubskaya were successfully micrografted on to either unrooted or previously rooted shoots of sour cherry rootstocks VP1 and Vladimirskaya. The grafted shoots were then cultured on half-strength Murashige and Skoog medium; IBA was added when unrooted rootstock shoots were used. The technique was more successful when rooted rootstock shoots were micrografted. It has been suggested that the technique may be of value in raising rootstock clones or cultivars which are extremely difficult to propagate using more conventional techniques. It is difficult to envisage, however, how such a complicated process can have any merits over conventional budding of the scions on to rooted rootstocks.

A better use of the micrografting technique is as a means of speeding up plant regeneration following heat therapy treatment of cherry cultivars to rid them of viruses and mycoplasmas. Apices are excised from the branches of heat-treated cherry cultivars and then top- or side-grafted on to micropropagated virus-free rootstock cuttings. The grafted cuttings are then grown for 5–6 weeks on agar before transplanting. Lateral micrografting during the root induction phase of micropropagation (cutting bases kept in the dark for 8–10 days) gave the best results.

6.5.3 Trees grown on their own roots

Several sour cherry cultivars may be propagated, quite easily, on their own roots from softwood or semihardwood cuttings. This technique has advantages in lowering the cost and sometimes also the time taken to produce young trees. However, most of the popular sour and all of the major sweet cherry cultivars are quite difficult to propagate in this way.

Improvements in micropropagating techniques have meant that it is now much easier to propagate scions of any cherry cultivar, either sweet or sour, on its own roots. Before embarking on this course of action, the cherry propagator should carefully weigh up the advantages and disadvantages of raising trees using micropropagation techniques. These are discussed earlier in this chapter, in Section 6.4.4.

Scions raised from cuttings

Many sour cherry cultivars are quite easy to root from either softwood or hardwood cuttings. Sour cherries root best under an intermittent mist or fogging environment, with the cuttings treated with IBA and inserted over bottom heat. Cuttings of the sour cherry cultivar Montmorency root very well under mist if taken just before active extension growth ceases and given a 2500 mg l^{-1} IBA 'quick dip' (10 s) before insertion. These cuttings take 4–6 weeks to root in a sharp sand medium. Successful rooting of softwood cuttings of the sour cherry variety Lyubskaya was shown to be negatively correlated with the degree of shoot lignification. Best rooting (80–96%) was obtained when cuttings were taken when shoots were growing very vigorously (0.4–0.8 cm day^{-1}) and when the lignified component of the stem accounted for only 10–20% of the shoot radius. In Russia this occurred between 10 and 20 June.

In Russia many cultivars of sour cherry are raised from softwood cuttings. Cuttings 30–40 cm long of the current season's shoots are treated with a rooting hormone and then inserted at a density of 300–400 m^{-2} under polythene protection. The cuttings are given bottom heat (26–30°C) and mist irrigation. Rooting generally takes 11–12 days.

Micropropagation of scions on their own roots

Techniques for the micropropagation of sour cherry cultivars are very similar to those described previously for micropropagating cherry rootstocks. In the former Yugoslavia, where the techniques are occasionally employed, 88% success in rooting and 90% establishment of the rooted plants has been achieved.

Research in Poland on the popular sour cherry cultivar Schattenmorelle used Murashige and Skoog medium and rooting was initiated by supplementing the medium with 2.0 mg l^{-1} IBA and 5.0 mg l^{-1} IAA. Shoots were retained for only 3 days on this medium before transfer to an auxin-free medium. This treatment initiated abundant rooting and minimal callus growth. Omission of the IAA and increasing the IBA concentration reduced the root density and increased callus production. Further work in Poland on the same variety showed that cultures could be initiated from whole vegetative buds or from actively growing shoot tips; both produced axillary shoots after 8 weeks' culture. Transfer of these shoots to a fresh medium produced more axillary shoots in a further 4–6 weeks. Comparisons of different cytokinins showed BA to be superior to zeatin, kinetin and their ribosides. Half-strength Murashige and Skoog or the rooting medium recommended by Druart (see above) proved best for rooting Schattenmorelle. In some cases addition of phloroglucinol to the medium improved rooting, which usually took approximately 5 weeks. The plants were then transplanted to a 3 : 1 : 1 (peat : pine bark : sand) mixture in pots.

Aftercare of Schattenmorelle micropropagules is also critical to the success of the technique. Plantlets rooted in half-strength Murashige and Skoog medium and then transferred to a peat : bark : sand mixture in pots made much better subsequent growth in the nursery if chilled at 4°C for 5 weeks approximately 2 months after potting.

Italian research has shown that the addition of proline to IBA-enriched rooting media is useful in stimulating successful micropropagation of several difficult-to-root sour cherry cultivars.

Research in Germany showed that the sweet cherry cultivars Hedelfinger and Sam and the sour cultivars Schattenmorelle, Boscha and Schwabische Weinweichsel could all be successfully micropropagated. Shoot proliferation rates were greater for the sour than for the sweet cultivars and rates for the sweet cultivars were much less than for the rootstock F.12/1.

Research at the Volcani Centre in Israel developed micropropagation techniques for seven different cultivars of sweet cherry and produced several hundred plants using these methods. Compact varieties, such as Compact Stella, have also been grown on their own roots following micropropagation, although records on the stability of these mutant clones after such treatments are lacking.

Whether there are any real advantages of growing cherries on their own roots is doubtful. French work has shown that sweet cherry trees raised on their own roots by micropropagation are slow to begin fruiting and very vigorous. Similarly, Italian researchers found micropropagated sour cherry trees were less productive than budded trees and also had very low production efficiencies.

6.6 Nursery Aftercare of the Tree

6.6.1 Management in the nursery

Supplementary nutrition of rootstocks and young trees in the nursery should not usually be necessary if soil fertility at the time of planting is good. It is wise to have the pH and nutritional status of the nursery soil checked prior to its use, as this will then allow any defects to be remedied prior to planting.

Supplementary irrigation will be necessary in many of the hotter and drier parts of the world. It is particularly important that the young developing scion shoot is not subjected to drought stress; this leads to a check in growth and reduced size at the end of the season. Overhead rain guns or sprinklers are most commonly used for irrigation in nurseries. It is vital to check the purity, pH and mineral status of the water applied, particularly if extracted from rivers or bore holes.

Control of pests and diseases is vital in the nursery and the young plants should be regularly monitored by experienced staff, checking for the first signs of any problem. Details of the pests and pathogens causing serious problems are not presented here as they differ from country to country, as do the approved control measures. It is essential that the propagator takes expert and up-to-date advice on the most appropriate control measures for his or her situation.

Cherry trees are rarely staked in nurseries. If a suitable sheltered site has been chosen, there should be no need for additional support for the growth of trees on Mazzard or Mahaleb rootstocks. A review of this policy may be necessary in the future, however, as more dwarfing rootstocks, which are often poorly anchored, become popular.

Induction of branching

Work at East Malling (Harrison-Murray and Mitra, 1985) has shown that the height of feathers (lateral branches) on sweet cherry scions may be raised if shoots are temporarily retained on the rootstock stems. The presence of these rootstock shoots has the effect of suppressing the development of the lowest feathers and speeding the elongation of the central leader. Stem thickening of the leader tip is reduced slightly by the practice, possibly as a result of delayed xylem differentiation just below the tip. It is thought that ABA may play a role in inhibiting the development of the lowest feathers. The technique has not been taken up by nurserymen, however, on account of its effect on stem shape. Unfortunately, unless bud guides are used, the presence of rootstock shoots leads to the formation of rather bent leader stems.

6.6.2 Tree lifting and storage

Nursery-grown cherry trees are allowed to defoliate naturally in the late autumn. They usually do this relatively rapidly and there is no need for any artificially stimulated defoliation. Cherry trees are rather tender compared with apples and pears, and use of chemical defoliants may cause damage to the trees.

In the UK, cherry trees are lifted from the nursery by first undercutting the roots mechanically and then hand-lifting the trees. In many other countries, the whole lifting process is mechanized. In the USA, either side tree diggers mounted on conventional tractors are used (Fig. 6.5) or tracked tractors are customized into tree diggers which drive over the trees, undercutting and lifting them (Fig. 6.6).

After lifting, any damaged roots and/or branches are trimmed off and the trees rapidly transported off the nursery field. This is very important if damaging root desiccation is to be avoided. In the UK, the trees are usually tied in bundles of five trees and either stored in a jacketed cold store or heeled-in close together outside. If heeling-in is the choice, it is important to make sure that the roots are well covered with soil and the soil is firm around the roots. In colder regions, trees must be cold-stored to avoid winter cold injury. Most stores are of the jacketed type, to avoid tree desiccation, and are maintained at 1–2°C above freezing and at 95–100% humidity. It is also vital to maintain good air circulation if the development of diseases such as oedema are to be prevented.

Fig. 6.5. A US-designed side tree lifter mounted on a conventional tractor (photograph L. Brandt, Washington State, USA).

Fig. 6.6. An over-tree lifter on a customized crawler tractor operating in the USA (photograph L. Brandt, Washington State, USA).

General Reading

Garner, R.J. (1979) *The Grafter's Handbook*, 4th edn. rev. Faber and Faber, London, 319 pp.
Hartmann, H.T. and Kester, D.E. (1983) *Plant Propagation: Principles and Practices*, 4th edn. Prentice-Hall, Englewood Cliffs, New Jersey.
Howard, B.H. (1987) Propagation. In: Rom, R.C. and Carlson, R.F. (eds) *Rootstocks for Fruit Crops*. Wiley Interscience, New York, pp. 29–77.

References

Ahuja, M.R. (1993) *Micropropagation of Woody Plants*. Forestry Science 41. Kluwer Academic, Dordrecht, the Netherlands.
Bernhard, R. and Claverie, J. (1986) Le bouturage d'extrémités semi-ligneuses, application à diverses espèces fruitières à noyaux. [Rooting of semiwoody tip cuttings: application to various stone fruit species.] In: *Cinquieme Colloque sur les Recherches Fruitières. Recueil des Communications*. CTIFL-SDIT, Pont-de-la-Maye, France, pp. 229–236.
Druart, P. (1985) Multiplication conforme de sujets porte-greffe et des variétés de cerisier par la culture *in vitro*. *Acta Horticulturae* 169, 319–328.
Fogel, H.W. (1958) Effects of duration of after-ripening, gibberellins and other pre-treatments on sweet cherry germination and seedling growth. *Proceedings of the American Society for Horticultural Science* 72, 129–133.
Garner, R.J. (1979) *The Grafter's Handbook*, 4th edn. rev. Faber and Faber, London, 319 pp.
Harrison-Murray, R.S. and McNeil, J.C. (1984) Practical alternatives to peat-based media. In: *Report of East Malling Research Station for 1983*. p. 81.
Harrison-Murray, R.S. and Mitra, S.K. (1985) Control of lateral branching in nursery trees. In: *Report of East Malling Research Station for 1985*. p. 98.
Hartmann, H.T. and Kester, D.E. (1983) *Plant Propagation: Principles and Practices*, 4th edn. Prentice-Hall, Englewood Cliffs, New Jersey.
Howard, B.H. (1971) Propagation techniques. *Scientific Horticulture* 23, 116–126.
Howard, B.H. (1986) Factors affecting the rooting response of fruit tree cuttings to IBA treatment. *Acta Horticulturae* 179, 829–840.
Howard, B.H., Skene, D.S. and Coles, J.S. (1974) The effects of different grafting methods upon the development of one-year-old nursery apple trees. *Journal of Horticultural Science* 49, 287–295.
Howard, B.H., Harrison-Murray, R.S., Vasek, J. and Jones, O.P. (1988) Techniques to enhance rooting potential before cutting collection. *Acta Horticulturae* 227, 176–186.
Hutchinson, J.F. and Zimmerman, R.H. (1987) Tissue culture of temperate fruit and nut trees. *Horticultural Reviews* 9, 273–349.
Jones, O.P. and Hopgood, M.E. (1979) The successful propagation *in vitro* of two rootstocks of Prunus: the plum rootstock Pixy (*P. insititia*) and the cherry rootstock F12/1 (*P. avium*). *Journal of Horticultural Science* 54, 63–66.
Kevers, C., Coumans, M., Coumans-Gilles, M.F. and Gaspar, T. (1984) Physiological and biochemical events leading to vitrification of plants cultured *in vitro*. *Physiologia Plantarum* 61, 69–74.
Quoirin, M. and Lepoivre, P. (1977) Etude de milieux adaptés aux cultures *in vitro* de Prunus. *Acta Horticulturae* 78, 437–442.
Webster, C.A. and Jones, O.P. (1989) Micropropagation of the apple rootstock M.9: effect of sustained subculture on apparent rejuvenation *in vitro*. *Journal of Horticultural Science* 64, 421–428.

7 Selecting the Orchard Site, Orchard Planning and Establishment

M. LONGSTROTH and R.L. PERRY

Department of Horticulture, Michigan State University, East Lansing, MI 48824-1325, USA

7.1 Site Selection

Many factors influence the health and longevity of cherry orchards, not least the choice of an appropriate site. Detrimental site factors may shorten orchard life, reduce tree productivity or increase the costs of maintaining production. In contrast, beneficial site factors can greatly increase productivity and/or reduce the risk of damaging disease infection. Orchardists have only one opportunity to make decisions relating to orchard establishment and mistakes made at this time may greatly reduce the orchard's profitability.

The proposed site's history and soil type often reveal its potential for cherry production. Climate and local environmental conditions, in particular weather conditions during bloom, fruit growth and harvest, determine growing potential. These environmental factors are often, in turn, directly related to the likelihood of disease or insect damage and associated effects upon fruit yield and quality and overall tree health. Growing conditions after harvest also influence tree health and the productive life of the orchard.

In choosing a suitable site climatic, local environmental and edaphic factors must all be considered.

7.1.1 Climatic conditions

Cherries are temperate fruits requiring both a warm growing season and a winter dormant period. They also require a relatively short, frost-free growing season to set and mature their fruit and a rain-free harvest period to avoid problems of fruit splitting.

Firstly, winter chilling is necessary in order to break winter dormancy, allowing growth to resume in the spring. Cherry shoot and flower/fruit growth occur simultaneously after bud burst. The trees rely heavily on stored food reserves from the previous year for bud burst and initial growth (Keller and

Loescher, 1989). Warm temperatures and good light levels are then needed to support cell division and photosynthesis. Finally, they also require good growing conditions after harvest, allowing the tree to accumulate reserves for next year's growth.

Climate, together with soil type, also determines whether the orchard will require supplementary water in the form of irrigation. This may be required all year, to meet the demands of both the trees and any orchard floor cover crops, or it may only be required during dry periods as a supplement to rainfall and soil moisture reserves at periods of peak demand. If irrigation water is not available, or is costly, this may preclude growing cherry orchards on dry but otherwise desirable sites.

Winter climatic conditions

The range and duration of winter temperatures are extremely important. They must be cool enough to allow the accumulation of chilling units (for dormancy breaking), but not so cold as to injure the trees (Table 7.1). The temperatures most useful for chilling are just above freezing (0–6°C); temperatures below freezing have no influence on the accumulation of chilling units. Richardson *et al.* (1974) have developed a model where temperatures from 2.5 to 9.1°C are assigned a value of one, temperatures from 2.4 to 1.5°C and those between 9.2 and 12.4°C are assigned a value of one-half, and temperatures below 1.5°C and between 12.5 and 15.9°C have no value. Long periods of time during dormancy with temperatures above 16°C may actually negate accumulated chilling units and increase the time required to complete rest. Depending on the variety, cherries require 400–1500 h of temperatures below 7°C (Westwood, 1978; Childers, 1983; Seif and Gruppe, 1985). In areas experiencing long winters, varieties with short chilling requirements may suffer damage if they begin growth during temporary midwinter thaws. Conversely, in areas experiencing short, mild winters, varieties with long chilling requirements may not accumulate sufficient chilling units to fully break rest and growth in the spring will be erratic.

The timing of bloom is often correlated to a cultivar's chilling requirement. Cultivars with a short chilling requirement and bloom early begin growth earlier in the spring, depending on the level of heat units accumulated after completion of rest, while cultivars with longer chilling requirements are still accumulating chill units.

Table 7.1. Chilling requirements of several cherry cultivars. Chilling units are determined using weighting values given in the text. (Adapted from Seif and Gruppe, 1985, and Anderson *et al.*, 1976.)

Species	Variety	Chilling units	Chilling hours at < 7°C
Prunus avium	Bing	900	900
	Emperor Francis	1100	1300
	Early Burlat	1100	1300
	Van	1150	1350
	Hedelfingen	1200	1400
Prunus cerasus	Montmorency	950	

Although the tree is most resistant to cold temperatures when dormant, extremely cold winter temperatures can cause considerable damage to the tree itself. Fully dormant sweet cherry trees can withstand temperatures as low as − 29°C (Proebsting, 1970). Although trees lose cold-hardiness during warm weather and regain it during colder periods, cold tolerance is lost much more quickly than it is regained. This means that trees losing cold-hardiness during winter warm spells become very susceptible to sudden drops in temperature. Sites susceptible to large fluctuations in winter temperatures are, therefore, not desirable. Some cultivars, particularly many of those planted in Russia and parts of eastern and central Europe, can withstand colder temperatures better than others and are preferred for areas with colder winters.

In areas of severe winter cold, the presence of snow cover on the ground is very important. The snow acts as an insulator, reducing heat loss from and freezing of the ground. Cherry roots (usually of the rootstock) are more susceptible to cold damage than the above-ground parts of the tree and rootstocks may differ in their sensitivity. Mazzard roots are killed by temperatures below − 11°C, while Mahaleb roots are slightly more cold tolerant, experiencing damage only if temperatures drop below − 15°C (Carrick, 1920). In research conducted by one of the authors (Perry, 1987), the hybrid rootstock Colt seemed to be less cold tolerant than either Mazzards or Mahalebs. The cold tolerance is also affected by interactions between the rootstock and scion. Cold-hardy rootstocks can increase the cold tolerance of both hardy and susceptible scions, while sensitive stocks, such as Colt, may reduce scion cold-hardiness (Howell and Perry, 1990).

Spring climatic conditions

Spring growth begins with the onset of warm temperatures, when both floral and vegetative buds begin to swell and lose their ability to tolerate cold temperatures. Severe spring frosts can kill buds, flowers and young fruits and at each advancing bud stage the lethal temperature rises (Table 7.2). The proportion of buds killed depends on the duration of the freeze and its lowest temperature. Not all the buds on a tree develop at the same rate. The most advanced buds

Table 7.2. Average critical freezing temperatures for cherry buds.

		Percentage damage to floral buds		
Stage	Description of stage of bud development	10% (Sweet)	90% (Sweet)	50% (Sour)
1	First swell	− 8.3	−15.0	− 17.0
2	Side green	− 5.5	− 13.0	− 13.0
3	Green tip	− 4.0	− 10.0	− 5.5
4	Tight cluster	− 3.3	− 8.3	− 5.0
5	Open cluster	− 2.8	− 6.1	− 4.0
6	First white	− 2.8	− 4.4	− 3.5
7	First bloom	− 2.2	− 3.9	− 3.5
8	Full bloom	− 2.2	− 3.9	−3.5
9	Postbloom	− 2.2	− 3.9	−3.5

are the most susceptible to freezing. Colder minimum temperatures expose more buds to their own individual critical threshold at which they freeze and die. Buds on the most exposed portions of the tree lose heat rapidly by radiation to the open sky and these freeze first. Buds in less exposed positions receive radiated heat from the ground and other portions of the tree and these buds survive longer. The longer the time that air temperatures are below the critical threshold, the greater the number of buds that are allowed to cool to their critical threshold and to freeze. Flower buds of the sour cherry variety Montmorency are very susceptible to frost damage soon after the buds begin to swell. At bloom, cherry blossoms can withstand $-2°C$ for only half an hour, and nearly all are killed at $-4°C$ (Ballard *et al.*, 1982). The frequency of frosts in the early spring and the average date of the last killing frost are, therefore, important statistics to consider before choosing a site.

The type of spring frost is also an important consideration. Radiation frosts, which are more common in drier climates, occur when the ground cools as a result of radiation loss to the sky on still nights with no cloud cover. The cold ground chills the air immediately above it and may result in damage to fruit buds in the lower portion of the tree. This is a sure sign of a radiation frost.

It is possible to reduce frost damage to trees during radiation frosts by orchard heating, by the use of orchard wind machines to mix the upper warm with the lower cold air layers or by water sprinkling during the frost, using the heat of fusion of ice formation to keep the buds above their critical temperatures (Ballard and Proebsting, 1972). Nevertheless, these measures begin to lose their effectiveness at temperatures below $-6°C$. Rising costs of oil for heating and water for sprinkling now make conventional frost protection methods very expensive. Locating orchards in frost-free sites, where frost damage should be minimal, is a much cheaper method of avoiding frost damage.

Advective or wind frosts are more common in areas with moist climates. These are caused by the movement of cold air masses into a region after the passage of cold fronts. Orchard heating is the only means of reducing losses caused by advective frosts. Unfortunately, even this is seldom effective because of the large amount of heat required to heat the orchard and maintain temperatures above that of the surrounding air. Regions where such cold air masses are common in the late spring after bloom should, therefore, be avoided, when choosing sites for cherry orchards.

Growing season climatic conditions

Depending on the variety, cherries require between 40 and 80 days from the time of bloom to fruit maturity. Rain, cold weather or wind during bloom can result in reduced activity by insect pollen vectors and poor pollen germination or growth, leading to poor fruit set. Rain during bloom can result in reduced pollination or infection of the blossoms by fungi (*Monilinia laxa*) or bacteria (*Pseudomonas syringae*). Rain before or during harvest can result in fruit splitting of sweet cherry and infection with brown rot (*Monilinia fruticola*) or other fungal fruit rots (*Alternaria* species, *Botrytis cinerea*, *Cladosporium herbarum* and *Penicillium expansum*). Foliar diseases, such as the cherry leaf spots (*Coccomyces hiemalis* and *C. lutescens*, *Cerospora circumscissa*, *Alternaria citri* and *Phyllosticta pruni-avium*) and powdery mildew (*Podosphera oxyacanthae*), cause leaf loss, reducing growth during the current season and in the following year. Moist conditions also favour bacterial (*Pseudomonas mors prunorum*) and fungal (*Valsa* or *Cytospora cinta*) stem cankers that can kill the tree.

Cherries need a period of favourable growing conditions following harvest during which the tree generates and stores reserves for the coming growing season. Climatic conditions that support only poor growth in this period will result in reduced reserves, which can, in turn, lower the tree's cold tolerance during the winter. Drought is the most common cause of poor growth in this period and sites likely to suffer this problem will require supplemental irrigation to sustain adequate growth.

Extremely warm temperatures during the growing season often result in twin fruits because the flower pistil doubles during flower formation (Iezzoni *et al.*, 1990).

7.1.2 Physical environmental considerations

Environmental factors largely determine a site's microclimate. The proximity and size of any nearby bodies of water and the topographic relief of the site and surrounding countryside directly influence growing conditions.

Proximity of bodies of water

Water has a large specific heat and it changes temperature slowly. Proximity to large bodies of open water will, therefore, moderate winter temperature extremes. In contrast, frozen lake surfaces cool the surrounding air, delaying the resumption of growth and bloom in the spring, so reducing the danger of spring frosts. The severity of spring frosts can also be reduced by the warming of the air above the water, as water retains its heat accumulated during the day longer than soil during the night. For these reasons, some cherry-growing regions, such as those in Michigan, USA, are located near large bodies of water.

Topography

Surface relief affects air and water drainage. Air drainage is especially important during the spring in determining susceptibility to radiation frosts. Surface and subsurface drainage of water is important at all times of the year, determining the water status of the soil and root system. Steep slopes favour erosion of bare soil, while soils on ridges and hilltops are also often poor choices on account of the potential loss of topsoil due to erosion and their exposure to wind which can damage fruits. Winds can also reduce canopy volume by inhibiting shoot growth and increasing water loss through evapotranspiration. Sloping land also affects the feasibility and efficiency of orchard equipment operation; some areas with excellent drainage may be too steep for safe operation of orchard equipment.

Air drainage. Cold air is heavier than warmer air, settling at ground level and draining down the slope to lower elevations. Orchards should not, therefore, be planted in areas where cold air is likely to accumulate, such as the bottoms or ends of valleys. More favoured frost-free sites are often located a short distance away on higher ground. Low areas, slopes and valleys should also be clear of tree lines, windbreaks and other features that block air movement and create frost pockets. Cutting openings in windbreaks or pruning off lower branches to allow air drainage through them will help alleviate this problem. Similarly, cherry orchards should not be located on level sites unless there is little chance of frost, as drainage of cold air from such sites will be minimal.

Water drainage. Surface relief allows drainage of excess surface water by runoff, essential in preventing soil waterlogging during periods of heavy rain or snow melt. Flat or low areas are usually wetter than surrounding sloped sites and, if such sites are selected, surface or subsurface drainage should be provided to correct the problem. Smoothing the land surface by filling in low spots and providing surface drainways reduce subsurface flooding by eliminating standing water. However, it should be borne in mind that excess surface runoff can result in surface erosion, leading to loss of topsoil, soil nutrients and rooting volume.

Edaphic conditions

A thorough understanding of a proposed orchard site's soil is essential if the grower is to avoid poor sites and take appropriate actions at marginal sites to improve potential tree longevity and productivity. Some soil problems may be obvious from visual inspection of the land: low spots, ravines, areas where the current crop's growth is reduced, or where the native vegetation changes. All these are clues that the soil type or profile is not uniform. Digging trenches or holes to evaluate the soil profile at the proposed site will give the grower useful information.

Soil conditions determine the volume of soil available for root growth, the nutrients available for tree growth and the need for irrigation or drainage. Soil texture and structure are two important variables that determine drainage and water-holding capacity of the soil and thus greatly influence rooting. Soils for cherries should, ideally, be of a well-buffered medium loam at least 1 m deep, well-drained but with good water-holding capacity.

Table 7.3. Sensitivity of *Prunus* rootstocks to waterlogging. (Adapted from Rowe and Beardsell, 1973, and Day, 1951.)

Sensitivity	Scientific binomial	Common name
Extremely sensitive	P. amygdalus	Almond
	P. armeniaca	Apricot
	P. mahaleb	Mahaleb
	P. persica	Peach
	P. davidana	Nemaguard
Sensitive	P. avium	Mazzard
Moderately sensitive	P. cerasus	English Morello
	P. salicina	Japanese Plum
Moderately tolerant	P. cerasifera	Myrobalan
	P. domestica	Brompton
	P. insititia	St Julian
Tolerant	P. cerasifera × munsoniana	Marianna

Cherry root systems are sensitive to poorly drained or wet soils (Table 7.3). In wet soil, the soil pores are full of water with little, if any, air present. Water supplies much less oxygen than air to the plant and oxygen diffuses slowly in the soil, since it diffuses through the channels between the soil particles. As the soil pores become filled with water, less and less oxygen, which is vital for root respiration, is available to the tree. An experiment conducted by Beckman (1990) with actively growing potted Montmorency trees on Mahaleb rootstocks demonstrates the sensitivity of cherries to waterlogging. Shoot growth of trees flooded for 2 or 4 days recovered to levels of unflooded control trees in 24 and 40 days, respectively. Trees flooded for 8 days or more never recovered. This experiment showed that even short-term flooding can have long-lasting deleterious effects on the tree. Additionally, wet soils encourage the development of *Phytophthora* root rots.

Mazzard rootstocks are often used in areas where poor drainage is known to be a problem, as they regenerate roots more quickly than Mahaleb, allowing them to utilize any unflooded portion of the soil (Roth and Gruppe, 1985; Beckman and Perry, 1987). Repeated flooding or high water-tables may, however, result in the development of a shallow root system above the habitually flooded soil layers. Such root systems are susceptible to drought stress in the drier season and may result in poor tree anchorage.

Soil texture and structure. Soil texture is determined by the size of the particles (sand, silt and clay) which make up the soil. The arrangement of these primary particles, known as the soil structure, determines the size of the soil pores. Soils with large particles, such as sandy soils, have larger pores and such soils do not readily retain water, as it drains away rapidly to lower layers within the profile. Sandy soils lack water-holding capacity and their lack of cation exchange capacity (CEC) means they readily lose nutrients through leaching. In the arid regions of the American West, where such soils often predominate, Mahaleb rootstock is preferred because it is considered to be more drought-tolerant than Mazzard. Under these conditions, trees propagated on Mahaleb rootstocks are usually more vigorous than trees with Mazzard roots.

The presence of clay and organic matter in the soil allows the consolidation of small primary soil particles together into aggregates or peds. The voids between these larger particles allow rapid water drainage, while some water is retained in pores inside the peds. Soils with a range of particle sizes are ideal, as they provide both small pores, which retain water, and larger pores, which favour drainage of water and through which oxygen can diffuse into the soil.

Fine-textured soils have a high proportion of clay and many small pores which drain slowly. These soils are easily waterlogged and can pose problems of survival for cherry trees. A grower should consider growing less sensitive tree species such as apple, pear, or plum on such soils. Otherwise, rootstock selection and soil amelioration by physical means are the only options. In this situation, rootstocks such as Mazzard or the Mazzard × Mahaleb hybrid M×M 2 should be used rather than Mahaleb (Perry, 1990).

Soil profile restriction. The soil's water-holding capacity depends on its profile characteristics as well as its texture. Abrupt soil texture changes can restrict subsurface water movement. The change from a sandy layer to a clay layer or from clay to sand results in what is known as a 'perched' water-table with excessive moisture in the soil above the layer, which restricts root growth.

Compacted layers, such as plough pans, fragipans or clay pans, also restrict root growth through them. The roots are unable to push aside soil particles, and large pores through which the roots might grow are absent. Because the root system cannot penetrate the layer, roots are concentrated above it. Such shallow root systems are more susceptible to drought stress during the growing season, because they have smaller soil volumes from which to extract water.

7.1.3 The site's previous crop history

Previous crops grown on the site will influence, favourably or unfavourably, the condition of the soil. Previous cherry plantings may result in 'replant' problems. The causes of 'replant' problems may be nematodes, soil fungi, residues of toxic minerals or other, often unknown, causes. While some of these causes may be alleviated by fumigation or choice of appropriate rootstock, affected sites are always best avoided wherever possible. A long history of field or row crops often suggests the presence of a plough pan, a compacted zone in the soil profile formed by the continued operation of heavy equipment or cultivation at the same depth, often when soils are too wet. Such impervious layers must be broken up, possibly by subsoiling, if subsequent poor drainage is to be avoided. Intensely farmed soils may also be depleted in one or more soil nutrients. This is more easily rectified by appropriate nutrition based on soil analysis. Crops such as potatoes, strawberries or alfalfa are hosts for *Verticillium*, which can also infect cherry. Soils should be tested for such organisms before selection for cherry culture.

Sites still in their natural state (virgin soils) often offer useful indications to soil conditions. For example, the presence of poplar (*Populus*), willow (*Salix*) or other water-loving species may indicate a high water-table and unfavourable wet soil conditions. *Pinus* species, in contrast, suggest deep, coarse, sandy soils. Forested sites often lack the A horizons of traditional agricultural soils. In many cases, these soils have a shallow O horizon composed of humus at the soil surface. Below the O horizon is a leached E horizon, which often lacks nutrients because clay and silt particles have been washed to lower levels in the profile (Fig. 7.1).

Previously forested or wooded areas and old or declining orchards are a potential source of many pests and diseases. *Armillaria mellea*, which causes Armillaria root rot or honey fungus, is especially deadly to cherries propagated on Mahaleb rootstock. Wild cherries growing in surrounding woodland or hedgerows are also often a reservoir for disease. In the USA, *Prunus virginiana*, or Choke cherry, is an alternate host for X disease, a potentially damaging disease caused by a mycoplasma. Old orchard sites where trees were lost because of wet soils should not be replanted with cherries until drainage problems have been corrected.

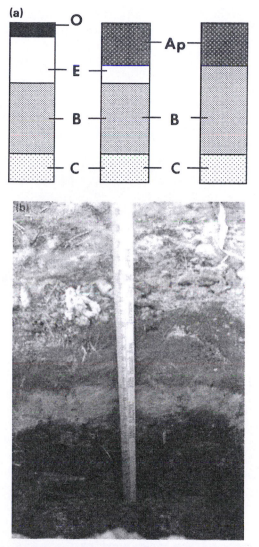

Fig. 7.1. (a) Comparison of a virgin forest soil (left), a recently ploughed forest soil (centre) and a typical agricultural soil (right). O, organic horizon, consisting of decomposing organic matter; E, zone of maximum leaching or eluviation of clays, organic matter and soluble minerals; Ap, the cultivated soil horizon; B, horizon of maximum accumulation of clays, contains less organic matter than A, but more organic matter than the E horizon; C, underlying horizon of least weathering. Except for the abrupt transition at the bottom of the cultivated soil, transitions between soil horizons are rarely as abrupt as shown here. (b) A recently ploughed forest soil. The ploughed layer is visible at the top of the profile. The leached E horizon is the light-coloured band beneath it. Below the E horizon is the dark B horizon, the zone of accumulation.

7.2 Preparation of the Site

Once a site has been selected, planning and preparations must be made for establishing the cherry orchard. Any site limitations in relation to its future cherry culture must be identified, so that strategies can be developed to avoid or minimize their negative impact on future productivity and profitability. A plan for the entire orchard needs to be developed.

7.2.1 Orchard planning

First, a map of the entire site should be prepared which includes its elevation and physical relief and the area and location of different soil types. This plan should also show the location of roads, buildings and the existing or planned irrigation system, as well as areas for filling sprayers, storing chemicals, loading and unloading equipment and handling the crop during harvest. This map will be useful in determining any areas to be avoided when planting trees. If the site is windy, suitable windbreaks should be planned which are orientated so as to allow air drainage in the spring. Ideally, such windbreaks should be established several years before planting so that the young trees may immediately benefit from shelter.

Next, plans for each individual orchard block should be developed. Generally, tree rows should run in a north–south direction, to provide maximum light penetration. If cold air drainage is likely to be important, tree rows should also run downhill rather than across the slope or on the contour. Tree spacing depends on expected tree vigour and the system of tree pruning and training planned. Vigour is determined by rootstock, scion vigour and soil fertility. For example, sweet cherry trees planted on Mazzard or Mahaleb rootstocks may require spacing 5–7 m apart in conventional training systems. In contrast, less vigorous tart cherry scions, such as Montmorency on the same rootstocks, can be planted 1–2 m closer. In eastern Europe, Morello tart cherries, of even less vigour than Montmorency, are grown in bush form at even closer between-tree spacings. New dwarfing rootstocks for the more vigorous sweet cherries which were bred in Europe are now being tested (Chapter 5). If these prove successful, they will permit the adoption of even closer tree spacings. Cherries grown on Mazzard, Mahaleb and some tart cherry stocks are traditionally grown as centre leader or open-centre trained trees without stakes. Use of dwarfing rootstocks and/or adoption of novel training systems in most cases requires trees to be supported by stakes or on some form of trellis system (Chapter 10).

7.2.2 Topography

Unfortunately, there is little that can be done to correct excessive slopes on a site. Cherry growers in Michigan find that slopes exceeding 9% limit the operation of orchard equipment, especially mechanical cherry harvesters. Earth movers have been used in Michigan to reshape sites by cutting or reducing the tops of hills while filling in valleys and depressions. This makes sites more uniform in topography, facilitating equipment operation, but less uniform in soil conditions. Trees established on these altered sites often express micronutrient deficiency symptoms and poor growth; this is hardly surprising, as the shallow A horizon is usually lost or buried and replaced by a coarse alkaline C horizon (Fouch, 1987). If such operations are to be successful, the A horizon should first be removed, the subsoil levelled and then the topsoil replaced evenly above.

Hilly sites will also require cover crops to control erosion. In these situations the cover crop may be planted first and a weed-free strip of soil (1–2 m wide) created for the tree row just prior to planting, by rotavating, burning out or using a contact herbicide. Land grading, i.e. creating ridges of soil on to which the trees are planted, may be necessary to provide surface relief and to avoid waterlogging in low-lying areas.

7.2.3 Windbreaks

Windbreaks should be established to shelter orchards planned for windy sites, such as the tops of hills, ridges and in windy valleys. In areas with prevailing winds, the windbreaks should be on the upwind side of the orchard, and should be established before the orchard is planted to provide protection for the young trees. Windbreaks should not compete with the orchard for light, water or nutrients. Local recommendations should be consulted to determine which species will provide adequate protection without excess maintenance, while avoiding competition with the orchard for light and soil moisture. In many parts of Europe, non-competitive *Alnus* species are preferred to more competitive types, such as *Salix* or *Populus* species. It is vital to avoid blocking cold air drainage by leaving gaps in windbreaks. In Australia, South Africa and Washington State, artificial windbreaks made of woven plastic screens, attached to poles, are used. These have the advantages of not competing for plant resources, they are quickly installed and their height may be increased as the orchard grows. They are, however, not as permanent and in the long term are more expensive.

7.2.4 Soil preparation

If possible, green manure crops should be grown for 1–2 years prior to orchard establishment. Crops such as white clover and Sudan grass improve soil tilth, soil organic matter and cation exchange capacity (CEC), all of which will be especially useful on sites with coarse, freely drained soils. Alfalfa, rape and other large, tap-rooted crops can be used to improve the porosity of fine-textured soils. Sites which have previously been planted to orchards should be cleared of old roots and debris, tilled and planted to cover crops at least 1 year, preferably several years, in advance of establishing a cherry orchard. The site should be deep ploughed if a plough pan is present. Prior to planting, soils should be cultivated to produce a good uniform tilth.

Soil pests and fumigation

Soil samples should also be taken to determine if plant parasitic nematodes such as *Pratylenchus*, *Xiphenemia* or *Meloidogyne* are present. An application of a soil fumigant may be recommended in some local areas to suppress nematode and other soil pathogen populations. Choice of fumigant will depend upon local pesticide legislation and on the precise identity of the pathogen. Cherry root systems are generally very susceptible to soil-borne diseases such as *Armillaria*,

Fig. 7.2. Portions of this Californian sweet cherry orchard have been replanted to replace trees killed by *Phytophthora* root rot.

Phytophthora (Fig. 7.2), *Verticillium* and *Agrobacterium*. Unfortunately, soil fumigants are relatively ineffective in suppressing these organisms, especially in the long term. Fumigation is generally recommended for orchards which are replanted with the same orchard crop, or one of its close relatives; fumigation of sites previously planted to stone fruits is recommended prior to planting cherries. In some cases, increased growth followed fumigation (Webster, 1984), in others the differences in growth were not significant (Miller *et al.*, 1990). Much will depend upon the causal organisms, the rootstock choice and the fumigant used.

Drainage problems

Depending on the situation, deep tillage or subsoiling can successfully break up hardpans. A permanent enhancement of root and water penetration can be made if there are differences in soil texture within the soil profile. Loose or coarse materials from one region are mixed with the dense soil materials in the profile. Young fruit tree growth has been improved in California sites that were 'slip ploughed' (a chisel plough fitted with a wing at the tip) to 1–2 m depth before planting (Wildman, 1984), and in Oregon sites where the tree holes were dug with a 'back-hoe' (Miller *et al.*, 1990). In France, 'ripping' (use of a multi-tine chisel plough) to at least 80 cm in depth has been found to enhance young tree growth (Lichou *et al.*, 1990). These techniques have only limited success in uniform soil profiles because of their short-term effect (Unger *et al.*, 1981).

Overall, little can be done to improve deep, fine-textured soils where poor internal water drainage will create damaging anaerobic conditions for cherry roots. Similarly, soils with high water-tables are very difficult to improve. If such soils must be used, then the only solution is the installation of drainage tiles and deep drainage ditches, and/or the use of raised beds or ridges. Drainage tiles, when properly engineered and installed, are indeed an effective means of alleviating soil profile and high water-table drainage problems (Cannell and Jackson, 1981; Hillel, 1982). However, not all systems continue to function properly over the life of the orchard.

The formation of raised beds or ridges may also be effective and costs considerably less to achieve than installation of drainage tiles. Raised beds are commercially and successfully used for the production of tree fruit crops in fine-textured and high water-table soils of Australia, New Zealand, South Africa and Germany. In southern Florida, over 125,000 ha of citrus are grown on raised beds where a permanent water-table exists less than 1 m from the surface. In Michigan, one study shows that Montmorency cherry on Mahaleb rootstock has performed well for 10 years when grown on wide (2 m) beds, 45 and 70 cm high, in comparison with a flat control treatment. This study was carried out on a clay loam soil with a 20–30-cm A horizon over a dense clay B horizon. More trees have survived and the cumulative yields per tree are slightly higher with these treatments than the control (Table 7.4). Preliminary observations on this plot indicate, however, that cherries on raised beds require irrigation to avoid excessive drought stress.

Table 7.4. Cumulative results (1981–1990) for Montmorency/Mahaleb cherries grown on 2-m-wide raised beds in Michigan. (Unpublished data from R.L. Perry.)

Treatment	Bed height (cm)	Mortality (%)	Cumulative yield (kg tree^{-1})	Trunk area increase (cm^2)
Flat	0	15	90	153
Low	45	2	121	165
High	70	0	108	162

Supplemental irrigation

Irrigation can benefit cherries grown on many of the more shallow-rooted rootstocks and orchards established on raised beds or in coarse (sandy) soils with little water-holding capacity. Local climatic factors will influence evapo-transpiration demands and the need for supplementary irrigation. In some situations, mulching is quite effective and economical in conserving soil moisture and should be considered as an alternative. Suitable mulches are black plastic, straw or other organic materials.

Mahaleb rootstocks should be considered as an alternative to Mazzard where drought conditions prevail, on account of their improved tolerance to water deficits.

The type of supplementary irrigation system chosen will be determined by availability of water, the slope of the site and system cost. Disease problems and the need for frost protection will also be important considerations. Overhead sprinkler irrigation can be used to provide frost control in the spring, but wetting the trees can also spread diseases such as bacterial canker (*Pseudomonas* species) and brown rot (*Monilinia* species). Trickle irrigation allows precise metering of water to the trees and can also be used to apply fertilizers with the irrigation water (fertigation).

Soil chemical factors

Soil samples should be taken from representative areas of the potential orchard site to determine the soil's pH and nutrient status. Different soils within the site should be sampled separately. The added cost of multiple tests is trivial compared with the useful information gained. Care should be taken not to plant young trees into soils heavily contaminated with residues of herbicides. 'Simofune' (simazine) and several other persistent herbicides are phytotoxic to young cherries.

Soil pH

Availability of many soil nutrients is influenced by soil pH. Many of the metallic elements become more available to plants as the soil becomes more acidic, whereas availability of a few elements, such as nitrogen, sulphur and potassium, increases as the soils become more alkaline. Phosphorus and many other elements are readily available at pH 6.0–6.5, which is generally considered optimal for cherries and other fruit trees. Trees grown on soils above pH 7.5 are susceptible to deficiencies of iron, zinc, boron and manganese. In these situations, acid-forming fertilizers or gypsum should be used, or foliar nutrient sprays, often containing the elements in a chelated form, can be applied after establishment.

There is evidence that rootstocks differ in their ability to extract minerals under different pH conditions. An example is the preference for Mahaleb in France and the western USA, where calcareous soils are used.

Deficiencies of micronutrients should be anticipated from the results of soil tests and appropriate fertilizers applied before the trees are planted, when the material can be fully worked into the soil.

Cation exchange capacity

CEC is a measure of the soil's ability to acquire and exchange cations, positively charged ions such as H^+, Ca^{2+}, K^+ and Mg^{2+}, from the soil solution. Soils with high CEC can provide, therefore, a greater reservoir of mineral nutrients for the trees. Soils with a high content of either clay or soil organic matter have high CECs, whereas sandy soils commonly have low CECs. Soils with low CEC are incapable of retaining satisfactory levels of nutrients, and trees on these soils are often weak and vulnerable to winter injury, diseases and insects. Incorporating organic manures or planting green crops such as alfalfa or annual grasses in the years before planting the orchard and incorporating them into the soil can increase soil organic matter and CECs.

7.3 Planting and Postplanting Care

Cherries should be planted as early in the spring as possible, as soon as the soil can be worked. If planted too late, little new growth will occur. In regions with mild winters, the trees may be planted during late autumn or early winter, allowing root growth and tree establishment during the dormant season.

Upon receipt from the nursery, cherry trees should be inspected for the quality of their shoot and root systems. Root systems which have dried during transit must be rehydrated and all roots should be inspected for evidence of crown gall.

Crown gall, caused by *Agrobacterium tumefaciens*, is commonly found on the roots of Mazzard seedlings, the Mazzard clone F.12/1 and Colt rootstocks. Preplanting treatments of the soil and root systems against crown gall are particularly important if maiden trees on these rootstocks are to be given a healthy start. If the roots have galls, or the grower suspects the presence of the pathogen on the roots, treatment of the tissue with a solution of *Agrobacterium radiobacter*, or commercial preparations available in the USA and known as Gallex or Galltrol, can help protect against further disease development. Growers should consult local advisory services for recommendations concerning this problem.

Trees should be planted with the graft union slightly above the soil surface, preferably facing the prevailing wind to help alleviate blowing out. The planting hole should be large enough to accommodate the root system after pruning back excessively long or damaged roots. If the sides of the planting hole are glazed (Fig. 7.3), the glazed soil should be removed or tree growth will be retarded (Miller *et al.*, 1990). If the trees were budded high in the nursery (30 cm or above), be careful to plant the trees so that the uppermost lateral root is no more than 5–10 cm below the soil surface. Problems have arisen in orchards where budded trees were planted with the graft union buried and the highest lateral root 15 cm below the surface. This is an old practice to protect the graft union from winter cold damage. In these orchards unwanted scion rooting may occur, and, especially with Mahaleb rootstocks which regenerate roots slowly, root initiation is inhibited deep in the soil by low oxygen levels. In these cases, distal roots may rot and the lower roots grow upwards toward the surface (Fig. 7.4). General care following planting should follow local recommendations.

7.3.1 Tree support

Sweet and sour cherries are traditionally planted without supporting stakes; trees on Mazzard or Mahaleb rootstocks are usually sufficiently well anchored to support themselves. However, with the increasing use of new dwarfing rootstocks, which generally produce trees that are much less well anchored, some form of tree support will need consideration. Often the support will be provided as part of a specialized training system, for example a trellis or hedgerow system supported by post and wires (Chapter 10). Alternatively, a single stake per tree may be preferred for central leader spindle-trained trees.

In all cases strong supports with many years' life are essential. Nothing is

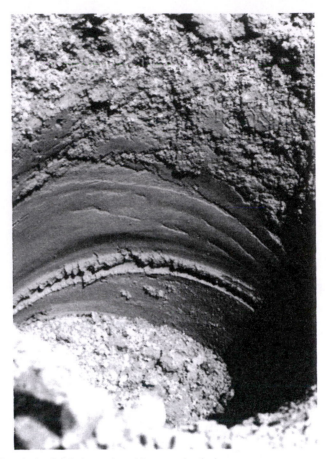

Fig. 7.3.   Glazed planting hole produced by a mechanical auger.

worse than loss of mature cropping trees due to stakes rolling and snapping in gales.

Sweet cherries are extremely sensitive to bacterial canker (*P. mors prunorum*) in certain situations and this problem may be exacerbated by the chafing of branches against wires or tying materials. Great care should be taken to minimize tying damage on susceptible cultivars.

7.4 Summary and Conclusions

The primary restrictions on cherry production are climatic. Cherries are grown profitably near the upper edge of their climatic range where winter temperatures are stable and snow covers the ground. Growth begins in the spring after the risk of late winter freezes is reduced. Planting trees when late winter temperatures can fluctuate can lead to damage if the trees finish rest and begin growth before the last of the extremely cold weather has passed. Furthermore, early spring frosts can reduce or even destroy the year's crop. Extremely warm temperatures

Fig. 7.4. An exposed root and trunk of a young, deep-planted Montmorency scion on Mahaleb rootstock.

in the late spring and early summer result in fruit doubling, an undesirable trait. Rain at bloom or during harvest can also destroy the crop.

Specific sites can offer advantages: proximity to large bodies of water will moderate temperatures, elevated sites provide air drainage, reducing the danger of spring frosts. Low sites with poor air and soil drainage should be avoided. The soil of the proposed site should be examined. Ideal soils for cherries are deep, well-drained soils that also have good water-holding capacity. Well-buffered soils with a pH of between 6 and 7 provide mineral nutrients for the tree. Restrictions to soil drainage, such as compacted layers, abrupt changes in texture or high water-tables, should be avoided or remedied before planting. A plan should be prepared for the orchard. Areas which are marginal for tree growth should be used for other purposes. If dwarfing rootstocks are used, tree supports may be necessary. The orientation to tree rows and the placement of windbreaks will be dictated by a site's topography. The irrigation system should be planned with the requirements of the site and a mature orchard in mind. Soil amendments such as lime or fertilizer should be added to the soil before the orchard is planted. Soil fumigation may be necessary.

The orchard should be planted as early in the spring as possible. The graft union should be just above the ground surface to avoid scion rooting. Make sure there is good contact between the soil and the roots of the tree, avoiding voids in the soil of the planting hole. Supplemental irrigation may be necessary to obtain optimum growth of the new trees. Local recomendations for tree care are often based on experience and should be heeded.

Care and planning in the selection of the orchard site and establishment of the orchard will result in a long-lived and profitable planting.

Acknowledgement

Acknowledgement is made to the Michigan Agricultural Experimental Station for their support of much of the research described in this review.

General Reading

Rendig, V.V. and Taylor, H.M. (1990) *Principles of Soil–Plant Interrelationships.* McGraw-Hill, New York.

References

Anderson, J.L., Ashcroft, G.L., Griffin, R.E., Richardson, E.A., Seeley, S.D., Walker, D.R., Hill, R.W. and Alfaro, J.F. (1976) Reducing fruit losses caused by low spring temperatures. In: *Final Report of the Utah Agricultural Experiment Station to Four Corners Regional Commission.* Logan, Utah.

Ballard, J.K. and Proebsting, E.L. (1972) *Frost and Frost Control in Washington Orchards.* Washington State University Extension Bulletin 634. Washington State University, Pullman, Washington State, USA.

Ballard, J.K., Proebsting, E.L. and Tukey, R.B. (1982) *Critical Temperatures for Blossom Buds: Cherries.* Washington State University Extension Bulletin 1128. Washington State University, Pullman, Washington State, USA.

Beckman, T.G. (1990) Flooding tolerance of sour cherries. PhD thesis, Michigan State University, East Lansing, Michigan, USA.

Beckman, T.G. and Perry, R.L. (1987) The effect of scion and graft union on root growth potential (RGP) of two seedling cherry rootstocks, *Prunus mahaleb* L. and *P. avium* L. 'Mazzard'. *Fruit Varieties Journal* 41, 8–13.

Cannell, R.Q. and Jackson, M.B. (1981) Alleviating aeration stresses. In: Arkin, G.F. and Taylor, H.M. (eds) *Modifying the Root Environment to Reduce Crop Stress.* American Society of Agricultural Engineers, St Joseph, Michigan, USA, pp. 141–192.

Carrick, D.B. (1920) Resistance of the roots of some fruit species to low temperature. *Cornell University Agricultural Experiment Station Memorandum* 36, 613–661.

Childers, N.F. (1983) *Modern Fruit Science*, 9th edn. Horticultural Publishers, Gainesville, Florida, USA.

Day, L.H. (1951) Cherry rootstocks in California. *California Agriculture Experimental Station Bulletin* 725, pp. 7–11.

Fouch, S.B. (1987) Causes of poor sour cherry tree growth on reshaped sites in Northwestern Michigan. MS thesis, Michigan State University, East Lansing, Michigan, USA.

Hillel, D. (1982) *Introduction to Soil Physics*. Academic Press, Orlando, USA, and London, UK.

Howell, G.S. and Perry, R.L. (1990) Influence of cherry rootstock on the cold hardiness of twigs of the sweet cherry scion cultivar. *Scientia Horticulturae* 43, 103–108.

Iezzoni, A., Schmidt, H. and Albertini, A. (1990) Cherries (*Prunus*). *Acta Horticulturae* 190, 111–173.

Keller, J.D. and Loescher, W.H. (1989) Nonstructural carbohydrate partitioning in perennial parts of sweet cherry. *Journal of the American Society for Horticultural Science* 114, 969–975.

Lichou, L., Edin, N., Tronel, C. and Saunier, R. (1990) *Le Cerisier*. CTIFL, Paris.

Miller, A.N., Lombard, P.B., Westwood, M.N. and Stebbins, R.L. (1990) Tree and fruit growth of 'Napoleon' cherry in response to rootstock and planting method. *HortScience* 25, 176–178.

Perry, R.L. (1987) Cherry rootstocks. In: Rom, R.C. and Carlson, R.F. (eds) *Rootstocks for Fruit Crops*. John Wiley & Sons, New York, USA, pp. 217–264.

Perry, R.L. (1990) Cherry rootstocks. *Compact Fruit Tree* 23, 22–27.

Proebsting, E.L. Jr (1970) Relation of fall and winter temperatures to flower bud behavior and wood hardiness of deciduous trees. *HortScience* 5, 422–424.

Richardson, E.A., Seeley, S.D. and Walker, D.R. (1974). A model for estimating the completion of rest for 'Redhaven' and 'Elberta' peach trees. *HortScience* 9, 331–332.

Roth, M. and Gruppe, W. (1985) The effects of waterlogging on root and shoot growth of three clonal cherry rootstocks. *Acta Horticulturae* 169, 295–302.

Rowe, R.N. and Beardsell, D.V. (1973) Waterlogging of fruit trees. *Horticultural Abstracts* 43, 533–548.

Seif, S. and Gruppe, W. (1985) Chilling requirements of sweet cherries (*Prunus avium*) and interspecific cherry hybrids (*Prunus* × ssp.). *Acta Horticulturae* 169, 289–294.

Unger, P.W., Eck, H.V. and Musick J.T. (1981) Alleviating plant water stress. In: Arkin, G.F. and Taylor, H.M. (eds) *Modifying the Root Environment to Reduce Crop Stress*. American Society of Agricultural Engineers, St Joseph, Michigan, USA, pp. 61–96.

Webster, A.D. (1984) The effects of fumigation and rootstock on the growth of young cherry trees planted in land previously cropped with cherries. *Journal of Horticultural Science* 59, 349–358.

Westwood M.N. (1978) *Temperate Zone Pomology*. Freeman, San Francisco, USA.

Wildman, B. (1984) *Deep Tillage to Improve California Soils*, Slide Set 84/124. California Cooperative Extension Service, University of California at Davis.

8 Flowering, Pollination and Fruit Set

M. THOMPSON

Department of Horticulture, Oregon State University, Corvallis, OR 97331-2911, USA

A panoramic view of a cherry orchard in full bloom is a glorious sight. Many city-dwellers make springtime pilgrimages to seek inspiration from this experience. For the orchardist, it is a time for renewed hope, but also the beginning of a period of heightened anxiety. A bountiful harvest depends upon the successful completion, during the next 2–3 months, of the annual reproductive cycle for this crop.

A working knowledge of floral biology and of the factors that influence the normal progression from flower initiation through flower development, fruit set and fruit maturation will prove valuable to all orchard management decision makers. Interruption of any of these stages can lead to crop reduction or even complete crop failure.

8.1 Flowering Habit of Sweet and Sour Cherries

In sweet cherries, most of the flowers are borne on long-lived (10- to 12-year) spurs on 2-year and older wood, while very few are borne near the base of 1-year shoots. Each bud, which is surrounded by several bud scales, contains two to four flowers and many buds are crowded at each spur. Thus, at full bloom the tree appears extremely floriferous with all branches literally roped with flowers.

Sour cherries also produce flower buds laterally, both on 1-year wood and on spurs, and there are two to four flowers per bud. Although they generally produce a higher proportion of flowers on 1-year wood than do sweet cherries, the relative amount of blossom borne on shoots vs. spurs varies considerably, both between cultivars and within a cultivar, depending upon cultural management. In one study, Meteor bore 30% of its fruit on 1-year wood, whereas Montmorency bore 68% in this location. In another report, Montmorency produced only about 35–45% of its fruit on 1-year shoots.

Sour cherry spurs are short-lived compared with those of sweet cherry. Most of the vigorous and productive fruiting spurs are concentrated on 2- to 3-year-old wood while a few less productive spurs persist on 4- to 5-year-old wood. On trees low in vigour due to stresses such as viruses, winter injury, etc., a high percentage of the buds on the very short 1-year shoots will form flower buds with the result that there are no, or very few, leaf buds to develop into spurs the next season. The result is that 2-year and older wood becomes barren and unproductive. For optimal spur production and sustained high yields of sour cherry it is necessary to maintain sufficient annual shoot growth to develop both flower and vegetative buds.

8.2 Flower Structure and the Process of Fertilization

Individual cherry flowers consist of the outer sepals, petals, stamens consisting of a filament and a pollen-bearing anther, and a pistil (Fig. 8.1). The pistil is comprised of the upper stigmatic surface, the style, and the ovary wherein lie a pair of ovules. One ovule (the primary ovule) develops into the seed while the other (the secondary ovule) aborts very early. At anthesis, or full bloom, each ovule lies in a small open space within the ovary (the locule) and is attached to the central part of the ovary (the placenta) by the funiculus. The ovule consists of two outer layers of tissue (integuments) which wrap around the ovule except for a narrow passageway (the micropyle) leading from the locule of the ovary to the inner tissue of the ovule (nucellus) within which the mature embryo sac

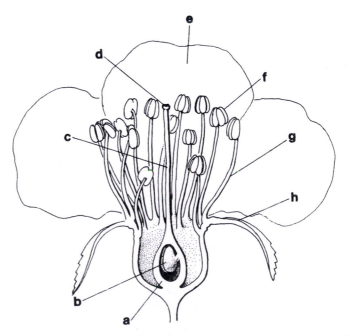

Fig. 8.1. Longitudinal section of a cherry flower: (a) ovary, (b) ovules, (c) style, (d) stigma, (e) petal, (f) anther, (g) filament, (h) sepal.

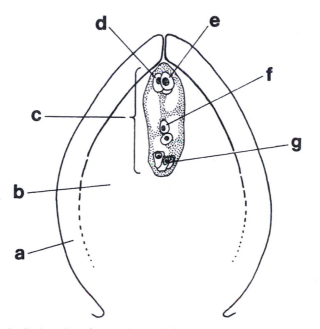

Fig. 8.2.   Longitudinal section of an ovule containing a mature embryo sac: (a) integument, (b) nucellus, (c) embryo sac, (d) synergid, (e) egg, (f) polar nuclei, (g) antipodals.

(megagametophyte) develops (Fig. 8.2). The embryo sac is an elongated structure which differentiates within the nucellus at the micropylar end of the ovule shortly before anthesis. It arises from a single cell (macrosporocyte), which is first distinguished by its larger size, larger nucleus and denser-staining cytoplasm. This cell undergoes meiosis (a series of two cell divisions, the first of which halves the number of chromosomes), which results in four megaspores. While three of the megaspores atrophy, one of them enlarges and undergoes three mitotic divisions, yielding eight haploid cells within the elongated embryo sac. Normally the embryo sac (or megagametophyte) has matured and is functional at anthesis. At this stage, three of the cells are clustered at the micropylar end (the egg cell and two synergid cells), two are centrally located (the polar cells) and three are located at the opposite end of the sac (the antipodals). The egg cell is destined to fuse with a sperm and develop into the embryo and eventually the mature seed. The two polar cells fuse with each other and with a second sperm cell and subsequently develop into a temporary nourishing tissue (the endosperm). One of the synergids plays an important, albeit temporary, role in that it is associated with the penetration of the pollen tube into the embryo sac. The three antipodal cells disintegrate very early and have no known function.

Differentiation of anthers and maturation of pollen occur shortly before anthesis. In Montmorency, meiosis in the anthers was not observed until bud break (green tip to half green). The anther consists of a wall of specialized cells surrounding the two lobes which contain the pollen sacs. Pollen differentiates from the mass of dense-staining, thin-walled sporogenous cells (microsporocytes) occupying the central parts of the anther lobes. Each microsporocyte undergoes meiosis to yield four microspores which have half the number of chromosomes of the original cell. As pollen matures from the microspore, the pollen wall differentiates and one mitotic division occurs, resulting in two cells. One of these divides again to form two sperm cells, while the other becomes a vegetative cell which has a role in growth of the pollen tube down the style and ovary. Thus, the mature pollen grain (the microgametophyte) consists of three cells surrounded by a complex, relatively thick and often elaborately sculptured wall with apertures for pollen tube emergence.

Following pollination, pollen germinates on the stigma. Pollen tubes grow down the style into the ovary, pass across the locule, enter the ovule through the micropyle and penetrate the embryo sac via one of the synergids. Once inside the synergid, the tip of the tube bursts and ejects the two sperms, one of which fuses with the nearby egg cell while the other moves toward, and fuses with, the polar nuclei.

Adequate fruit set in cherries is absolutely dependent upon normal flower development, effective pollination and successful fertilization with subsequent seed development.

8.3 Floral Initiation

Although bud burst and flowering in spring provide a dramatic, visible manifestation of the annual reproductive process, floral initiation, the invisible, microscopic changes which convert a vegetative bud into a floral bud, actually commences the previous summer. The exact date of initiation depends upon the cultivar and the physiological condition of the tree, which, in turn, depends upon the weather and cultural practices.

Great precision in determining the anatomical changes associated with floral initiation of Montmorency sour cherry has been achieved with scanning electron microscopy (Diaz *et al.*, 1981). Very subtle alterations can be detected with this method because the entire shoot apex can be viewed in three dimensions and at very high magnifications. Changes in the shoot apex (the meristem) within the bud become evident about 4 weeks after anthesis. The first evidence of transformation of a vegetative bud to a floral bud is a broadening and flattening of the rounded meristem. Then two to four (usually three) small lateral protuberances, representing primordial bracts which subtend each flower, emerge on the periphery of the meristem. By about 7 weeks after anthesis, individual flower primordia are evident in the axil of each bract. Over the summer, sepal primordia first differentiate in a pentagonal whorl and are followed, in inward succession, by a whorl of petals, whorls of stamens and finally the pistil primordium which appears in the centre. By the time of leaf fall, all the floral parts are in evidence, albeit in an immature stage (Fig. 8.3).

Although morphological development ceases with the onset of dormancy, certain physiological processes continue during midwinter (Felker *et al.*, 1983). Prominent among these are starch accumulation and enlargement of nuclei and nucleoli. After the chilling requirement is satisfied and as temperatures rise in

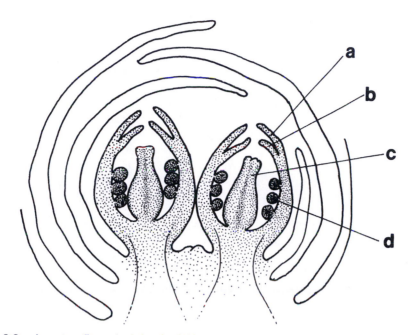

Fig. 8.3.   Immature flower bud showing initial stages of all floral parts at the end of September: (a) sepal primordium, (b) petal primordium, (c) pistil primordium, (d) stamen primordium.

spring, buds swell and floral parts mature rapidly, reaching their final form at anthesis.

8.3.1 Cultural factors influencing floral initiation

The onset of floral initiation is triggered by internal signals believed to involve a balance of endogenous hormones and various assimilates. However, certain orchard management practices can influence both the time of floral initiation and the number of buds initiated. For example, a tree with very low vigour will initiate too many flowers. Severe pruning that stimulates excessive vigour will both delay floral initiation and decrease the amount of blossom produced. Irrigation practices that prolong vigorous vegetative growth also delay the onset of floral initiation. Different rootstocks (see Chapter 5), apart from their influence on vigour, can influence both the age at which flowers first appear on a young budded tree and the amount of flowering in subsequent years. Growth regulating chemicals (Chapter 11) can also influence floral initiation in cherry.

8.4 Factors Affecting Flower Development

8.4.1 Summer temperatures

Unusually hot weather during the critical early stages of floral initiation may cause abnormal flower development, including pistil-like or petal-like appendages which replace anthers on the end of filaments and double pistils, which may accompany these abnormal flowers or occur in otherwise normal flowers. Thus, hot weather during the previous summer can cause varying percentages of unmarketable, double fruit. This problem occurs in both sweet and sour cherry and the amount varies with the season, the cultivar and the growing region. The problem is particularly severe in an unusually warm season in the warmer growing regions. Bing and Napoleon are particularly prone to heat-induced double and malformed fruits. In Utah, where summers are very warm, Bing may have 40–80% double fruits. Cultivars such as Angela, Sweet Ann, and Utah Giant have been selected for this region because they are not subject to this problem.

8.4.2 Inadequate winter chilling

Normal bud break and blossoming are dependent upon completion of a specific amount of chilling followed by temperatures warm enough to stimulate growth. Insufficient chilling can be a serious problem in regions with mild winters.

Chilling requirement is commonly expressed as the number of hours the plants are exposed to temperatures between 0 and 7°C after the onset of rest, which occurs in late October–November in the northern hemisphere. Chilling requirements for cherry cultivars generally range between 750 and 1400 h (Seif and Gruppe, 1985). In general, sour cherries tend to have somewhat higher chilling requirements than sweet cherries.

Once the chilling requirement is satisfied, bud break occurs after a specific amount of heat has accumulated. Both the base temperature (the temperature below which there is little or no flower development) and the total heat units required to achieve anthesis vary among cultivars.

Insufficient chilling reduces fruit set due to delayed and erratic blooming, abscission of some buds and poor synchronization of the main crop cultivar with the pollinizer cultivar. In regions with frequent mild winters, cultivars with the lowest chilling requirements should be chosen. Alternatively, application of dormancy-breaking agents may be effective. Hydrogen cyanamide has been successful in advancing bloom of insufficiently chilled Sam sweet cherry trees by as much as 18 days (Snir and Erez, 1988). Buds of this cultivar were most responsive to the treatment immediately after 350 h of chilling had been accumulated. In regions prone to spring frosts, growers are advised to consider cultivars with high chilling requirements because they flower somewhat later in the spring.

8.4.3 Freeze damage

In some cherry-growing regions, low-temperature damage to flower buds is the most important limiting factor for yields. Freeze damage may occur in autumn before buds are acclimatized, during the winter 'rest' period (endodormancy), during late winter after the chilling requirement is satisfied but before de-acclimatization takes place (ecodormancy), and especially during deacclimatization as the buds swell and progress towards anthesis. Flower bud resistance to low temperature changes rapidly and predictably in response to both temperature changes and stage of floral development.

Deep supercooling – a mechanism of hardiness

During the period of flower bud endodormancy, buds survive low temperatures by avoiding freezing through the process of deep supercooling.

During subfreezing temperatures, water in the buds remains liquid even though there are ice crystals in the woody tissues immediately subtending the buds. This barrier to ice crystal growth (the ultimate cause of tissue death) is established in early autumn and can be induced by cool temperatures. Supercooled water is unstable. When crystallization starts, it progresses rapidly and destructively throughout a primordial flower, but it does not propagate from one flower to another. Thus, after a damaging freeze, some buds may appear dark brown (dead) while others are green (alive). Even within a bud, one of the flowers may be dead, while the others are alive. Partial injury to an individual flower is rare.

When the ability to deep-supercool is acquired in autumn, the temperature at which 50% of the florets are killed (LT_{50}) is $-18°C$ to $-20°C$. This is lower than that which kills woody tissues. Thus, very early freezes may kill shoots that bear living flower buds. By late November, the wood becomes hardier than the buds and the LT_{50} declines to $-20°C$ to $-22°C$. This value, termed the minimum hardiness level (MHL), remains constant throughout endodormancy, even when temperatures rise above freezing. During long periods of subfreezing temperatures, the LT_{50} slowly decreases by $1-2°C$ per day, and may reach as low as $-35°C$. This level of hardiness is more a function of the duration of subfreezing temperatures than the low temperature reached on any given day. However, subsequent periods with temperatures above freezing raise the LT_{50} very rapidly, at $1-2°C$ per hour, but only back to the MHL. These relationships have been incorporated into a model that estimates the LT_{50} from measured air temperatures (Andrews and Proebsting, 1987).

Deacclimatization and frost prediction

With sufficient chilling to complete endodormancy, warm temperatures start the developmental sequence that leads to anthesis. During this deacclimatization period, prediction of bud hardiness becomes more complex (Andrews *et al.*, 1983; Andrews and Proebsting, 1987). At first, buds enlarge without showing other external changes, the MHL rises to near $-18°C$ and the ability to deep-supercool is lost. As visible development proceeds, individual buds lose hardiness, from MHL of $-18°C$ to $-5°C$. During this period, buds vary considerably in their level of hardiness. Also, the ability to deep-supercool may be reacquired temporarily during periods of freezing temperatures. Progressively, more and more buds lose resistance so that, shortly before anthesis, all the buds are killed at $-2°C$ to $-3°C$, the freezing-point of water in tissues.

Orchard management to minimize freeze damage

Healthy trees have the best potential to resist flower bud damage during winter. Other than utilizing optimal orchard practices to maintain tree health and abundant flower buds, little can be done to induce greater cold-hardiness within dormant flower buds. Frost protection with wind machines or orchard heaters is often effective during winter freezes, provided that one knows what the critical temperatures are for operating these systems. The computer model referred to above, coupled with laboratory tests of resistance, provides the necessary information.

During spring, wind machines and water, applied either over or under the trees, and sometimes supplemented by orchard heaters, are the preferred methods of frost protection. Critical temperatures for each of nine stages which reflect the decreasing hardiness levels during spring bud development have been established for Bing sweet cherry in Washington (Table 8.1) (Proebsting and Mills, 1978) and for Montmorency sour cherry in Michigan (Dennis and Howell, 1974). These can be used as a guide for activating frost protection systems. However, the best frost protection method is to establish the orchard on a frost-free site.

Table 8.1. Mean values for bud freeze-kill temperatures (°C) in controlled freezing tests of Bing sweet cherry at Prosser, Washington State, USA. (From Proebsting and Mills, 1978.)

Bud stage	T_{10}	T_{50}	T_{90}
Dormant	−35 to −14.3*		
First swell	−11.1	−14.3	−17.2
Side green	−5.8	−9.9	−13.4
Green tip	−3.7	−5.9	−10.3
Tight cluster	−3.1	−4.3	−7.9
Open cluster	−2.7	−4.2	−6.2
First white	−2.7	−3.6	−4.9
First bloom	−2.8	−3.4	−4.1
Full bloom	−2.4	−3.2	−3.9
Postbloom	−2.1	−2.7	−3.6

*Wide range, depending upon the time of year.

8.5 Pollination and Fruit Set

An abundant crop of cherries is dependent upon the successful completion of a sequence of reproductive events. Failure or deficiency of any aspect of this process can result in crop reduction or total loss. The first requirement is the availability of an adequate source of viable, compatible pollen. Secondly, there must be an effective transfer of pollen when stigmas are receptive. Thirdly, pollen tubes must grow down the style and enter the ovule during the period when embryo sacs have matured and ovules are viable. And, finally, double fertilization and subsequent growth and development of the embryo and endosperm must occur to provide the necessary stimulus for fruit development.

8.5.1 Pollen source – incompatibility

The phenomenon of incompatibility prevents inbreeding and promotes out-crossing in natural populations of many plant species. However, in orchard management it is an obstacle to efficient pollination because of the problems associated with pollinizer trees and the necessity of introducing pollinators into the orchard to transfer pollen. In sweet cherries, incompatibility is genetically controlled by a single gene with many alternative forms (multiple alleles), designated S_1, S_2, S_3, etc. Any pollen tube bearing an allele in common with either of the two alleles in the somatic tissue of the pistil fails to achieve fertilization because its growth is inhibited part way down the style. Obviously, pollen from the same plant is rejected (self-incompatibility), but fertile pollen from any other cultivar sharing an allele is also rejected (cross-incompatibility). Compatibility relationships in sweet cherry are always reciprocal.

Cultivars have been classified into groups according to their compatibility relationships, those bearing the same two S alleles being placed in the same group (Tehrani and Brown, 1992). All cultivars within each group are cross-incompatible and cannot be planted as pollinizers for each other, whereas members of one group are all compatible with cultivars in all other groups. Thus far, a total of 13 groups have been defined, each having a specific pair of the six alleles which have been identified. All cultivars which do not fit into any of the known groups, and thus must have a new, undetermined combination of S alleles, are placed in group O and designated as universal donors because they are compatible with members of all known groups. Further research is necessary to assign specific S alleles to cultivars in this heterogeneous O group. Examples of cultivars in each group are given in Table 8.2.

A major breakthrough in sweet cherry incompatibility was reported by Lewis and Crowe (1954). Using X-irradiation of pollen, they induced a mutation, now designated S' (occasionally S_F = self-fertility), which destroys the inhibitory reaction and permits pollen to function on its own pistil. Also, trees bearing this self-fertility allele are compatible with all other cultivars. Thus, they are con-sidered as universal pollen donors. This trait is relatively simple to incorporate into new cultivars through breeding because it is transmitted to half of the progeny when the non-mutated S allele in each of the two parents is different, or to all of them if it is the same. Stella, the first self-compatible cultivar, which was released by Agriculture Canada's Summerland Research Station in 1971, has S_3S_4 alleles but the S_4 is mutated to the 4' type. In controlled pollination experiments, it has been demonstrated to be fully self-fertile, that is, fruit set was equally high with self-pollen or pollen from another cultivar. Subsequently, Stella has been used in breeding programmes to incorporate self-compatibility into new cultivars such as Lapins, Sunburst, New Star, Sylvia and Starkrimson. Other self-compatible advanced selections from breeding programmes at Sum-merland, British Columbia, and at the Horticultural Research Institute of Ontario at Vineland are currently under evaluation by cooperators in Europe, United States, Australia and Asia. Self-fertile cultivars offer distinct advantages: the entire orchard can be planted to one high-value cultivar, thus simplifying orchard operations as well as marketing; yields are more reliable (even in years when

Table 8.2. Pollen incompatibility groups in sweet cherry and their respective S alleles. (Taken largely from Knight, 1969.)

Group	Cultivars
Group I (S_1S_2)	Baumann's May, Bedford Prolific, Bigarreau van Piringen, Black Circassian, Black Downton, Black Eagle, Black Tartarian A, Black Tartarian B, Carnation, Early Rivers, Knight's Early Black, Leicester Black, Ronald's Heart, Roundel Heart, Sparkle, Tillington Black
Group II (S_1S_3)	Abundance, Belle Agathe, Black Cluster, Black Elton, Black Heart B, Caroon B, Frogmore Early, Gil Peck, Jubilee, Kristin (= NY 1599), Lamida, Maiden's Blush, Merton Bigarreau, Merton Bounty, Merton Crane, Merton Favorite, Semis de Burr, Shrecken, Sodus, Van, Venus, Victoria Black, Waterloo, Windsor
Group III (S_3S_4)	Bing, Büttners Röte, Emperor Francis, Lambert, Merton Marvel, Mezel Nos 1 and 2,* Napoleon (= Royal Ann), Ohio Beauty, Star, Vernon
Group IV (S_2S_3)	Allman Gulrod, Kassin's Frühe Hertz, Kentish Bigarreau, Late Amber, Ludwig's Bigarreau, Merton Premier, Sue, Velvet, Victor, Viva, Vogue, West Midlands Bigarreau, Weston's Amber, Yellow Spanish
Group V (S_3S_5)	Late Black Bigarreau, Turkey Heart
Group VI (S_3S_6)	Early Amber, Elton Heart, Gold (= Stark's Gold), Governor Wood, Merton 42, Merton Heart, Ohio Beauty, Rival, Turkish Black
Group VII (S_4S_5)	Black Republican, Bradbourne Black, Burlat, Hooker's Black, Mezel No. 3,* Moreau
Group VIII (S_2S_5)	Büttners Späte Röte Knorpelkirsche, Noir de Schmidt, Peggy Rivers, Poolse, Schmidt
Group IX (S_1S_4)	Chinook, Giant, Hudson, Merton Late, Merton Reward, Rainier, Red Cluster, Red Turk, Ursula Rivers, Viscount, Yellow Glass
Group X (unknown)	Bigarreau Jaboulay (= Lyons), Black Tartarian D, Ramon Oliva, Rodmersham Seedling
Group XI (unknown)	Cryall's Seedling, Guigne D'Annonay (= Annonay), Knight's Bigarreau
Group XII (unknown)	Caroon A, Monstreuse de Mezel, Newington Late Black, Noble
Group XIII (S_2S_4)	Ulster
Group O (universal donor)	Beeve's Heart, Belle D'Orleans, Bigarreau Gaucher, Black Oliver, Black Tartarian E, Bowyer Heart, Cleveland Bigarreau, Dikkeloen, Florence, Goodneston Black, Guigne très Précoce, Hedelfingen (Vineland Clone), Heinrich's Riesen, Knight's Early Black G, Malling Black Eagle, Merton Glory, Mumford Black, Noir de Guben, Norbury's Early Black, Nutberry Black, Ord, Smoky Heart, Strawberry Heart, Valera, Vic, Vista, Wellington A, Wellington B, White Heart, Zweit Frühe Wadenswil

*A cultivar which falls into more than one group indicates mistaken identity, that is, more than one clone has been represented by the same name.

bee flight is limited due to poor weather conditions because set is not entirely dependent upon bees); and they can be used as universal pollen donors for any self-incompatible cultivar.

Compatibility relationships are less clear-cut in sour cherry. Many sour cherry cultivars have proved to be fully self-compatible, e.g. Schattenmorelle, Fanal, English Morello, Early Richmond, Suda and several cultivars released from breeding programmes in Hungary and Romania. Reports for Montmorency are contradictory. This cultivar is usually considered to be fully self-compatible. However, in some experiments, cross-pollination has resulted in higher fruit set. Recent evidence that pollen of both Montmorency and Meteor grows more slowly in their own styles than does pollen of other cultivars supports the concept of partial self-incompatibility for these two cultivars (Lansari and Iezzoni, 1990). However, on a practical basis, Montmorency is effectively self-compatible because full crops, and even oversetting in some cases, occur in solid commercial plantings.

Self-incompatibility in sour cherry has been reported for several cultivars, e.g. Ostheim, Crisana, Pándy, Tschernokorka, Lyubskaya, Pamyat, Vavilova, Plodorodnaya Michurina, Körös, Ottawa 391, Homer, Kentish Red and Chase Morello. Only a few cases of cross-incompatibility have been observed. There is a report of an interincompatible group which includes Montmorency, Bruin Waalse and Rode Waalse. Other reports of cross-incompatibility are Ottawa 391 × Körös, Ottawa 391 × Ostheim, Tschernokorka × Meteor and Tschernokorka × Ostheim (Redalen, 1984). The tetraploid nature of sour cherry, as compared with the diploid sweet cherry, probably accounts for the more complex inheritance and expression of incompatibility.

8.5.2 Pollinizers

Since most major sweet cherry cultivars are self-incompatible, requiring cross-pollination to obtain fruit set, important decisions must be made when establishing a new orchard. A cultivar selected for use as a pollinizer should have fruit with the highest possible economic value while fulfilling its other requirements. It must produce abundant, viable pollen with proven cross-compatibility with the main crop cultivar. Bloom time of both cultivars must be reliably synchronous so that pollen is available as soon as flowers begin to open. Maximum pollination efficiency is achieved by planting equal numbers of each cultivar and alternating their locations down the rows so that each tree is surrounded by a compatible pollinizer. However, because it is unlikely that the two cultivars will have equal value, it is necessary to balance the desire to provide for maximum pollination with the economics of planting the minimum number of less valuable pollinizer trees. Therefore, the common practice is to plant a pollinizer tree at every third position in every third row, a design that provides the minimum acceptable number of pollinizers but ensures that every main crop tree is adjacent to a pollinizer. This proximity is necessary because bees forage primarily on two, or possibly three trees per trip and they tend to move to the nearest trees. When trees are planted in a hedgerow, at closer spacing within the row than between rows, it is important to distribute pollinizer trees along the rows because bees tend to forage on the closest trees, i.e. along the rows.

8.5.3 Pollination and pollinators

Assuming that an adequate number of effective pollinizer trees is available, the next step is to ensure that sufficient pollen is transferred in a timely manner. Poor pollination is a common cause of less-than-optimum fruit set, especially in crops which require cross-pollination and bees to effect pollen transfer. A simple method to determine if this factor is responsible for poor fruit set is to provide supplementary pollination. Flowers on some limbs can be hand-pollinated and fruit set compared with that on open-pollinated limbs. If fruit set is significantly increased on hand-pollinated limbs, it can be concluded that inadequate pollination is contributing to reduced fruit set in the orchard.

The introduction of honey bees into the orchard is essential for fruit set on self-incompatible cultivars. However, even for self-fertile cherries, bees will increase fruit set by improving the distribution of pollen within the flower or between flowers on the same tree. Bees should be moved into the orchard when about 10% of the flowers have opened (Mayer *et al.*, 1986). If brought in before cherry blossoms are open, their foraging habit may become established on another species. If bees are brought in later, some of the cherry flowers may have become non-functional. Flowers on all competing vegetation, such as weeds, or ground covers, should be eliminated so that foraging bees will focus on the cherry trees. All toxic pesticide sprays should be avoided immediately before and during the period when bees are foraging in the orchard.

A grower must be confident about the quality of hives placed in the orchard. In some fruit-growing regions, colony-strength regulations are in place that stipulate minimum standards for hives regarding size of comb, presence of an active laying queen, number of adult bees, amount of brood (eggs, larvae and pupae), sufficient bee food in the form of honey or its equivalent and minimum levels of disease. A good, strong colony suitable for orchard pollination consists of 20,000–30,000 adult bees.

To assess the quality of the hive, its contents can be examined by an experienced person or one can observe flight activity from the hive. A good pollinating unit will have over 100 bees per minute entering the hive entrance under warm (above 18°C), sunny conditions with winds less than 16 km h^{-1}. Nearly equal bee activity in front of all hives indicates uniform-strength colonies. Another method of determining if there are adequate bees present to effect pollination is to count the number of bees working the trees. When weather conditions are optimal for bee flight, slowly move around a tree and count the number of bees working the bloom for 1 min. An average of 25–35 bees for each of ten mature cherry trees counted indicates that sufficient pollinators are present.

The number of hives recommended varies from three to five strong hives per hectare, depending upon local weather conditions during the pollination period. Ideally, hives should be distributed uniformly throughout the orchard. However, for more efficient handling, they may be placed in groups of four to six and the distance between groups adjusted depending upon the need. It is not advisable to place hives in larger groups at greater distances because the foraging ranges of bees in adjacent groups will not overlap sufficiently.

In regions subject to adverse weather conditions during pollination, it is critical to have luxury levels of bees available to take advantage of short periods suitable for bee foraging. Bees exhibit very little activity below 12°C but, as temperatures rise to 20°C, there is a rapid increase in flight and foraging. Winds greater than 25 km h^{-1} will slow or altogether stop bee flights. Rain inhibits bee flight and high humidity prevents anthers from dehiscing and shedding pollen. Also, light intensities falling below 6000 foot-candles (1200 μmol m^{-2} s^{-1}) will reduce foraging behaviour.

If an orchard exhibits chronic oversetting with consequent reduction in fruit size and tree growth, it may be advisable to limit the number of bee colonies or their time in the orchard in order to reduce the number of flowers pollinated.

8.6 Plant Nutrition and Fruit Set

Nutritional levels which produce an optimum balance between vegetative and reproductive growth are important for sustained, annual high fruit production. The objective of fertilizer management is to produce healthy trees which produce moderate tree vigour and an abundance of strong, well-developed flowers. Late-summer leaf analyses provide a good measure of the nutritional status of the vegetative phase of the tree. However, there is increasing evidence that there may be different optimal nutritional levels for the reproductive phase, levels which are not detectable by the usual leaf analyses (Chaplin and Westwood, 1980).

8.6.1 Boron

Boron (B) seems to be in particularly high demand in floral tissues during the time of fruit set. In several fruit species, including both sour and sweet cherry, foliar applications of B just before leaf fall have resulted in greatly increased boron levels in flower buds and increased fruit set (Hanson, 1991).

B applied to trees in the autumn moves, along with other metabolites, from the leaves into the adjacent buds, where elevated levels are maintained and expressed in flowers at anthesis. Although the mechanism of B action is not understood, the increased B content in flower buds often improves fruit set. Furthermore, a positive response to B may occur in trees not considered deficient according to midsummer leaf level standards.

8.6.2 Nitrogen

With apple, a summer application of nitrogen after the termination of shoot elongation influenced flower quality the next spring (Williams, 1965). Externally, flowers appeared 'stronger' and stigmas remained receptive for 13 days as compared with 7 days for controls. Internally, embryo sacs continued to enlarge after those of controls stopped growing, ovule longevity was almost twice as long and the growth rate of fertilized eggs and embryo sacs was faster. Based on vegetative symptoms, trees selected for the experiments were not deficient in nitrogen but, nevertheless, the extra nitrogen had a positive effect on ovule quality and longevity.

More research is needed on cherries to define more precisely the nitrogen and perhaps other nutrient requirements of floral tissues for optimal fruit set. It may be that the midsummer leaf level standards for good vegetative growth do not accurately reflect this requirement.

8.7 Postpollination Failures

Although reduced fruit set is commonly attributed to poor pollination, post-pollination failures, while difficult to interpret, can have equally devastating effects. Specifically, these failures result from disruption of any of the reproductive events following pollen transfer.

8.7.1 Stigma receptivity

At anthesis, the stigmatic surface in cherries appears wet with a cellular secretion and the papillate cells are turgid and greenish in colour and appear receptive. Within 1–2 days these thin-walled cells collapse, the secretion is no longer evident, the surface turns brownish, and stigmas no longer appear receptive. However, this appearance is misleading because even 9–10 days after anthesis it has been observed that pollen germinates and tubes grow as rapidly in the styles as on the day of anthesis (Stösser and Anvari, 1983). Thus, stigma receptivity and stylar senescence are not often limiting factors for fruit set of cherry.

8.7.2 Pollen tube growth rate

Research directed towards understanding postpollination failure has focused on the various factors affecting pollen tube growth rate, ovule longevity and the effective pollination period (EPP). These parameters are known to be influenced by temperature, nutrition and genetics. It is well established that, within the temperature range (about 5–25°C) at which pollen tubes are growing in flowers on orchard trees, their rate of growth increases as temperature rises. Under continual warm conditions, cherry pollen tubes may reach the embryo sac in 2 days. However, in one field study involving Napoleon flowers, pollen tubes reached the micropyle by the fourth day after pollination when the mean postbloom temperature was 13.7°C but not until the sixth day at 10.6°C (Guerrero-Prieto et al., 1985). Cultivar differences in pollen tube growth rates and in the response of growth rate to temperature, are well documented in many species. When comparing growth of Bada, Rainier and Corum sweet cherry pollen in Napoleon styles at four temperatures, all grew at the same rate at 12.3 and 16.4°C, but Bada tubes grew the fastest at lower temperatures (9.9 and 7.3°C): 3.0 mm day^{-1} for Bada as compared with 1.4 mm day^{-1} for Corum and Rainier. However, during the 2 years of this study, when postbloom temperatures in the orchard averaged 6.6 and 12.6°C, fruit set was equally high with pollen of all three cultivars, indicating that the slower-growing pollen tubes of Corum and Rainier in the cooler year was not a limiting factor for fruit set. More research is needed to determine if the more rapid growth of Bada pollen tubes under cool temperatures actually translates into increased fruit set in seasons with marginal temperatures for fruit set.

8.7.3 Ovule viability and longevity

The development of a high percentage of normal, functional embryo sacs is essential for a high fruit set. Thus far, no studies of cherry cultivars have found a high enough proportion of inherently abnormal embryo sacs to account for poor fruit set. Although Montmorency has been reported to have 25–40% incomplete, degenerating or abnormally immature (presumably non-functional) embryo sacs at anthesis, this amount is not sufficient to fully account for the low fruit set observed in some years.

Longevity of normally developed ovules and embryo sacs may be a more important limiting factor for fruit set than defective embryo sacs. Reports on the duration of ovule viability in cherries vary considerably because of different methods of assessing viability, different temperatures during anthesis and post-anthesis and different cultivars being studied. Two methods are used to assess ovule viability. One involves preparing serial sections through ovules collected periodically after anthesis and microscopically examining the embryo sac and ovular tissues. Absolute determinations are difficult with this method because it depends, to some extent, upon subjective interpretations by the investigator. The other method, considered to be more accurate, involves periodic dissections of intact ovules, squashing them on a slide, staining with aniline blue and making observations with a fluorescence microscope. With this stain, viable ovules either fluoresce very faintly or not at all, whereas non-functional ovules fluoresce intensely as a result of the interaction of cellular degeneration products with the stain. At anthesis, this latter method clearly distinguishes the secondary ovule from the primary within each pistil, and is useful to determine the longevity of the primary ovules. As time elapses after anthesis and ovules remain unfertilized, or if fertilization does occur and embryo or endosperm degenerates early, ovule senescence can be observed with this fluorescent stain.

The effect of temperature on ovule longevity is well known in many species, with warmer temperatures hastening senescence and cooler temperatures prolonging it. With a constant temperature of 20°C a high percentage of ovules of three sweet and two sour cherry cultivars began to senesce within 1–2 days after anthesis, whereas at 5°C senescence did not begin until 5 days after anthesis (Postweiler *et al.*, 1985). In another study, under field conditions with mean postbloom temperatures of 10°C, Napoleon sweet cherry ovules appeared to be viable 12 days after anthesis (Guerrero-Prieto *et al.*, 1985).

Other cultivar differences in ovule longevity have been reported. In sweet cherry cultivars, Schmidt had particularly advanced embryo sacs at anthesis and very early degeneration (beginning before anthesis) when compared with Bing, Windsor and Hedelfingen (Eaton, 1962). This very short ovule viability of Schmidt probably contributes to chronic low yields in this cultivar. In a field study of sour cherries, ovules of Köröser Weichsel and Kelleriis 14 began to senesce 4 days after anthesis whereas, under the same environmental conditions, only 8 and 9% of Schattenmorelle ovules were senescent at 6 and 10 days after anthesis, respectively (Stösser and Anvari, 1982).

A relationship has been shown between ovule longevity and fruit set on Schattenmorelle sour cherry trees growing in the same orchard in the same year. Tree-to-tree variability in fruit set, as well as reduced fruit set on low vs. high branches on the same tree, was associated with variation in ovule longevity. These differences may relate to mineral nutrition or perhaps to carbohydrate resources. Variation in nutritional status among trees in an orchard can occur due to site differences in soil composition or to differential fertilizer distribution or uptake. Variations due to location within the canopy may be related to preferential allocation of nutrients to the upper part of the tree but are more likely to be due to reduced carbohydrate reserves associated with excessive shading.

Effective pollination period

The concept of effective pollination period (EPP) has been developed to express the period during which pollination must occur in order to achieve fruit set. EPP represents the period of ovule longevity minus the period of time required for pollen tubes to grow down the style and effect fertilization. The longer the EPP, in units of days after anthesis, the more prolonged the pollination period may be and still achieve fertilization and fruit set. On the other hand, with a very short EPP, pollination must occur immediately after anthesis.

At optimum temperatures for fruit set, pollen tubes grow fast enough and ovules remain viable long enough for fertilization to occur in a high percentage of ovules. While cool temperatures prolong the viability of ovules, they also reduce the growth rate of pollen tubes, but not necessarily to the same extent. At very low temperatures, pollen tubes may grow too slowly to reach the ovule before it senesces. Although moderately cool temperatures do generally prolong the EPP, this does not always result in higher fruit set. When temperatures are very warm during and after anthesis, the EPP is very short because of reduced ovule longevity. Under these conditions, it is extremely important to have pollinators in the orchard as soon as the first flowers open so that pollen can be transferred within as short a period as possible.

In the orchard, the EPP is determined by comparing fruit set from controlled pollinations made on successive days after anthesis. The upper limit of the EPP is the day before there is a significant decrease in fruit set. In Prosser, Washington State, the EPP for Bing, Lambert and Chinook sweet cherries in different years ranged from 4 to 7 days (Toyama, 1980). In Corvallis, Oregon, it was 4 to 5 days for Napoleon (Guerrero-Prieto *et al.*, 1985). This duration was also reported for sweet cherries in Stuttgart, Germany (Stösser and Anvari, 1983). In Michigan, the EPP for Montmorency was from anthesis to 4 days postanthesis in a warm year and from 2 to 4 days postanthesis in a cool year (Furukawa and Bukovac, 1989). Under more extreme weather conditions than those prevailing in these studies, and considering other cultivars, the EPP for cherries may be as short as 1–2 days or as long as 1 week.

8.7.4 Premature fruit drop

Final fruit set, which varies considerably from year to year, depends upon the percentage of flowers that remain on the tree to develop into mature fruits. Depending upon the species, the cultivar and especially the abundance of flowers, good cherry crops result from about 20–65% set. In cherries, there are three periods when flowers or fruit are lost. The exact timing of these drops varies with the cultivar, the year and the growing conditions.

A detailed 3-year study made of drops in the self-pollinated Montmorency and Early Richmond sour cherries in Wisconsin serves as an example of this phenomenon (Bradbury, 1929). The average total drop was 64% for Montmorency and 70% for Early Richmond.

The first drop occurred about 2–2.5 weeks after full bloom. Virtually all first-drop fruit had pollen tubes in the ovarian cavity, indicating that pollination had occurred. In years of poor bee activity, especially in cross-pollinated cherries, the first drop will also include some unpollinated flowers. Dissection of ovaries revealed two shrunken ovules as compared with developing flowers which had one shrunken and one larger, functional ovule. The shrunken primary ovules in first-drop fruits contained non-functional embryo sacs, either defective ones or normal ones which had lost viability before fertilization occurred.

The second drop occurred about 1 week later, and sometimes overlapped the first drop. Ovules in these drops were of three types: those with aborted embryos; those having vestiges of degenerated or defective embryo sacs; and those that were unpollinated.

The third drop (sometimes called the 'June' drop whether it occurs in June or early July) occurred about 3 weeks after the second. Over 90% of these aborting fruits contained embryos, but their smaller-than-normal size indicated that growth had been arrested (Fig. 8.4).

The exact cause of the heavy loss of flowers and young fruit in cherry is

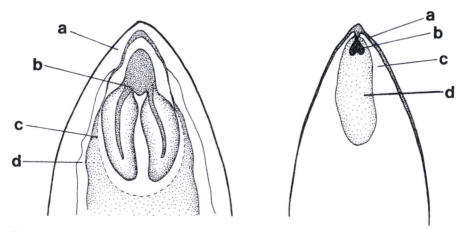

Fig. 8.4. Diagrams of ovules in a developing fruit (left) and an ovule in an aborting fruit of the June drop: (a) integument, (b) embryo, (c) nucellus, (d) endosperm. (Redrawn from Bradbury, 1929.)

unknown. Competition for nutrients or environmental stresses such as moisture deficiency, winter injury, viruses, mechanical damage or defoliation have been suggested as contributing factors. Evidence supporting the concept of competition for nutrients as a causative factor is as follows. First, when flowers are thinned, either by hand or by loss from freeze damage, a larger percentage of remaining flowers set fruit. Secondly, the percentage of flowers lost in the first drop affects the percentage lost in the second; that is, a heavy first drop tends to be followed by a lighter second drop, and a light first drop tends to be followed by a heavier second. Also, lighter-than-usual first and second drops may be followed by a heavier-than-usual third drop. Occasional excessive third drops have also been observed in sweet cherries. This annual variation in the percentages lost at each of the three drops suggests an influence of weather, either by having a direct effect on fruit growth or by indirectly influencing uptake or mobilization of nutrients to the developing fruit.

Although poor fruit set is commonly a limiting factor for efficient cherry production, too much fruit set can also be a problem. When trees overbear, there results a serious reduction in fruit size and quality. The challenge for the grower is to manage all cultural factors, so far as possible, so that they interact to produce the maximum crop that the tree can bear without sacrificing fruit quality in the current season and good production and tree growth in subsequent years.

References

Andrews, P.K. and Proebsting, E.L. Jr (1987) Effects of temperature on the deep supercooling characteristics of dormant and deacclimating sweet cherry flower buds. *Journal of the American Society for Horticultural Science* 112, 334–340.

Andrews, P.K., Proebsting, E.L. Jr and Gross, D.C. (1983) Differential thermal analysis and freezing injury of deacclimating peach and sweet cherry reproductive organs. *Journal of the American Society for Horticultural Science* 108, 755–759.

Andrews, P.K., Proebsting, E.L. Jr and Lee, G.S. (1987) A conceptual model of changes in deep supercooling of dormant sweet cherry flower buds. *Journal of the American Society for Horticultural Science* 112, 320–324.

Bradbury, D. (1929) A comparative study of the developing and aborting fruits of *Prunus cerasus*. *American Journal of Botany* 16, 525–542.

Chaplin, M.H. and Westwood, M.N. (1980) Relationship of nutritional factors to fruit set. *Journal of Plant Nutrition* 2, 477–505.

Dennis, F.G. and Howell, G.S. (1974) Cold hardiness of tart cherry bark and flower buds. *Michigan State University Agricultural Experiment Station Research Report* 220.

Diaz, D.H., Rasmussen, H.P. and Dennis, F.G. Jr (1981) Scanning electron microscope examination of flower bud differentiation in sour cherry. *Journal of the American Society for Horticultural Science* 106, 513–515.

Eaton, G.W. (1962) Further studies on sweet cherry embryo sacs in relation to fruit setting. In: *1962 Report of the Horticultural Experiment Station and Products Laboratory*. Vineland, Ontario, Canada, pp. 26–38.

Felker, F.C., Robitaille, H.A. and Hess, F.D. (1983) Morphological and ultrastructural development and starch accumulation during chilling of sour cherry flower buds. *American Journal of Botany* 70, 376–386.

Furukawa, Y. and Bukovac, M.J. (1989) Embryo sac development in sour cherry during the pollination period as related to fruit set. *HortScience* 24, 1005–1008.

Guerrero-Prieto, V.M., Vasilakakis, M.D. and Lombard, P.B. (1985) Factors controlling fruit set of 'Napoleon' sweet cherry in western Oregon. *HortScience* 20, 913–914.

Hanson, E.J. (1991) Sour cherry trees respond to foliar boron applications. *HortScience* 26, 1142–1145.

Knight, R.L. (1969) *Abstract Bibliography of Fruit Breeding and Genetics to 1965*. Commonwealth Agricultural Bureau, Technical Communication No. 31. East Malling, Maidstone, Kent.

Lansari, A. and Iezzoni, A. (1990) A preliminary analysis of self-incompatability in sour cherry. *HortScience* 25, 1636–1638.

Lewis, D. and Crowe, L.K. (1954) The induction of self-fertility in tree fruits. *Journal of Horticultural Science* 29, 220–225.

Mayer, D.F., Johansen, C.A. and Burgett, D.M. (1986) *Bee Pollination of Tree Fruits*. Pacific Northwest Cooperative Extension Bulletin 0282. Washington State University Cooperative Extension Service, Oregon State University Extension Service and University of Idaho Cooperative Extension Service, Pullman, Washington, 10 pp.

Postweiler, K., Stösser, R. and Anvari, S.F. (1985) The effect of different temperatures on the viability of ovules in cherries. *Scientia Horticulturae* 25, 235–239.

Proebsting, E.L. Jr and Mills, H.H. (1978) Low temperature resistance of developing flower buds of six deciduous fruit species. *Journal of the American Society for Horticultural Science* 103, 192–198.

Redalen, G. (1984) Cross pollination of five sour cherry cultivars. *Acta Horticulturae* 149, 71–75.

Seif, S. and Gruppe, W. (1985) Chilling requirements of sweet cherries (*Prunus avium*) and interspecific hybrids (*Prunus* × ssp.). *Acta Horticulturae* 169, 289–294.

Snir, I. and Erez, A. (1988) Bloom advancement in sweet cherry by hydrogen cyanamide. *Fruit Varieties Journal* 42, 120–122.

Stösser, R. and Anvari, S.F. (1982) On the senescence of ovules in cherries. *Scientia Horticulturae* 16, 29–38.

Stösser, R. and Anvari, S.F. (1983) Pollen tube growth and fruit set as influenced by senescence of stigma, style and ovules. *Acta Horticulturae* 139, 13–22.

Tehrani, G. and Brown, S.K. (1992) Pollen-incompatibility and self-fertility in sweet cherry. *Plant Breeding Reviews* 9, 367–388.

Toyama, T. (1980) The pollen receptivity period and its relation to fruit setting in stone fruits. *Fruit Varieties Journal* 34, 2–4.

Williams, R.R. (1965) The effect of summer nitrogen applications on the quality of apple blossom. *Journal of Horticultural Science* 40, 31–41.

9 Cherry Nutrient Requirements and Water Relations

E.J. Hanson[1] and E.L. Proebsting[2]

[1]Department of Horticulture, Michigan State University, East Lansing, MI 48824-1325, USA; [2]Irrigated Agriculture Research and Extension Center, Washington State University, Prosser, WA 99350-9687, USA

9.1 Introduction

Cherries, although of several distinct fruit crop species, are less widely grown than many other temperate fruits. Thus, the production methods currently in vogue for cherries have evolved from information and experience gained with these other crops. Our knowledge of the water and nutrient requirements of cherries certainly fits this pattern.

Cherries are the first of our deciduous fruit crops to mature each summer, the trees having only 2 months from flowering to develop their fruits to maturity. Fruit growth, shoot growth and flower bud initiation all occur at the same time, often competing as sinks for the tree's substrates. Consequently, after harvest the rest of the growing season is spent with no fruit on the tree. Cherry fruit are very small in comparison with most other fruits, making it essential that numerous flower buds are formed, set and mature into fruits for commercially economic yields to be achieved. Large fruit are required of sweet cherries for the most lucrative markets and, as hand or chemical thinning is either impractical or impossible, optimum supplies of water and nutrients are essential if fruit size is to be maximized.

Characteristics such as these put the two cherry species in a very interesting class by themselves. Discussion of water relations and nutrient requirements for cherries must, therefore, be read in the context of the unique characteristics of the two species and of the paucity of scientific information that bears directly on them.

9.2 Mineral Nutrition of Cherries

9.2.1 The nutritional requirements of cherries

Fertilization programmes for cherries inevitably vary according to region and are largely determined by local climate and soil conditions. The tree's requirements for some nutrients have been studied in some detail and are generally well understood. The requirements for other nutrients have only been studied to a limited extent and our understanding of the cherry tree's requirements for these elements is incomplete.

How a tree responds to applications of specific nutrients depends primarily on the nutrient status of the tree at the time of application, the tree's health and the efficiency of uptake. Perhaps the most universal basis for evaluating tree nutrient status is by tissue analysis, usually leaf analysis. Nutrient concentrations in cherry leaves are an accurate indicator of whether nutrition is limiting tree performance and whether trees are likely to respond to nutrient applications. Tissue analysis also provides the grower with a means of anticipating nutrient shortages before crop performance is adversely affected. The commonly accepted sufficiency values for various mineral nutrients, expressed as percentage dry weight or parts per million (ppm) of leaf tissue dry weight are shown in Table 9.1.

Soil testing also provides information that is useful in developing cherry fertilization programmes. Soil pH should be monitored through soil testing, since pH influences the availability of most nutrients to plants. However, direct measurements of soil nutrient levels are generally a crude indicator of orchard nutrient status.

Some nutrient deficiencies or excesses cause specific symptoms to develop which are of diagnostic value to growers. However, symptoms are easily misdiagnosed, and they usually indicate that a disorder is relatively advanced and that crop performance has already been adversely affected. For these reasons, growers need to be familiar with the symptoms of common nutritional problems, rather than relying solely on symptomatology to guide fertilization practices.

In this chapter, each nutrient element will be discussed separately. Current knowledge of the effects of deficiencies and toxicities on tree appearance will be discussed, followed by information on optimum tissue nutrient levels and common fertilization techniques.

Table 9.1. Standard nutrient levels (% dry weight or ppm) in the leaves of sweet and sour cherry. (Values compiled from Huguet, 1984, and Shear and Faust, 1980.)

Nutrient	Deficient	Sufficient	Excessive
N (%)		2.2–3.4	> 3.4
P (%)	< 0.08	0.16–0.4	> 0.4
K (%)	< 1.0	1.0–3.0	> 3.0
Ca (%)		0.7–3.0	
Mg (%)	< 0.24	0.4–0.9	> 0.9
S (%)		0.13–0.8	
B (ppm)	< 20	25–60	> 80
Cu (ppm)		5–20	
Fe (ppm)		20–250	
Mn (ppm)	< 20	20–200	
Zn (ppm)	< 10	15–70	

Nitrogen

Most orchard soils do not supply adequate nitrogen (N) to meet the requirements of cherry trees. Thus, annual applications of N-containing fertilizers are required in most orchards for optimum growth and yields. N-deficient trees are generally low in vigour and productivity. Since extension shoot length is reduced, trees may produce abundant spurs and flower buds. Leaves on deficient trees are pale green in colour and may develop a reddish tint late in the season. Older, basal leaves typically exhibit the earliest and most severe symptoms, although all leaves may be affected on severely deficient trees. Affected leaves are also smaller and likely to drop prematurely, while fruits may also be smaller and earlier to mature on N-deficient trees.

Although heavy N applications are seldom directly toxic to plants, excessive rates may adversely affect trees by stimulating vigorous, late-season growth which is subject to winter injury. Fruit maturity is also delayed on heavily N-fertilized trees. Excessive rates of N have been observed to both increase and decrease cherry firmness, although N fertilizers may affect fruit firmness indirectly by either increasing cherry size or delaying maturity.

There is some disagreement concerning deficient, optimum and excessive leaf N levels for cherries. However, as a general rule, N levels in cherry leaves should be maintained between 2.2 and 3.4%. Higher levels may be occasionally recommended for young trees to encourage rapid growth and establishment of the tree canopy. Mature trees may benefit from slightly lower leaf N levels if continued vigorous growth is not desired.

The rates of N fertilizer required annually to maintain optimum leaf N levels are influenced by numerous factors. Mature cherry orchards typically receive 50–130 kg N ha^{-1}. Highest rates are required on light-textured sandy soils in humid regions, where leaching losses are likely to be highest. Heavy soils that are high in organic matter supply the most native N, so lower fertilizer rates are required on these soils.

Orchard floor management practices also influence N availability and the requirements of cherry trees. If vigorous sod (grass) covers are maintained, these compete with the trees for N and higher rates may be needed to supply sufficient N to the trees. Clean cultivated orchards may, in contrast, require lower N rates to achieve similar N levels in the trees. Standard inorganic N fertilizers, irrespective of type, all appear equally effective as sources of N for cherries.

N fertilizers are typically broadcast over the soil surface in the spring before growth begins. Autumn fertilization or combinations of autumn and spring applications may be equally effective. Foliar N sprays are not commonly applied to cherries and, although some studies indicate that foliar urea sprays may increase cherry yields, others report no benefits. Trees very low in N might be most likely to respond to such sprays. Cherries are also very easily injured by biuret, a by-product of urea production, so sprays of high-biuret urea may prove toxic to cherry leaves.

Cherries can be fertilized effectively by injecting N into trickle irrigation systems (fertigation). This technique appears to improve the efficiency of fertilizer use, since similar leaf N levels and yields can be achieved by injecting only half of the N rates required when fertilizer is broadcast over the soil surface. Efficiency is probably improved because injection allows growers to control fertilization timing and rates and to supply N directly to the root system, while reducing leaching losses common with broadcast applications.

Careful planning is required for effective fertilizer injection through trickle irrigation systems. The irrigation system must provide uniform rates of water application throughout the orchard, since non-uniform systems will result in both inadequate and excessive applications of N. Various types of injectors are available to introduce the fertilizer solution into the irrigation lines. Calcium nitrate, ammonium nitrate and urea are soluble in water and commonly used as N sources for injection. The annual fertilizer requirement is usually split into three or more applications, beginning at bloom and ending 4–8 weeks later.

Phosphorus

Cherry trees seldom respond to applications of phosphorus (P) under field conditions. Since deficiencies are uncommon, symptoms of P deficiency on cherries are not fully characterized. General symptoms on other tree fruit crops include weak, slender shoot growth and dark green to purplish-coloured leaves. Root growth is usually reduced substantially by P deficiency.

Leaf P concentrations in midsummer are normally between 0.16 and 0.40%. Concentrations decrease through the season, so samples taken earlier are normally higher and late-season samples are lower in P. Levels of P associated with deficiency in cherry leaves have not been firmly established, but are likely to be near to 0.08% and below.

There is some evidence that newly planted trees on soils low in phosphorus may respond to P applications by producing more rapid initial growth. However, it is generally believed that P applications are of little value in orchards as long as soil P is sufficient to maintain orchard floor species.

Potassium

Potassium (K) deficiencies occur periodically in cherries in most production areas of the world. The first symptoms of K deficiency are a slight upward curling of the leaf margins. The undersides of leaves then turn bronze in colour and the margins of leaves eventually become necrotic or 'scorched'. Leaf curling and scorching are usually most severe on the basal leaves of the current season's growth, but may also appear on terminal leaves. Both shoot extension growth and leaf size are reduced. Symptoms are most severe on trees carrying a heavy crop of fruit, since the fruits accumulate relatively large amounts of K at the expense of the other parts of the tree. Leaves may drop prematurely from severely affected trees.

Leaf analysis provides a useful measure of the K status of trees and the likelihood that applications will be of benefit. Sufficient leaf K concentrations are between 1 and 3%, whereas trees with leaf levels < 1.0% are generally considered deficient. Leaf K levels are influenced by the sampling time and crop load. Highest levels would be expected in leaves sampled early in the season or from trees carrying a light crop load.

K deficiency can be corrected readily by K applications to orchard soils. Rates of 100–200 kg K_2O ha^{-1} are usually sufficient to ameliorate the problem.

Applications normally increase leaf K levels and eliminate symptoms within 1 year. Foliar sprays of K may be of some value in stone fruits grown on soils that fix large quantities of K. Sprays of KNO_3 applied after bloom appear most effective.

Excessive applications of K should be avoided. High soil K levels can prevent trees from absorbing sufficient quantities of magnesium (Mg) or calcium (Ca) and may induce deficiencies of these elements.

Calcium

Ca deficiency has not been reported in cherry orchards. Symptoms induced experimentally on young trees include light brown to yellow markings on leaves. Leaves may become tattered with numerous holes, and shoot growth is minimal. Cherry leaves normally contain 0.7–3.0% Ca. Cherry fruits usually contain approximately 12.0 mg Ca 100 g^{-1} fresh weight of flesh.

Fruit Ca has been implicated in determining sensitivity to rain-induced cracking at harvest time and in fruit postharvest potential. High crop loads and sprays of gibberellic acid applied to improve fruit set or firmness may reduce fruit Ca levels, while sprays of growth retardants that reduce shoot growth and gibberellin biosynthesis may increase fruit Ca.

Sprays of Ca, commencing 10 days before harvest, are applied in some sweet cherry production areas in an attempt to reduce rain-cracking. Unfortunately, the results of these sprays are very inconsistent (Chapter 12).

Magnesium

Although not often recorded, Mg deficiency is occasionally a problem in UK sweet cherry orchards. The symptoms of deficiency include an interveinal browning and necrosis, starting first on basal (older) leaves. This browning may start in the middle of leaves or progress inward from the margins. Bright red or yellow coloration typically precedes and may border necrotic areas and in severe cases affected leaves may fall prematurely.

Healthy cherry leaves typically contain 0.4–0.9% Mg. Leaf Mg levels below 0.24% are associated with the appearance of deficiency symptoms on young, container-grown cherry trees. Mg deficiency is often exacerbated on soils very high in K and is particularly severe on sweet cherries grown in containers of medium which is comprised of peat and sand with no loam content. Symptoms of Mg deficiency are often much more severe on trees on Mazzard than on Colt rootstock.

Sulphur

Sulphur (S) deficiency in cherries (and tree fruits in general) is uncommon. However, shortages were identified in orchards in Washington State, USA. Symptoms are very similar to N deficiency, and include pale yellow, small leaves, and poor shoot growth. Necrosis of leaf margins may occur on severely affected trees. S concentrations less than 100 ppm in immature tip leaves indicate a deficiency. Deficiency can be corrected by applications of gypsum or ammonium sulphate or the use of irrigation water containing more than 0.7 ppm of S.

Boron

Boron (B) deficiency or toxicity can potentially occur in many cherry orchards. Deficient trees exhibit little shoot growth. Some buds may fail to open in the spring, whereas others may open and then shrivel and die. Shoots may grow for some time and then tips cease growth and die. Leaves are distorted in shape, with irregular serration. Leaves may cup or roll in a downward direction and feel thick and leathery. These vegetative symptoms typically develop only if leaf B concentrations are lower than 20 ppm.

B applications may improve fruit set and yield in sour cherry trees containing B levels as high as 30 ppm and exhibiting no visible deficiency symptoms. This is probably partly explained by B's beneficial influence on pollen germination and growth down the style. Flower clusters have a large demand for B during blossoming if fruit set is to be fully effective.

In one report, cracking of sweet cherries in Oregon was reduced by ground applications of B. The test trees may have been marginally deficient in B, since applications also improved foliage colour. However, leaf B levels were not reported, so it was impossible to confirm tree B status. Sweet cherries commonly crack when rain occurs just prior to harvest, but this characteristic is generally independent of tree B status. Whether cherry cracking is influenced by B deficiency is, therefore, unclear.

Cherries are relatively sensitive to excessive B levels. Toxicity symptoms include a dieback of twigs, accompanied by gumming. Severe toxicity may induce gumming along main limbs and trunks. Leaves are normal in shape and size but necrotic zones may develop along main veins. Flower buds may fail to open and few fruit are set. Leaf B levels above 100 ppm are unusually high and toxicity symptoms can be expected when levels are 140 ppm or higher.

Foliar sprays or ground applications of B-containing fertilizers can be used to correct deficiencies. Fertilizer borate is granular and well suited for bulk blending and ground applications. Solubor is a useful B source for foliar or ground sprays. Rates of 1–2 kg B ha^{-1} will correct deficiencies. Annual applications of 0.5–1 kg B ha^{-1} are an effective maintenance programme for orchards in low-B regions.

Copper

The copper (Cu) requirements of cherry trees have not been studied in detail. Cherry leaves typically contain 5–15 ppm Cu. No published descriptions of Cu deficiency symptoms on cherries occurring naturally or induced experimentally could be found. Cu-deficient apple and pear trees develop a condition called wither tip, in which the terminal leaves turn yellow and eventually wither and fall. The tips of twigs may die also.

Work in Tasmania, Australia, has recently suggested that Cu may be involved in determining sensitivity to rain-induced cracking. Also, research prior to 1950 showed that sprays of between 0.01 and 0.1% Cu sulphate applied early in the season reduced the incidence of cracking at harvest. Whether this is a direct effect on fruit sensitivity or an indirect effect by influencing fruit size and maturity is not known.

Iron

Iron (Fe) deficiency occurs periodically in cherries produced in all areas of production but is most common in arid regions. Terminal leaves on deficient trees turn chlorotic to bright yellow between the veins. Veins remain green and stand out against the rest of the leaf. Although terminal leaves are first affected, symptoms may progress in a basal direction to include older leaves. Tissue along the margins of severely affected leaves may die. Leaf Fe concentrations may not correlate closely with the severity of symptoms, so a definitive deficiency level has not been established. Leaves from healthy trees may contain 20–250 ppm Fe.

Fe deficiency is common on alkaline soils (pH > 7.0) and may be more severe on poorly drained soils or where root growth and drainage are restricted by compacted soil horizons. Since Fe chlorosis is essentially caused by high soil pH, the condition is most effectively corrected by reducing soil pH. Improved soil drainage alone will often correct Fe chlorosis when drainage is inadequate. Various Fe chelates provide some benefit when used as repeated foliar applications, although these are usually an expensive remedy. Unfortunately, soil applications of Fe-containing fertilizers are not effective corrective measures.

Manganese

Manganese (Mn) deficiency has been reported in several cherry production regions. Symptoms include interveinal chlorosis, similar to that caused by zinc (Zn) deficiency. Chlorosis starts at the margins of leaves and progresses inward between the main veins. Leaves are typically smaller and shoot growth is inhibited in proportion to the severity of the leaf symptoms. In mild instances, symptoms may be most apparent on spur leaves, although both spur and shoot leaves are affected in severe cases. Fruit yields and quality may be reduced severely, with fruit from affected trees being small and lacking in juice, although well coloured and firm.

Deficiencies of Mn are most common where soil pH is relatively high (> 7.0), because Mn availability in alkaline soils is limited. Mn deficiency may be confused with shortages of Fe or Zn, since shortages of these nutrients induce somewhat similar symptoms to Mn shortage and are also prevalent on alkaline soils. Symptoms are likely to develop when leaf Mn levels are less than 20 ppm. Sufficient leaf Mn levels range from 20 to 300 ppm.

Mn deficiency can be corrected by foliar sprays of Mn sulphate, Mn-containing fungicides or various chelated Mn products available commercially. Sprays applied after petal fall and again several weeks later appear to be most effective. Direct injection of Mn solutions into trunks and large limbs has corrected deficiencies experimentally, but such treatments are likely to be too labour-intensive for commercial use. Application of Mn fertilizers to the soil may not be effective if Mn availability is limited by high soil pH.

Zinc

Zinc (Zn) deficiency occasionally occurs in cherries. Shortages cause a reduction in leaf size and irregular mottling and chlorosis of leaves, and leaves may drop prematurely. Shoots typically fail to elongate normally. Shortening of the internode distance between leaves results in tufts or 'rosettes' of leaves at the tip of shoots. Affected limbs may be confined to certain portions of the tree or distributed uniformly throughout the tree. Buds on the previous year's growth may fail to open, resulting in lengths of shoots which are devoid of leaves. Fruit size and soluble solids content are severely reduced.

Zn deficiency is most common where soil pH is relatively high, but shortages have also been reported on slightly acidic soils. Leaf Zn concentrations in normal trees range from 15 to 70 ppm. Some work suggests that trees with leaf Zn concentrations as high as 19 ppm may be marginally deficient. Deficiency of Zn is likely if leaf levels are less than 10 ppm.

Deficiencies are usually corrected by Zn sprays or soil applications. Zn sulphate or chelated Zn compounds are generally effective. Dormant sprays of Zn sulphate appear more effective than applications during the growing season, whereas chelated Zn materials are preferred for soil applications. Annual or alternate-year applications are usually needed where a deficiency exists.

9.2.2 Influence of rootstocks on nutrient status

Common cherry rootstocks have been observed to influence nutrient levels in the leaves of the scion cultivar. Perhaps the most consistent effects have been on leaf K concentrations. Cherry trees on *Prunus avium* (Mazzard) accumulate higher leaf K levels than trees on *Prunus mahaleb* rootstocks. Mazzard rootstocks may also result in higher leaf Ca, B, N and Mn concentrations. Studies in Norway showed that leaf N was much higher on trees of three sweet cherry cultivars if worked on F.12/1 (Mazzard) compared with Colt, while leaf Ca and Mg levels were higher on the Colt trees.

Rootstocks may directly affect the nutrient status of the scion if they vary in efficiency of nutrient uptake or impart indirect effects on leaf nutrient concentrations by altering cropping levels or tree vigour. The mechanisms of how cherry rootstocks influence leaf nutrient levels are not clear. However, rootstock selection may influence the nutrient status of trees and the fertilization requirements of orchards and should be given appropriate consideration when planning a new orchard.

9.2.3 Tissue analysis

The nutrient status of cherries is commonly assessed by measuring nutrient concentrations in trees. Although different tissues and sampling times may be used, some standard recommendations have been established. Whole leaves (including petioles) are typically sampled in midsummer. Specific sampling times may vary by region and are typically defined in relation to a phenological stage of development. Each sample requires an adequate number of leaves to accurately represent the sampling area. Typically, 100 leaves are collected from the middle of the current season's shoots. Leaves should be sampled from all sides of trees from shoots approximating the average length in the orchard. Leaves are usually washed briefly, dried, ground and analysed. Mature orchards may benefit from sampling every 3–4 years. More frequent sampling may be useful in new orchards where little previous cropping history is available.

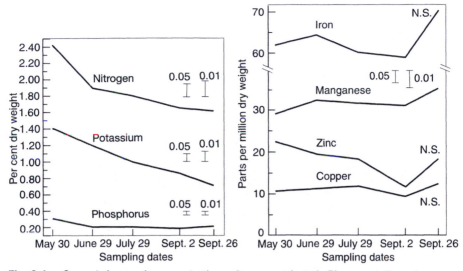

Fig. 9.1. General changes in concentrations of some nutrients in Bing sweet cherry leaves under Utah conditions (from Christensen and Walker, 1964). N.S., not significant.

Interpretation of leaf nutrient levels is based on deficient, sufficient and toxic concentrations of nutrients in cherry leaves (Table 9.1).

Recommended levels for some nutrients are based on considerable research and are likely to be reasonably accurate. Research on other nutrients is limited, so standard levels should be viewed as only tentative suggestions. Values in Table 9.1 are generalized from several sources, so levels for specific locations may vary slightly. When nutrient concentrations fall below the deficiency level, deficiency symptoms are likely to develop and nutrient applications are justified.

Samples taken at other times during the season can be generally evaluated by considering how leaf nutrient concentrations change with time (Figs 9.1 and 9.2). For example, early-season samples are likely to contain higher concentrations of N, P and K and lower levels of Mg.

9.3 Water Relations of Cherries

The water requirements of temperate-zone fruit trees have been reviewed (Landsberg and Jones, 1981; Chalmers *et al.*, 1983) and, although these reviews do not cover cherries specifically, on the broad scale cherry trees are similar to apples and peaches. Principles involving the soil, the plant and the atmosphere describe and determine fruit tree water relations. While the soil and atmosphere (climate) present conditions that differ little from species to species, as described in Chapter 7, the cherry tree under cultivation has some interesting and somewhat different characteristics on which this discussion will dwell.

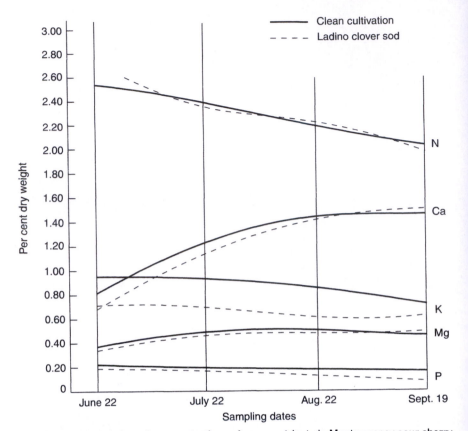

Fig. 9.2. General changes in concentrations of some nutrients in Montmorency sour cherry leaves under Pennsylvania conditions (from Smith *et al.*, 1961).

9.3.1 Soil and atmospheric effects on a tree's water balance

Soil is the reservoir for moisture that is available for tree growth. The depth and texture of the soil determines its capacity to hold water. The volume of soil explored by tree roots defines the amount of water that is available from a given soil. Available soil moisture, i.e. the water that can be extracted by the tree, is held at water potentials in the range between -0.1 and -1.5 megapascals (MPa), which approximate field water-holding capacity and the permanent wilting percentage, respectively.

The atmosphere near leaf surfaces develops very low water potentials on warm, sunny days, with values well below -100 MPa. This strong potential gradient through the tree drives water movement from the soil to the top of the tree and out to the atmosphere. Evaporation at the leaf surface, through the stomatal apertures, causes a potential difference between the stomata and the roots that pulls water to the top of the tree. The water in capillary columns in the xylem tissues is under tension and these fine columns of water are usually sufficiently cohesive to remain unbroken as the water is pulled to the top.

The water that moves from the roots through the tree and is evaporated at the leaf surface, transpiration, is near zero at night and reaches maximum values during midday. The rate of transpiration is controlled by the size of the stomatal apertures (stomatal resistance), the evaporating power of the atmosphere, the soil moisture content and the resistance to water movement throughout the tree. Since the resistance of the tree is relatively constant and since soil moisture, especially under irrigation, changes slowly, the rate of transpiration is determined mostly by stomatal resistance and by the weather. Transpiration, which is usually measured in grams of water lost per unit leaf area per hour, ranges diurnally from near 0 at night to over $1.0 \text{ g dm}^{-2} \text{ h}^{-1}$ during the day. This is associated with leaf water potentials from near -0.5 at night to -1.5 to -2.0 MPa during the day. Leaf water potentials on clear days might reach -2.5 to -3.0 MPa when the available soil moisture is depleted to near the permanent wilting percentage. It is the stomatal closure on such days that prevents much lower values. Osmotic adjustment in the leaf and other tissues serves to minimize wilting and other injury under these extreme conditions.

9.3.2 Water needs of the cherry tree

Sweet cherries, if irrigated to their full requirements, may use over 100 cm of water ha^{-1} in a season in the USA. Twenty-five-year-old cherry trees in Washington State were trickle-irrigated from a system controlled by a sensor/valve placed under the tree. When soil moisture was drawn below -0.6 MPa, the valve opened to allow irrigation until enough water reached the sensor site to raise the soil moisture back above -0.6 MPa. When the sensor was placed 75 cm deep and 100 cm from the trickle line, the trees demanded and received over a metre of irrigation. However, when the sensor was placed 30 cm deep directly below the emitter, the trees only used 76 cm water ha^{-1} with apparently equal horticultural performance by the trees.

Cherries, especially during the first half of the season before the crop matures, are irrigated to avoid plant water deficits that might reduce fruit growth rates and produce small cherries. This calls for irrigating before the soil reaches its maximum allowable depletion, usually in the range of 40–60% of the total available water in the soil.

9.3.3 Irrigation techniques

Irrigation may be applied in one of several ways. It can be channelled down furrows close to the tree or broader areas beneath the trees may be temporarily flooded. Other alternatives are to apply the water through sprinklers, usually positioned beneath the trees close to the soil surface, or to use a trickle system which slowly emits water through narrow capillaries or nozzles placed on the soil. Irrigation by furrow, flood or sprinkler wets the entire soil surface and, if enough water is applied, refills the whole root volume. Trickle-irrigation, in contrast, may be applied at one or a few points, may wet only a portion of the root volume and should be applied much more frequently than traditional methods. The consequences of these differences to successful cherry cultivation have not been determined.

Plants respond to drought stress by closing stomata and reducing growth.

Closed stomata reduce gas exchange, thereby reducing both transpiration and photosynthesis. Reduced terminal growth and fruit growth are the most obvious manifestations of early drought stress. Efforts to control excessive terminal growth while minimizing effects on fruit growth are successful on peaches and pears when irrigation is controlled properly. This practice, labelled regulated deficit irrigation (RDI), is not likely to be successful on cherries because the fruit matures very early in the summer concurrently with the most active period of terminal growth. Since large cherries are the most desirable in the market place, cherry trees are generally irrigated to minimize plant water deficits. The objective is usually to maintain leaf water potentials as high as is practical, given the fairly low values that are reached on warm days even with soil moisture near field capacity.

Cherries are harvested in May to July or early August in the northern hemisphere, depending on the regional climate. In most areas, the hottest part of the summer occurs after harvest. It is easy to overlook cherry orchard irrigation during this time, when carbohydrate reserves are being produced and flower buds for next year's crop are differentiating and developing. High evaporative demand and less rigid attention to irrigation needs often lead to very low plant water potentials. Cherry trees respond to low water potentials by dropping leaves and by quite substantial osmotic adjustment. Interestingly, the leaves that remain on droughted trees have been observed to photosynthesize at respectably high rates.

9.3.4 The effects of drought on tree growth and cropping

An experiment conducted in central Washington State, USA, explored responses of cherries to severe drought that might be imposed by failure of irrigation water supplies. This region receives less than 200 mm of rain a year, but only about 50 mm during the growing season. Trickle-irrigated trees subjected to low irrigation rates throughout the season reduced soil moisture to near the permanent wilting percentage except close to the emitters, where the soil moisture was near field capacity throughout the summer. The volume of wetted soil was proportional to the rate of irrigation. The average midday leaf water potential for trees irrigated at 100% of evaporation was -1.4 MPa compared with -1.6 and -2.2 at 50% and 15% of evaporation, respectively. Trees that were trickle-irrigated at 50% of evaporation received only 45 cm-ha ha^{-1} during the main irrigation season, 1 June to 1 September. These trees produced normal yields without a reduction of fruit size in the year of the drought and in subsequent years.

The very dry treatment (15% of evaporation) began on 3 August of the previous year. By early September defoliation began on the smallest interior spur leaves. Wilting was observed in mid-September and, by late October, relatively few leaves remained on the trees. These leaves remained green, failed to abscise and were killed by low temperatures. Normal trees retained green leaves until late October, and then dropped all their foliage by mid-November.

The soil remained dry throughout the winter with very little precipitation. Bloom was delayed 1–2 days on the droughted trees and the flowers appeared to be slightly smaller than flowers from normally irrigated trees. Vegetative growth was short with small leaves and the trees wilted by early June. They remained wilted throughout the summer, dropping a few leaves each week, under the irrigation regime of 15% of evaporation. Small branches began to die back by late July, continuing throughout the summer. By September the slow defoliation had stopped. The remaining leaves stayed green and abscised several days later than those on adequately irrigated trees.

In the following year, when normal irrigation was restored, bloom was normal although branches continued to die back and temporary wilting was observed in June. Also, mild interveinal and marginal chlorosis was observed but this did not become a significant problem. The best regrowth was from the interior part of the tree. Shoots on exterior branches grew very little, many remained weak and some died back. After harvest, an appreciable number of short, spur-like shoots resumed growth that extended 10–20 cm. After removing dead and weak limbs, the trees looked normal but with a smaller leaf canopy. South-facing surfaces of the major scaffold branches were killed by sunburn. Ten years later the trees were essentially normal except for the sunburned portions of the scaffolds.

The fruit was smaller in the year of the drought and, in the following year, the yield was reduced about 20%. After 3 years, the trees were once again producing normally.

By droughting part of the root zone, these trees were growing with a reduced effective root volume even though part of the root zone was at field capacity. Effective root:shoot ratios were, therefore, also reduced.

9.3.5 Effects of modified root:shoot ratios on tree growth

Adjusting root:shoot ratios provides a somewhat unexplored avenue to aid in cherry tree vigour control and management. Shoots and roots tend to maintain a functional equilibrium. Although the value of the ratio will differ on different soil types and for trees of differing age, mature trees grown on the same site with consistent management will start to maintain a constant value by either increasing or reducing shoot and/or root growth. We capitalize on this property by reducing the size of the top through pruning to promote more vigorous growth of the shoot portion.

Recent research, mostly on peaches, has shown the possibility of controlling excessive top growth by restricting the volume of soil available for root exploration. This technique seems to be particularly appropriate to sweet cherry, a very large tree with no satisfactory size-controlling rootstocks yet available. Root volume might be controlled by wetting only a portion of the soil or it might be controlled by building root-restricting beds using a physical barrier to root growth to limit root volume.

Trials in their fourth year at Prosser, Washington State, show that sweet cherry trees, despite their variability, show the expected size control with restricted root volume. The most fruitful trees are growing in containers in the range of 0.2 to 0.5 m^3. Both smaller and larger containers have produced less fruitful trees. In the first year after planting, the trees in the smallest containers 'recognized' that they were restricted by closing their stomata as early as mid-June. Research to develop beneficial and practical means to control cherry tree size through root restriction is needed.

9.3.6 Water relations and cherry fruit cracking

Sweet cherry fruit crack readily when wet (Chapter 12). Water is absorbed through the relatively porous, often fractured, cuticle, possibly partly driven by an osmotic potential gradient. High soil moisture or high atmospheric humidity does not cause the most severe cracking although trials in Britain and New Zealand indicate that it may exacerbate the problem. Cracks may occur in concentric rings around the pedicel, at the distal end, usually associated with the stylar scar, and on the sides of the fruit. Cracked fruit are culled, representing direct losses of up to 30–50% of the crop. Beyond that, when damage exceeds some value, perhaps 20–30%, it is not economical to run the crop over the packing line, effectively eliminating the whole crop. Even when the cracked cherries are sorted out, some still find their way into the pack, providing a source for decay, and others with fractured cuticle and soft flesh often deteriorate during shipping.

Control of cherry cracking depends on multiple strategies. A rain-free site is best. Lacking that, growers approach the problem by growing less susceptible, usually softer-fleshed cultivars, by growing cultivars having a succession of ripening periods, thus spreading the risk of untimely rain, and by trying to remove the water following light rains by blowing the trees with air-blast sprayers, helicopters or wind machines. Some have employed antirainmakers, who believe they can prevent rain by excessive seeding of rain clouds with silver iodide. (Dancing and praying have also been tried.) Research dating back at least 60 years has indicated that Ca will reduce the incidence of cracking but the practice has not been generally adopted, probably because of erratic results, unsightly residues and a tendency toward smaller fruit on treated trees. In some parts of the world, shelters to ward off the rain are employed commercially. This practice is expensive, is not fully effective and appears to be justified only where very high quality fruit is grown for a very high-priced market.

9.3.7 Conclusions

In summary, the water relations of cherries are a critical production factor. Early maturation of the crop presents an opportunity to modify mid- to late-season irrigation practices in ways not appropriate for trees of other species that are still maturing crops. Modern irrigation technology permits water to be applied daily or less frequently, at a single point or over the entire orchard floor, at the rate used by the crop or some other rate, over the tree or under the tree, with or without mineral nutrients. All these options and their interactions have physiological implications that have not been explored thoroughly. Given present understanding, cherry orchard irrigation should be based on traditional principles but with an open mind toward exploring new concepts.

General Reading

Anderson, P.C. and Richardson, D.G. (1982) A rapid method to estimate fruit water status with special reference to raincracking of sweet cherries. *Journal of the American Horticultural Society* 107, 441–444.

Chalmers, D.G., Mitchell, P.D. and van Heek, L. (1981) Control of peach tree growth and productivity by regulated water supply, tree density and summer pruning. *Journal of the American Horticultural Society* 106, 307–312.

Evans, R.G., Proebsting, E.L. and Kroeger, M.W. (1986) Soil moisture-sensitive valves for determining crop water use. *American Society of Agricultural Engineers Paper* 86–2572.

Faust, M. (1989) *Physiology of Temperate Zone Fruit Trees.* John Wiley & Sons, 338 pp.

Glenn, G.M. and Poovaiah, B.W. (1989) Cuticular properties and postharvest calcium applications influence cracking of sweet cherries. *Journal of the American Horticultural Society* 114, 781–788.

Kozlowski, T.T. (1968) Diurnal changes in diameters of fruits and tree stems of Montmorency cherry. *Journal of the Horticultural Society* 43, 1–15.

Leece, D.R. (1975) Diagnostic leaf analysis for stone fruit 5. Sweet cherry. *Australian Journal of Experimental Agriculture and Animal Husbandry* 15, 118–122.

Proebsting, E.L. Jr, Middleton, J.E. and Mahan, M.O. (1981) Performance of bearing cherry and prune trees under very low irrigation rates. *Journal of the American Society for Horticultural Science* 106, 243–246.

Swietlik, D. and Faust, M. (1984) Foliar nutrition of fruit crops. *Horticultural Reviews* 6, 287–355.

Tvergyak, P.J. and Richardson, D.G. (1979) Diurnal changes of leaf and fruit water potentials of sweet cherry during the harvest period. *HortScience* 14, 520–521.

Westwood, M.N. and Wann, F.B. (1966) Cherry nutrition. In: Childers, N.F. (ed.) *Nutrition of Fruit Crops.* Rutgers University Press, Bridgton, New Jersey, pp. 158–173.

References

Chalmers, D.J., Olsson, K.A. and Jones, T.R. (1983) Water relations of peach trees and orchards. In: Kozlowski, T.T. (ed.) *Water Deficits and Plant Growth*, Vol. 7. Academic Press, New York, pp. 197–232.

Christensen, M.D. and Walker, D.R. (1964) Leaf analysis techniques and survey result on sweet cherries in Utah. *Proceedings of the American Society of Horticultural Science*, 85, 112–117.

Huguet, C. (1984) Cherries. In: Martin-Prevel, P., Gagnard, J. and Gautier, P. (eds) *Plant Analysis as a Guide to the Nutrient Requirements of Temperate and Tropical Fruit.* P. Lavoisier Publishing Co., New York, pp. 279–298.

Landsberg, J.J. and Jones, H.G. (1981) Apple orchards. In: Kozlowski, T.T. (ed.) *Water Deficits and Plant Growth*, Vol. 6. Academic Press, New York, pp. 419–469.

Shear, C.B. and Faust, M. (1980) Nutritional ranges in deciduous tree fruits and nuts. *Horticultural Reviews* 2, 142–165.

Smith, C.B., Fleming, H.K. and Kardos, L.T. (1961) Leaf composition and performance of sour cherry trees as influenced by fertilizer and soil management. *Pennsylvania Agriculture Experiment Station Bulletin* 683.

10

Tree Canopy Management and the Orchard Environment: Principles and Practices of Pruning and Training

J.A. FLORE[1], C.D. KESNER[1] and A.D. WEBSTER[2]

[1]*Department of Horticulture, Michigan State University, East Lansing, MI 48824-1325, USA;* [2]*Horticulture Research International, East Malling, West Malling, Kent ME19 6BJ, UK*

10.1 Introduction

Pruning, training, tree shape and orchard spacing are all management decisions that the orchardist has control of to influence cropping precocity, productivity and fruit quality. Recent interest in growing trees which produce quickly following planting, crop abundantly and consistently thereafter and give rapid economic returns has stimulated interest in higher intensity culture (more trees per hectare) for cherry. As there are currently no dwarfing rootstocks for cherry that have reached commercial acceptance, the orchard manager must use other tools (pruning, training, bending and regulation of cropping) to assure a balance between vegetative and reproductive growth.

The objective of efficient canopy management is to achieve full canopy development at an early tree age, to sustain fruit production over the projected life of the orchard, to produce high-quality fruit and to facilitate the efficient use of equipment and hand labour. Knowledge of the fundamentals of the tree's fruiting habit, of pruning and training techniques and of the importance of light interception and the influence of shade, can provide a basis for choice of tree size, shape and spacing, time and degree of pruning and a training system for optimum production and efficiency. In this chapter we will explore these fundamentals and how they apply to cherry orchard canopy management, both now and in growing systems of the future.

10.2 Fruiting Habit and Vegetative Growth

Sweet and sour cherries have simple buds (only leaves or flowers, not both) with one bud per node and buds may have up to five flowers per bud. Fruit are borne in a lateral (axillary) position on 1-year-old shoots or on spurs that are formed on shoots 2 years old or older. The percentage of fruit borne on spurs and shoots varies with tree age (a higher proportion of spur-borne fruits as the tree ages) and with variety; the sour cherry variety Montmorency, for example, has fewer spurs than Meteor. Sweet cherry spurs may remain productive for up to 10 years if good light levels are maintained in the trees. Spurs on sour cherries usually have a shorter productive life.

Trials in Britain have shown that spur buds generally set much better than buds formed on the base of 1-year-old shoots. The effective pollination period of these axillary buds is usually lower than for spur buds. If an axillary bud produces a flower, it cannot initiate vegetative growth and it will be 'blind' thereafter. Some sour cherry varieties, such as Schattenmorelle, have a greater tendency to produce blind wood than other varieties. Flower formation in Montmorency is inversely related to vigour, and, if vigour is too low (average length of new extension shoots less than 15 cm year^{-1}), all laterals will produce flowers, resulting in all blind wood except for the terminal bud the following year (Tukey, 1927).

This blind wood effect, common on sour cherry varieties, should not be confused with the long lengths of wood often noted on sweet cherry trees that have no visible growth below the terminal bud of vigorous shoots. Here apical dominance is so great that the apical bud inhibits the growth of buds below it on the shoot. In this case dormancy of lateral buds can be released by heading the shoot, by bending it to a more horizontal position or through the use of appropriate growth-regulating chemicals.

Sweet and sour cherries differ in their apical dominance; sweet cherries are strongly acrotonic (i.e. apically dominant), while the terminal buds of sour cherry trees are weaker and less dominant. This trait differs greatly among varieties, especially in sour cherry. Strong apical dominance in sweet cherry is characterized by vigorous upright growth and strong inhibition of lateral buds below the terminal. Sour cherry does not have such strong apical dominance. Indeed, the lateral shoots directly under the terminal are themselves extremely vigorous and can be dominant over the terminal shoot.

10.3 Types of Pruning and Their Influences on Tree Growth and Cropping

Several experiments conducted during the early part of the twentieth century indicated that pruning delayed fruiting, reduced yields and reduced tree size, but locally invigorated shoot growth (Chandler, 1923; Roberts, 1924; Tukey, 1927; Shoemaker, 1929; Crane, 1931). This led to the common opinion that pruning was a dwarfing process, that it reduced yields, delayed fruiting and should only be done to form the tree during its early years, with limited corrective pruning sufficient thereafter. Recent studies have shown that these generalizations are an oversimplification and that specific responses will depend upon the time and degree of pruning, the variety, tree age, orchard site and specific training system used.

For the purposes of this discussion we will define pruning terms as follows.

Summer pruning:	the selective removal of shoots or branches during the growing season.
Dormant pruning:	the same as summer pruning, except that cuts are made during the dormant season before active new growth has begun.
Summer tipping or pinching:	the removal of the apical bud of shoots or spurs during the growing season.
Hedging:	the indiscriminate removal of all branches within a plane, in either the summer or the dormant period.
Heading cut:	a cut made anywhere below the terminal bud on a shoot, but not the total removal of the shoot.
Thinning cut:	total removal of the shoot or branch at its insertion point on a larger branch or scaffold.

10.3.1 Summer or dormant season pruning and its severity

In many parts of the world where sweet cherries are grown, the trees may be severely damaged by bacterial canker (*Pseudomonas mors prunorum* and *P. syringae*) and silver leaf disease (*Chondrostereum purpureum*). Pruning during the late autumn, winter and early spring has been shown to increase the risk of wood infection from these organisms. German research (Stosser, 1984) has shown that when cherries are pruned in late spring there is a flush of sap flow in the shoot which blocks the damaged vessels and other wood cells with gums and phenolics within 10 days. In favourable warm conditions this is followed by phellogen activity and the growth of a corky callus over the wound within 4–5 weeks. Shoots pruned in winter show no healing until late the following spring and remain vulnerable, therefore, to disease infection.

Because of these risks many sweet cherry growers in Europe delay pruning their trees until after flowering in May when the risks of disease infection of the wounds is reduced. If winter pruning cannot be avoided, the risks of disease infection following pruning may be reduced if the wounds are painted immediately with a mercuric-based pruning paint. Unfortunately, many of these paints have recently been withdrawn by manufacturers in Europe and are now unavailable to cherry orchardists.

Summer pruning is considered more dwarfing than dormant pruning, and the earlier it is accomplished in the season the greater the dwarfing effect on the whole tree (Flore, 1992). However, in one sense it is also invigorating, as shoot growth is stimulated locally, directly behind the cut. Reduction in total growth is directly related to the degree of pruning, and stimulation of terminal growth is inversely related to severity of pruning.

Crane (1931) conducted an experiment over a period of 4 years on young sweet and sour cherries to determine the effect of time and severity of pruning on vegetative and reproductive growth. Eight pruning treatments were selected which differed in their severity (light or corrective only, moderate or heavy) and in their timings (dormant or at two different times during the summer, or in combination with one another). The trees receiving light dormant season pruning produced the most bloom (blossoming), and as the severity of dormant pruning increased the amount of bloom decreased. Summer pruning decreased bloom and this effect was independent of timing. Trees which were both heavily dormant- and summer-pruned had the lowest total bloom of all. Yields followed the same trend as bloom and trees light-dormant-pruned had the highest yields. Yields decreased with the severity of pruning, and both early and late summer pruning also reduced yields. In all cases light pruning resulted in more growth than heavy pruning. There was also a strong variety difference; Montmorency produced much less growth than the sweet cherry variety Schmidt when both were pruned in the summer at the same time and with the same severity.

The dramatic effect on growth reduction from summer pruning is likely to be a result of removing a substantial amount of the leaf area and with it the potential to produce photosynthate for the rest of the tree. If done too late in the season, for instance after harvest, further shoot growth can be stimulated and the tree may not acclimatize to cold temperature and will be more vulnerable to early autumn freezing of tissue. Also, pruning early in the dormant season does not allow the cut to heal before spring, and generally results in more winter injury than late spring pruning. Summer pruning on peach has been shown either to increase damage from cold or to have no effect (Flore, 1992); no similar studies have been conducted on cherries. Roberts (1917) showed that more buds were winter-injured on trees with minimal shoot growth compared with those with much greater growth. The implication is that pruning should not be carried out to the extent that it severely reduces tree growth. On the other hand, pruning to increase the amount of light in the interior of the tree can have a positive effect on shoot hardiness.

10.3.2 Summer tipping or pinching

More recently, summer tipping or pinching has been used as a method to control growth, stimulate bud break and influence flower bud initiation. In theory, this should achieve the pruning objectives without the excessive removal of leaf material and associated reduction of photosynthesis. Therefore, with summer tipping, dwarfing is kept to a minimum and tree size reduction is lower than that achieved with summer pruning, especially in the early years of growth.

In experiments with sweet cherry (Van and Merton Glory) Webster and Shephard (1984) found that tipping reduced mean shoot length and tree size and that timing of tipping had little effect on the degree of size reduction. However, tipping later in the season induced more new shoots than early tipping. Fruit set on 1-year-old wood was increased, but the effect on spurs was either positive or negligible. Soluble solids and sometimes fruit size were increased if tipping was carried out just before harvest. The two varieties did not always respond the same way in the same year. Looney (1989) indicated that the systematic cutting back of the current year's shoots in June and July had little effect on cropping or fruit quality, but it reduced set in the third year. He concluded that the main effect of summer pruning was in controlling tree size.

10.3.3 Summer hedging

Unless coupled with substantial dormant pruning, most fruit trees do not respond positively to repeated summer hedging. This is because of the excessive growth stimulated at the top and on the sides of the tree. The exception seems to be the sour cherry Montmorency when grown under Michigan, USA, conditions. In response to grower interest in intensification of planting and in increasing early production, Kesner and colleagues conducted a number of studies (Kesner *et al.*, 1981; Kesner and Nugent, 1984) and reported favourable results when summer hedging was conducted in the following manner. The sides of trees were hedged with a double sickle bar to remove one-third to one-half of the current season's growth. This operation was conducted at the end of pit hardening, i.e. 40–47 days after petal fall and just before final swelling of the fruit. At this time spur growth has stopped and terminal leaf emergence has slowed. As a consequence, there is little extension regrowth of shoots and some stimulation of spur formation along the shoot. If hedging is done during this period, fruit size is sometimes increased, ripening is more uniform, flower bud initiation is increased, there are more flower buds per limb cross-sectional area and fruit set in the following year is slightly increased and is more uniform on all sides of the tree. When trees are hedged during this period of time, no adverse effects on cold acclimatization or winter hardiness are noted. Figure 10.1 shows a mature, mechanically hedged orchard of Montmorency sour cherries growing in Michigan, USA.

Tipping can also retard leaf ageing, as demonstrated by higher leaf photo-

Fig. 10.1. A typical mature orchard of Montmorency sour cherries growing in Michigan, USA. The trees are pruned by mechanical cutter bars.

synthesis and transpiration rates, greater chlorophyll concentration and later leaf drop (Flore, 1992).

Summer hedging is a satisfactory method of controlling Montmorency tree growth; total terminal shoot length of each limb and the number of spurs and short shoots are increased. The tree shape which has proved most successful for hedging is an inverted V in the direction of the row, with rows orientated north–south. When compared with rectangular shaped trees, these triangular-shaped trees produced less fruit in the early years, possibly because their canopy volume was lower. However, this cropping effect reversed as trees became larger, because of greater tree-to-tree shading with the rectangular shapes (Flore and Layne, 1990). Hedging increases the density of the foliage on the outside of the tree, causing internal shading. Therefore, supplementary, but minimal, dormant pruning to remove a few major branches every 1–2 years is recommended to facilitate light penetration into the interior of the canopy.

10.3.4 Double sectorial pruning

Research in Hungary (Brunner, 1990) has advocated the double sectorial system of pruning for sweet cherry high-density systems. The technique involves heading back obliquely orientated laterals to an upper (inside) facing bud and then removing the erect shoot that emerges from this bud either 1 or 2 years later. The side-shoots formed below this shoot are reported to be ideally wide-angled and fairly horizontal in their growth. The advantages reported for this technique in Hungary are precocious cropping, reduced tree height and reduced labour for shoot bending and tying. Figure 10.2 shows how wide-angled laterals are reported to develop if trees are pruned using the double sectorial system of pruning.

10.3.5 Shoot bending

Shoot bending is an effective method of reducing shoot vigour on pome fruit species such as apple. Bending towards or below the horizontal reduces shoot extension and may increase flower induction. While the technique may also prove effective with cherries, timing and degree of bending are much more critical. Bend too early and the shoot tips curve upwards and continue growth, bend too severely and gummosis is stimulated. Bending too late in the season also often results in branch breakage.

Lateral shoot (feather) bending in the nursery is particularly useful if scaffold branches which have wide angles with the central leader are desired. Proprietary clips or clothes-pegs are often utilized to improve feather branch angles (Fig. 10.3).

10.3.6 Techniques of root manipulation

Root pruning of sweet cherries has also been considered in Italy as a method for reducing excessive vigour and is currently being examined in trials in Italy and Britain. Results have proved disappointing in Italy but have shown slightly more promise in British trials.

A technique called correlative mechanical pruning has been developed in

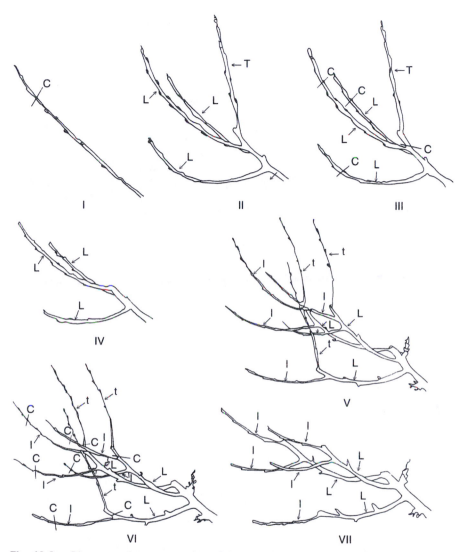

Fig. 10.2. Diagrammatic representation of the procedures and consequences of double sectorial pruning of sweet cherries over a 3-year period. L and l = lateral and sub-lateral shoots, T and t = terminal and sub-terminal shoots and C = position of spring pruning cuts. (Modified from Brunner, 1986.)

Hungary (Brunner, 1990) for use with cherries. This involves mechanical pruning of the hedge canopy in the spring but with the addition of root pruning, when necessary, to maintain the desired root:shoot balance. The heavy equipment needed to carry out these operations necessitates wide row spacings and the technique is not widely popular for this reason.

Perhaps more promising are techniques in which the soil available to the root system is limited by planting the trees within restrictive membranes which are buried in the soil. The membranes allow water and soil solutes to move into the restricted root zone but prevent the roots from growing out. This technique

Fig. 10.3. Wide-angled young lateral branches on Regina sweet cherry trees acheived by use of clothes-pegs and small branch spreaders.

has shown early promise in US trials (Chapter 9) but further trials will be needed before the root restriction strategy for control of growth and cropping can be fully appraised.

10.3.7 Pruning effects on cherry fruit size

High crop loads with fewer than two leaves per fruit have been shown to inhibit maturity and reduce fruit size in sour cherry (Flore and Layne, 1992) and fruit size has also been shown to be directly related to leaf area per fruit for sweet cherry (Roper *et al.*, 1987). Patten *et al.* (1986) showed that fruit size of sweet cherries was reduced on later-flowering clusters and that later flowering was linked with flower location on the shoot; the more basipetal the bud the more delayed the flowering.

Tukey (1927) noted that severe pruning increased the size of sour cherries but greatly reduced total yield. Because large sweet cherries bring a premium price, indirect crop reduction, by shoot thinning before bloom or some more direct method of reduction of crop load, has often seemed to be worth while.

However, total yield is always reduced when crop load is reduced, even when fruit size is increased. In a crop density study on sour cherry, using hedging to control tree size, Flore and Layne (1990) found that fruit ripening was delayed in a high-crop year in the highest-density planting. This probably resulted from the removal of too much leaf area during the hedging process. It appears that hedging or summer pruning to control tree size must be used with caution in years when the crop load is light, since excessive leaf removal may affect maturity of the crop and carbohydrate reserves in the tree.

10.4 Pruning Sweet and Sour Cherries

10.4.1 Sweet cherries

Although occasionally influenced by environmental conditions, the majority of sweet cherry varieties are characterized by their strong apical dominance. Failure to prune such trees results in strong growth of terminal shoots with inhibition of bud break for long distances behind the growing point. This can result in branches that extend to 75–100 cm without any lateral branching or spur formation. For this reason, trees are generally headed at planting and their lateral branches also headed annually during the formative years to ensure bud break and lateral shoot development along the main branch. In some cases branch bending towards the horizontal, bud notching or use of growth regulators such as Promalin® are necessary to promote bud break.

Research in New Zealand and the USA (Jacyna and Brown, 1987; Southwick, 1989) demonstrated that painting specific vegetative buds on 1-year-old shoots of young sweet cherries with a formulation containing the gibberellins 4 and 7 (GA_{4+7}) and benzyl adenine (BA) (Promalin®) stimulated growth extension from these buds. It was suggested that the technique could be used to develop trees with well-spaced and balanced branching, by stimulating the growth of appropriately positioned vegetative buds.

Traditionally sweet cherries were, and largely still are, trained as open-centre, central-leader or modified central-leader trees. To form a modified central-leader system, probably the most commonly used tree shape for sweet cherries, the following technique is recommended.

Year of planting

Obtain well-feathered trees from the nursery that have branches that are evenly spaced and with wide crotch angles. Poor crotch angles provide a place for water to freeze during the winter months, which can result in splitting of the bark, and trees with narrow branch angles form scaffold branches that are not structurally as strong as those with wider 90° angles. If there are less than three good laterals, remove them all at planting time and head the tree leader to approximately 20–30 cm above the height where the lowest scaffold branch is desired. Branches will be formed from buds just below the cut.

Second year

Do not prune young trees until the danger of frost is completed. Then choose three or four well-spaced scaffold limbs with wide crotch angles that are 15–25 cm apart on the central leader. Head these scaffolds to approximately 20–30 cm above the height where you want the next set of laterals to be produced. Without this heading, the next set of branches may be spaced too far apart. The central leader should also be headed back in the second year.

Third year

Choose two to four more well-spaced scaffolds on the central leader, as in year two; try to have six to eight in all. Try to avoid leaving one scaffold limb directly above another. Head the leader and the scaffold branches, as in year two. Once a satisfactory tree canopy structure is established, head the leader only slightly, to the weakest side lateral, preferably one facing the wind.

After this point, only corrective pruning is needed. On older trees, as spurs age they start to produce lower-quality fruit. If this happens, it may be necessary to reinvigorate the branch by making a few severe thinning cuts to stimulate new growth. Productivity will go down in the year of treatment but will return in 2–3 years with higher-quality fruit.

10.4.2 Sour cherries

The sour cherry has less apical dominance than the sweet cherry and a greater tendency towards production of blind wood and poor branch angles. Also, lateral shoots just below the leader have a tendency to become dominant and can 'choke out' the leader. In Michigan, 95% of the sour cherries grown commercially are mechanically harvested. To facilitate mechanical harvesting the industry has rapidly moved from an open-centre form to a modified central-leader system for Montmorency. The following is recommended for a modified central-leader system of pruning.

Year of planting

The objective is to develop six to eight major scaffold branches on the leader. If there are not at least three good laterals (feathers) on the nursery tree, it should be whipped (all laterals removed), otherwise these will become too dominant. Laterals should be evenly distributed around the leader and 10–15 cm apart; generally the tree does not need to be headed. However, if the leader is excessively long and thin, such that the first scaffold formed would be higher than desired, it should be headed to a viable lateral vegetative bud, preferably one that faces into the prevailing wind. When new shoots are 10–15 cm long special clips or 'spring' clothes-pegs may be placed above them to force them to grow at a 90° angle. These should be removed 4–6 weeks later when the shoots begin to lignify.

Second year

After the danger of frost has passed, choose four well-spaced scaffold limbs at least 10 cm apart and evenly distributed around the leader. Remove any strong limb that is equal in diameter to the leader, as it will otherwise become too vigorous and compete with the leader. Also remove the uppermost laterals within 10–15 cm of the terminal, as they will also become very vigorous and compete with the leader. Lateral limbs that develop above the chosen limbs should be cut back, leaving 10–15-cm 'stubs' rather than complete removal. These stubs will force the limb directly below to form a wide crotch angle, and will have foliage that should provide valuable photosynthate for the growth of the tree during the season. These short stub limbs should then be removed the following spring. If four good scaffolds cannot be found or if those present are not in the proper position, the tree should be 'whipped' again, leaving a short stub on the leader. Usually there will be a vegetative bud just below the insertion point of the lateral branch, which, after cutting, will break and form a branch with a very good (wide) angle.

Third year

Choose two to four additional laterals to provide a total of six to eight. Unwanted laterals should again be cut to 10–15-cm 'stubs' and then removed the following spring, as in year two. Remove all unwanted laterals, including the uppermost limbs, that compete with the leader.

Fourth year

By this time the basic tree structure is set, and only correlative pruning should be done. Main scaffold branches should be thinned out, and limbs that block light penetration into the lower part of the tree should be removed.

10.5 Pruning and Training in Relation to Light Interception and Shade

Light interception has been directly related to dry matter production for a number of annual and perennial plants (Monteith, 1977; Cannell *et al.*, 1987; Palmer, 1988). In high-density sour cherry systems, preliminary results (Flore and Layne, 1990) also indicate that there is a direct relationship between yield and light interception. Shading can have a profound effect on the vegetative and reproductive growth of a tree. Shading occurs not only within a canopy but also between trees and this problem can be especially acute in high-density orchards. In orchard design and canopy management, the orchardist has the objective of developing the tree canopy as rapidly as possible to maximize light interception, but must also train and prune trees so that shading does not adversely affect fruit quality or yield. Studies on the effect of light or shade on vegetative growth and yield, coupled with a characterization of light levels in different canopies at different times of the year, provide a scientific basis for orchard design and canopy management.

Numerous studies have been conducted on apple, but until recently little has been done on sour cherry (Flore, 1980; Flore and Kesner, 1982; Flore *et al.*, 1983; Flore and Layne, 1990, 1992) and almost none on sweet cherry. However, studies by Roper and Kennedy (1986) have shown that sweet cherry leaves become photosynthetically competent early in their development and this may supplement the reserve carbohydrates used for growth early in the season. This indicates that in early spring, when young leaves have limited leaf area and temperatures are cool, the sweet cherry canopy is still able to provide, efficiently, photosynthates for the demands of both young fruits and developing leaves and shoots. Also, the research of Patten and Proebsting (1986) suggested that shading sweet cherry trees reduced fruit set and that the relationship between light and fruit colour or soluble solids content at harvest was logarithmic, with both variables greatly reduced by light levels below 10–15% of full sun in Washington State.

In general, trees or shoots grown in the shade have larger, thinner leaves which appear greener and have more chlorophyll per unit area. They also have shoots that are longer and thinner. If shading is very severe the area per leaf will actually decrease because of lack of carbohydrate. Even after full leaf expansion, leaf development is influenced by shade. Photosynthetic rates under increasing light intensity are higher for leaves grown under full sun than those grown in the shade. Under artificial shade or in canopies there is a dramatic decrease in flowering at levels below 20% of full sunlight. Shading also retards fruit maturity as indicated by poorer colour and higher fruit retention force. Fruits formed in the interior of the tree canopy are often greener and harder to harvest than those on the periphery of the tree canopy, where exposure to sunlight is better. Light, therefore, is a major factor associated with uniformity of ripening and can affect harvest time, especially in high crop years when leaf-to-fruit ratios are less than two. Both wood and bud hardiness decrease with shade.

10.5.1 Light distribution in canopies

Leaf emergence and development are dependent upon temperature and are directly related to degree-day accumulations in cherry (Eisensmith *et al.*, 1980). Leaf emergence on shoots and spurs occurs at approximately the same time as flower bud break. Spur leaves emerge rapidly and are usually fully developed within 20–30 days after bud break. They do not resume growth once emergence has stopped. Shoots on terminals continue to develop new leaves until fruit harvest; they seldom resume growth once it has stopped, unless some severe cause of stress is removed or after summer pruning or hedging.

Leaf development is most rapid during the first 30 days after bud break and is generally complete by harvest, 60 days later. This leads to rapid canopy development and shading within the canopy. As a consequence, light interception and shading occur very rapidly and light in the interior of the tree can be less than 20% of values above the canopy within 30 days of bud break. At full canopy development, in trees that are trained to a central leader, most of the fruiting zone is located as a peripheral mantle, which averages approximately 1.0 m thick around the outside of the tree (Fig. 10.4a). Light levels in this area range from 100% of full sunlight at the surface to less than 10% at the interior. The thickness of this mantle will depend upon the age and size of the tree and upon the pruning practices employed. This zone is thicker at the top of the tree and narrower at the bottom. Hedging trees for 6 years and the consequent

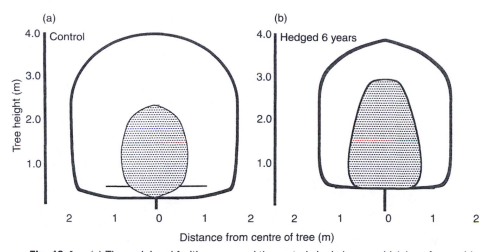

Fig. 10.4. (a) The peripheral fruiting zone and the central shaded zone, which is unfavourable to fruit development, on a typical central-leader-pruned sour cherry tree. (b) Similar fruiting and shaded zones of the canopy on a tree which was hedge-pruned mechanically for 6 years.

development of increased shoot and leaf density in the canopy can change the relative proportions of shaded and adequately illuminated tree canopy (Fig. 10.4b).

The reduced fruiting zone in the lower levels of the tree is not only affected by internal tree shading, but by shading from trees in adjacent rows. The extent of this depends upon the shape of the tree, the width of the clear alleyway, the height of the tree and row orientation. Shading in the lower level of the tree decreases as tree height decreases and as clear alleyway width increases.

Trees that take the shape of an inverted V have better light penetration than bush or rectangular-shaped trees. At latitudes far from the equator, north–south row orientation results in less row-to-row shading than east–west orientation, if tree height is greater than the alleyway width. This is because of the angle of incidence of the sun in relation to the row orientation in the summer as it moves from east to west. If trees are too tall or clear alleyway width is too narrow, one row will shade the lower portion of the adjacent row for most of the day.

Generally, minimum clear alleyway width is determined by the width of the equipment used by the fruit grower, while tree height is determined partly by the rootstock/scion combination in relation to the soil fertility. Pruning, hedging and tree shape can be used as tools to keep the tree within a prescribed height and shape. Under Michigan conditions, for north–south row orientation, it has been found that the height of bush or rectangular-shaped trees should not exceed twice the clear alleyway width, while height of triangular-shaped trees should not exceed three times the clear alleyway width.

10.6 Pruning and Training Systems for the Future

Although most cherry trees grown throughout the world are pruned and trained as vase-shaped or central leader-shaped standard trees of one sort or another, interest has also been shown in the use of more novel pruning and training systems. The objective of most of these non-traditional systems of pruning and training is to increase tree density per hectare and improve cropping precocity and efficiency. Although they were largely developed for apples, some systems have been tested for cherries in several parts of the world. Among these are the Tatura system (Ende and Jerie, 1985), the Swiss four-wire system (Zbinden and Widmer, 1980) and various modifications of spindle systems. Another system, the palmate, has, in contrast, been extensively used for many years in Italy. Unfortunately, sweet cherry trees growing on the traditional Mazzard or Mahaleb rootstocks adapt poorly to these intensive systems of planting unless their natural vigour is reduced in some way.

One method of achieving this is to grow trees on shallow soils in areas experiencing low summer rainfall and to control growth by judicious use of irrigation. This practice has been used successfully, when coupled with the Tatura trellis system of training developed in Victoria, Australia (see below).

Although withholding water supplies early in the season (regulated deficit irrigation (RDI); see Chapter 9) has shown promise in controlling the excessive growth of peaches and pears on the Tatura system, cherries ripen earlier in the growing season and severe water deprivation at the time of most active shoot growth could lead to reduced fruit size.

10.6.1 The Tatura trellis system

The Tatura trellis system for sweet cherries can be described as follows (Ende and Jerie, 1985). Pine poles 100–125 mm in diameter and 3.6 m long are erected, with 0.8 m of their length buried in the ground, in a V shape with an angle of 60° to the horizontal. This leaves a gap of approximately 1.8 m between the tops of the poles. These V frames are erected every 10 m down the proposed cherry rows and two, high-tensile steel wires (2.8 mm diameter) are attached to the poles on each side, one at the top and the other 1.8 m lower on the pole. Temporary wires (1.6 mm in diameter), five in number, are attached to the poles between the two permanent wires for training purposes. All wires are tensioned and tied back to a buried anchor at the end of each row.

Year of planting

Training in the first year involves selecting two shoots from each tree in spring, one on each side. These are trained to a string which is stretched from the tree base to the top wire on each side of the V canopy. Several methods of holding the shoot to the string have been tried with success.

Second year

In the second year Promalin® (a proprietary mixture of GA_{4+7} and BA) mixed with acrylic paint is applied to buds on the 1–1.5-m-long shoots formed in the first year. The lateral shoots induced by the treatment are tied to the adjacent wires, firstly at an angle of 30–45° from the main leader and then vertically to fill in space on the trellis.

Third year and later

In the third year Promalin® is again applied to 1-year-old wood of the leaders and strong secondary shoots. By this time the 3-year-old shoots and the basal sections of some of the leaders should be producing flowers. Training continues with the aim of filling all the space on the trellis. Varieties differ in their branching habit and hence in their adaptability to trellis training.

Summer pruning is usually undertaken twice in year three and involves removing shoots which are growing inside the V framework and which are of no use for filling the canopy. The tree tops are also pruned, as by this time they should have reached the top 2.5-m height of the V canopy. Thereafter, the tree tops should be trimmed in early summer, and after harvest new long laterals are also trimmed to prevent the canopy becoming too thick.

Tatura plantings with up to 3000 trees ha^{-1} have yielded extremely well in Australia, and more than 17 different sweet cherry varieties have now been

Fig. 10.5. Mature sweet cherry trees trained on a Tatura trellis system growing on a shallow soil in Victoria, Australia.

tested on this system. In New Zealand, where soils are deeper and summer rainfall more frequent than in Australia, control of sweet cherry tree growth on the Tatura trellis can prove more of a problem. One solution is to use the chemical growth retardant paclobutrazol (Chapter 12). Figure 10.5 shows the ideal development of cherry trees on a Tatura trellis system planted on the shallow and semiarid soils of Victoria, Australia. Figure 10.6 shows young trees trained to the same system on the highly fertile soils of New Zealand. In this instance treatment with paclobutrazol was necessary to control excessive shoot vigour.

Fig. 10.6.   Young sweet cherry trees trained on a Tatura trellis system on a deep fertile New Zealand soil. Excessive tree vigour is controlled by use of the growth retardant paclobutrazol.

10.6.2 The Swiss four-wire system

Swiss research (Zbinden and Widmer, 1980; Zbinden, 1983) showed that, by combining two summer prunings, first in June and again in late July or early August, with shoot bending, sweet cherry varieties grown on F.12/1 rootstock could be successfully maintained within their allotted hedge canopy space. A feature of this system was that several main branches were trained at an oblique angle to the leader, using four horizontal wires spaced on posts. The leader was regularly weakened by pinching back to allow plenty of light to the tree centre. This system of training has not been widely adopted outside Switzerland, its country of origin.

10.6.3 Spindle systems

Spindle systems of training, such as the slender spindle, the north Holland spindle, the central axe or one of the many other variations developed, are likely to become more popular with sweet cherry producers, once dwarfing rootstocks become more readily available. The principles and practices of the technique are similar to those described for apple (Barritt, 1992) except that timing of pruning and training is altered to account for the greater disease sensitivity of the cherry.

Zahn's (1989) research in northern Germany recommends the use of a type of free spindle training system for sweet cherries. This system advocates the retention of a strong central leader in the tree with only minimal pruning in the early years after establishment. The aim is to produce a natural pyramid-shaped tree. While this often gives excellent tree shape with varieties that branch freely, with others, such as Stella and Lapins, which are reluctant to branch naturally when young trees, the system leads to poor tree shape. More severe leader heading is required with these varieties if they are to develop sufficient primary branches.

10.6.4 The palmate system

The palmate system was, for a time, a very popular method of training sweet cherries in Italy (Nicotra and Fideghelli, 1972). However, problems of excessive tree vigour on many sites meant the system was less than ideal for sweet cherries. Adoption of size-controlling rootstocks, such as the CAB series of *Prunus cerasus* selections, may stimulate renewed interest in this system for Italian-grown cherries.

All of these intensive systems are characterized by high numbers of trees per hectare. They also need tree supports (posts and canes) and either individual stakes or larger supports for erecting trellises. Growth control is achieved by a combination of techniques. Shoot bending, tipping and summer pruning are used in some systems; in others plant growth-regulating chemicals or regulated deficit irrigation (RDI) techniques are used. The objective is to maximize light interception, minimize shading in tree canopies and develop a full canopy as soon as possible after planting. Tests have shown these systems to be very productive in early years, but the economics and the long-term yield of these systems have still not been fully examined.

The major constraint on the successful development of intensive systems of pruning and training for cherries, in particular sweet cherries, has been the lack of suitable methods for controlling the inherently strong vigour of the trees. With the release of dwarfing rootstock clones bred in Belgium and Germany (Chapter 5), this problem should be diminished in the future and one can anticipate the development of successful trellis and hedgerow systems for sweet cherries.

References

Barritt, B.H. (1992) *Intensive Orchard Management.* Washington State Fruit Commission. *Good Fruit Grower* 211, 1–2.

Brunner, T. (1986) Hajlitas metszessel. [Bending by pruning.] *Kerteszet es Szoleszet* 35 (8), 6–7.

Brunner, T. (1990) *Physiological Fruit Tree Training for Intensive Growing.* Akadémiai Kiadóo és Nyomda, Budapest.

Cannell, M.G.R., Milne, R., Sheppard, L.J. and Unsworth, M.H. (1987) Radiation interception and productivity of willow. *Journal of Applied Ecology* 24, 261–278.

Chandler, W.H. (1923) Results of some experiments in pruning fruit trees. *Cornell University Agricultural Experimental Station Bulletin* 415.

Crane, H.L. (1931) Effects of pruning on the growth and yield of cherry trees. *West Virginia Agricultural Experimental Station Bulletin* 240.

Eisensmith, S.P., Jones, A.L. and Flore, J.A. (1980) Predicting leaf emergence of 'Montmorency' sour cherry from degree-day accumulations. *Journal of the American Society for Horticultural Science* 105, 75–78.

Ende, B. van den and Jerie, P. (1985) The Tatura trellis – a new look at growing sweet cherries. *Compact Fruit Tree* 18, 95–101.

Flore, J.A. (1980) Influence of light interception on cherry production and orchard design. *Annual Report, Michigan Horticultural Society* 111, 161–169.

Flore, J.A. (1992) The influence of summer pruning on the physiology and morphology of stone fruit trees. *Acta Horticulturae* 322, 257–264.

Flore, J.A. and Kesner, C. (1982) Orchard design for stone fruit based on light interception. *Compact Fruit Tree* 15, 159–165.

Flore, J.A. and Layne, D.R. (1990) The influence of tree shape and spacing on light interception and yield in sour cherry (*Prunus cerasus* cv. Montmorency). *Acta Horticulturae* 285, 91–96.

Flore, J.A. and Layne, D.R. (1992) Orchard design and pruning in stone fruit. *Proceedings of the Oregon Horticultural Society* 82, 11–18.

Flore, J.A., Howell, G.S. and Sams, C.E. (1983) The effect of artificial shading on cold hardiness of peach and sour cherry. *HortScience* 18, 321–322.

Jacyna, T. and Brown, G. (1987) Earnscleugh canopy – the answer for cherries. *Orchardist of New Zealand* 62 (3), 13–15.

Kesner, C.D. and Nugent, J.E. (1984) Training and pruning young cherry trees. *Michigan State Cooperative Extension Bulletin* E-1744.

Kesner, C.D., Hansen, C.M. and Fouch, S. (1981) Tree training and mechanical pruning of tart cherry in high density planting. *Compact Fruit Tree* 14, 135–139.

Looney, N.E. (1989) Effects of crop reduction, gibberellin sprays and summer pruning on vegetative growth, yield and quality of sweet cherries. In: Wright, D.J. (ed.) *Manipulation of Fruiting.* Butterworths, Borough Green, Sevenoaks, Kent, pp. 39–50.

Monteith, J.L. (1977) Climate and the efficiency of crop production in Britain. *Philosophical Transactions of the Royal Society, Series B* 218, 277–294.

Nicotra, A. and Fideghelli, C. (1972) Prova di confronto tra i sistemi di allevamento del ciliego a Marchand ed a palmetta. *Annali dele Instituto Sperimentale per la Frutticoltura* 3, 289–298.

Palmer, J.W. (1988) Annual dry matter production and partitioning over the first five years of a bed system of Crispin M27 apple trees at four spacings. *Journal of Applied Ecology* 14, 539–549.

Patten, K.D. and Proebsting, E.L. Jr (1986) Effect of different artificial shading times and natural light intensities on the fruit quality of 'Bing' sweet cherry. *Journal of the American Society for Horticultural Science* 111, 360–363.

Patten, K.D., Patterson, M.E. and Proebsting, E.L. Jr (1986) Factors accounting for within tree variation of fruit quality in sweet cherries. *Journal of the American Society for Horticultural Science* 111, 356–360.

Roberts, R.H. (1917) Winter injury to cherry blossoms. *Proceedings of the American Society for Horticultural Science* 14, 105–110.

Roberts, R.H. (1924) How to prune the young cherry tree. *Wisconsin Agricultural Experimental Station Bulletin* 370.

Roper, T.R. and Kennedy, R.A. (1986) Photosynthetic characteristics during leaf development in 'Bing' sweet cherry. *Journal of the American Society for Horticultural Science* 111, 938–941.

Roper, T.R., Loescher, W., Keller, J. and Rom, C. (1987) Producing big firm fruit studied at Cherry Institute. *Good Fruit Grower* 38, 33–36.

Shoemaker, J.S. (1929) Pruning cherry trees. *Fruits and Gardens* 27 (1), 8–9.

Southwick, S.M. (1989) New approaches to training sweet cherries in California. *Compact Fruit Tree* 22, 106–109.

Stosser, R. (1984) Uber den Wendverschluss bei Obstgeholzen in Beziehung zu Schnittmassnahmen. [Wound healing in fruit trees in relation to pruning.] *Erwerbsobstbau* 26 (6), 141–143.

Tukey, H.B. (1927) Responses of the sour cherry to fertilizers and to pruning in the Hudson River Valley. *New York Agricultural Experimental Station Bulletin* 541.

Webster, A.D. and Shephard, U.M. (1984) The effects of summer shoot tipping and rootstock on the growth, floral bud production, yield and fruit quality of young sweet cherries. *Journal of Horticultural Science* 59, 175–182.

Zahn, F.G. (1989) 10 Jahrige Erfahrung mit Starkenbezogener. *Baumbehandlung* 174–191.

Zbinden, W. (1983) Erfahrungen mit Verschiedenen baum Formen in Susskirschenanbau. *Schweizerische Zeitschrift für Obst und Weinbau* 119, 407–418.

Zbinden, W. and Widmer, A. (1980) Erfahrungen mit dem Sommercchnitt bei Steinobst. *Schweizerische Zeitschrift für Obst und Weinbau* 116, 492–497.

11 Principles and Practice of Plant Bioregulator Usage in Cherry Production

N.E. Looney

Agriculture and Agri-Food Canada Research Centre, Summerland, BC V0H 1Z0, Canada

11.1 Introduction

Plant growth-regulating chemicals, referred to here as plant bioregulators (PBRs), are widely used in tree fruit horticulture to facilitate plant propagation, control aspects of tree growth and development and achieve specific improvements in cropping and fruit quality. This chapter discusses PBR products and practices that have been developed for sweet and/or sour cherry production in various parts of the world.

Some of these practices are widely used by nurserymen or orchardists in many countries, while others are either restricted in their availability or have very limited applications. These latter, and even some others still under development at the time of writing, are included because they demonstrate a potential that might be explored by others.

For example, we know that climate has a large bearing on PBR effectiveness and those interested in introducing cherries to a new production region may want to explore some of these practices. Furthermore, cultivars often differ in their responsiveness to various PBR techniques, so new opportunities for PBR usage may arise as new cultivars are introduced.

PBRs are natural or synthetic chemicals that either augment or duplicate a natural plant growth substance (i.e. plant hormones such as ethylene, abscisic acid or various natural gibberellins) or they accelerate or inhibit a normal physiological process that involves natural growth substances (Table 11.1).

Simply put, PBR chemicals exhibit useful biological activity in plants and some (the natural plant growth substances) are essential for normal plant growth and development. By regulating specific growth and development stages or events, PBRs adapt and adjust healthy plants to their physical environment. They permit the completion of horticulturally important processes as varied as seed germination, shoot growth, flowering and fruiting.

While much is made of the preference for natural product usage and

279

Table 11.1. Some plant bioregulators important in fruit production and their effects on sweet cherry or sour cherry.

Natural growth substance 'family'*	Commercial chemicals	Use in cherry production
Auxin Indoyl-3-acetic acid	Indoyl-3-butyric acid (IBA)	Rooting of cuttings
	Naphthaleneacetic acid (NAA)	Controlling sucker growth Rooting of micropropagules
	Naphthaleneacetamide (NAAm)	Improving fruit set of sweet cherry
Cytokinins Zeatin and others	6-Benzyladenine (6-BA)	Encouraging shoot proliferation in micropropagation systems Promoting feathering of nursery stock and spur development of bearing trees (when combined with GA_{4+7})
	Thidiazuron	Regenerating shoots in genetic transformation systems
Ethylene	2-Chloroethylphosphonic acid (ethephon)	Facilitating mechanical harvesting Defoliation of nursery stock Improving bud cold-hardiness and delaying anthesis to avoid frost damage
Gibberellins Many GAs known to occur in cherry	Gibberellins $A_4 + A_7$	Promoting feathering of nursery stock (together with 6-BA)
	Gibberellic acid (GA_3)	Delaying fruit coloration and improving fruit quality of sweet cherry Delaying fruiting of sour cherry and treating the 'yellows' virus Achieving parthenocarpic fruit set (in combination with auxins and/or cytokinins)
	Paclobutrazol and other triazole inhibitors	Suppressing shoot elongation by inhibiting GA biosynthesis
	Daminozide	Advancing sweet and sour cherry fruit ripening

*The commercial chemicals listed in the next column either mimic the action of these natural growth substances or are known inhibitors.

'sustainable' agriculture production systems, it is important to understand that PBRs, natural or synthetic, exhibit negligible activity in other biological systems. The target is the tree or the fruit, not something living on the tree. PBRs are not pesticides.

11.2 PBR Usage in Nursery Stock Production

Nursery stock production is an exacting business with little margin for error. Each of many operations, from propagation of the rootstock to delivery of the finished tree, requires very specialized knowledge and skills. Not surprisingly, therefore, nursery operators have proved to be very interested and innovative participants in the development and introduction of new PBR technology.

11.2.1 Clonal propagation

Promotion of callus and root development on woody stem cuttings by synthetic auxins such as indole-3-butyric acid (IBA) was discovered before 1940 and remains important today. In cherry propagation, IBA-based technology has long been used to propagate such popular rootstock clones as Mazzard F.12/1 and Colt and will most certainly be applied to other rootstock clones in the future. The basal end of large winter cuttings taken from virus-free 'mother' trees is treated with a powder formulation containing 0.25–0.50% IBA to encourage root initiation. Rooting takes place in a carefully constituted and temperature-controlled rooting medium (Howard, 1981). More details are provided in Chapter 6.

Some nurseries also possess the capability to multiply rapidly cherry rootstock clones and scion cultivars of sweet and sour cherry by rooting tiny cuttings taken from shoot cultures proliferating *in vitro*. The scion cultivars propagated in this manner are those known to perform satisfactorily as 'own-rooted' trees. Using techniques now well established for woody plants (Jones and Hopgood, 1979), cultures capable of producing thousands of plants arise from a single shoot tip collected in the field.

The success of this procedure depends on the culture medium containing 1 mg l^{-1} (1 ppm) or less of 6-benzyladenine (6-BA) or perhaps another PBR with cytokinin activity to induce shoot proliferation (Borkowska and Opilowska, 1988). Other PBRs required for success are gibberellic acid (GA$_3$) and IBA (or another synthetic auxin) at about 0.1 mg l^{-1} each to maintain healthy growth in the absence of roots (Snir, 1983). Shoots excised from these cultures are first transplanted to a medium free of cytokinin and then one free of IBA. After 2–3 weeks the rooted microcuttings are removed to a shaded greenhouse and, after acclimatization for about 10 days, are transplanted into compost and grown to a size suitable for field planting.

Commercial use of tissue culture propagation of cherry is limited to specific cases where rapid propagation of a new cultivar or a new rootstock clone is desired. However, this practice may increase in importance, since own-rooted trees of commercial sour and sweet cherry cultivars, produced by tissue culture, are reported to be as productive as grafted trees (Quamme and Brownlee, 1993) and may prove less expensive than traditional budded or grafted trees. Furthermore, the growing interest in size-controlling rootstocks for sweet cherry may determine that rootstock liner production by tissue culture will increase in importance.

Shoot regeneration from immature cotyledons of sour cherry has been achieved using thidiazuron (N-phenyl-N-1,2,3-thidiazol-5-ylurea), a synthetic cytokinin with greater biological activity than 6-BA (Mante *et al.*, 1989). Looking to a future where crop improvement and genetic engineering may be synonymous terms, this technology will surely prove important to those interested in plantlet regeneration from genetically transformed cells and tissues.

11.2.2 Feathering agents

Many orchardists prefer to purchase planting stock with well-developed lateral branches ('feathers') positioned where scaffold branches would be desired on the mature tree. This is difficult to achieve on sweet cherry trees grown in the nursery for one growing season after budding. However, success in increasing the number of lateral branches on 1-year planting stock has been reported from the use of Promalin (a proprietary mixture containing 1.8% (w/w) of 6-BA and 1.8% (w/w) gibberellins $A_4 + A_7$ (GA_{4+7}); Abbott Laboratories, North Chicago, IL 60064) applied when tree height in the nursery row reaches about 40 cm (Cody *et al.*, 1985). Sprays containing as little as 0.5 g l^{-1} of the active ingredients of Promalin (i.e. 500 mg l^{-1} Promalin contains 500 mg l^{-1} 6-BA plus 500 mg l^{-1} total gibberellins) significantly increased branch number and also widened branch attachment angle on Bing sweet cherry trees (Table 11.2).

Table 11.2. Effect of Promalin treatment and mechanical heading on number, mean length and mean attachment angle (°) of lateral branches on 1-year Bing sweet cherry trees (adapted from Cody *et al.*, 1985).

Treatment*	Laterals tree^{-1}	Mean lateral length (cm)	Mean attachment angle (°)
Untreated	1.0	21.2	15.6
Promalin (500 ppm)	4.1†	39.1†	40.4†
Promalin (1000 ppm)	5.0†	29.6	41.9†
Promalin (2000 ppm)	7.1†	38.6	45.7†
Mechanical heading	2.3	56.3†	29.2†

*Trees were sprayed once or mechanically headed when average tree height was 41 cm.
†Value differs significantly from the untreated control at the 5% level of probability.

Based upon the success of treatments applied to young trees in the orchard (Miller, 1983), it is possible that even better quality planting stock of sweet cherry could be obtained by applying Promalin just before a second year in the nursery after budding. Such planting stock would require wider spacing in the nursery and would, of course, be more expensive. However, if such trees begin cropping as early as expected the added costs may be justified.

The technique is to apply Promalin (about 5000 mg l^{-1} of the two active ingredients) in a latex-based paint to the part of the tree where branching is desired. To be effective this treatment must be applied before bud burst.

11.2.3 Defoliating nursery stock

In regions where winters are cold enough to present a high risk of low-temperature injury, nursery operators prefer to lift their planting stock in late autumn and store the trees for early spring delivery. However, the cultural practices used to produce the large-caliper planting stock desired by most customers also delay the onset of natural autumn defoliation. Thus, soil freezing may occur before natural defoliation occurs and tree digging can be completed. Hand stripping of leaves, particularly on the 'feathered' trees desired by many customers, is a tedious and expensive operation but may be required to avoid mould build-up and other sanitation-related problems in cold storage.

Unfortunately, the search for a PBR-based technique to hasten defoliation of nursery stock without adverse side-effects to shoots and buds has failed to identify a completely satisfactory material. Some western North American nurseries have achieved reliable defoliation using a mixture of 1% Dupont WK surfactant (based on the dodecyl ether of polyethylene glycol; Dupont Chemical Co.) and 100 mg l^{-1} 2,3-dihydro-5,6-dimethyl-1,4-dithiin-1,1,4,4-tetroxide (Harvade; Uniroyal Chemical Co.). Most commonly, a single spray is applied in mid-October but with some cultivars a second spray, applied 1 week later, may be required (Larsen and Lowell, 1977). Dupont WK plus 100–200 mg l^{-1} ethephon has also proved reasonably effective. Defoliation occurs within 2–3 weeks of treatment. Terminal bud damage and tip dieback have been observed with some fruit tree species (i.e. pears and peaches) but sweet cherry appears to be more resistant to such injury.

It is worthy of mention that defoliation too early in the autumn, with or without chemicals, is a dangerous practice if the trees are expected to survive autumn planting in a cold climate or even to grow satisfactorily after cold storage. Normal acclimatization to low temperatures and maximum accumulation of storage reserves require the presence of leaves well into the mid- to late-autumn period.

Thus, the ideal situation is to produce planting stock where spring lifting is the normal procedure and to plant cherry orchards in the spring in those more continental regions where winter injury is a normal concern. This would eliminate the need for either manual or chemical defoliation.

11.3 Influencing Tree Growth and Development in the Orchard

Aside from the important usage of GA$_3$ to suppress flowering of young trees and to maintain productivity of mature sour cherry trees infected with the 'yellows' virus, most PBR practices aimed at influencing tree growth and development of cherry have been developed for sweet cherry. The need to hand-harvest sweet cherries, especially those intended for the fresh market, and the inherent large size of sweet cherry trees have led many researchers to look for PBR solutions to the excessive tree size problem.

11.3.1 Promoting lateral branching of sweet cherry

Miller's (1983) success with Promalin to improve lateral branching has led to widespread use of this technique in Australia and New Zealand to improve canopy architecture and achieve early production on young trees in high-density plantations (Jacyna *et al.*, 1989). This technique is also gaining favour with producers in western North America (T.J. Facteau, Oregon State University, personal communication). Promalin (0.5–1.0% a.i.) applied in a brown latex paint to 1-year-old branch sections on young trees, preceding or at the time of bud burst, greatly increases the number of axillary buds that develop into spurs or lateral shoots along the treated branch section. The result is greater flowering and fruiting of treated trees in subsequent years (Table 11.3). Significantly, treatments applied after bud burst inhibit flowering the following season.

Table 11.3. Effect of Promalin-containing paint applied at time of bud burst to 1-year branch sections of 3-year-old Napoleon cherry trees in 1979 and again in 1980 on spur and lateral shoot development in 1980/81 and yield per tree in 1981 and 1982 (adapted from Miller, 1983).

Treatment	Laterals and spurs per limb 1980	Laterals and spurs per limb 1981	Yield 1981 (kg tree^{-1})	Yield 1982 (kg tree^{-1})
Control	4.5	8.1	1.9a	29.1a
Promalin 0.5%	17.0	23.9	6.3b	36.9b
Promalin 1.0%	18.8	21.9	8.1b	42.2c

Means followed by different letters differ significantly at $P = 0.05$.

11.3.2 Promoting growth of young sour cherry trees

Commercial sour cherry production in the Great Lakes region of the USA and Canada is largely based on the Montmorency cultivar and involves mechanical harvesting with 'shake and catch' equipment. Relatively wide tree spacing (often 6 m × 6 m or greater) is necessary to accommodate this equipment and small trees cannot be safely subjected to mechanical shaking.

Therefore, management of young sour cherry orchards in this region focuses on encouraging tree growth and developing strong scaffold branches. Precocious flowering and fruiting is undesirable because it slows tree growth (Fig. 11.1) and, since sour cherry flower buds have no vegetative component, can lead to a serious lack of spurs and lateral shoots on 2-year and older wood. Early flowering also permits the spread of pollen-transmitted viruses, such as cherry yellows, which seriously reduce tree vigour and future productivity.

Flowering of Montmorency sour cherry can be reduced by 80% or more by a high-volume spray containing 50–100 mg l^{-1} GA$_3$ (Stang and Wiedman, 1986) applied 2–3 weeks after full bloom in the second, third and even fourth seasons

Fig. 11.1. Effect of gibberellic acid sprays on growth of Montmorency sour cherry trees. Right-hand tree was treated with 50 mg l^{-1} GA$_3$ 2 weeks after full bloom in 1979, 1980 and 1981. Both trees were planted in 1977 and photographed in 1982. (Courtesy of Prof. E.J. Stang, University of Wisconsin.)

after planting. A higher rate of GA$_3$ (up to 100 mg l^{-1}) may be required to significantly reduce flowering of low-vigour trees and those growing under obvious stress (e.g. drought, weed competition, etc.). It has also been suggested that there are long-term productivity benefits from a lower rate of GA$_3$ (25 mg l^{-1} or lower) applied the year before commercial cropping is scheduled to commence. The reason given is that overcropping in the first commercial production year, combined with the stress of mechanical shaking, may lead to excessive vigour reduction and the start of a 'blind wood' problem.

11.3.3 Ameliorating the symptoms of sour cherry yellows disease

As mentioned above, the 'yellows' virus is known to reduce tree vigour and limit productivity of Montmorency sour cherry trees in the major production regions of eastern North America. Infected trees produce too few vegetative buds to maintain the leaf-to-fruit ratio needed to achieve good fruit size and quality (Fig. 11.2). Furthermore, unless vegetative buds develop on 1-year wood sections, there can be no spur renewal and cropping potential declines precipitously.

New York State producers are advised to apply 10–25 mg l^{-1} GA$_3$ 2–3 weeks after full bloom to trees suffering the effects of the 'yellows' virus (Parker *et al.*, 1969). The higher rate is used on trees with very low vigour. The aim is to induce about three vegetative buds per terminal shoot since fewer than three will depress the long-term cropping potential of the tree.

11.3.4 Suppressing shoot growth and tree size

Because hand-picking is a normal requirement and protection from birds and rain would be a highly desirable practice, excessive tree vigour and large tree size are a particular problem with sweet cherries grown for the fresh market. This being the case, researchers worldwide have searched for a PBR that could be used to control tree vigour. The best results to date have involved chemicals in the triazole group of plant growth retardants such as paclobutrazol (PBZ)

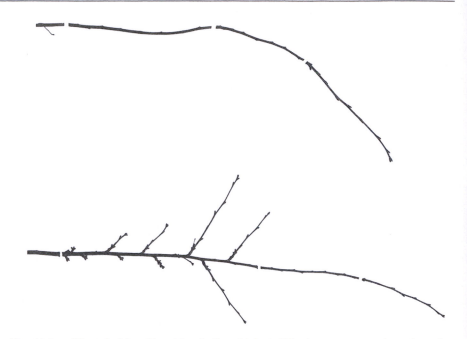

Fig. 11.2. Effect of gibberellic acid on 'yellows'-infected Montmorency sour cherry branches photographed in the autumn of 1963. Top: branch from untreated tree with each of 3 years' extension growth (1961, 1962 and 1963) separated with a white tape. Bottom: branch from tree treated with 50 mg l^{-1} GA$_3$ 3 weeks after bloom in 1961. (Courtesy of Prof. L.J. Edgerton, Cornell University.)

and uniconazole (UCZ) (Davis *et al.*, 1988). Both are highly effective inhibitors of gibberellin biosynthesis.

Paclobutrazol usage on sweet cherry

Since the early 1980s, many studies have shown that PBZ (Cultar, Zeneca Inc.) can profoundly suppress shoot elongation of sweet cherry. PBZ, a potent gibberellin biosynthesis inhibitor, is effective when applied as a foliar spray (sprays containing 500–1000 mg l^{-1} active chemical have proved effective), as a soil drench (amounts ranging from milligrams to grams per tree) and, perhaps most successfully, when metered to trees through a drip irrigation system (Fig. 11.3). Absorption by the root system is highly efficient but, as discussed below, soil applications are not always a practical alternative. UCZ (Sumagic, Sumitomo Chemical Company) is believed to be interchangeable with PBZ in field usage situations (Davis *et al.*, 1988).

PBZ-treated trees crop normally, may even exhibit greater flowering and fruiting efficiency, and fruit quality is little different from that on untreated trees (Webster *et al.*, 1986; Belmans and Keulemans, 1987; Looney and McKellar, 1987; Leonard *et al.*, 1988). Fruiting spurs on PBZ-treated trees are reported to be more productive and have a longer life expectancy (Lauri, 1993). This latter effect is probably due to better light conditions within the canopy of PBZ-treated trees.

Fig. 11.3. High-density sweet cherry orchard treated with paclobutrazol (Cultar) via the drip irrigation system. (Courtesy of Dr M.C.T. Trought, Marlborough Research Centre, Blenheim, New Zealand.)

Typically, PBZ treatment advances full bloom a few days but cropping has not been adversely affected by this phenomenon except where synchronization with pollinizers was disrupted.

The effectiveness of whole-tree sprays relies on absorption by non-lignified stem tissues and xylem transport to the apical meristem (Lever, 1986). Thus, thorough coverage of the tree is important. Foliar absorption does occur but without phloem transport there can be no direct influence on shoot growth. Eventual transport to the shoot tip might occur following decomposition of fallen leaves and movement of the chemical into the soil solution but it is not known whether such recycling is significant.

Important characteristics of PBZ that influence its use as a soil treatment are: (i) its relatively slow rate of decomposition (the half-life of PBZ in soil often exceeds 1 year); (ii) the fact that soil type and texture can greatly influence its availability to the tree; and (iii) its very high biological activity. Work with cherry seedlings grown in solution culture has demonstrated that a few parts per billion of PBZ or UCZ in the culture medium fully controls shoot elongation (Fig. 11.4).

Thus, the theoretical amount of chemical needed in the root zone to control growth of an orchard tree for several years can be measured in milligrams, perhaps even micrograms. In practice, however, since soil type and irrigation practice can profoundly influence the amount of chemical that enters the tree, the amount of chemical needed to achieve the desired amount of growth control can be difficult to predict. Soils high in clay or organic matter reversibly bind PBZ, resulting in reduced mobility and lower amounts of chemical in the soil

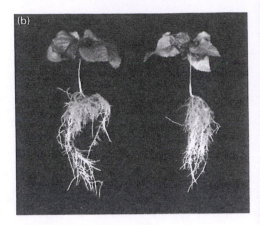

Fig. 11.4. Influence of paclobutrazol (PBZ) on growth and development of sweet cherry seedlings grown in continuously aerated Long Ashton full nutrient solution for 4 weeks. (a) control seedlings; (b) solution contained 5 ppb (1.7×10^{-8} M) PBZ; and (c) solution contained 50 ppb PBZ.

solution. Thus, the suppression of shoot elongation following ground applications may not appear as soon as expected. This leads to the temptation, even the necessity in severe cases, to increase the dosage to several grams of chemical per tree. The result can be excessive growth suppression in the subsequent year and sometimes for a period of several years. With sandy soils, PBZ is more freely available and annual applications and lower dose rates (as little as 100 mg tree^{-1}) are suggested. Here the concern may be leaching of the chemical into the groundwater.

Since spur renewal depends on shoot growth in sweet cherry, the long-term cost of excessive PBZ application rates will most certainly be reduced productivity (Facteau and Chestnut, 1991).

Effects of paclobutrazol on sour cherry

In tests conducted in Utah, sour cherry trees responded to PBZ in a manner similar to that of sweet cherry (Walser and Davis, 1989). Soil surface applications of less than 1 g PBZ tree^{-1} suppressed shoot growth for up to 3 years with no adverse effects on cropping or fruit quality. However, the observation that mid-winter cold-hardiness of flower buds was reduced by PBZ treatment was cause for concern.

Furthermore, where the sour cherry 'yellows' virus is endemic and therapeutic GA_3 sprays are required to maintain a healthy balance between floral and vegetative buds, it would seem unwise to apply PBZ.

11.4 PBR Effects on Flower Initiation, Flower Bud Hardiness, Time of Flowering and Fruit Set

While flowering and cropping can be adversely affected by canopy architecture and poor flowering often occurs in poorly illuminated sections of older trees (see Chapter 10), mature trees of both commercial cherry species usually initiate enough flowers to produce a full crop. However, serious crop reductions can result from autumn and winter low-temperature events that damage immature flower parts in dormant buds. Spring frosts are even more likely to damage cherry flowers since developing flower parts become increasingly susceptible to freeze injury. Finally, even a full display of healthy flowers does not guarantee an adequate crop if conditions for pollination, fertilization and fruit set are problematic.

These are important and widespread concerns and the search continues for PBRs that will improve cropping reliability.

11.4.1 Promoting flower initiation

Facteau and Rowe (1979) carefully compared the effects of two shoot growth-suppressing chemicals, daminozide and ethephon, on flower initiation on young Napoleon sweet cherry trees. They concluded that ethephon sprays (50 or 100 mg l^{-1} ethephon) applied 4 and 8 weeks after full bloom did promote flower initiation on wood grown prior to treatment (i.e. on 2-year and 3-year-old branch sections). However, percentage fruit set was considerably lower than on same-age branch sections on untreated trees and the overall cropping benefit was negligible. Daminozide (up to 4000 mg l^{-1} 2 weeks after full bloom) not only failed to increase flower number, but also reduced fruit set.

Daminozide may have a more beneficial effect on sour cherry. Unrath *et al.* (1969) observed increased flower bud initiation with as little as 1000 mg l^{-1} in one experiment and 4000 mg l^{-1} in another.

PBZ, a much more potent growth suppressant, is also reported to increase the number of flower buds on branch sections present at the time of root zone applications (Facteau and Chestnut, 1991). This increase in flower number resulted in higher yields, even though percentage fruit set declined at the highest PBZ level.

Because of the persistence of PBZ, treated trees produce a greater proportion of the total crop on mature spurs on older branch sections. Thus, information about the long-term productivity of these spurs is of great importance in evaluating the costs and benefits of this technology. The findings of Lauri (1993) are encouraging in that spurs on PBZ-treated trees proved to be more productive over a 3-year period than comparable spurs on untreated trees.

11.4.2 Effects on flower bud hardiness and time of anthesis

The greatest success in regulating flower bud development to improve cropping of cherry has come from autumn applications of ethephon to sweet cherry (Proebsting and Mills, 1976). A treatment of 300 mg l^{-1} ethephon, applied in mid-September, is used by Washington State producers wishing to delay anthesis by 2–3 days the following spring. This delay can result in improved cropping in seasons where flowers on untreated trees are damaged by a spring frost event. Probably more importantly, ethephon treatment also improves midwinter hardiness of flower buds and can result in significantly improved bud survival in some test winters.

The use of this technique, however, is limited by the fact that the absolute effect on hardiness is relatively small and ethephon treatment can result in reduced fruit size the following season, an effect not totally explained by increased crop load.

The usefulness of autumn-applied ethephon for delaying anthesis and improving winter hardiness of sour cherry flower buds has not been widely investigated. Dennis (1976) found that ethephon concentrations in excess of 1000 mg l^{-1} were required to delay bloom of Montmorency sour cherry and at this rate of application fruit set was reduced and excessive gumming was observed.

Other PBR chemicals can have the opposite effect on flower bud hardiness of cherry. For example, autumn applications of GA$_3$ resulted in significant winter injury to the cambium layer of both sweet and sour cherry (Dennis, 1976) and Proebsting and Mills (1974) found that 100 mg l^{-1} GA$_3$ applied in the autumn reduced hardiness of sweet cherry flower buds and cambium tissues.

11.4.3 Influencing fruit set

While a great deal of effort has been directed toward the development of PBR treatments to enhance fruit set of sweet cherry, the overall results have been disappointing. The most promise has come from mixtures of several PBRs, usually involving GA$_3$ and chemicals with cytokinin and auxin activity, applied at petal fall (Modlibowska and Wickenden, 1982). Percentage fruit set is improved in seasons where natural fruit set is very poor (presumably because of frost injury to flower parts or weather too cold for adequate pollination and/or fertilization). However, the fruit set benefit is often too small to produce a commercial crop and can be accompanied by reduced flowering in the following season (Webster *et al.*, 1979).

A technique available to some European producers involves the use of naphthaleneacetamide (NAAm) to avoid premature fruit drop of sweet cherries. This procedure, based upon research by Schumacher (1966), uses 70 mg l^{-1} NAAm applied just after full bloom.

11.5 Facilitating Mechanical Harvest

The sour cherries produced in North America, largely Montmorency, have been commercially harvested by 'shake and catch' equipment for decades. However, the use of this practice expanded rapidly in the early 1970s when ethephon was found to improve significantly harvest efficiency (Bukovak *et al.*, 1969; Looney and McMechan, 1970). This ethylene generator, applied about 1 week before shaking, substantially reduces fruit removal force. A higher proportion of the crop is removed and less mechanical energy is required to achieve this end. Lower energy inputs mean less damage to trunks and scaffold branches.

Like daminozide (Unrath *et al.*, 1969; Looney and McMechan, 1970), ethephon treatment also results in some advancement of colour and sugar levels (Anderson, 1969), although ethephon cannot be used to advance significantly the harvest season without deleterious effects on other aspects of fruit quality (Christiansen and Grauslund, 1979). Since ethephon-treated fruit separate cleanly from the stem during mechanical harvest, they are believed to be less prone to invasion by fungal pathogens that can infect fruit before it is processed. Furthermore, since less juice escapes via the stem scar, higher processing yields have been observed (Table 11.4).

Table 11.4. Effects of ethephon treatment on efficiency of fruit removal, fruit quality, leaf removal and 'gummosis' on mechanically harvested Montmorency sour cherries (adapted from Looney and McMechan, 1970).

Treatment*	Percentage of crop removed	Leaf removal†	Percentage of fruit acceptable to processor‡	Evidence of gummosis
Year one				
Control	60.6	100	37.3	None
Ethephon 500 mg l⁻¹	87.9	89	76.7	Slight
Ethephon 1000 mg l⁻¹	98.0	541	81.7	Appreciable
Year two				
Control	69.9	–	86.3	None
Ethephon 250 mg l⁻¹	92.9	–	95.8	None
Ethephon 500 mg l⁻¹	92.2	–	95.2	None

*Ethephon sprays applied to runoff 1 week before harvest.
†As percentage of the average number of leaves removed from the control trees.
‡In year two the fruit was held in cold water prior to grading.

Ethephon spray concentrations of 250–500 mg l⁻¹ have consistently proved to be safe and effective but higher concentrations can result in excessive leaf removal during shaking. More importantly, excessively high rates of ethephon can cause 'gumming' on older branch sections (Wilde and Edgerton, 1975).

11.6 Influencing Fruit Ripening and Quality

Two PBRs have been tested rather extensively for their specific effects on cherry fruit maturation and fruit quality at harvest. Daminozide treatment advances harvest date by several days, an effect that has been important to producers in some regions, and GA$_3$ has the opposite effect. However, substantial improvements to fruit quality have made the use of GA$_3$ an important practice in western North America.

11.6.1 Effects of daminozide

It is interesting to note that while butanedioic acid mono-(2,2-dimethylhydrazide) (daminozide; Alar-85, Uniroyal Inc.) delays the onset of ripening of apples, it behaves more like ethephon when applied to cherry. Sour cherries treated with daminozide 2 weeks after bloom ripen earlier and more evenly and exhibit better quality (Unrath *et al.*, 1969), even after subsequent ethephon treatment and mechanical harvesting (Looney and McMechan, 1970).

Sweet cherries treated with 1000–2000 mg l^{-1} daminozide shortly after full bloom ripen earlier, exhibit higher sugar and anthocyanin levels and have smaller stones relative to the weight of flesh (Schumacher and Fankhauser, 1971). However, treated fruit may deteriorate more quickly in storage and is reported to be less suitable for processing (Drake *et al.*, 1978).

11.6.2 Effects of gibberellic acid

It is now established practice in western North America to treat red/black sweet cherries intended for the fresh market with GA$_3$ at the straw-yellow stage of fruit development (at the start of the final fruit swell; in most seasons in British Columbia about 3 weeks before commercial harvest). High-volume sprays containing between 10 and 20 mg l^{-1} GA$_3$ slow the rate of red colour development which in turn extends or delays the harvest period by as much as a week, with a 3–4-day delay being more common (Proebsting, 1972; Looney and Lidster, 1980). This delay in harvest maturity can benefit producers in those regions where high-quality late-ripening cultivars are preferred. For example, by treating Lapins and Sweetheart cultivars with GA$_3$, British Columbia producers continue to harvest fresh market fruit well into August.

However, the most important benefits are a reliable increase in fruit size of about 10%, improved fruit firmness and a substantial reduction in poststorage losses due to 'pitting', a disorder characterized by sunken and discoloured tissues and associated with bruising during harvest and handling (Porritt *et al.*, 1971) (see Chapter 17).

Interestingly, white cherries treated with GA$_3$ in the same manner provide a superior product for canning (Proebsting *et al.*, 1973). They retain better texture and are less prone to discoloration in the can or bottle.

It is also worthy of mention here that GA$_3$ treatment to sweet cherries may reduce fruit susceptibility to rain-cracking. The author has observed instances where treated blocks have escaped serious damage from a rainstorm event while adjacent untreated blocks were largely destroyed. While there is little direct evidence, it is commonly felt that this effect is due to delayed fruit maturity. In other words, had the rainstorm event come a few days later, the untreated block might have been harvested while the GA$_3$-treated block would be both susceptible and vulnerable. None the less, the use of GA_3 treatment to 'spread the risk' of rain damage is of economic importance to some producers.

Despite these important benefits, GA_3 usage to improve fruit quality is limited to a few regions of the sweet cherry-producing world. The problem in some regions is that cherry trees treated with GA_3 in the manner described may produce too few flowers the following season to ensure a full crop.

Another limitation to the adoption of this practice relates to marketing strategy. Delayed fruit maturity is not seen as a desirable effect by producers in southern Europe or California, where earliness in the market is a major asset.

References

Anderson, J.L. (1969) The effects of Ethrel on the ripening of Montmorency sour cherries. *HortScience* 4, 92–93.

Belmans, K. and Keulemans, J. (1987) Paclobutrazol, effect on growth, yield and fruit quality of sweet cherry cv Hedelfinger. *R. Mededelingin van de Faculteit Land-bouwwetenschappen, Rijksuniversiteit Gent* 52, 1271–1274.

Borkowska, B. and Opilowska, M. (1988) Influence of BA and other cytokinins on proliferation and metabolic status of sour cherry cultured *in vitro*. *Fruit Science Reports* 15, 147–156.

Bukovak, M.J., Zucconi, F., Larsen, R.P. and Kesner, C.D. (1969) Chemical promotion of fruit abscission in cherries and plums with special reference to 2-chloro-ethylphosphonic acid. *Journal of the American Society for Horticultural Science* 94, 226–230.

Christiansen, P.F. and Grauslund, J. (1979) Changes in contents of important constituents during ripening of *Prunus cerasus* L., cv. 'Stevnsbär'. *Tidsskrift for Planteavl* 83, 95–99.

Cody, C.A., Larsen, F.E. and Fritts, R. (1985) Stimulation of lateral branch development in tree fruit nursery stock with GA_{4+7} + BA. *HortScience* 20, 758–759.

Davis, T.D., Steffens, G.L. and Sankhla, N. (1988) Triazole plant growth regulators. *Horticultural Reviews* 10, 63–105.

Dennis, F.G. (1976) Trials of ethephon and other growth regulators for delaying bloom in fruit trees. *Journal of the American Society for Horticultural Science* 101, 241–245.

Drake, S.R., Proebsting, E.L. Jr and Nelson, J.W. (1978) Influence of growth regulators on the quality of fresh and processed 'Bing' cherries. *Journal of Food Science* 43, 1695–1697.

Facteau, T.J. and Chestnut, N.E. (1991) Growth, fruiting, flowering, and fruit quality of sweet cherries treated with paclobutrazol. *HortScience* 26, 276–278.

Facteau, T.J. and Rowe, K.E. (1979) Growth, flowering, and fruit set responses of sweet cherries to daminozide and ethephon. *HortScience* 14, 234–236.

Howard, B.H. (1981) Propagation by leafless winter cuttings. *Plantsman* 3, 99–107.

Jacyna, T., Wood, D.E.S. and Trappitt, S.M. (1989) Application of paclobutrazol and Promalin (GA_{4+7} + BAP) in the training of 'Bing' sweet cherry trees. *New Zealand Journal of Crop and Horticultural Science* 17, 41–47.

Jones, O.P. and Hopgood, M.E. (1979) The successful propagation *in vitro* of two rootstocks of *Prunus*: the plum rootstock Pixy (*P. insititia*) and the cherry rootstock F12/1 (*P. avium*). *Journal of Horticultural Science* 54, 63–66.

Larsen, F.E. and Lowell, G.D. (1977) Tree fruit nursery stock defoliation with harvest aide chemical and surfactant mixtures. *HortScience* 12, 580–582.

Lauri, P.E. (1993) Long-term effects of (2RS, 3RS)-paclobutrazol on vegetative and fruiting characteristics of sweet cherry spurs. *Journal of Horticultural Science* 68, 149–159.

Leonard, W.F., Brown, G.B. and Harris, G.W. (1988). A review of paclobutrazol use on New Zealand stonefruit crops. In: *Proceedings of the 41st New Zealand Weed and Pest Control Conference*, Wellington, pp. 275–279. Weed and Pest Control Society, Wellington, New Zealand.

Lever, B.G. (1986) 'Cultar' – a technical overview. *Acta Horticulturae* 179, 459–466.

Looney, N.E. and Lidster, P.D. (1980) Some growth regulator effects on fruit quality, mesocarp composition, and susceptibility to postharvest surface marking of sweet cherries. *Journal of the American Society for Horticultural Science* 105, 130–134.

Looney, N.E. and McKellar, J.E. (1987) Effect of foliar- and soil surface-applied paclobutrazol on vegetative growth and fruit quality of sweet cherries. *Journal of the American Society for Horticultural Science* 112, 71–76.

Looney, N.E. and McMechan, A.D. (1970) The use of 2-chloroethylphosphonic acid and succinamic acid 2,2-dimethyl hydrazide to aid in mechanical shaking of sour cherries. *Journal of the American Society for Horticultural Science* 95, 452–455.

Mante, S., Scorza, R. and Cordts, J.M. (1989) Plant regeneration from cotyledons of *Prunus persica*, *P. domestica* and *P. cerasus*. *Plant Cell, Tissue and Organ Culture* 19, 1–11.

Miller, P. (1983) The use of Promalin for manipulation of growth and cropping of young sweet cherry trees. *Journal of Horticultural Science* 58, 497–503.

Modlibowska, I. and Wickenden, M.F. (1982) Effects of chemical growth regulators on fruit production of cherries. I. Effects of fruit-setting hormone sprays on the cropping of cvs Merton Glory and Van cherry trees. *Journal of Horticultural Science* 57, 413–422.

Parker, K.G., Edgerton, L.J. and Hickey, K.D. (1969) Gibberellin treatment for yellows infected sour cherry trees. *Farm Research* 29 (4), 8–9.

Porritt, S.W., Lopatecki, L.E. and Meheriuk, M. (1971) Surface pitting – a storage disorder of sweet cherry. *Canadian Journal of Plant Science* 51, 409–414.

Proebsting, E.L. Jr (1972) Chemical sprays to extend sweet cherry harvest. *Washington State University Extension Multilith* 3520.

Proebsting, E.L. Jr and Mills, H.H. (1974) Time of gibberellin application determines hardiness responses of 'Bing' cherry buds and wood. *Journal of the American Society for Horticultural Science* 99, 464–466.

Proebsting, E.L. Jr and Mills, H.H. (1976) Ethephon increases cold hardiness of sweet cherry. *Journal of the American Society for Horticultural Science* 101, 31–33.

Proebsting, E.L. Jr, Carter, G.H. and Mills, H.H. (1973) Quality improvement in canned 'Rainier' cherries (*P. avium* L.) with gibberellic acid. *Journal of the American Society for Horticultural Science* 98, 334–336.

Quamme, H.A. and Brownlee, R.T. (1993) Early performance of micropropagated trees of several Malus and Prunus cultivars on their own roots. *Canadian Journal of Plant Science* 73, 847–855.

Schumacher, R. (1966) Rötelbekämpfung bei Jungbäumen. *Schweizerische Zeitschrift für Obst- und Weinbau* 102, 230–232.

Schumacher, R. and Fankhauser, F. (1971) Beeinflussung der Fruchtentwicklung von Süsskirchen durch den Hemmstoff 2,2-dimethylhydrazid der Bernsteinsäure (Alar). *Schweizerische Zeitschrift für Obst- und Weinbau* 107, 756–762.

Snir, I. (1983) A micropropagation system for sour cherry. *Scientia Horticulturae* 19, 85–90.

Stang, E.G. and Wiedman, R.W. (1986) Economic benefit of GA_3 application on young tart cherry trees. *HortScience* 21, 78–79.

Unrath, C.R., Kenworthy, A.L. and Bedford, C.L. (1969) The effect of Alar, succinic acid 2,2-dimethyl hydrazide, on fruit maturation, quality and vegetative growth of sour cherries, *Prunus cerasus* L., cv 'Montmorency'. *Journal of the American Society for Horticultural Science* 94, 387–391.

Walser, R.H. and Davis, T.D. (1989) Growth, reproductive development and dormancy characteristics of paclobutrazol-treated tart cherry trees. *Journal of Horticultural Science* 64, 435–441.

Webster, A.D., Goldwin, G.K., Schwabe, W.W., Dodd, P.B. and Pennell, D. (1979) Improved setting of sweet cherry cultivars, *Prunus avium*, L. with hormone mixtures containing NOXA, NAA or 2,4,5-TP. *Journal of Horticultural Science* 54, 27–32.

Webster, A.D., Quinlan, J.D. and Richardson, P.J. (1986) The influence of paclobutrazol on the growth and cropping of sweet cherry cultivars. I. The effect of annual soil treatments on the growth and cropping of cv. Early Rivers. *Journal of Horticultural Science* 61, 471–478.

Wilde, M.H. and Edgerton, L.J. (1975) Histology of ethephon injury in 'Montmorency' cherry branches. *HortScience* 10, 79–81.

12 Rain-induced Cracking of Sweet Cherries: Its Causes and Prevention

J. Vittrup Christensen

Institute of Pomology, Kirstinebjergvej 12, 5792 Aarslev, Denmark

12.1 Introduction

Fruits of several species have a tendency to split during or just after rain, but none seem to be quite as sensitive to rain-induced fruit cracking as sweet cherries. In many cherry-producing areas of the world fruit cracking is the major problem limiting profitable cherry production. In some years and with some varieties the problem is extremely severe, with up to 90% of the fruits cracking. However, it has been shown that as little as 25% cracked fruit can render harvesting uneconomic on account of the escalating costs of harvesting and handling of the fruit (Looney, 1985). Furthermore, as cracked fruits are very much more susceptible to the entry of fungus rots (such as *Botrytis*), just a few cracked fruits may destroy the entire crop of susceptible varieties. Although in cherries grown for juice or other products fruit splitting is not so destructive, it still leads to crop weight loss and poor fruit quality.

12.2 The Causes of Fruit Cracking

According to the earliest reports in the literature, cherry fruit cracking was thought to be due to excessive water uptake through the root system or through the fruit skin. However, subsequent research (Sawada, 1931; Verner and Blodgett, 1931; Christensen, 1976) showed that only water absorption through the skin leads to cracking. Furthermore, splitting was considered to be a very local phenomenon, which occurs only close to the wet part of the fruit.

More recent observations, by J.A. Cline in the UK and M. Trought in New Zealand (personal communications), indicate that, although most fruit cracking is indeed attributable to direct and possibly localized water uptake through the fruit skin, a smaller proportion of fruits crack even when fully protected by rain shelters.

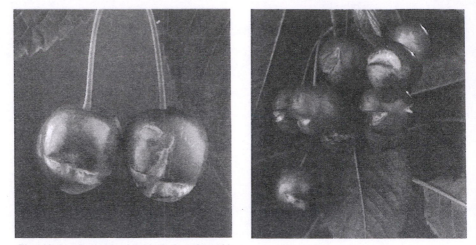

Fig. 12.1. Examples of rain-induced cracking.

12.2.1 Types of cracking and their possible causes

Cherries exhibit three different, well-defined types of cracking: circular or semicircular cracks around the stem end, similar fine cracks at the apical end and, the most injurious kind, in the form of long, irregular, and often deep splits on the sides of the fruit (Fig. 12.1).

The small cracks at the base and top end of the fruit will often occur at a very early stage, many days prior to fruit maturity. These cracks are seldom detrimental to the commercial value of the cherries and, as they often appear at an early stage in fruit development, usually cork over or seal before the fruit ripens. In contrast, the large, deep, side cracks very seldom heal; on the contrary, they are frequently followed by severe attacks of grey mould (*Botrytis* and other mould species).

In a study of the mechanism of cracking, Sawada (1934) stated that the prevalent types of cracking are mainly determined by the shape of the fruit. He argued that cracking occurs primarily at right angles to the curvature of the fruit, and noted that other fruits and vegetables, such as grapes, plums, cabbage and carrots, were affected in a similar way.

Verner (1937) noted that most often cherries cracked first at their apex. By sampling fruits which were divided into sections from the apical to the stem end, he found a sugar concentration gradient from the stem end to the apex. He concluded from this that the higher osmotic concentration of juice at the apex accounted for a more rapid water absorption through the skin, resulting in an earlier formation of cracks in this part of the fruit. He further noted that upon cessation of rain, a drop of water usually adhered to the apex of the cherry, thereby subjecting that portion of the fruit to a longer period of potential water absorption.

Studies on another crop, the closely related plums (Uriu *et al.*, 1962), have added to our knowledge by showing that apical cracks and ordinary lateral splitting are two independent phenomena, as irrigation affected the two cracking processes differently. The stage of maturity had a great influence on the development of lateral splits, whereas apical splits appeared immediately after irrigating in areas with a severe water deficit, regardless of the degree of ripeness. When soil moisture conditions were favourable throughout the whole period of fruitlet growth, very few apical splits occurred.

A similar result was recorded on sweet cherries by the author (Christensen, 1972b), who showed in a study of 32 varieties that the incidence of apical cracks did not correlate with the variety's susceptibility to more severe lateral or flesh cracking. Some varieties were susceptible to stem-end cracking and others to apical cracking. Belmans *et al.* (1989) confirmed these results, showing, for instance, that Early Rivers mostly cracked at the apex but Hedelfinger only at the stem end.

12.3 Measurements of Cracking

In most experiments on cherry fruit cracking it is important to have a well-defined expression for the intensity of cracking. The method of determination need not necessarily be the same for all types of studies.

12.3.1 Cracking measurements in the orchard

The most obvious method of determining the varietal tendency to cracking in cherries is to determine the percentage of cracked fruits under orchard conditions. However, this method can only furnish an accurate standard of comparison for cherry varieties which have an identical ripening time, as the cracking sensitivity is greatly dependent on the stage of development of the fruit and the climatic conditions.

As it is difficult to find even two varieties which are almost identical in this respect, it is evident that the cracking percentage under orchard conditions in a single year can just as easily reflect the distribution of accidental showers during the season, as indicate the varietal susceptibility to cracking.

A reliable determination of relative varietal susceptibility to cracking by this method requires measurements to be taken over several years with a representative range of climatic conditions, as the ripening season may stretch over a period of 6–7 weeks depending upon the year. An accurate standard of comparison can be obtained only in a climate with daily rainfall during the whole maturing period. Fortunately, from the grower's viewpoint, such conditions are rarely experienced in any of the major cherry-growing areas.

However, this method of cracking measurement is appropriate if the aim is merely to study the efficacy of preventative measures. One major difficulty is that rain in the sensitive period is vital if such trials are to yield any useful results. Unfortunately, simulating rain by oversprinkling with water does not give the same results as rainwater. The water droplet size, temperature and mineral content of applied water are rarely similar to those of rainwater.

12.3.2 Estimating cracking sensitivity by immersion in water

The simplest way to determine cracking sensitivity is to immerse the fruits in water and measure the percentage of the total number of fruits cracked after a specified period of time. This space of time may be varied according to the cracking resistance of the fruits and the aims of the experiment. The method is especially suitable when an approximate expression of the influence of mineral salts on the formation of cracks is desired. However, the method has the obvious shortcoming that it takes no account of the speed of crack formation. For instance, if the time interval is fixed at 6 h, then the measurement gives no indication of whether the cracking occurred in the first or last hour.

12.3.3 The cracking index

The imperfections in the previously described methods have long been recognized by cherry researchers seeking more exact expression of cracking sensitivity. Verner and Blodgett (1931) developed a laboratory procedure for determining the susceptibility of cherries to cracking which took account of some of these shortcomings. Fifty cherries, free of blemishes, were immersed in water under controlled temperature conditions for a period of 10 h. At each 2-h interval, all cracked cherries were counted and discarded. A cracking index, expressing the cracking intensity, was arrived at by multiplying the number of cracked cherries at each time of recording by a weighting factor, which was highest for the early recordings and least for the last recording. All samples used were collected in the early hours of the morning, since it was found that the time of picking could influence the cracking index.

This procedure was standardized in 1957 and the cracking index calculated as the percentage of the maximum reading obtainable. The method of determining the cracking index is described in Appendix 12.1 and a typical example is illustrated in Table 12.1.

Table 12.1. Cracking index.

Hours submerged	2	4	6	8	10
Factors for weighting	9	7	5	3	1
Number of cracked fruit	3	7	17	9	2
Weighted values	27	49	85	27	2
Total weighted values					190
Maximum possible value					450
Cracking index: (190/450) × 100					42

Distilled water held at a constant temperature should be used for calculating the cracking index, as only small amounts of cations in the water may considerably modify the incidence of cracking. As the susceptibility to cracking may differ greatly from year to year, the cracking index can be of value in the evaluation of a new variety only when the index of that variety is directly compared with that of a variety whose tendency to crack is already known and when the two varieties have been grown under comparable conditions. Such comparisons should, furthermore, be made on several different days in each of several years.

The technique presupposes, without much supporting evidence, a close correlation between the cracking index and the susceptibility to cracking under natural conditions in the field. Nevertheless, the now widely used Verner cracking index for determining cracking susceptibility offers, in principle, a good method for expressing the incidence of cracking under laboratory conditions. However, it does not provide a full description of the rhythm of cracking during the recommended 10-h period of immersion.

Another shortcoming of this method is the long immersion period required. A working day of 11–12 h will often be impracticable for researchers during the whole of the season, and a reduced time of determination is, therefore, desirable. A further problem with the method is that the cracking indices measured show a considerable annual variation. In comparisons of varieties spanning the whole ripening season, only differences of more than 20 points in the index may prove statistically significant. However, for varieties with about the same ripening period and where one or more determinations are conducted on the same day, the chances of obtaining a more precise significant result are much better. A single year's results should, in any case, be judged with care, and ideally the average of 2–3 years' determinations should be used as a way of grouping varieties.

12.3.4 Determination of cracking under artificial rain

Determination of cracking by immersion of fruits in water has some weaknesses. In experiments evaluating treatments to alleviate cracking, the conditions of fruit wetting by immersion diverge so much from natural orchard conditions that the results may be misleading or the experiment rendered impossible. For instance, wetting agents cannot be tested effectively as immersion in water offers no possibility of the water being rejected by the treatment, i.e. the effects of the treatment(s) on the amount of free water on fruits and leaves cannot be measured. Furthermore, techniques other than immersion in water are needed to stimulate different wetting conditions, such as showers of different length.

Laboratory facilities that simulate artificial rain offer extensive opportunities for experimentation during the short cherry season. To obtain results quickly using such equipment, it is preferable that room and water temperature are maintained at 25–30°C. However, the effect of preventive chemical treatments is usually weaker under natural conditions, as the fruits will not be covered as effectively with field sprays as in the laboratory sprays.

12.3.5 Varietal comparisons of cracking resistance

Varietal differences between cherries in their tendency to crack are considerable, and the cracking susceptibility is often, therefore, indicated in textbooks alongside varietal descriptions. Unfortunately, these references are usually of a highly subjective nature and often indicate the susceptibility to cracking only as high, medium or low, making comparisons with other references difficult and problematic.

Little importance can be attached to just a few tests of the cracking susceptibility of one particular cherry variety. But on some varieties sufficient information has been accumulated to allow a fairly reliable assessment of their tendency to crack. Bing, for instance, which is probably the most widely grown cherry in the USA, has so often been described as highly susceptible to cracking that further determination would appear superfluous. On the other hand, on account of these many tests Bing does lend itself very well to serving as a standard for comparison with other varieties of unknown sensitivity.

The use of the cracking index as a description of varieties' cracking susceptibilities has increased somewhat during recent years, even though observations of cracking on the trees are still more frequently quoted.

Appendix 12.2 gives cracking determinations made in Idaho, Oregon, Denmark, the Netherlands, Belgium, Norway and Spain on a range of varieties. These results frequently show differences in the order of magnitude and also in the order of susceptibility to cracking between locations. It is remarkable, for instance, that Van has the lowest index in Oregon while elsewhere it is mostly very susceptible. As mentioned earlier, the cracking index is most useful for expressing the order of resistance within varieties ripening at the same time and in tests conducted in the same year. In the Danish results the differences in indices needed to be as high as 20 points before genuine varietal difference in susceptibility could be confirmed.

12.4 Factors Influencing Cracking

12.4.1 Rate of water absorption

If one accepts the premise that cracking of sweet cherries is largely associated with water (rain) uptake by the fruit itself and not by the rest of the tree, it is important to consider the influence of both the speed and the amount of water taken up on cracking, the factors which might influence these processes and also the sensitivity to cracking.

Christensen (1972a) and Wade (1988) showed that water uptake by sweet cherries is linear with time up to 20 h. The rate of uptake varied between varieties, and fruit weight increased from 0.14% to 0.70% h^{-1}. However, it was also shown that the rate of water absorption only explained part of the fruit's susceptibility to cracking. In a study with 26 cultivars, Christensen (1972a) found that only 36% of the cause could be attributed to this factor. Verner (1937) proposed that capacity for expansion, as well as rate of water absorption, should be taken into account when ranking cracking sensitivity. On this basis he divided varieties into four groups on the basis of their susceptibility to cracking.

Group 1 – severe-cracking variety: rapid absorption and low capacity for expansion.

Group 2 – moderate-cracking variety: rapid absorption and high capacity for expansion.

Group 3 – moderate-cracking variety: slow absorption and low capacity for expansion.

Group 4 – slight-cracking variety: slow absorption and high capacity for expansion.

Verner did not try to define skin elasticity, but merely used it as an expression for the fact that some varieties absorbed more water than others before cracking.

12.4.2 Threshold for water uptake

Few studies demonstrate the threshold for water uptake, i.e. the amount of water taken up before cracking is triggered. One (Verner and Blodgett, 1931) showed that fruits increased 2–5% in volume before cracking, and Sawada (1931) suggested that there was a correlation between this threshold and varietal cracking tendency. In other studies, by Kertesz and Nebel (1935), the weight increase of fruits was as high as 18% before cracking occurred, while, in a study of 26 varieties over 3 years, Christensen (1972a) showed that water uptake varied from as low as 0.44% to 2.04% weight increase prior to cracking. A multiple correlation between rate of water uptake, cracking index and cracking threshold for water uptake gave a coefficient of 0.684, higher than that of any simple correlation. Out of 11 major deviations from a linear correlation between rate of water uptake and cracking index, eight of them could be ascribed to their cracking threshold (Table 12.2).

These results were only partly in accordance with Verner's (1937) four categories of fruit types. Ohio Beauty is the variety which is closest to the ideal in respect of cracking susceptibility. It has a very slow rate of uptake but a high threshold for water uptake before cracking – hence a low cracking index. In contrast, Bing has a very rapid rate of uptake and a rather low cracking threshold. This variety has a high cracking index and is also known to have a high tendency to crack in the field.

Rate and threshold of water uptake also vary considerably between years and growing sites; they are also influenced by several other external factors, some of which will be discussed later.

12.4.3 Fruit characteristics

If it is accepted that cracking is at least in part associated with the amount or speed of water taken up and the ability of the skin to adjust to this, then physical characteristics of the fruit and especially the skin itself must be influential in these processes and the cracking response. Skin anatomy, strength, elasticity/plasticity, cuticle integrity and stomatal function and density may all have a role to play in water uptake and splitting. Also of possible importance are the osmotic potential of the fruit juices and the water-absorbing and retaining capacity of the fruit pulp.

Table 12.2. Rate of water uptake and threshold for cracking (from Christensen, 1972a).

Variety	Rate of water absorption (% h^{-1})	Cracking index	Threshold for cracking (% wt increase)
Early Rivers	0.14	29	0.61
Erianne	0.15	54	0.64
Sam	0.17	39	0.48
Knauff	0.19	62	0.44
Ohio Beauty	0.20	26	1.18
Merton Premier	0.21	66	0.55
Starking Hardy Giant	0.22	57	0.49
Kunze	0.23	44	0.63
Seneca	0.23	36	0.89
Merton Favourite	0.27	50	0.95
Ostedgard	0.29	57	0.67
Black Tartarian	0.29	51	0.89
Snarkle	0.29	55	0.79
Kassins	0.30	42	1.12
Victor	0.31	64	0.69
Frühe Französissche	0.32	50	0.91
Merton Heart	0.35	38	1.02
Merton Bounty	0.36	81	1.06
Star	0.41	62	0.99
Merton Glory	0.44	56	0.63
Van	0.49	71	1.11
Gil Peck	0.52	92	1.06
Sodus	0.58	58	2.04
Bing	0.60	81	0.85
Heinrich Riesen	0.60	65	1.40
Napoleon	0.70	57	1.47

Fruit size

It is generally supposed that large fruits of cherries are more prone to cracking than smaller ones. However, it is essential to distinguish between differences in fruit size between varieties and differences between fruits within a single variety.

Several studies (Tucker, 1934; Sekse, 1987) have shown that large-fruited cultivars have a greater tendency to crack than small-fruited ones. In contrast, other studies, comparing many more varieties (Zielinski, 1964; Christensen, 1975) found no relationship between cracking susceptibility and varietal-determined fruit size. The relationship between cracking susceptibility and fruit size is, however, much greater within a variety than between varieties. This was confirmed by the author (Christensen, 1975), who in five experiments found an increase in the cracking index from 45 in small fruits to 67 in large fruits within the same varieties.

Studies by Bullock (1952) and Way (1967) showed that fruit on heavily cropping trees tend to crack less than fruit of the same cultivar on a tree carrying a light crop. The very high correlation between cracking susceptibility and fruit size may account for these observations, as fruits from heavily cropping trees are generally smaller.

Fruit firmness

There are considerable varietal differences in fruit firmness. It is a general assumption that cultivars with firm-fleshed fruits have a greater tendency to crack than do soft-fleshed ones. As cracking of a cherry fruit is caused by excess uptake of water resulting in bursting of the skin, it may seem logical that firm-fleshed cultivars are more susceptible to cracking than are soft-fleshed ones.

Although Kertesz and Nebel (1935) stated that cherries prone to cracking were generally those classed as firm cultivars and those less prone to cracking were soft-fleshed ones, Tucker (1934) found no correlation between firmness and a tendency to crack. Also, in a study by the author of varietal firmness in relation to susceptibility to cracking, it was not possible to find any significant correlation (Christensen, 1975).

Anatomy and strength of the fruit skin

Several investigators have tried to associate varietal differences in cracking susceptibility with the anatomy and integrity of the fruit skin. Tucker (1934) measured fruit skin thickness but found no direct relationship with cracking susceptibility. Nevertheless, he did find a correlation between the estimated unit area of skin per gram of soluble solids in the fruit and the cracking susceptibility, possibly indicating that the more skin a fruit produces, the less is its tendency to crack.

Studies by Kertesz and Nebel (1935) showed no relationship between the size of the epidermal cells and the cracking susceptibility, but a positive correlation between the thickness of the inner wall of the epidermis and the cracking susceptibility. However, they found no other differences in the skin structure which might have influenced permeability of the skin, nor any other anatomical differences influencing the cracking susceptibility, beyond the distinctions which separate firm- and soft-fleshed fruit and which could be observed without the use of a microscope.

Nobody has yet been able to measure accurately the elasticity of the skin. Although studies (Levin *et al.*, 1959) indicated that the skin modules of elasticity were less in a longitudinal than in a transverse direction, use of this technique for comparing varieties is not recommended. Rootsi (1960) concluded that the elasticity of loose detached pieces of skin depended greatly on their degree of hydration. In an investigation of cracks in plums he measured the pressure necessary to pierce the skin with a probe 1 mm in diameter. On the basis of these investigations, he concluded that an increase in hydration of the skin, possibly attributable to changes in the fruit composition when ripening or to admission of water from the outside, played a decisive part in the formation of cracks. Improved methods of measuring skin elasticity, plasticity and strength should be used in future studies (Vincent, 1990).

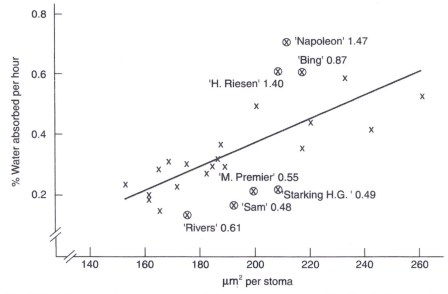

Fig. 12.2.   Rate of water absorption in relation to size of stomata: ⊗ = threshold, r = 0.679
(Christensen, 1976).

Stomata in the fruit skin

It can be argued that rate of water uptake into fruits should be influenced
by stomatal density and function. However, plant physiologists have often
demonstrated that water absorption into leaves is mostly through epidermal cells
and that the absorption through stomata is insignificant. Nevertheless, Sawada
(1931) found that at least part of the water absorption into cherry fruit was
through the stomata.

In a study of 33 varieties, the author (Christensen, 1972c) demonstrated
considerable differences between varieties in number and size of fruit stomata.
The density of stomata did not appear to have any significant influence on the
rate or threshold amount of water taken up before cracking occurred. A high
correlation with the size of the stomata was demonstrated (0.679) with rate of
water uptake and also with both rate and threshold water uptake combined
(0.827). These results are shown in Fig. 12.2, where the major deviations from
the regression line are indicated with their cracking threshold. Two of the
three varieties exhibiting the largest positive deviations had very high cracking
thresholds and the third (Bing) a medium high threshold. Bing did have the
second highest total area of stomata of all the varieties tested. The varieties with
the largest negative deviations were among those with the lowest threshold for
cracking capacity.

The correlation of water uptake with size of stomata is in accordance with
Poiseuille's law concerning the flow rate of liquids, i.e. flow rates increase
exponentially with increasing size of orifice.

Glenn and Poovaiah (1989), studying the variety Bing in the USA, found no
water-damaged tissue near stomata even when water damage occurred in regions

adjacent to stomata. On this evidence they contended that it was unlikely that stomata played an important role in the development of cracking.

Cuticular properties

The surface of the cherry fruit is covered by a thin cuticular membrane. Fractures have been shown to occur in this membrane as early as 1 month after flowering and it has been suggested that fruit cracking may, in many cases, start from such fractures. Glenn and Poovaiah (1989) noted that water uptake through the fruit skin caused separation of the cuticle from the cell wall and the associated swelling in the epidermal cell wall region resulted in cuticular fracturing, which generally preceded fruit cracking. They also observed an occurrence of preharvest cuticular fractures on fruits, which markedly increased the rate of water absorption. They suggested that the annual variation in the incidence of preharvest fractures was caused by environmental or cultural factors.

Although Glenn and Poovaiah did not find the exact cause of the formation of preharvest cuticular fractures in fruit that had not been rain-damaged, their findings seem to be an important step forward in solving the enigma of big annual variations in cracking susceptibility.

In support of this hypothesis, Belmans et al. (1990) found a very strong correlation between thickness of cuticle and the cracking resistance of 13 varieties (Fig. 12.3). Nevertheless, Bangerth (1968) could find no consistent correlation between the fruit cracking susceptibility and the amount or composition of the cuticle.

Even if investigations on the anatomy of the fruit skin are not entirely convincing in providing a complete explanation of the varietal and climatical variation in cracking susceptibility, there is no question that the surface constituents of the skin do, to a great extent, influence both the rate of water absorption and the cracking threshold for water uptake prior to splitting.

Osmotic concentration

As water absorption in plant cells is predominantly an osmotic phenomenon, it is understandable that cherry investigators have attempted to associate differences in the amounts of water absorbed into cherries with variations in the osmotic concentration of the fruit juice.

Early studies in the USA (Verner and Blodgett, 1931) showed that the rate of absorption of water was proportional to the osmotic concentration of the fruit juice, which showed a rapid increase with the degree of maturity of the fruit. After 2 h immersion in water, 5% of the fruits of the Bing variety with 14–16% sugar content showed cracking, while in the same variety with 20–22% sugar 21% were cracked. The rather simplistic conclusion was that cracking was directly affected by the osmotic concentration of the fruit juice. Further studies by Sawada (1931) at about the same time tended to support this conclusion but similarly took no account of the fact that fruits' increased sensitivity as they ripened may have had nothing to do with the parallel changes in osmotic potential.

Indeed, other investigations (Tucker, 1934) showed no correlation between the sugar content, which varied from 15.2 to 24.5%, and the tendency to crack

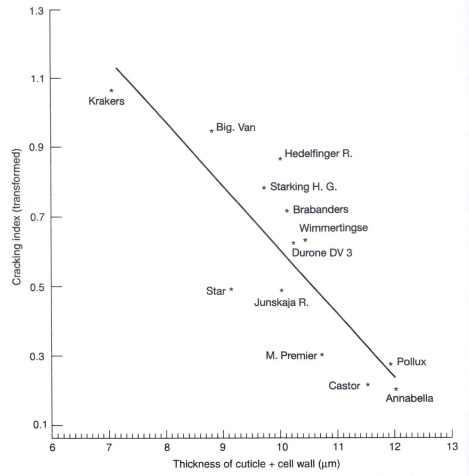

Fig. 12.3. Correlation between the cuticle of fruits and cracking index (transformed): $r = -0.85$; $y = 1.8 \times x + 180.8$ (Belmans *et al.*, 1989).

of a range of varieties to crack. The conclusion was that, while the sugar content could be of some importance, other varietal factors influenced the cracking tendency to a greater extent.

By determining the sugar content in localized portions of the fruit, Verner (1937) recorded a sugar concentration gradient from the base to the apex of the fruit. Verner argued that this helped to explain the fact that most cherries crack first at the apex, where the osmotic concentration of the juice is particularly high and absorption of water therefore takes place more quickly. However, cracking in the apex can just as easily be explained by cuticular imperfections in the region or the tendency for water drops to persist longer in this area before drying.

Bullock (1952) cast further doubt on the hypothesis that cracking is strongly influenced by osmosis. He found that although the tendency to crack increased with sugar content of the fruit from 17 to 21% in fruits of nearly uniform maturity, at concentrations of sugar between 21 and 24% the tendency to crack decreased gradually. Also, testing 35 cherry varieties, the author (Christensen, 1972c) found no or very weak correlations between the refractometer readings and the cracking index.

Comparatively large deviations from the normal increase in sugar content with advancing maturity may be explained by the ability of cherry fruits to absorb water during periods of high humidity. Such absorption of water affects the sugar concentration. In periods with dry, warm weather, the fruits may have little juice and therefore a high sugar content in relation to the stage of maturity.

In conclusion, the very slight correlation found between the cracking index and the fruit sugar content indicates that osmotic effects account very little for cracking susceptibility of the fruits. This refutes the earlier conclusions of Sawada (1931), Verner and Blodgett (1931) and Bullock (1952). In periods with relatively constant weather conditions, it is very difficult indeed to establish whether the increasing fruit sugar content, or other fruit characteristics associated with advancing maturity are responsible for the increased cracking susceptibility. Nevertheless, it is still a common contention that cracking susceptibility is in some way related to osmotic influences, with increased rates of water absorption caused by high osmotic pressures. But, as shown in research by the author (Christensen, 1972c; Fig. 12.4), there was no strong relationship between the varietal refractometer values and the rate of water absorption. Therefore, it

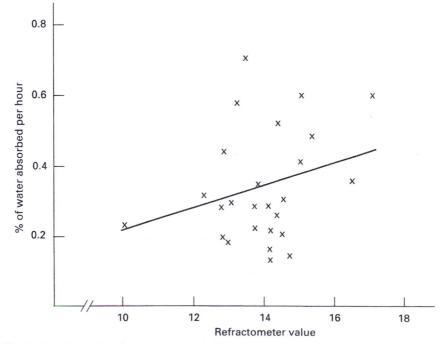

Fig. 12.4. Water absorption in relation to refractometer value: 26 varieties, 1968–1970 (Christensen, 1972c).

seems likely that a high sugar content and an increasing susceptibility to cracking are two independent characteristics which both increase with advancing maturity.

Water-retaining capacity of the fruit pulp

Kertesz and Nebel (1935), after investigating the fruit pulp from a number of varieties, hypothesized that the swelling due to absorption of water by the colloids present was mainly responsible for cracking susceptibility. They determined the water-retaining capacity of the pulped cherries by measuring, at different intervals, the amount of liquid which passed from the pulp through filter-paper. The results showed that pulp of three sweet cherry varieties retained much more water than that of the sour cherry Montmorency. They concluded from this that the water-retention capacity and swelling of this colloidal pulp was important in determining cracking susceptibility.

However, in a study comparing the water-retention force of the fruit pulp of 30 varieties of sweet and three of sour cherries, the author (Christensen, 1972c) showed no relationship between the retention force and the varietal cracking index, even though he recorded very big differences between varieties.

While in theory it is reasonable to suppose that the content of colloid in the fruit pulp may in some way affect the varietal cracking susceptibility, the methods used by Kertesz and Nebel (1935) are scientifically unsound. The pulping may, for instance, have caused an increased enzymatic activity, so that the results obtained may have been an expression of the varietal pectic-decomposing enzymes rather than their water-retention force.

Cracking in relation to the growth stage of the fruit

The stages of fruit development and the growth rhythm of cherry fruits are briefly described elsewhere (Chapter 8). The susceptibility to cracking in relation to the development stage of the fruit is of importance for the grower in making logical decisions concerning protective measures. This knowledge is also important in studies of the mechanism of cracking and for determination of cracking indices. Severe cracking is not usually observed until late in the third growth phase (Christensen, 1973), and sensitivity then increases rapidly until harvest.

Van is one variety which deviates from this norm, as it shows considerable cracking as early as the beginning of the third growth phase. This result matches commercial experience, that Van in some years cracks while the fruits are still green (Fig. 12.5).

Although most cultivars show an increasing susceptibility to cracking with increasing maturity, there are instances of deviation from this relationship. The reason for this is thought to be due to climatic conditions influencing the fruit's sensitivity to cracking before fruit sampling has begun.

The considerable reduction in cracking susceptibility observed in some cultivars after normal harvest time is probably due to a decrease in the turgor pressure of the fruit brought about by overripeness (Fig. 12.6). Verner and Blodgett (1931) showed that fruits with low turgor had a cracking index 30% less than that of normal fruits.

However, despite these occasional deviations from the norm, most research

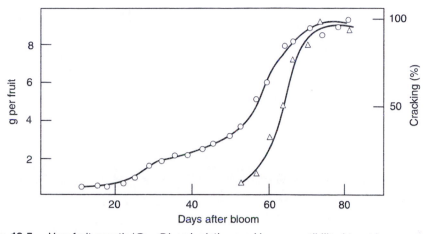

Fig. 12.5. Van: fruit growth (O—O) and relative cracking susceptibility (△—△) (Christensen, 1973).

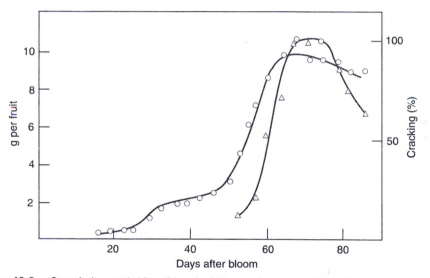

Fig. 12.6. Sam: fruit growth (O—O) and relative cracking susceptibility (△—△) (Christensen, 1973).

indicates that future work on the prevention of cracking should be concentrated on the latter half of the third growth phase. The relatively short period with maximum cracking susceptibility indicates that the time for assessing the varietal cracking index should be chosen with care.

12.5 Environmental Influences on Cracking

12.5.1 Influence of temperature

Several researchers have shown that temperature has a considerable influence on the rate of fruit cracking. In general there was a linear increase in cracking with temperature increase from 10 to 40°C (Fig. 12.7).

The proportionality between temperature and fruit cracking suggests that water uptake is possibly a physical process. Indeed, according to Poiseuille's law, the rate of flow of liquids through vessels with small diameters is directly proportional to temperature. But, as temperature also affects many other factors such as the permeability of the cell walls and biochemical processes in the cells, an unambiguous explanation of this temperature effect cannot be given. The influence of temperature on cracking indicates that the risk of cracking under orchard conditions is considerably greater during and just after rain on hot days (e.g. after thunder showers) than during cold periods (during the night or on cool days). This knowledge is of importance when deciding on the need for remedial measures to be taken for the prevention of cracking.

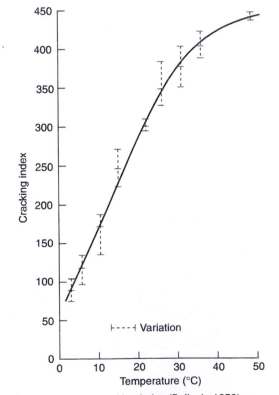

Fig. 12.7.   Effect of temperature on cracking index (Bullock, 1952).

12.5.2 Orchard environmental conditions

Examination of the many trials which have compared the cracking susceptibility of varieties, using cracking index tests, suggests that there is a considerable variation in susceptibility not only from year to year but also from one location to another. Unfortunately, variations in the published cracking indices may be, and probably often are, to a great extent the result of the different techniques employed. Inconsistency in the temperature of the water used for immersion and in the definition of a crack and, not least, differences in the turgor of fruits at the time of sampling all contribute to the variable results. Fruits usually have a high turgor in the early morning, which then decreases throughout the day. An example is shown in Table 12.3. The results show clearly that cracking susceptibility decreases markedly during the hottest parts of dry days. On the sampling day in the Venus trial b, a short shower fell between 8.00 and 10.00 a.m., which resulted in a large increase in the index at 10.00 a.m. The fruits dried quickly, however, and the test at 12.00 a.m. showed the normal decrease in susceptibility with decreasing turgor. From these data, it is clear that cracking susceptibility also varies during the day.

The influence of climatic conditions during and just after rain was studied by Levin *et al.* (1959) under laboratory conditions. Finding that significant air circulation reduced cracking of wetted fruits, they proposed drying the trees and the surrounding air by means of a helicopter or, alternatively, heating the air. Many growers shake the trees just after rain to reduce cracking and others put their air-blast sprayers through the orchard but run them without sprays, i.e. with just the high-velocity air output.

Despite several attempts, using simple and multiple correlations relating meteorological data and cracking indices, the author has failed to demonstrate a strongly significant influence of climate on cracking resistance. Such relationships are only recorded in experiments where several of the climatic conditions are controlled.

Over a short time period a single climatic factor may have quite different effects, depending on other contributory factors. For instance, high temperatures if combined with wind and dry weather reduce turgor and thereby the risk of cracking. In contrast, these same high temperatures on humid and rainy days

Table 12.3. Cracking resistance in relation to time of sampling.

	% Cracked fruits after 4 hours' immersion				
	Venus		Büttners		Winston
Time of harvest (a.m.)	a*	b*	a*	b*	a*
7.00	41	23	67	81	95
8.00	29	22	61	81	95
10.00	15	57	53	82	83
12.00	7	11	45	75	60
LSD_{95}	7	6	5	6	4

LSD_{95} = least significant difference, $P = 0.05$.
* Denotes separate and distinct trials.

increase the turgor of the fruit and increase the rate of water uptake, and consequently the risk of cracking. Risk of cracking is always highest on cloudy days or in periods with little wind and after warm nights with dew-wet fruits.

As well as these short-term climatic effects, results also indicate a long-term climatic effect on the fruit's cracking susceptibility. It has long been a subject for discussion whether cherries grown in a warm climate, such as California, are more or less susceptible to cracking than fruits grown in a cool climate, such as Norway. It is reasonable to expect that such climatic differences may influence cuticular properties, the incidence of preharvest skin fractures, skin thickness or elasticity, anatomy of the skin or size of stomata. However, studies comparing fruits from climatically different locations have not yet been carried out.

12.6 Prevention of Cracking

There are many reports in the literature suggesting that sprays of mineral salts, fungicides and other chemicals reduce cherry cracking. However, in many cases it has not been possible to reproduce these results. Also, many of the chemicals used had deleterious secondary effects, left residues on the fruits or damaged the leaves, such that they were not adopted in commercial practice.

Any compound with a positive effect in reducing fruit cracking may be expected to have at least one of the following characteristics.

1. Causes a delay or reduces the amount of water uptake into the fruit.
2. Increases transpiration of free water from the fruit surface.
3. 'Improves' the skin of fruit (strength, elasticity/plasticity or cuticular properties).

Some of the substances reported to alleviate cherry cracking are described below.

12.6.1 Sprays of calcium salts

It was in 1936 that it was first noted that some orchards, sprayed 5–6 weeks before harvest with Bordeaux mixture, produced fewer cracked fruits (Foster, 1937). These observations were rapidly followed by research and, in experiments comparing Bordeaux mixture (a mixture of copper sulphate and lime with water) and calcium hydroxide, both showed equally beneficial effects on cracking. It was concluded from these experiments that calcium was the effective chemical ion in these compounds or mixtures. However, the sprays left visible deposits on the fruits and therefore were of no practical use.

Since those early experiments many more tests have been undertaken to evaluate calcium as a remedy against cracking. Experiments on Bing in the early 1950s showed that immersion of the fruits in solutions of calcium acetate or calcium hydroxide (0.01 M) had only a small effect on their rate of cracking (Bullock, 1952). Nevertheless, spraying with the same salts (0.02–0.2 M solutions)

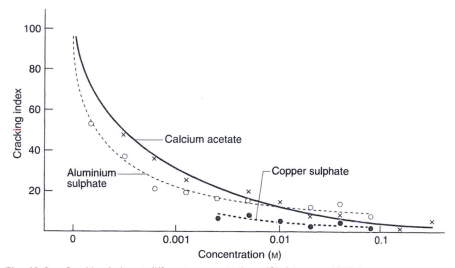

Fig. 12.8.   Cracking index at different concentrations (Christensen, 1972d).

8 and 18 days before harvest reduced the cracking index by 10–60%, even though the chemical deposits had to be removed with acetic acid. Subsequently, other researchers (Ono *et al.*, 1954) showed that calcium-caseinate sprays also greatly decreased cracking and these did not leave residues on the fruit.

In contrast, Ackley (1956) and Bangerth (1968) both found that calcium reduced cracking of fruits immersed in water but had only a slight effect when sprayed on the trees. These studies demonstrated that only a small proportion of the spray solution was deposited on the fruits and that calcium transportation from leaves to fruits was minimal. Laboratory studies by the author in Denmark (Christensen, 1972d) showed that calcium acetate (in a 0.000313 M solution) halved the cracking index compared with distilled water (Fig. 12.8).

Nevertheless, many other experiments testing calcium acetate showed no significant benefits and the only reliable treatment involved spraying with 5% solutions every hour during rain; this effectively reduced the percentage of cracked fruits. However, too much deposit was left on the trees and the treatment did not prove popular. In other Danish experiments spraying with 0.8% calcium nitrate just before or during artificial rain simulation gave a significant reduction in cracking in five out of eight experiments (Christensen, 1976).

In Washington State, USA, the local advisory service has recommended a weekly application of calcium chloride, starting 3–4 weeks before harvest. However, work in neighbouring British Columbia, in Canada (Looney, 1985), found the effect of 0.35% calcium chloride sprays to be minimal in 2 years' experiments and it was suggested that a more effective calcium source should be sought. Work in Montana (Callan, 1986) showed that three sequential preharvest sprays of calcium hydroxide consistently reduced cracking more than sprays with calcium chloride.

Although there is evidence of calcium having beneficial effects in reducing cracking and although it is recommended in some locations, these treatments do not seem to have been a great success in commercial practice. A major problem is the deposits left on the fruits by the quite concentrated sprays used. One possibility may be to modify rainwater into a weak calcium solution by overhead sprinkling with calcium solutions during rain, beginning perhaps 10–15 days before harvest. In immersion trials Danish tap water reduced cracking by 85% in comparison with distilled water and US results (Ackley and Krueger, 1980) showed that irrigation with normal water gave less cracking than irrigation with deionized water. This emphasizes the effect of even small amounts of dissolved salts in reducing fruit cracking.

The possible reasons for the beneficial calcium effect have been extensively studied and it is believed to be attributable to two major causes.

1. A direct effect of calcium on decreasing the rate of uptake of free water on the fruit.
2. A more long-term effect on the anatomy of the skin.

A reduction in the rate of water absorption is certainly evident, as shown in Table 12.4 (Christensen, 1972d).

Table 12.4. Water absorption per hour as a percentage of fruit weight.

Variety	Distilled water	0.1% Calcium acetate
Gil Peck	0.55	0.28
Merton Glory	0.35	0.08
Primavera	0.50	0.18

It is interesting to note, also, that the beneficial effects of calcium are only rarely obtained following spraying under orchard conditions 2–4 weeks before harvest; much better results are achieved by spraying during rain.

Calcium may also have a less transient effect on cracking by influencing cell permeability and the imbibition of water by the fruit colloids and by its ability to bind itself to pectins in the cell wall. According to Lidster *et al.* (1978), the cuticle and epidermal cells are very permeable to calcium, which also penetrates the fruit flesh very quickly.

In their studies on the cuticular properties of the cherry fruit, Glenn and Poovaiah (1989) showed that calcium maintains or strengthens the cell wall. However, it should be noted that Bangerth (1968) could find no correlation between the natural calcium content in fruits of eight varieties and their cracking index.

12.6.2 Sprays of aluminium, copper, other metallic salts and boron

Studies in the USA comparing the effects of different metal salts on cracking (Bullock, 1952) showed that aluminium salts had a stronger effect in reducing cracking than calcium. Sprays of 0.01–0.1 M solutions of aluminium sulphate 18 and 8 days before harvest reduced cracking considerably. Three other studies have confirmed these results. However, none of the investigators found the treatments acceptable under orchard conditions, due to the visible deposits left on the fruits and the risk of the sprays scorching the leaves.

Effective prevention of cracking has also been achieved with sprays of 0.1% copper sulphate and some tests have indicated that it was the copper, rather than the calcium, in sprays of Bordeaux mixture which gave beneficial effects. However, as with aluminium, the problem with copper sprays is the deposit left on the fruits and the risk of phytotoxic damage to leaves.

Other metal salts have received little attention in connection with prevention of cracking. US tests (Bullock, 1952) showed no effect of either iron chloride or nitrate and Danish experiments (Christensen, 1972d) similarly found no effect of sprays of magnesium, zinc, manganese or iron sulphates. Nevertheless, in a study of the effect of metabolic inhibitors on the cracking process, both silver and mercuric salts drastically reduced cracking.

The rate of cracking may be reduced by approximately 50% by immersion of fruits in a 0.25% borax solution (Powers and Bollen, 1947). However, sprays of 0.5 or 1.0% borax, while significantly reducing cracking on one variety, had no effect on several others (Knoppien, 1949). As with sprays of many of the metallic salts, these boron treatments caused serious damage to leaves.

12.6.3 Treatments with fungicides

In several experiments testing the efficacy of fungicides, investigators have noted effects on the incidence of cracking. Cation and Friday (1959) compared several fungicides sprayed at 2-week intervals beginning at petal fall and obtained remarkable results, as shown in Table 12.5. However, Christensen (1972d and 1976) could find no effect of Captan, Benlate, Thiram and Phaltan, either in immersion tests or applied in the orchard.

12.6.4 Treatments with wetting agents

Experiments in the early 1970s (Kampe, 1972) showed that wetting agents had a considerable effect by reducing cracking on several varieties and in several years, when sprayed 10 days before harvest. However, in experiments on picked fruits placed under artificial rain, Christensen (1976) found a considerable increase in cracking when the wetting agents were sprayed before rain, but a 25% reduction when spraying was delayed until immediately after 1 hour of rain.

Wetting agents may, therefore, be expected to have contradictory effects on cracking, depending on whether they are sprayed before or after rain. Their normal function is to reduce surface tension, so that water drops on the fruits are broken down to a thin water layer covering the whole fruit surface. Water may then be absorbed from the whole surface instead of at discrete locations. It is argued that spraying after rain has a beneficial effect because the water on the fruit evaporates much faster if present as a thin layer instead of drops. Experiments have shown that fruits sprayed with a wetting agent just after rain were dry after 15–20 min, while unsprayed fruits were still wet 1.5–2 h later. A very large single drop of water was observed on the fruits at the time of spraying.

Table 12.5. Effect of fungicide sprays on fruits of Windsor sweet cherry (from Cation and Friday, 1959).

Materials	Cracked fruits (%)
1. Copper–sulphur–lime	13.0
2. Phaltan–sticker	29.0
3. Sulphur–glyodin	40.0
4. Grower's programme	41.5
5. Thylate	67.0
6. Cyprex (phthalate)	71.0
7. Captan–sticker	72.0
8. Ferbam–sulphur–glyodin	74.0
9. Actidioneglyodin	77.0
10. Control (insecticide only)	82.0
11. Cyprex	88.0
12. Glyodin	93.0
13. Cyprex (borate)	96.0

12.6.5 Treatments with antitranspirants

Sprays or dips of hydrophobic compounds are often used to prevent or reduce transpiration of cherries. Davenport *et al.* (1972) applied antitranspirants to the whole fruit surface in a laboratory experiment on Californian cherries and reduced water uptake to 50% of control. Although Italian trials confirmed these Californian results, similar trials in Norway could achieve no reliable results.

Despite this inconsistency, the promising results in California and Italy indicate the possibility of using antitranspirants as cracking-preventive sprays, possibly for cherries grown in the hotter, drier production areas.

12.6.6 Treatments with plant growth regulators

In Bullock's (1952) comprehensive work on cherry cracking, he also studied the effect of sprays of the auxin, α-naphthalene acetic acid (NAA). He demonstrated a decrease of cracking when spraying with a 0.1–1.0% NAA solution 30–35 days before harvest. When applied later, however, the cracking was markedly increased.

Gibberellic acid has been shown to decrease fruit pitting, but Looney (1985) observed in two out of three experiments an additive and useful effect of the sprays in reducing cracking.

12.7 Conclusions

Much knowledge has been accumulated concerning the mechanism of cracking in cherries. Unfortunately, many results are contradictory and a final and convincing explanation of the varietal and climatic variation in cracking susceptibility is still lacking. As stated by Wade (1988):

The interaction between metabolism and cracking has far-reaching implications on cracking and the mode of action of chemical treatments which affect cracking incidence in the orchard. The possibility that chemicals which affect metabolism may also affect cracking should be considered in orchard management.

Efforts through the years to find methods for the efficient prevention of cracking have not succeeded. Many experiments have given positive results using different chemical sprays, but none seem to be safe and reliable under orchard conditions, particularly when the current demand is for a clean product without deposits on the fruit. The most promising method seems to be to spray with a weak solution of calcium salts during rain using an overhead sprinkling system (Christensen, 1972d).

The best method of reducing the risk of cracking is to choose the most resistant varieties and to minimize risk of total crop loss by growing several varieties in the same orchard. The high value of cherries in some countries may make covering the crop with rain shelters a viable option (see Chapter 16).

References

Ackley, W.B. (1956) Fruit cracking of sweet cherries. *Washington State University Progress in Agriculture and Home Economics Research, Progress Report No. 9.*

Ackley, W.B. and Krueger, W.H. (1980) Overhead irrigation water quality and the cracking of sweet cherries. *HortScience* 15 (3), 289–298.

Bangerth, F. (1968) Zur ursache des Platzen von Kirschenfrüchten. In: *ISHS Symposium on Cherries.* Institut für Obstbau de Universität Bonn, Germany, pp. 198–201.

Belmans, K., Keulemans, J. and Bronchart, R. (1989) Sensibilitée variétale à l'éclatement chez les cerises douces. *Revue de l'Agriculture* 42, 155–162.

Belmans, J., Keulemans, T., Debarsy, T. and Bronchart, R. (1990) Influence of sweet cherry epidermal characters on the susceptibility to fruit cracking. *Proceedings of the International Horticulture Congress* xxiii, 637.

Bullock, R.M. (1952) A study of some inorganic compounds and growth promoting chemicals in relation to fruit cracking of Bing cherries at maturity. *Proceedings of the American Society for Horticultural Science* 59, 243–253.

Callan, N.W. (1986) Calcium hydroxide reduces splitting of 'Lambert' sweet cherry. *Journal of the American Society for Horticultural Science* 111 (2), 173–175.

Cation, D. and Friday, J. (1959) Cracking and root control of sweet cherry. *Plant Disease Reporter* 43, 394–395.

Christensen, J.V. (1972a) Revner i kirsebaer I. Rytme og hastighed af frugernes vandoptagelse i relation til revnetilbojelighed. [Cracking in cherries. I. Fluctuation and rate of water absorption in relation to cracking.] *Danish Journal of Plant and Soil Science* 76 (1), 1–5.

Christensen, J.V. (1972b) Cracking in cherries III. Determination of cracking susceptibility. *Acta Agriculturae Scandinavica* 22, 128–136.

Christensen, J.V. (1972c) Cracking in cherries IV. Physiological studies of the mechanisms of cracking. *Acta Agriculturae Scandinavica* 22, 153–162.

Christensen, J.V. (1972d) Revner i kirsebaer V. Nogle salte og kemikaliers virkning pa revnetilbojeligheden. [Cracking in cherries V. The influence of some salts and chemicals on cracking.] *Frukt og Baer Oslo* 37–47.

Christensen, J.V. (1973) Cracking in cherries VI. Cracking susceptibility in relation to the growth rhythm of the fruit. *Acta Agriculturae Scandinavica* 23, 52–54.

Christensen, J.V. (1974) Numerical studies of qualitative and morphological characteristics of 41 sweet cherry cultivars. *Danish Journal of Plant and Soil Science* 78, 303–312.

Christensen, J.V. (1975) Cracking in cherries VII. Cracking susceptibility in relation to fruit size and firmness. *Acta Agriculturae Scandinavica* 25, 301–312.

Christensen, J.V. (1976) Revnedannelse i kirsebaer. [Cracking in cherries.] *Danish Journal of Plant and Soil Science* 80, 289–324.

Christensen, J.V. (1977) Evaluation and numerical studies of qualitative and morphological characteristics of 49 sweet cherry cultivars. *Danish Journal of Plant and Soil Science* 81, 148–158.

Christensen, J.V. (1984) Evaluation of qualitative characteristics of 48 sweet cherry cultivars. *Danish Journal of Plant and Soil Science* 88, 277–285.

Davenport, D.C., Uriu, K. and Hagan, R.M. (1972) Antitranspirant film: curtailing intake of external water by cherry fruit to reduce cracking. *HortScience* 7 (5), 507–508.

Foster, W.R. (1937) Cracking of cherries. *Scientific Agriculture* 17, 550.

Glenn, G.M. and Poovaiah, B.W. (1989) Cuticular properties and postharvest calcium applications influence cracking of sweet cherries. *Journal of the American Society for Horticultural Science* 114 (5), 781–788.

Kampe, W. (1972) Ensatz von Netzmitteln gegen das Aufplatzen der Kirschen. *Obst und Garten* 91, 161–162.

Kertesz, Z.I. and Nebel, B.R. (1935) Observations on the cracking of cherries. *Plant Physiology* 10, 763–772.

Knoppien, P. (1949) Het Scheuren van kersen. *Mededelingen van Directeur van Tuinbangewassen* 12, 77–78.

Levin, J.H., Hall, C.W. and Deshmukh, A.P. (1959) Physical treatment and cracking of sweet cherries. *Quarterly Bulletin of the Michigan Agricultural Experiment Station* 42, 133–141.

Lidster, P.D., Porrett, S.W. and, Tung, M.A. (1978) Texture modification of 'Van' sweet cherry by postharvest calcium treatments. *Journal of the American Society for Horticultural Science* 10, 527–530.

Looney, N.E. (1985) Benefits of calcium sprays below expectations in B.C. tests. *Goodfruit Grower* 36 (10), 7–8.

Ono, T., Oyaizu, W. and Suzuki, K. (1954) Studies on the reduction of cracking in sweet cherries (1). *Journal of Horticulture, Japan* 22, 239–243.

Powers, W.L. and Bollen, W.B. (1947) Control of cracking of fruit by rain. *Science* 105, 334–335.

Rootsi, N. (1960) Magnadprocess och Sprickbildning hos plommon. In: *Sveriges Pomologishe Förening. Arsskrift for 1959.* Stockholm, pp. 117–127.

Sawada, E. (1931) Studies on the cracking of cherries. *Agriculture and Horticulture* 6, 864–892.

Sawada, E. (1934) Physical consideration of the mechanism of the cracking of sweet cherries. *Sapporo Natural History Society Transactions* 13 (3), 365–376.

Sekse, L. (1987) Fruit cracking in Norwegian grown sweet cherries. *Acta Agriculturae Scandinavica* 37, 325–328.

Tabuenca, M.C. and Cambra, M. (1982) Susceptibilid ad al agriemento de los frutos de destintas variedases de cerezo. *An Aula Dei* 16, 95–99.

Tucker, R. (1934) A varietal study of the susceptibility of sweet cherries to cracking. *University of Idaho Agriculture Experimental Station Bulletin* 211, 1–15.

Uriu, K., Hansen, C.J. and Smith, J.J. (1962) The cracking of prunes in relation to irrigation. *Proceedings of the American Society for Horticultural Science* 80, 211–219.

Verner, L. (1937) Reduction of cracking in sweet cherries following the use of calcium sprays. *Proceedings of the American Society for Horticultural Science* 36, 271–274.

Verner, L. and Blodgett, E.C. (1931) Physiological studies of the cracking of sweet cherries. *University of Idaho Agriculture Experimental Station Bulletin* No. 184.

Vincent, J.F.V. (1990) Fracture properties of plants. *Advances in Botanical Research* 17, 235–287.

Wade, N.L. (1988) Effect of metabolic inhibitors on cracking of sweet cherry fruit. *Scientia Horticulturae* 34, 239–248.

Way, R.D. (1967) Cherry varieties for New York State. *Cornell Extension Bulletin* 1197. New York State College of Agriculture.

Wertheim, S.J. and Scholtens, A. (1987) Da barstindex by zoete kers. *Fruitteelt* 77, 14–15.

Zielinski, B.Q. (1964) Resistance of sweet cherry varieties to fruit cracking in relation to fruit and pit size and fruit colour. *Proceedings of the American Society for Horticultural Science* 84, 98–102.

Appendix 12.1

Based upon the work of Verner and Blodgett (1931) and Verner (1937), the author has modified and refined the cracking index (Christensen, 1972b). The modified procedure is as follows.

1. Fifty uniform, well-developed fruits are collected, preferably in the morning and in the shortest possible time (maximum 1 h).

2. The fruits are quickly immersed in 2 l distilled water held at a constant temperature.

3. Cracked fruits are counted in a good light and discarded after 2, 4 or 6 h of immersion.

4. The cracking index is calculated as:

$$\frac{(5a + 3b + c)\,100}{250}$$

where *a, b* and *c* are numbers of cracked fruits after 2, 4 and 6 hours, respectively, and 250 is the maximum obtainable.

5. The annual index is the average of three determinations made at 2–3-day intervals around the time of optimum maturity.

6. The index of a variety should be the average of determinations made over 2–3 years.

Using this technique the indices are usually a little lower than those determined using the original 10-h period (Verner and Blodgett, 1931), especially in tests of varieties with high cracking resistance. However, the range, measured between highest and lowest, is much the same (Christensen, 1972b).

Appendix 12.2: Cracking Indices as Determined in 7 Independent Studies

Variety	Denmark	Idaho, USA	Oregon, USA	Norway	Nether-lands	Belgium	Spain
Abels Späte	67						
Adriana	14						
Alfa	66						
Alfheim	50						
Allers Späte	31						
Alma	52						
Almaen Bigarreau	50						
Alman Gulröd	54						
Altenburger Melonen	53						
Ambrunés							3
Ampenar	65						
Angela	51						
Annabella	42				24		
Annonay	62						
Arslama	97						
Asdonkse	32						
Bada	25						
Badasconer	69						
Badeborner	79						
Balsgard 20406	44						
Balsgard 20414	67						
Balsgard 21441	59						
Barbara	72						
Basler Adler	80						
Belvitsa	43						
Beta	33						
Bianca	30						
Bianca di Verona	20						
Bigarreau Esperen	76		13				
Bing	75	66	49				49
Black Elton	60						
Black Giant	84						
Black Oliver	16						
Black Republican	75	20					
Bladorozova	28						
Bleylhs Braune	77						
Boambe de Cotnari	44						
Boitzeburger	73						
Burbank	74						
Burlat	88						59
Büttners Rote Knorpel	50	15		68			
Castor						1	8
Centennial			58				
Cerna Edra	37						
Chinook	95		41				
Coes Transparante	6						

Variety	Denmark	Idaho, USA	Oregon, USA	Norway	Nether-lands	Belgium	Spain
Compact Lambert	96						
Compact Stella	58						
Corazón Pichón							34
Corum	46		19				
Cristobalina							72
Daiber							43
Delta	83						
Dikkeloen	59						
Donnissens Gule	70						
Dorona Vara Vinola I	73						
Drogan Yellow	81						
Drögsperyds Medeltidiga	60						
Drögsperyds Tidiga	82						
Durella di Cesena	74						
Durona de Cesena	78					40	
Duroncino di Cesena	38						
Eagle		6					
Elton	60						
Early Rivers	30						7
Ebony	80		44				
Erianne	54						
Ermstaler	44						
Flamentiner	50						
Flamingo SRIM	12						
Forli	65						
Frans Meyling	47						
Frogmore	46						
Fromms Schwarze Herzk	74						
Frühe Bernstein	58						
Frühe Rote Meckenheimer						22	
Frühe von Dobitschen	56						
Früheste der Mark	70						
Gamma	75						
Gardebo	69						
Garrafal de Lérida							33
Gaucher	85						
Gemella	69						
Geisepitter	68						
Gil Peck	92						
Gold	89						
Grafiona	64						
Greening	59						
Grosse Germersdorfer	45						
Grosse Schwarze Herz	63					75	
Guillaume	60						16

Variety	Denmark	Idaho, USA	Oregon, USA	Norway	Nether-lands	Belgium	Spain
Hatif de Burlat	87				45	40	
Hative de Bale	50						
Hative de Berny	55						
Haumüllers Mitteldicke	36						
Hedelfinger	82					68	18
Heinrichs Riesen	65						
Holmabär	51			22			
Hoskins	59						
Hudson	34						
Inspekteur Löhnis	39				33		
Intekaer	67						
Jaboulay	75						57
Jubilee	92						
Junskaja						41	
Karesova	52						
Kassins Hjerte	42						
Knauff	71						
Knuthenborg	29						
Kozerska	44						
Krakers						77	
Kristin	38			24			
Kunze	50						
Lambert	87	35	44				12
Lamida	96		12				
Lampé							7
Lapins	41						
Larian	52						
Lucien Kirsche	34						
Macmar	57		43				
Magda	51						
Mahongnibär	40						
Maibigarreau	42						
Marmotte	74						60
Meikers					18		
Merla	46						
Mermat	70						
Merpet	80						
Merton Bigarreau	86						
Merton Bounty	57						
Merton Crane	66						
Merton Favourite	52						
Merton Glory	63			92			
Merton Heart	39						
Merton Late	16						
Merton Marvel	44						
Merton Premier	73				23	29	

Variety	Denmark	Idaho, USA	Oregon, USA	Norway	Netherlands	Belgium	Spain
Merton Reward	55						
Mollar de Cáceres							43
Mona	64						
Montagnola	54						
Moreau	89						28
Morette di Cesena	91						
Moserkirsche	84						
Napoleon	57	46	32			38	42
Noble	75						
Noir de Guben	58						
Nordwonder	59						
Ohio Beauty	26						
Oksehjerte	75		36				
Oktavia	63						
Oregon		16					
Ostads Röda	42						
Ostergard	66						
Pater van Mansfelt	39						
Pico Colorado						5	
Pico Negro						6	
Pollux				38	40		
Posnanska	52						
Primavera	63						
Producta						30	
Querfurter Königskirsche	84						
Rainbow Stripe	90						
Rainier	75		20	72			
Ramon Oliva	96						
Ranna Ljaskovska	39						
Rebekka	65						
Reverchon	48					13	
Rivan	66						
Rotterts Braune Riesen	84						
Roundel	57						
Salmo	89						
Sam	40		17	43		32	
Sanborn		19					
San Giovannino Grossa	48						
Schauenburger	35						
Schmahlfelts Schwarze	71						
Schmidt	22		35	50			
Schneiders Späte	68					39	
Schrecken	81						
Schumacher	53						

Variety	Denmark	Idaho, USA	Oregon, USA	Norway	Netherlands	Belgium	Spain
Sekunda	80						
Seneca	36						
Sigrid	54						
Sodus	8		18				
Sort Spansk	82						
Sort Tartarisk	51	59					
Souvenir des Charmes	87						
Spalding	75		39				
Spanish Yellow			38				
Sparkle	55						
Spitze Braune	51						
Star	62					18	
Starking Hardy Giant	69					67	22
Starks Royal Purple	81						
Stella	72						
Sue	26		10	26			
Sumflets Bunte	62						
Summit	71						
Sunburst	57						
Sysebär	60						
Taleguera Brillante							43
Tardif de Vignola	63						
Teickners Schwarze	65						
Tigre							24
Tolleiv	82						
Ulrichs Braune	88						
Ulster	47			26			
Uriasa de Bistrita	36						
Valera	86						
Valeska	49						
Van	74		12	92	41	66	27
Vega	72			85			
Velvet	78						
Venus	77				48		
Vernon	90		39				16
Vic	63						
Victor	63		18	24			
Vignola							25
Villareta							61
Viola	55						
Viscount	51						
Vista	72			89			
Vittoria	33						
Viva	80						
Vogue	84						
Wal Purgis	79						
Waterhouse		11					
Wellington	73						
Werder				29			

Variety	Denmark	Idaho, USA	Oregon, USA	Norway	Netherlands	Belgium	Spain
Wills-Dallas			27				
Wimmertingse						47	
Windsor	84						
Wolska Wczesna	66						
Zweitfruhe	66						

The following papers were used in compiling the above table. Denmark: Christensen (1974, 1977, 1984); Idaho, USA: Tucker (1934); Oregon, USA: Zielinski (1964); Norway: Sekse (1987); Netherlands: Wertheim and Scholtens (1987); Belgium: Belmans *et al.* (1989); Spain: Tabuenca and Cambra (1982).

IV CROP PROTECTION

13 Orchard Floor Vegetation Management

E.J. Hogue and N.E. Looney

Agriculture and Agri-Food Canada, Research Centre, Summerland, BC V0H 1Z0, Canada

13.1 Introduction

Orchard floor management is a comprehensive topic and includes what is found in the fruit crops production and protection literature under such headings as soil management, cover crops and intercrops, weed control, cultivation, etc.

The concept of orchard floor management encompasses all of the vegetation in the orchard other than the crop plant. It can include fallowing or fumigation to prepare the orchard site for planting or replanting to fruit trees, the control of competitive plants in the vicinity of the newly planted trees, the seeding, fertilizing and mowing of temporary or permanent cover crops, and the control of problem weeds in the tree row or the alleyway of established orchards. Increasingly, it must encompass discussions about crop and non-crop plant interactions with regard to pests, diseases and orchard climate. Cherry orchard vegetation management has also been discussed by Lichou *et al.* (1990) and Caprile (1992).

13.2 The Historical Development of Orchard Floor Management Systems for Cherry

Traditionally, sweet cherries were grown on vigorous rootstocks and these strong-growing trees in low-density plantation systems were relatively unaffected by weed competition or even by specific replant disease problems. Being large, free-standing trees they were often planted in fence rows, on headlands and on hillsides and left to compete with grasses and forbs (broad-leaved weeds) until they established sufficient canopy to shade out competing vegetation. However, near the end of the nineteenth century, previously sodded orchards began to be tilled and cover crops came into use (Slate, 1976).

For example, Hedrick (1915) discussed the changing practices in cherry orchards of New York. Cultivation was recommended until August, when a cover crop was sown. He reports that cultivation was especially important on light soils and could double the yield compared with trees grown in sod, particularly in dry years. An added advantage of tillage in an area subject to brown rot (*Monolinia fructicola*) was the burying of mummified fruit and hence the reduction of spores to infect next year's crop.

The interest in more intensive orchard management also led to experiments with mulching. In sour cherries, Judkins (1949) reported a long-term comparison of sod mulching (grassed floor, mulched trees) and a system of spring tillage followed by soybean and autumn rye cover crops. Eleven years after planting the mulched trees were slightly larger but yields in both systems were comparable. Mulching of tree fruit crops with all types of organic materials was widespread by the mid-twentieth century (Hogue and Neilsen, 1987).

Fogle *et al.* (1973), recognizing that it was essential to eliminate competition from young sweet cherry trees, reported that a combination of mechanical and chemical control practices was the most suitable for this crop. Thus, the age of herbicides had arrived and it quickly became common practice to rely on an ever-increasing array of chemicals to control competing vegetation – first, only in the area surrounding the tree, and later to treat the entire tree row.

Today the grassed alley and herbicided row (Fig. 13.1) remains the most common method of managing the cherry orchard floor. However, various

Fig. 13.1. Young sweet cherry plantation managed with a grass alleyway and a herbicided tree row. Summerland, British Columbia, Canada.

Fig. 13.2. High-density sweet cherry plantation managed by summer tillage. Note the Spanish Bush training system used in this Ebro Valley (Spain) orchard.

problems with herbicide use have emerged and caused growers to re-examine previous vegetation control methods and devise new ones. Furthermore, overall tillage (Fig. 13.2) remains important in several major cherry-growing regions.

This chapter will introduce and discuss the range of objectives that can be addressed with orchard floor vegetation management technology. The reader will then be introduced to some specific technologies that have proved useful.

13.3 The Objectives of Orchard Floor Vegetation Management

The overall objective of orchard floor vegetation management is to provide the crop plant with the soil fertility and other growing conditions that will allow adequate vigour and the production of high yields of quality fruit. However, by choosing the appropriate system, a number of more specific objectives can be addressed. Orchard floor vegetation management can be a key ingredient for the successful re-establishment of fruit trees in old orchard soils. It is commonly used to provide the competition needed to prevent excessive late-season growth that can adversely affect winter hardiness. It can be used to improve traffic-carrying ability and orchard accessibility to people and equipment. It can help prevent soil degradation by reducing wind and water erosion. The long-term fertility of a soil can be protected by maintaining organic matter levels. It can even have an influence on the temperature of the orchard.

There are other environmental, ecological and orchard sanitation aims.

Providing habitat for beneficial insects and eliminating diseased fruit and wood can be addressed by orchard floor management activities. Herbicide rotation strategies are designed to avoid the development of herbicide resistance and the resulting build-up of problem species. Grass and weeds can become a significant hazard for workers moving ladders during harvest.

Finally, it is essential that the orchard floor management system be compatible with a range of other orchard activities and it must, of course, be cost-effective.

13.3.1 Site preparation for successful orchard establishment

The main concern in preparing the orchard site for planting, from the vegetation management standpoint, is the presence of weeds that will be potentially troublesome in the young orchard. Very persistent perennial plants, such as field bindweed (*Convolvulus arvensis*) should be controlled before planting. Such weeds can climb into small or low-growing trees and are difficult to control once the orchard is in place. Each production area has its set of troublesome weeds and some may prove very difficult to control. For most such weeds, chemical or cultural means, or a combination of both, have been devised to provide satisfactory control.

Depending on the previous crop, fallowing the land for an entire growing season will serve to reduce weed populations and may improve soil structure and fertility. Better still, several years of cover cropping with cereals and green-manuring with annual legumes will certainly accomplish the aim of reducing weeds and improving soil structure and fertility. A variation on this management option is to establish a permanent cover crop in the spring 1 year before planting and herbiciding row strips in the autumn with glyphosate. Choosing a grass mix that will quickly establish a dense cover will help to suppress both annual and perennial weedy species.

Where there is a known cherry replant problem caused by nematodes, soil pathogens or other biotic factors, soil fumigation may be the practice of choice for site preparation (Pitcher *et al.*, 1966). Furthermore, fumigation properly carried out will control most, if not all, annual and perennial weeds.

13.3.2 Managing tree vigour for maximum productivity

The yield of fruit trees is often related to tree vigour and, in turn, tree vigour is influenced by cultural practices discussed elsewhere in this book. In sweet cherries, the use of vigorous clonal rootstocks, such as Mazzard F.12/1, to produce large trees which are rather slow to come into bearing, has minimized the need for careful control of competing vegetation around the tree. However, the introduction of size-controlling rootstocks plus training methods to advance cropping and more intensive plantation systems will intensify the need for more careful management of orchard floor vegetation in cherries.

The first 3–4 years after planting are critical. During this period the tree must make a good start toward filling the allotted space and yet become fruitful enough to produce an economic yield. Thus, removal of all competing vegetation in the immediate area is essential during this period (Fogle *et al.*, 1973). A vegetation-free strip extending 60 cm into the alley (1.2 m in total width) is normally considered sufficient.

In view of the apparent relationship between tree vigour and the proximity of competing vegetation, there has been some hope that the width of the weed-free strip could be used to control tree growth (where desired) without adversely affecting cropping. Unfortunately, this does not appear to be the case. Weeds appear to be as competitive with flowering and fruiting as they are with shoot growth.

On the other hand, trees can be kept free of competition during the early part of the growing season and then left to 'weed over' to create both moisture and nutrient stresses to prevent late-season growth. Trees managed in this fashion are likely to be less vulnerable to early-winter freeze damage.

In the course of managing orchard floor vegetation for vigour and high yields, however, there must be concern for the long-term fertility of the soil and the possible detrimental effects of residual herbicides and some tillage operations. This is particularly true in intensive management systems where the objective is to replace the plantings after short intervals to maximize return from new cultivars or to take advantage of the fruiting characteristics of young trees.

13.3.3 Ecological and environmental objectives

The orchard floor vegetation management system most appropriate for any particular cherry orchard can depend upon such widely divergent factors as the slope of the terrain or the vegetation adjoining the orchard site. A strong slope may eliminate tillage as an option, while adjoining vegetation favouring the build-up of populations of harmful insects, such as aphids and thrips, may dictate the need for clean cultivation from time to time.

Although orchard floor vegetation can harbour harmful insects (Meagher and Meyer, 1990) and diseases (e.g. quackgrass (*Elytrigia repens*) is a host for *Phytophthora cactorum* (R. Utkhede, personal communication)), some mixes of plant species can in fact provide an environment that will favour parasite and predator populations that can reduce the levels of harmful insects (Holliday and Hagley, 1984). Such mixes are likely to vary from region to region, so local information should be sought.

Organic mulches are considered to be among the best orchard floor management systems (Hogue and Neilsen, 1987) and yet Merwin *et al.* (1992) found that straw mulch promoted *Phytophthora* crown or root rots. Thus, local knowledge about disease problems can also influence the choice of orchard floor management system.

Soil conservation must be an overriding consideration. Tilled cherry orchards, especially those on light soils, can suffer severe exposure to erosion. Soils devoid of vegetation following the use of persistent herbicides may also be at risk. Hipps *et al.* (1990) found that such an orchard with a 2° slope eroded significantly.

Finally, loss of soil structure can also be a long-term concern. Kenworthy (1954) reported that newly planted sour cherry trees grew poorly on an on old orchard site where mulching had not been part of the management. Tree growth was satisfactory where mulching had occurred. He attributed the beneficial effects of mulching to improved soil structure, greater moisture-holding capacity and better pH and nutrient availability. Others have related higher soil organic matter levels to improvements in nutrient exchange capacity, water absorption, soil structure and the maintenance of a healthy soil flora and fauna.

13.3.4 Compatibility with other orchard management activities

Various other orchard management activities will be influenced by the orchard floor management system. Here are a few examples.

1. Tillage can result in dust on the fruit and unpleasant working conditions during dry weather. Heavy rains, especially on heavier tilled soils, can delay spraying, harvesting and other operations.
2. Certain cover crop or permanent sod species (e.g. white clover (*Trifolium repens*)) become very slippery when wet. Operating machinery, especially on a slope, can be hazardous.
3. The choice of cover crop affects the rodent control programme. Pocket gophers are attracted by alfalfa, dandelion and other plants with fleshy roots. Mice and voles find food and cover in tall vegetation and organic mulches.
4. Orchard nutrition is usually influenced by the orchard floor vegetation management system (e.g. Merwin and Stiles, 1994). Manures, straws and hay mulches contribute nitrogen (N) and other elements and fertilizer applications should be reduced accordingly. On the other hand, with low carbon : nitrogen (C : N) ratios materials, such as fresh sawdust, tie up soil nitrogen and require high rates of applied N to prevent N deficiency and poor growth and yields. Eliminating competing vegetation with a herbicide treatment is often equivalent to an N application.

13.3.5 Cost-effectiveness

While effectiveness is measurable and well documented for most orchard floor vegetation management options (Hogue and Neilsen, 1987), the cost of the different options can be highly variable from location to location and, at a given location, variable over time. For instance, straw for mulching may be readily available for a farmer producing both cherries and cereals and the only cost would be the labour involved in application. More commonly, the straw for mulching must be purchased and transported for long distances. Thus, the costs of this option have increased steadily over the years. Conversely, several herbicides widely used in cherry orchards have not only declined in price over the years but are now being used much more effectively, i.e. at lower rates. Thus, the herbicide option is often more cost-effective.

However, while it is relatively easy to evaluate such things as tree vigour and the costs of labour and materials, the long-term effects of an orchard floor management choice, such as the loss of organic matter, may be much more difficult to quantify. There are important concerns about the long-term productivity of orchard soils and the factors that can contribute to soil deterioration are still being documented. Still, the choice between an option that contributes

Table 13.1. The advantages and disadvantages of tillage as a vegetation management technique.

Advantages	Disadvantages
Simple, straightforward operation	Promotes wind and water erosion
Conserves soil moisture	Promotes organic matter breakdown
Destroys habitat for voles, harmful insects, mites and nematodes	Increases leaching of nutrients
Improves fertilizer availability to crop	Damages trunks and roots
Clean, tidy appearance	Can promote root diseases and winter injury
	Creates orchard travel problems following heavy rains
	Frequent tillage is costly
	Causes compacting of subsoil
	Creates dust problem during harvest

Table 13.2. The advantages and disadvantages of controlling vegetation with herbicides.

Advantages	Disadvantages
Conserves soil moisture	Choice of chemicals can be problematic
Effective and economical	Requires careful, precise application
Avoids potential root and trunk injury by tillage implements	Can require special climatic conditions for application
Destroys habitat for voles, harmful insects, mites and nematodes	Soil residues and/or chemical drift can be phytotoxic
	Resistant weed population can build up
	Soil water absorption may be impeded
	No organic matter added to soil
	Earthworm populations may decline

Table 13.3. Advantages and disadvantages of using mulches in the tree row.

	Advantages	Disadvantages
Organic mulch	Adds organic matter to soil, improves soil structure	Requires regular reapplication
	Conserves soil moisture	Can be expensive
	Adds nutrients	Can unbalance nutrition
	Controls most weeds	Will not control all perennial weeds
	Insulates soil	Provides cover for voles
		May be source of weed seeds
Inorganic mulch	Conserves soil moisture	Costly
	Requires no attention during life of mulch (up to 6 years)	Difficult to apply
	Controls all weeds	Can be cover for voles
	Increases soil temperature	Disposal at end of life can be a problem
	Adapted to low branch system	

to higher short-term profits and a more costly option with long-term benefits will always be a difficult one to make.

In the following sections we will look at some alternative technologies for managing vegetation in the tree rows and in the orchard alleyways. There are

Table 13.4. The advantages and disadvantages of permanent cover crops in fruit tree orchards.

Advantages	Disadvantages
Increases organic matter	Competes for soil moisture
Improves soil structure	Can compete for N
Improves water penetration	Harbours voles and encourages gopher
Recycles nutrients	populations
Prevents leaching	Can harbour harmful insects, mites,
Moderates pH changes	nematodes
Prevents erosion	Increases blossom freeze hazard
Increases availability of some nutrients	Requires frequent mowing
Good base for machinery travel	
Moderates summer daytime orchard	
temperatures	
Provides soil insulation	
Provides habitat for beneficial insects and	
predators	

advantages and disadvantages to every approach. Some specific strengths and weaknesses of tillage (Table 13.1), herbicides (Table 13.2), mulching (Table 13.3) and cover crops (Table 13.4) are listed for the reader's consideration.

13.4 Managing the Tree Row

The tree row can be managed with tillage, with residual and non-residual herbicides and with the use of mulching materials. While trees can be planted into a living sod, there are few good reasons to do so. The key to good growth and cropping is the elimination of competition in the tree row.

13.4.1 Tillage

Tillage, whether in the alleyway or immediately around the trees, can be destructive of soil structure and organic matter. When performed close to the trees there is the added risk of damage to roots, trunks and low branches. This is in spite of the development of a wide range of new tillage equipment that can be guided around tree trunks and regulated to till at more precise depths. None the less, tillage can be an effective means of keeping the tree row free of competing plants. It may be the only viable option where herbicides or mulches are unavailable, unacceptable or uneconomical. Tillage is discussed in greater detail in relation to managing the orchard alleyway.

13.4.2 Herbicides

Herbicide usage varies widely from country to country. Regulations regarding the purchase and use of agricultural chemicals are more restrictive in some

Table 13.5. Chemicals recommended for weed control in sweet and sour cherries in some cherry-growing areas.

Herbicide	Washington, Oregon and Idaho	France	Canada
Postemergence			
Alloxydim-sodium	–	N,E	–
Amitrole	–	E	–
(2,4-dichlorophenoxy)acetic acid	E*	N,E	–
Fluazifop	N,E	N,E	–
Glufosinate	–	E	–
Glyphosate	N,E	E	N,E
Monocarbamide dihydrogensulfate	N	–	–
Oxadiazon	–	E	–
Oxyfluorfen	N,E	–	–
Paraquat	N,E	–	N,E
Paraquat + diquat**	–	N,E	–
Sethoxydim	N	N,E	–
Preemergence			
Butralin	–	N,E	–
Dichlobenil	E	–	E
Diuron + simazine†	–	E	–
Isoxaben	N	N,E	–
Linuron	–	–	E
Napropamide	N,E	–	–
Norflurazon	E	–	–
Oryzalin	N,E	N,E	–
Oxyfluorfen	N,E	–	–
Pendimethalin	N	–	–
Propyzamide (Pronamide)	E	–	–
Simazine	E‡	E	–
Total no. of recommendations	23	21	6

*N = recommended in newly planted orchard; E = recommended in established orchard.
**Paraquat and diquat combined in a formulated mix.
†Diuron plus simazine formulated with oil.
‡Sour – Oregon, Washington, Idaho; sweet – Oregon.

countries and cost and availability issues also contribute to this situation. To illustrate this issue of availability, Table 13.5 lists the chemicals available to producers in three countries in 1993. Sweet and sour cherry growers in the USA had 23 possible recommendations of products for use in newly planted and established orchards. Producers in the adjacent Canadian province of British Columbia had only six.

These great differences in product registrations profoundly influence the number of weed control options available to producers, but climate, soil and the weed species to be controlled are also important and are often restrictive variables in herbicide usage. They determine which specific chemicals can be used and the following general guidelines have universal relevance.

Age and sensitivity of trees

Newly planted trees are very susceptible to damage by glyphosate contacting immature bark or small amounts of spray drift on the foliage. Paraquat will also damage newly planted trees when directed on to immature bark, but is considered safer than glyphosate, because small amounts of spray drift will only retard growth. Young cherry trees with a shallow, undeveloped root systems are more likely to be damaged by residual herbicides leached into the soil by heavy rains or irrigation. Thus, the label of many such chemicals specifies that they be used only on established plantings.

Soils

Soil type is a major consideration when choosing a herbicide. For instance, all residual herbicides can prove unsafe on very light, sandy soils low in organic matter. In fact, for most residual herbicides, soil organic matter content is the main component determining their safety.

Irrigation and rainfall

Orchards in regions that experience frequent and heavy rainfall or those in drier areas irrigated regularly with under-tree sprinklers have a high potential for leaching. This must be taken into consideration when choosing residual herbicides. Oryzalin, for instance, is strongly adsorbed to soil particles and resists leaching. In contrast, terbacil is considerably more mobile in soils.

Predominant weeds

Annual weeds are usually easier to control than perennial weeds. Low rates of a contact herbicide may be sufficient to control a chickweed (*Stellaria* spp.) while a high rate of translocated herbicide, glyphosate, will be required to control field bindweed (*Convolvulus arvensis*). The weed species may also dictate the timing of sprays. Country- or region-specific information on species susceptibility and timing of sprays is detailed on product labels and in weed control publications.

Judicious use of chemicals

Herbicide usage in orchards expanded rapidly when growers recognized the attractive cost/benefit ratio for this practice. But, unfortunately, perhaps inevitably, the long-term costs soon became apparent. These include the build-up of residues of persistent herbicides that can limit future cropping options, the leaching of chemicals into the water-table and the appearance of herbicide-resistant weed species. To avoid such problems, growers are urged to use the minimum amount of residual herbicide required to achieve control, to avoid whole-season weed control strategies and to use herbicide combinations and rotations to avoid the development of herbicide resistance. Weed control alternatives such as mulching can be used to reduce, even eliminate, reliance on herbicides. Implementation of these various practices has reduced the amount of chemicals used and improved long-term weed control.

13.4.3 Mulches

Although experimental trials with mulching fruit trees date back to the early 1900s, this practice is much older than that and its benefits are generally acknowledged. Kenworthy (1954) reported that straw and alfalfa hay mulches tripled sour cherry yields when compared with unfertilized sods of chewing fescue or Kentucky bluegrass. Even when the sod was fertilized, this mulch increased yield by 50%. On the other hand, a sawdust mulch did not increase yield, even with added N, indicating that the choice of mulch material is very important.

Organic mulches beneficially affect the growth of trees by improving soil moisture under the mulch, which in turn allows for shallow rooting and feeding in the fertile, well-aerated surface soil. However, obtaining suitable mulching materials is often problematic and application and reapplication can be expensive. Thus, not withstanding the obvious advantages, mulching is not widely used in modern orchards.

Perhaps the greatest interest in mulches has come from growers wishing to produce fruit without the use of pesticides and inorganic fertilizers. The favourite mulch materials in these situations are those which provide sufficient N, such as legume hays or barnyard manure–straw mixtures.

An interesting new concept that has found favour in parts of the USA is the use of a killed-sod mulch in the orchard row (Welker and Glenn, 1988). Trees are planted directly into a heavy sod that has been killed with a non-residual herbicide before planting. This sod provides protection from other vegetation, does not compete with the trees and persists for the time needed to establish the young tree. It is eventually replaced with a herbicided strip or another kind of mulch.

13.4.4 Sod or living mulch

A solid stand of almost any species in the immediate vicinity of a newly planted cherry tree will result in competition and poor tree vigour. Even after a tree is established, allowing grasses and most other plants to grow around the trunk will result in competition for moisture and nutrients. This seems to occur even if the trees are irrigated and receive high rates of fertilizers. Thus, even though soil structure and overall fertility may be improved by such vegetation (see Table 13.4), the effects on tree vigour and yield are often too serious to be tolerated in the modern orchard.

However, there is growing interest in identifying plant species that will provide a non-competitive cover in the tree row. This 'living mulch' approach is being developed successfully for annual crops but it appears to be more difficult to implement in perennial crops like tree fruits. Unfortunately, species that do not compete with the trees are also uncompetitive with other aggressive 'weed' species. The exception to this may be plants which possess allelopathic properties against such weeds. (Allelopathy is the term used to describe the deleterious effect of one plant on another through the production of specific chemicals (Rice, 1974).) This allelopathy, however, must not be directed against the crop species!

13.5 Managing Orchard Alleyways

13.5.1 Sod alleyways

The term sod implies an established cover crop (as opposed to a green manure crop), but beyond that does not specify the density, the vigour or the species composition. Cover crops have been evaluated in various production areas and under different climatic and cultural conditions (Butler, 1986; Miller *et al.*, 1989) and the species or mixture of species used to establish the sod can vary widely.

The tally of benefits and possible problems (see Table 13.4) indicates that there are more advantages to sod alleyways than disadvantages. Furthermore, the advantages are very important in that many contribute to the maintenance of soil fertility, a prime concern to all orchardists. The disadvantages, on the other hand, can be minor and easily overcome by another orchard management practice. Voles can be controlled by various means. Irrigation may be required to maintain soil moisture levels in grassed orchards.

13.5.2 Tillage

Tillage, as the sole method of managing the alleyway, offers few real advantages and can cause some serious problems, particularly to the orchard soil (see Table 13.1). However, where water is a limiting factor, tillage, at least during the dry season, is frequently the method of choice for level orchard sites. Furthermore, in newly planted orchards where competing vegetation cannot be controlled by herbicides, overall tillage may be the best choice.

The deleterious effects of tillage can be reduced to some extent. For example, infrequent, shallow tillage is better than frequent, deep tillage and organic matter can be replaced by manuring. However, a problem like erosion may be serious enough to discourage the use of tillage for inter-rows except on a temporary basis.

13.5.3 Tillage combined with green manure crops

Summer tillage combined with the autumn planting of a cover crop is still widely used in some production regions. In fact, winter annuals may reseed themselves each year to provide a cheap and satisfactory cover crop. Cover crops serve to stabilize the soil and maintain or increase soil organic matter. When legumes are used alone or as part of the cover crop mix, the opportunity exists to add N to the soil.

The choice of cover crop and the time of planting are often dictated by climate, soil factors and other cultural practices. It is seeded after harvest when competition with the fruit trees is no longer an important issue and when most orchard activities are completed for the season. For cherries, this is early enough in the autumn to allow good establishment of the cover crop before winter sets in.

The cover crop should be left to grow long enough the next spring to contribute maximally to soil organic matter but not to the point that it competes unduly with the main crop. A heavy cover crop may prove difficult to handle, but thorough incorporation into the soil is not required. Tillage equipment that cuts and chops the cover crop during incorporation is usually preferred.

13.5.4 Overall herbicides and chemical mowing

The use of chemicals to control alleyway vegetation is a relatively recent development and provides an alternative to tillage (Robinson, 1983). Non-irrigated orchards have responded well to this treatment (Atkinson and White, 1980). Although there are some clear disadvantages (see Table 13.2), they are not necessarily of a serious nature. It is possible that this option may be less detrimental to the orchard soil than tillage.

Residual herbicides, such as simazine, oryzalin or oxyfluorfen, used in combination or rotation can be very effective for long-term vegetation control. Non-residual herbicides will also do the job but frequent applications will be required. Alternating between residual and non-residual chemicals can prevent the build-up of chemical residues and of weedy species resistant to herbicides.

More recently, the concept of chemical mowing has been suggested (Bell and William, 1987; Merwin and Stiles, 1994). Turf regulators (Atkinson and Crisp, 1986) or sublethal rates of certain postemergence herbicides are used to suppress sod growth. These chemicals are applied in early spring or autumn, when sod growth is the most active.

13.6 Overall Systems

Concerns about soil conservation, sustainability of production and orchard accessibility have resulted in permanent sod being the most popular alleyway management system for cherry orchards. When defined to include such variations as tillage in the year of planting, the use of a relatively narrow sod strip and sods treated for growth control (i.e. chemical mowing), most orchards could be managed using permanent sod alleyways.

The choice of a complementary tree row treatment is more difficult.

Worldwide, the grassed alley and herbicided row are probably the most commonly used in cherry orchards, indicating their benefits and cost-effectiveness. A wide range of chemicals enables growers to control most species of invading plants on a range of soil types and in most climatic and cultural conditions (see Table 13.5). This combination provides the grower with a system by which tree vigour can be controlled and orchard soil fertility maintained. Welker and Glenn (1989) demonstrated that peach tree productivity was increased as the size of the vegetation-free area around each tree was increased from 0.36 to 9 m^2. Glyphosate was used to kill the tall fescue sod, and diuron and terbacil were used to keep the squares clean from then on. The mixture of these residual chemicals, each at 1.12 kg ha^{-1} once a year, was sufficient to maintain the treated area vegetation-free. The success of this trial demonstrates the principles of the grassed alley and herbicided tree row system. The sod alley provides a stable travel area, with all the benefits to the soil of permanent vegetation. The vegetation-free area can be regulated to provide the tree vigour desired.

The combination of mulching and a grassed alley has advantages over the grassed alley and herbicided row during the establishment years of an orchard. This is especially true where the tree branch structure starts near the soil so that the use of herbicides such as glyphosate in directed application can be hazardous to newly planted trees. Other situations where weed control with mulches may be preferred are on sandy soils, where it is not advisable to use residual herbicides, on slopes where tree rows run down the slope and can be subject to erosion when left bare, or in 'organic' orchards. With this system it may still be necessary to use herbicides periodically or in spot treatments to control perennials that grow through the mulch.

Similarly, the grassed alley and tilled row combination is useful in newly planted orchards, when many herbicides are hazardous to use. However, tillage may be difficult when trees have low branches and it should not be attempted where the slope is too steep. In an established orchard, this system may be considered as an alternative to mulched rows in non-chemical production, when mulching materials are either difficult to obtain or too expensive to use.

References

Atkinson, D. and Crisp, C.M. (1986) The use of plant growth regulators for the control of orchard swards. *Acta Horticulturae* 179, 427–430.

Atkinson, D. and White, G.C. (1980) Some effects of orchard soil management in the mineral nutrition of apple trees. In: Atkinson, D., Jackson, J.E., Sharples, R.O. and Waller, W.M. (eds) *Mineral Nutrition of Fruit Trees*. Butterworth, London, pp. 241–254.

Bell, S.M. and William, R.D. (1987) Managing orchard floor vegetation in the Pacific Northwest. *Pacific Northwest Extension Publication* 313.

Butler, J.D. (1986) Grass interplanting in horticulture cropping systems. *HortScience* 21, 394–397.

Caprile, J. (1992) Orchard floor management options. In: Southwick, S.M. (ed.) *Proceedings of a California Sweet Cherry Production Workshop*. University of California Cooperative Extension Publication, Davis, California, pp. 44–48.

Fogle, H.W., Snyder, J.C., Baker, H., Cameron, H.R., Cochran, L.C., Schomer, H.A. and Yang, H.Y. (1973) *Sweet Cherries: Production, Marketing, and Processing*. Agricultural Handbook No. 442. Agricultural Research Service, US Department of Agriculture, Washington DC.

Hedrick, U.P. (1915) *The Cherries of New York*. New York Agricultural Experiment Station, Geneva, New York State, 371 pp.

Hipps, W.A., Hazelden, J. and Fairall, G.B.N. (1990) Control of erosion in a mature orchard. *Soil Use and Management* 6, 32–35.

Hogue, E.J. and Neilsen, G.H. (1987) Orchard floor vegetation management. *Horticultural Reviews* 9, 377–430.

Holliday, N.J and Hagley, E.A.C. (1984) The effect of sod type on the occurrence of ground beetles (*Coleoptera: Carabidae*) in a pest management apple orchard. *Canadian Entomologist* 116, 165–171.

Judkins, W.P. (1949) Sites and soil management for sour cherries. *Ohio Farm Home Research* 34, 167–169. (Cited from Teskey, B.J.E. and Shoemaker, J.S. (eds) *Tree Fruit Production*, 2nd edn. AVI Publishing Co, Westport, Connecticut, p. 270.)

Kenworthy, A.L. (1954) *Effect of Sods, Mulches and Fertilizers in a Cherry Orchard on Production, Soluble Solids and on Leaf and Soil Analyses.* Technical Bulletin 243. Michigan State College, Agricultural Experiment Station, East Lansing, 39 pp.

Lichou, J., Edin, M., Tronel, C. and Saunier, R. (1990) *Le Cerisier.* CTIFL, Paris, 361 pp.

Meagher, R.J. Jr and Meyer, J.R. (1990) Effects of ground cover management on certain abiotic and biotic interactions in peach orchard systems. *Crop Protection* 9, 65–72.

Merwin, I.A. and Stiles, W.C. (1994) Orchard groundcover management impacts on apple tree growth and yield, and nutrient availability and uptake. *Journal of the American Society for Horticultural Science* 119, 209–215.

Merwin, I.A., Wilcox, W.F. and Stiles, W.C. (1992) Effects of ground cover management on the development of *Phytophthora* crown and root rots of apple. *Plant Disease* 76, 199–205.

Miller, P.R., Graves, W.L., Williams, W.A. and Madson, B.A. (1989) *Cover Crops for California Agriculture.* Leaflet 21471. University of California, Oakland, California, 24 pp.

Pitcher, R.S., Way, D.W. and Savory, B.M. (1966) Specific replant diseases of apple and cherry and their control by soil fumigation. *Journal of Horticultural Science* 41, 379–396.

Rice, E.L. (1974) *Allelopathy.* Academic Press, New York.

Robinson, D.W. (1983) Herbicide management in apple orchards. *Scientific Horticulture* 34, 12–22.

Slate, G.L. (1976) Development of fruit growing in the American states and Canadian provinces: New York. In: Upshall, W.H. (ed.) *History of Fruit Growing and Handling in the United States and Canada, 1860–1972.* Regatta City Press Ltd., Kelowna, British Columbia, pp. 99–106.

Welker, W.V. Jr and Glenn, D.M. (1988) Growth responses of young peach trees and change in soil characteristics with sod and conventional planting systems. *Journal of the American Society for Horticultural Science* 113, 652–656.

Welker, W.V. Jr and Glenn, D.M. (1989) Sod proximity influences the growth and yield of young peach trees. *Journal of the American Society for Horticultural Science* 114, 856–859.

14   Cherry Diseases: Their Prevention and Control

G.I. MINK[1] and A.L. JONES[2]

[1]Irrigated Agriculture Research and Extension Center, Washington State University, Prosser, WA 99350-9687, USA; [2]Department of Botany and Plant Pathology, Michigan State University, East Lansing, MI 48824-1312, USA

14.1 Diseases Caused by Viruses, Phytoplasmas, and Other Virus-like Agents

Most of the cherry diseases now known to be caused by viruses, phytoplasmas or undefined virus-like agents were described long before any of the causal agents were known. Names that were originally applied to these diseases frequently emphasized a prominent symptom. Most of these original names have been retained, but they usually give no information about the causal agent. Consequently, names of diseases discussed in this section can be quite confusing. They confuse even those who study them for a living! Adding to this confusion over names is the fact that several cherry diseases once thought to be distinct, and thus given different names, are now known to be caused by variants of a single agent (Mink, 1992). For example, Table 14.1 lists seven diseases caused by *Prunus* necrotic ringspot virus (PNRSV) and six diseases caused by prune dwarf virus (PDV). One name appears under both viruses. As if this were not confusing enough, PNRSV and PDV are frequently found in the same trees which compounds problems of visual diagnosis. Finally, despite the fact that many significant advances have been made in the technology used to detect and identify viruses and phytoplasmas, the specific causal agents of several virus-like diseases of cherry have not yet been determined (Gilmer *et al.*, 1974; Ogawa *et al.*, 1995).

So what does all this mean? Why should anyone other than plant pathologists care about what agent causes which cherry disease? First, for those interested in developing control strategies for these diseases, it is critical to understand that viruses and phytoplasmas cannot be eliminated from infected trees in the field by any economically practical means. Second, the spread of viruses and phytoplasmas into and within orchards cannot be controlled by chemical sprays. Consequently, strategies to control these diseases involve efforts to exclude or eradicate the causal agents. Such activities must be based on knowledge of the causal agents, their probable vectors and external factors that influence field spread.

Table 14.1. Cherry diseases that spread via pollen and are associated with the presence of *Prunus* necrotic ringspot (PNRSV) or prune dwarf (PDV) ilarviruses.

Disease name	Causal virus	Type of symptom*
Lace leaf	PNRSV	S
Necrotic leaf spot		S
Prunus necrotic ringspot		S
Rugose mosaic		C
Shot hole		S
Stecklingberger		C
Tatter leaf		S
Blind wood	PDV	C
Chlorotic ringspot		C
Narrow leaf		C
Shot hole		S
Sour cherry yellows		C
Tatter leaf		S

*S = shock symptoms appear the first year virus invades tissue; symptoms do not appear on those limbs in subsequent years even though the virus is present; C = chronic symptoms appear annually.

The following sections present information about virus and virus-like diseases that affect both sweet and sour cherries. The diseases are grouped according to the manner in which they are spread.

14.1.1 The diseases, their causal agents and how they spread

Diseases spread via pollen

The most common viruses that infect cherry trees worldwide are the two that utilize seed and pollen as their primary means for distribution. These are PNRSV and PDV. Both are classified as ilarviruses because of their tripartite genome and, while they are serologically and biochemically distinct, their biological properties overlap to the extent that both viruses are found commonly in the same orchard and often in the same trees. They can be introduced into new orchards through infected nursery stock or by virus-contaminated pollen from other orchards. Once established in an orchard, both viruses spread from tree to tree in association with pollen. In western North America, long-distance spread of PNRSV has been linked to movement of commercial beehives used to pollinate stone fruit orchards in California and Washington State. While the exact mechanism for pollen-associated orchard spread is not yet known, thrips have been shown to play a mediating role in this process.

In both sweet and sour cherries, PNRSV and PDV cause a variety of symptoms. Most of the diseases caused by these viruses derive their names from the most conspicuous symptoms (see Table 14.1). New infections by either virus often begin with a 'shock reaction', which consists of chlorotic or necrotic rings and arcs that appear on young leaves in early spring. Centres of the necrotic spots subsequently fall out, giving leaves a 'shot-hole' appearance. With some virus–cultivar combinations, this effect can be severe, leaving only skeletonized leaves on affected limbs. This condition is descriptively called lace leaf or tatter leaf. Shot hole, lace leaf, or leaf tattering may occur on a few small twigs of one branch or throughout an entire tree during the initial year of infection. These symptoms gradually fade as the season progresses. Fruit on symptomatic branches are seldom affected during the shock phase. While shock symptoms can appear on new limbs of the same tree over a period of years, they rarely, if ever, reappear on the same portions of the same limbs in succeeding years.

Chronic symptoms of either virus can be highly variable, depending upon the cultivar, the particular virus isolate and several ill-defined environmental factors. Recurrent leaf symptoms include chlorotic spots or blotches, enations, twisting or malformation and, in severe cases, significant reduction in size. Tree vigour and size can be reduced over time. Isolates of PNRSV that induce rugose mosaic disease (see Table 14.1) cause delays in fruit ripening that vary from a few days to several weeks. In orchards where fruits are picked for the fresh market, the fruits on rugose mosaic diseased trees are usually immature at optimum harvest time and therefore seldom picked. Some rugose mosaic-inducing isolates affect fruit shape as well as making the crop totally unmarketable.

In sour cherries, PDV causes sour cherry yellows disease, which gradually reduces tree vigour and productivity. In sweet cherries some PDV isolates produce narrow leaves but do not adversely affect fruit size or maturity. However, chronically infected sour and sweet cherry trees produce few permanent fruiting spurs. This lack of fruit spurs gives branches a bare, willowy appearance. These long bare spaces are referred to as 'blind wood'. Although this condition may reduce the total fruit obtained from an infected tree, it is partially offset by a parallel phenomenon where there is a substantial increase in the amount of fruit produced on 1-year wood. These fruits are exposed to good light conditions and often display high quality. Thus, while PDV may reduce tree yield, overall fruit quality may be increased.

The complex interactions among cherry cultivars, virus strains and environment make accurate diagnosis of the pollen-borne cherry disease difficult. Most strains of both PNRSV and PDV can be transmitted during early spring to herbaceous plants using rub-inoculation techniques. From these plants the viruses can be purified and studied in detail. However, these isolation and purification techniques are time-consuming and not suitable for routine diagnosis. In recent years, serological techniques have been used to assist field identification. One such technique, enzyme-linked immunosorbent assay (ELISA), is now widely used in both diagnostic and regulatory activities (Mink and Aichele, 1984). However, use of ELISA as an orchard diagnostic tool is severely limited by the fact that many isolates of PNRSV can be found in some orchards which cause no detectable effects on common cherry cultivars. In such cases it is not possible to identify, on the basis of serological results alone, which infected trees are likely to develop symptoms.

Diseases spread by airborne vectors

Five cherry diseases are known or strongly suspected to be spread by airborne vectors. These comprise two virus diseases, one disease caused by a phytoplasmas and two diseases whose causal agents have yet to be determined (Table 14.2).

Table 14.2. Diseases known or suspected to be spread by airborne vectors.

Disease name	Causal agent	Vector
Little cherry	Little cherry virus	Apple mealybug
Mottle leaf	Mottle leaf 'virus'	Eriophid mite
Rusty mottle	Unknown	Unknown
Twisted leaf	Unknown	Unknown
X disease	Phytoplasma	Leafhoppers

Little cherry disease (LCD). This has been reported from various parts of Europe, Japan, Canada and the USA. It has been especially devastating in the Kootenay region of southeastern British Columbia, Canada, where it spread rapidly during the 1940s. The only known vector is the apple mealybug *Phenacocus aceris* (Signoret) (Raine *et al.*, 1986). The rate of field spread is directly related to vector density. Flowering cherry cultivars are symptomless hosts.

In the first year, fruit on infected portions of the tree reach only about half normal size. They appear triangular when viewed from the bottom. Their slightly shrivelled skin gives them a leathery look. By harvest time, fruit of dark-coloured varieties remain pink to bright red and have an insipid flavour. Leaves on infected trees tend to be somewhat lighter green than those on non-infected trees. With time, tree growth is reduced. However, expression of fruit symptoms becomes less pronounced over the years, making visual diagnosis difficult for chronically infected trees.

Although filamentous particles up to 1500 nm in length have been consistently associated with the disease, there is, as yet, no serological method for detecting the virus in woody plants (Ragetli *et al.*, 1982). Methods utilizing fluorescent and electron microscopy have been used to detect certain disease-related abnormalities. However, none of these have proved useful for routine diagnosis. The virus can be detected biologically by leaf symptoms that develop on the sweet cherry cultivar Sam or the indicator clone Canindex 1, following graft inoculation (Hansen, 1985). Graft indexing is time-consuming but remains the most accurate method to verify diagnosis of LCD. Molecular methods for rapid detection have been developed (Eastwell and Bernardy, 1993) but require further evaluation before routine application can be recommended.

Mottle leaf disease (MLD). This is found mainly in western North America. Little is known about the causal agent. The only known vector is the cherry bud mite *Eriophyes inaequalis*. *Prunus emarginata* and peach are symptomless hosts. *P. emarginata* is also a host for the vector.

Most cherry cultivars exhibit similar symptoms. In the spring expanding leaves develop light green to whitish chlorotic areas with diffuse margins between many of the secondary veins. Affected leaves become irregularly shaped, frequently with shredded margins. Reduced terminal growth and shortened internodes cause infected trees to have a rosette-like appearance.

Fruit on severely affected trees may be smaller and may ripen later than fruit on non-infected trees. However, these fruits are not misshapen. Fruit on trees with mild leaf symptoms may appear normal.

Rusty mottle disease (RMD). This has been reported only in British Columbia,

Canada, and in the US states of Idaho, Montana, Oregon and Washington. So far nothing is known about the causal agent except that it is spread naturally by an as yet unknown vector. All sweet cherry cultivars tested have been susceptible to graft inoculation.

Leaf symptoms appear in spring as discrete chlorotic spots or rings on the lowest leaves of rapidly growing shoots. Spots and rings gradually appear on most expanded leaves. Just prior to harvest, many of the oldest leaves become bright yellow and subsequently develop mottles that include shades of brown, orange and green over a yellow background. Most of the mottled leaves soon drop. Considerable defoliation can occur during and after harvest. Chronically affected trees exhibit weak growth and decline progressively. Fruit on newly infected trees ripens normally while that on chronically infected trees may be smaller and lighter in colour than fruit on non-infected trees.

Twisted leaf disease (TLD). This has been found in the same region as RMD. Natural hosts are choke cherry (*P. virginiana*) and hybrid plums (*P. salicina* × *P. simonii*). While the disease spreads rapidly in some orchards, nothing is known about the vector. Recently, a rod-shaped virus approximately 800 nm long, has been transmitted from diseased cherry trees to *Nicotinia occidentalis* by rub inoculation. A similar virus has been transmitted from apricot trees expressing symptoms of apricot ring pox and apricot pit pox diseases. Graft transmission studies suggest that all three diseases are caused by closely related agents.

In early spring a pronounced downward twist develops in the leaf midrib near the petiole attachment of expanding leaves. Small elliptical, necrotic lesions can often be found on the lower side of the midrib where the twist occurs. Affected shoots become stunted and lower leaves begin to drop by midseason. Fruit on severely affected trees may ripen late and be somewhat misshapen.

X disease. This was first reported in California where it is called cherry buckskin disease. It was subsequently found in most western states where it is called western X, and in many eastern states, where it is referred to as eastern X disease. There appear to be symptomatological differences among isolates from different regions. X disease is caused by phytoplasmas that are transmitted by various leafhopper species. The most important natural reservoir host is wild choke cherry. Other economic hosts include peach, nectarine and Japanese plum.

Symptoms in sour and sweet cherry are similar. They occur mainly on fruit of trees propagated on Mazzard rootstocks. In the early stages some fruits on affected limbs are smaller than normal, ripen more slowly, are often distinctly pointed and have a bitter taste. It is not uncommon for normal and symptomatic fruit to occur interspersed on the same branch, often in the same cluster. Symptomatic fruits remain immature long after non-affected fruit have matured. Over a period of years the pronounced difference in size and maturity becomes less obvious. Chronically infected trees on Mazzard rootstock appear unthrifty but display few diagnostic symptoms. Trees propagated on Mahaleb rootstock usually collapse and die suddenly during midsummer of the year of infection.

For trees on Mazzard rootstocks, diagnosis of X disease can be difficult, as many of the fruit symptoms resemble those caused by little cherry virus. Diseased trees can be indexed biologically by inoculating 1-year-old cherry trees growing on Mahaleb rootstock. Infected indicator trees die within one growing season. Recently, complimentary deoxyribonucleic acid (cDNA) probes and monoclonal antibodies have been developed to supplement visual and biological diagnoses.

Diseases spread by soil-borne vectors

All of the diseases discussed in this section are caused by viruses and, with one exception, the viruses all belong to the nepovirus group (Table 14.3). Nepoviruses are nematode-transmitted and have a bipartite genome. Purified preparations also contain a third component: hollow protein shells that lack ribonucleic acid (RNA). The viruses involved include arabis mosaic (AMV), cherry leaf roll (CLRV), cherry raspleaf (CRLV), raspberry ringspot (RRSV), strawberry latent ringspot (SLRV), tomato black ring (TBRV) and tomato ringspot (TomRSV). Of these AMV, CLRV, RRSV, SLRV and TBRV are found predominantly in Europe. Both CRLV and TomRSV are found exclusively in North America. Although CLRV causes black line disease of walnut in California, this virus has not been reported to occur in cherries in the USA.

The European nepoviruses cause a variety of symptoms on cherries, including chlorotic leaf spots, leaf curling, leaf enations, reduced shoot growth, shoot dieback, bark splitting, gumming and decline. Leaf enations and decline may be intensified when trees are double-infected with PDV.

CRLV-infected trees exhibit numerous enations on the undersurface of leaves. Affected leaves become distorted and are reduced in size. The virus enters through roots but progresses through trees very slowly. Hence, most symptoms appear on lower branches and may be restricted to a few branches for many years.

TomRSV in sweet or sour cherries results in an overall unthrifty appearance. Leaves appear slightly wilted and develop light yellow or red hues. Bud break is delayed in spring. Distinct pitting develops in the woody cylinder and is most apparent on the lower trunk. Trunk sections exhibiting pitting develop abnormally thickened, spongy bark. In some cultivar/rootstock combinations, pitting may be limited to the rootstock portion (Forer *et al.*, 1984).

Tomato bushy stunt virus (TBSV). This belongs to the tombus group of viruses, characterized by isometric particles of 30 nm diameter and containing one single-

Table 14.3. Cherry diseases spread by soil-borne vectors.

Disease name	Causal virus	Vector
Decline	Tomato ringspot	Nematode
Fruit pitting	Tomato bushy stunt	Soil
Leaf enation	Arabis mosaic	Nematode
	Cherry leaf roll	Nematode
	Raspberry ringspot	Nematode
	Strawberry latent ringspot	Nematode
Leaf roll	Cherry leaf roll	Nematode
Raspleaf	Cherry raspleaf	Nematode
Xylem aberration	Tomato bushy stunt	Soil

stranded, messenger-sense RNA. While it appears to spread through soil, TBSV is not a nematode-transmitted disease. It infects cherries in both North America and Europe but has no known vector. The most conspicuous symptoms are severe depressions and distortion of fruit. Fruit symptoms may be accompanied by necrotic spots near leaf midribs, on fruit pedicels and on small twigs. These tissues often twist at the necrotic spots. The virus moves slowly through trees and can be isolated only from tissues exhibiting symptoms.

Diseases spread by grafting

Virus-like diseases that are spread by grafting represent a heterogeneous group of disorders, about which relatively little is known. The causal agent has been determined for only one of the 15 diseases listed in Table 14.4. Green ring mottle virus (GRMV), which infects but remains symptomless in sweet cherry cultivars and causes green rings and spots on leaves of sour cherry cultivars, has been shown to be a closterovirus, with flexuous particles measuring 1000–2000 nm in length and 5–6 nm in width (Zagula *et al.*, 1989). Although many closteroviruses can be transmitted by aphids, there is no evidence that GRMV is spread other than by grafting. This virus may occur without symptoms in many sweet cherry orchards planted in western North America before the establishment of virus certification programmes.

The other diseases listed in Table 14.4 have one or more characteristics that suggest they may be caused by viruses or phytoplasmas but their aetiology remains obscure. For the most part, the diseases are obscure. They are found only in limited geographic areas and then usually in relatively few trees. Their economic impact is marginal at most. Consequently, they attract little attention from growers or scientists.

Viruses isolated from cherry trees without symptoms

Three viruses known to cause diseases in other crops have been isolated from cherry trees. These are apple chlorotic leafspot virus, cucumber mosaic virus and tobacco mosaic virus. So far, these viruses have not been associated with specific diseases of cherry.

Non-infectious disorders that resemble virus diseases

Two non-infectious disorders occur commonly in certain sweet cherry cultivars and are frequently confused with virus diseases. These are cherry crinkle leaf and deep suture diseases. Both are assumed to be 'genetic' disorders because they are perpetuated by propagation, but not transmitted through grafts. Cultivars such as Bing and Black Tartarian are prone to both disorders. Trees propagated from healthy-appearing trees may or may not develop symptoms. Trees propagated from diseased shoots are nearly always diseased.

Crinkle-affected leaves are misshapen. The edges are deeply serrated and have large indentations on one or both sides. Distorted leaves have areas of light green mottle. Fruit are small, pointed, and have a raised suture. Trees affected with deep suture have leaves shaped much like those on crinkle trees but without the green mottle. The upper leaf surface has a pebbly appearance. Affected fruits

Table 14.4. Cherry diseases spread primarily by grafting.

Disease name	Symptoms	Distribution
Albino	Leaf roll; leaf casting; dieback; small, white fruit; tree death	Oregon
Amasya cherry disease	Leaf spots; defoliation; dieback; small, tasteless fruit	Turkey
Bark splitting (Montmorency cherry)	Narrow, brown stripes on young shoots; longitudinal splits in bark; gumming; tree death	Oregon
Black canker	Black cankers and swollen areas on limbs	Oregon, Washington State, British Columbia, Australia
Blossom anomaly	Green areas on petals; phyllidy; pointed fruit	Washington State
Boron rosette	Delayed flowering; flower abscission; narrow, chlorotic leaves	Oregon
Freckle fruit	Small, immature fruit with numerous, brown necrotic flecks	Oregon
Green ring mottle (Montmorency cherry)	Yellow or green rings on leaves; leaf drop; distorted fruit with necrotic spots	North America
Gummosis (Montmorency cherry)	Tip dieback; stem cankers; pronounced gumming	Oregon, Washington State
Line pattern	Chlorotic lines in leaves; deformed leaves; fruit spots	Eastern Europe
Necrotic rusty mottle	Angular, brown, necrotic spots on leaves; leaf drop; dieback	Western USA
Rough fruit	Fruit spots; deformed fruit	Bulgaria, Iran
Rusty spot	Leaf spots; shot-holes; curled, distorted leaves	New Zealand
Short stem	Leaf roll; misshapen fruit; short fruit stems	Montana, Oregon
Spur cherry	Leaf epinasty; bark necrosis; severe dwarfing	Washington State

have a pronounced depression along the suture. Fruits on both crinkle- and deep suture-affected trees have a distinct pointed shape.

14.1.2 Prevention and control

As mentioned above, there are no cures that can be applied commercially in the orchard for those cherry diseases caused by viruses or mycoplasmas. Consequently, with very few exceptions, when an orchard tree becomes infected, it remains so for the rest of its life. For those diseases spread only by grafting, subsequent spread can be prevented by tree removal or by avoiding propagation. For diseases that are spread by vectors, the infected trees become inoculum reservoirs for further spread. Attempts to use chemical sprays to suppress vector activity and thus to control spread have had only limited success and then only in highly specialized situations. Consequently, decades before it was fashionable to seek alternatives to chemical control, those who worked with virus and virus-like disease of fruit crops devised control strategies that integrate many different activities, some of which occur well away from the orchard. The basic premise is to exclude pathogens in one way or another from orchards. To be effective, exclusion activities must function simultaneously at several governmental levels. Some of the activities involved are summarized below.

Quarantine

Plant quarantine activities are promulgated by government agencies at both national and regional levels. Their purpose is to exclude pests or pathogens from areas where they do not already occur. At the national level, they restrict importation of plant materials and seeds from foreign countries where exotic pathogens or strains are known to occur. Individual states, provinces or regional districts may impose quarantines against trees produced in other areas of a country. Since quarantine regulations and needs are reviewed periodically, no attempt is made here to itemize specific regulations covering the movement of cherry tissues.

Plant introduction

Even though quarantine regulations can restrict importation of materials from foreign countries, they seldom prevent it entirely. For example, in the USA provisions exist for importation of cherry budwood through the National Plant Introduction Station at Beltsville, Maryland, or, in some cases, through the Interregional Project 2 (IR-2) facility at Prosser, Washington State. At either location, propagules are made, indexed for viruses or other transmissible agents, exposed to high temperatures to eliminate pathogens, retested and subsequently released. In general, this may take 1–3 years. Similar plant introduction protocols are used in many European countries. However, specific methods used vary greatly among countries.

Interregional Project 2. IR-2, a US government project recently renamed National Research Support Project 5 (IR-2/NRSP-5), is a good example of a national research support project whose objectives include development, maintenance and distribution of budwood and seed of deciduous fruit tree clones that are free of viruses, viroids and phytoplasmas. Clones of commercial *Prunus*, *Malus*, *Pyrus* and *Cydonia* are propagated, heat-treated to remove virus and virus-like agents and released under one of four programmes; (i) patented clones; (ii) restricted clones; (iii) sponsored clones; and (iv) public clones. A fee is charged for materials processed under programmes i–iii. An annual distribution list identifying those clones available to researchers, State certification programmes, and the public is provided upon request.

Certification programmes

At least ten US states, two Canadian provinces and several foreign countries have government-regulated activities, collectively referred to as nursery stock improvement programmes. These programmes usually obtain their élite propagation materials from the IR-2 project or similar programmes in other countries. These materials are increased by participating nurseries under guidelines developed and monitored by the appropriate state or federal agency. Certified trees are sold to growers.

Orchard activities

Once trees are planted in an orchard, they are closely monitored by the grower and possibly observed periodically by farm advisers and horticultural extension agents. When unusual symptoms appear that resemble virus diseases, these individuals often consult with researchers who specialize in the diagnosis of cherry virus diseases. Once testing verifies the presence of a virus or phytoplasma, the recommendation is almost always the same: remove the tree. However, growers universally tend to regard tree removal as a method of last resort. If fruit quality or yield is obviously reduced or if tree growth is adversely affected, the tree may be removed. If the effects are mild, the infected tree may remain for years and serve as a reservoir for further spread.

14.2 Diseases Caused by Fungi and Bacteria

Knowledge concerning those diseases of cherries caused by fungi and bacteria has increased rapidly in recent decades. Several new pathogens have been identified, new information on the biology and epidemiology of well-known pathogens has emerged and new or improved methods of control have been introduced. The economic importance of these diseases can be serious across a cherry-growing region or confined to individual orchards, with the method of transmission and the tissues affected being important factors. Some diseases affect the roots while others are confined to leaves and fruits. Those fungi and bacteria that attack roots are commonly referred to as soil-borne pathogens and are more likely to be of local importance.

Although information from around the world is included in this section, we have drawn heavily on research concerning fungal and bacterial diseases of cherries conducted in North America.

14.2.1 The fungal diseases of cherries

Cherry leaf spot

In North America, leaf spot occurs throughout the Great Lakes and mid-Atlantic states and provinces. In this region, leaf spot is the most important fungal disease of sour cherry and second to American brown rot in importance on sweet cherry. It is also a major disease of cherries in Europe. It occurs in both nursery and orchard plantations.

Symptoms first appear on the upper surface of leaves as small, circular, purple spots that eventually turn brown. Although infected leaves turn yellow, the area around lesions on sour cherry leaves may remain green, giving the leaves a mottled appearance. During wet weather, white to light pink spore masses are evident on the lower surface of infected leaves in the centre of the spots. The yellow leaves eventually abscise. Severe defoliation of trees between the petal fall stage of bud development and harvest can cause uneven ripening and poor fruit quality. It reduces tree vigour and winter-hardiness of the flower buds and the wood.

Leaf spot is caused by *Blumeriella jaapii* (Rehm) Arx (syn. *Coccomyces hiemalis* Higgins). Cup-like, ascus-bearing fruiting bodies (apothecia) are produced in leaves on the ground that were infected while on the tree during the previous growing season. Saucer-shaped fruiting bodies (acervuli) bearing conidiophores and conidia are produced on the lower surface of leaves infected in the current growing season.

Infection takes place through mature stomata on the lower surface of leaves and on fruit pedicels, but rarely on fruit. Infection of leaves by ascospores discharged during rain from apothecia on the orchard floor may start at about the petal fall stage of crop development. Infection is governed by the duration of wetting and the temperature during the wet period (Eisensmith and Jones, 1981).

Leaves are initially resistant to infection, probably due to the lack of stomata, but once they unfold they are highly susceptible. As the leaves age they become less susceptible to infection. Lesions appear in 5–15 days, depending on temperature. Except when the weather is very dry, acervuli containing masses of conidia will be visible on the bottom of the leaf about the time lesions become visible on the top. Conidia are dispersed by splashing rain to other leaves where secondary infection can occur repeatedly until leaf fall in autumn.

Breeding programmes aiming for high-quality cherry cultivars resistant to infection by *B. jaapii* are in progress in some East European countries, but the commercial acceptance of these cultivars in other regions is yet to be determined. In most cherry-growing regions, leaf spot is controlled with fungicides applied in calendar-based schedules beginning at petal fall. Spraying alternate row middles on a 7-day schedule, rather than spraying every row middle on a 10-day schedule, results in increased efficiency of fungicide usage (Jones *et al.*, 1993). Control is also improved by including a postharvest spray.

Sweet cherry is less susceptible to leaf spot than sour cherry (Sjulin *et al.*, 1989).

American brown rot

Brown rot is a common disease of sweet cherries, whereas sour cherries are somewhat resistant. Brown rot also occurs on apricots, nectarines, plums and peaches. American brown rot occurs throughout the cherry-growing areas of the New World (North America, Australia, New Zealand and parts of South America). It also occurs on cherries in South Africa and northeastern Japan.

Brown rot attacks blossoms and fruit, with fruit infections being the most destructive. Infected blossoms wilt, turn brown and persist into summer. Infected fruit may rot before or after harvest. Initially, small, circular, light brown spots develop on the surface of the fruit and expand rapidly under favourable conditions, destroying entire fruit in a few hours. Rotted fruit may fall to the ground or persist as mummies on the tree. Under wet and humid conditions, ash-grey tufts bearing conidia (sporodochia) of fungus develop over the surface of the lesions. The presence of conidia on lesions is the most obvious characteristic of brown rot.

Monilinia fructicola (G. Wint.) Honey (anamorph *Monilia* sp.) causes American brown rot. Apothecia are fleshy, cup-shaped, light brown and about 3 mm across at maturity. Asci are eight-spored, hyaline and ellipsoidal to ovoid. Conidia are produced in sporodochia in moniliid chains without disjunctors.

Sources of inoculum for blossom blight of cherries are overwintering brown rot-infected mummies, including the peduncles and apothecia from mummies on the orchard floor. Overwintering brown rot twig cankers are rare in cherry orchards compared with other stone fruit crops. The importance of mummies as a source of inoculum for blossom infection depends on having sufficient rainfall and favourable temperatures just before or during bloom to stimulate sporulation of the fungus. Although apothecia were not observed in sour cherry orchards during a 3-year study in New York State (Wilcox, 1989), they have been observed infrequently in Michigan sour cherry orchards. Conidia from the mummies are carried by splashing or wind-driven rain, and ascospores discharged from the apothecium are carried by wind to blossoms. Infection of the blossoms is highly dependent on duration of wetness and temperature.

Brown rot activity increases as the fruits start to mature (Northover and Briggs, 1990). Infection may occur directly through the cuticle or through natural openings in the fruit. Wounded fruits are infected much more readily than non-wounded fruits. Infected mummies constitute a significant source of inoculum, along with any blossoms or fruit infected during the current growing season, provided that the weather is favourable for sporulation of the fungus during the period of fruit ripening. External sources of inoculum are stone fruit crops that ripen before or simultaneously with cherry. Since decay of the fruit and spore production can occur in a few days, the disease is able to build up rapidly.

Environment is the key factor that governs the rate of disease development. Warm, wet, humid weather is particularly favourable for brown rot. The hours of wetting necessary for blossom infection decreases from 18 h at 10°C to 5 h at 25°C (Wilcox, 1989). Infection occurs more slowly at temperatures above 27.0°C and below 12.7°C, but may continue at temperatures as low as 4.4°C. Mature fruit can decay in 2 days under optimum conditions for disease development.

Brown rot is controlled using sanitation practices to reduce the amount of fungal inoculum, combined with protective fungicide programmes. Host resistance is not a major component of disease management programmes. However, immature sweet cherries are significantly more susceptible to infection than immature sour cherries (Brown and Wilcox, 1989). Because of their high susceptibility and the potential for the fruit to crack during rain near harvest, effective fungicide spray programmes are essential for sweet cherry production in many regions.

Other brown rot diseases

European brown rot, caused by *M. laxa* (Aderh. & Ruhl.) Honey, is considered an Old World disease because of its widespread distribution in the cherry-growing areas of Europe (Byrde and Willetts, 1977). In eastern North America, it is endemic on sour cherry from western New York State, across southern Ontario and into Michigan and Wisconsin. Outbreaks of the disease are rare compared with American brown rot. They have been more severe on English Morello and Meteor than on Montmorency sour cherry cultivars.

This disease attacks blossoms and spurs but rarely attacks fruit. Twigs are killed back up to 30 cm from blossom infection. Conidia are produced on the dead spurs, twigs and cankers. The fungus overwinters in cankers, and spores are produced the following spring around bloom time if adequate moisture is present. The main method of control is bloom sprays with effective fungicides. Pruning out infected spurs and branches helps to control the disease.

Powdery mildew

Mildew is a problem on sour cherry in both eastern and western North America and on sweet cherry in the west. It reduces the growth of young trees in the nursery and orchard, causes early defoliation and disfigures the fruit of susceptible cultivars of sweet cherry. A high incidence of mildew infection on sweet cherry fruit can result in rejection of the fruit for fresh market sale. A high incidence of the disease on leaves of sour cherry increases harvesting costs because of the need to separate excess leaf debris from the fruit during mechanical harvesting.

Powdery mildew is caused by the fungus *Podosphaera clandestina* (Wallr.:Fr.) Lev. (syn. *P. oxyacanthae* (DC.) de Bary). Mildew on young leaves appears as white, felt-like patches of mycelium and spores (conidia). Lesions spread rapidly, engulfing the entire leaf. Numerous small, brown to black, spherical bodies (cleistothecia) with dichotomously branched appendages develop in the felt-like patches as the season progresses. Severely infected leaves exhibit upward rolling, turn stiff and brittle with age and may drop prematurely. Defoliation may be particularly severe at harvest. Flower and fruit infections of sour cherry are rare, but infection of fruit of sweet cherry is common in Washington State. Lesions on fruit are particularly noticeable at harvest.

The fungus overwinters as cleistothecia on senescent cherry leaves on the orchard floor or trapped in bark crevices (Grove and Boal, 1991). Several researchers have speculated that the fungus overwinters as mycelium in buds but there is no direct evidence to support this theory. Primary infections occur from ascospores discharged from the cleistothecia at about the time of bud burst and continue until after the bloom period. With the appearance of primary mildew colonies, conidia are produced for initiating secondary infections. Secondary spread to fruit occurs 4–6 weeks before harvest, while secondary spread to leaves occurs throughout the summer months. Spread is favoured by warm, humid weather.

Powdery mildew is controlled largely with fungicides applied during the growing season. Wettable sulphurs applied frequently are commonly used on sour cherry to suppress the disease. Some, but not all, sterol-inhibiting fungicides are effective for mildew control. The most effective compounds in this group are strongly suppressive of conidial production and eliminate established infections, as well as preventing new infection. Although sulphur applications have traditionally been initiated with the onset of symptom development, the discovery by Grove and Boal (1991) that cleistothecia are the source of primary inoculum raises the question whether earlier sprays would improve the level of control by preventing primary infections.

Alternaria rot

A decay of sweet and sour cherry fruit caused by *Alternaria* sp. is a problem when fruit become overripe before harvest or when cracking or other injury exposes the flesh to infection. Lesions are circular to oblong and slightly sunken, later becoming firm, flattened and wrinkled, and often dark green to black because of abundant sporulation by the pathogen. Lesions may cover 30–50% of the fruit. The spores (conidia) are dark, with both cross and longitudinal septa, and are typically produced in chains. In cold storage, lesions start in open stem cavities, in cracks in the skin and flesh or in bruised areas and are often covered with fluffy, grey to white strands of the fungus.

Alternaria fruit rot can be avoided by harvesting the fruits before they become overmature. No fungicide control measures have been devised for this disease of cherry. It is known that benzimidazole fungicides may favour *Alternaria*, while dicarboximide and some other fungicides may suppress *Alternaria*.

Cytospora canker

In the USA, outbreaks of Cytospora canker have been reported on sweet cherry in New York, Oregon and Washington States. The disease is rare on sour cherry. Affected trees exhibit a dieback of the apical branches which continues downward, eventually affecting the major scaffold limbs. Infections appear as bark cankers, gummosis and hardwood discoloration.

The causal organisms are *Cytospora cincta* Sacc. (teleomorph *Leucostoma cincta* (Fr.:Fr.) Fr.; syn. *Valsa cincta* (Fr.:Fr.) Fr.) and *C. leucostoma* (Pers.:Fr.) Fr. (teleomorph *L. persoonii* Hohn.; syn. *V. leucostoma* (Pers.:Fr.) Fr.). The fungus produces a compound fruiting body consisting of a central pycnidium with hyaline conidia. Perithecia containing eight-celled, hyaline ascospores may surround the pycnidium in the same stoma. Perithecia are much less common than pycnidia and are produced in older stroma only.

Cytospora gains entry into cherry through winter-injured tissue and then advances beyond the winter-injured tissue (Kable *et al.*, 1967). It also gains entry through canker caused by *Pseudomonas syringae* pv. *syringae*, but this infection site is less important than winter injury. In pathogenicity studies, *C. cincta* was more virulent on sweet cherry than *C. leucostoma* (Regner *et al.*, 1990). Old stone fruit orchards, including sweet cherry, are a source of inoculum for infecting young orchards (Spotts *et al.*, 1990).

As Cytospora canker commonly follows winter injury, the disease can be avoided by maintaining sweet cherry trees in good vigour but with maximum hardiness. Cytospora canker was also reduced by applying white paint to trunks, but varying the level of nitrogen fertilization or applying sprays containing benomyl did not reduce the incidence of the disease (Spotts *et al.*, 1990).

Phytophthora root and crown rot

Prior to the work of Mircetich and Matheron (1976), sweet cherry trees with terminal dieback, weak growth and necrotic roots and/or crown tissue were diagnosed as suffering from wet feet. In California, *Phytophthora megasperma* Drechsler, *P. cambivora* (Petri) Buisman, *P. drechsleri* Tucker, *P. crytogea* Pethyb. & Laff. and an unidentified *Phytophthora* sp. were identified as causal agents of Phytophthora root and crown rot of sweet cherries (Mircetich and Matheron, 1976). In addition, isolates of *P. cinnamomi* Rands and *P. citricola* Sawada from other deciduous fruit and nut crops in California were pathogenic on Mahaleb cherry in greenhouse tests and were considered a potential threat to cherry (Wilcox and Mircetich, 1985). Then *P. megasperma*, *P. cryptogea*, *P. cambivora*, *P. syringae* (Kleb.) Kleb., *P. cactorum* (Lebert & Cohn) Schroet. and an unidentified *Phytophthora* sp. were identified as the causal agent of root and crown rot of sour cherry (Bielenin and Jones, 1988).

Affected trees exhibit poor terminal growth, sparse and chlorotic foliage and progressive decline over several seasons. More frequently, the trees collapse and die suddenly in late spring or summer following years with excessively wet weather the previous autumn or spring. Roots of affected trees exhibit decay, often extending to the crown and up to the graft union but seldom extending much above the graft union. Occasionally, only the crown will exhibit decayed tissue.

Phytophthora is primarily a soil-borne fungus that thrives when the soil is saturated or nearly so. The fungus produces zoospores within sporangia when soil moisture levels are high. When free or standing water is present, the zoospores swim or are moved passively by moving water. Once close to trees, zoospores actively move to the cherry roots and crown.

Soil water management is the most important method for avoiding Phytophthora root and collar rot. Selecting sites with good internal drainage usually eliminates future problems with this disease. If soil drainage is not excellent, it can be improved by installing drainage tiles or by planting trees on ridges. Trees on Mazzard rootstock are less susceptible to Phytophthora root and crown rot than trees on Mahaleb rootstock.

Armillaria root rot

Armillaria root rot is a widespread problem on Montmorency sour cherry trees planted on sandy soils in eastern North America. It is less common on sweet cherry trees grafted on to Mazzard rootstocks. The pathogen is also common in western North America and Europe on forest trees but outbreaks on cherry are rare.

Infected cherry trees on Mahaleb rootstock exhibit poor terminal growth for 1–2 years before collapsing suddenly in full leaf in midsummer. Trees on Mazzard rootstock decline for several years before collapsing. Trees usually die out in a circular pattern. A fan-shaped, fungal mat is present between the bark and the wood, and dark brown to black thread-like structures or rhizomorphs are often found on dead roots. Rhizomorphs enable the pathogen to survive many years in the soil. Mushrooms may arise in the autumn around the base of trees killed 1–2 years earlier.

Species of *Armillaria* recovered from the roots of cherry trees include *A. ostoyae* (Romagn.) Herink., *A. mellea sensu stricto* (Vahl ex Fr.) Kummer, North American group III of *Armillaria*, and *A. bulbosa* (Barla) Kile & Watling (Proffer *et al.*, 1987). Before 1979, these species were classified as *A. mellea*. Lumping several species together may explain the variability in pathogenicity observed by early researchers.

The best control is to plant less susceptible fruit crops, such as apple or pear, in soils infested with *Armillaria*. Fumigation is only moderately effective in controlling the disease.

Verticillium wilt

Verticillium wilt is a sporadic disease that can seriously affect cherry trees. Diseased trees are rarely killed, but they may be stunted and unproductive for several years or they may recover after one or two growing seasons.

The first symptom of infection is usually flagging of the leaves on one branch or perhaps the entire tree. Wilting usually begins in the middle of the summer, followed by browning and curling of the leaves and, ultimately, defoliation. Sometimes the leaves simply turn yellow or brown and drop. Symptoms usually develop on the lower part of the shoots and then progress upward, leaving a few green leaves at the tips of the shoots. In the second year these shoots may die back from the tip. Brownish or greyish streaks develop in the sapwood of infected trees. This discoloration may be observed in the current season's shoots, but is usually limited to the larger branches.

Both *Verticillium dahliae* Kleb. and *V. albo-atrum* Reinke & Berthier have been reported to cause the disease, but *V. dahliae* is the most common pathogen. *V. dahliae* is distinguished from *V. albo-atrum* by the formation of microsclerotia in culture and absence of growth at temperatures above 30°C. Both fungi can survive in the soil for several years in the absence of a susceptible host. Inoculum can build to very high levels on susceptible weeds and on such crops as strawberries, raspberries, tomatoes, peppers and eggplants. The fungi infect the roots of trees planted in infested soil.

The most effective control measure is to avoid planting orchards in soils where inoculum of the fungus has built up previously on susceptible crops. Preplant soil fumigation with a broad-spectrum fumigant that is active against fungi helps reduce inoculum levels. Adequate moisture to maintain good vigour will reduce the incidence of the disease and encourage the recovery of diseased trees.

Other fungal diseases

Eutypa dieback, caused by *Eutypa lata* (Pers : Fr.) Tul. & C. Tul., occurs widely on *Prunus* species, particularly apricot, and grapevine (*Vitis*) throughout semiarid regions of the world. In Europe, it is part of the disease complex on stone fruits called 'apoplexy.' Symptoms include cankering around wounds exhibiting exposed sapwood, gumming, and dieback of branches above cankers in summer with collapsed leaves remaining attached for several months on affected, sometimes swollen, limbs. Light to dark brown sapwood necrosis extends upwards and downwards from the cankers. Infected sweet cherry is the most important source of inoculum for infecting grapevines in California (Munkvold and Marois, 1994). Treating large pruning wounds with a fungicide, pruning when ascospore inoculum levels are low, and sanitation practices aimed at reducing inoculum levels in orchards help control the disease.

A strain of *Apiosporina morbosa* (Schwein.:Fr.) Arx, the cause of black knot of European and Japanese plums, is economically important on Montmorency sour cherry in southern Ontario, Canada (Northover and McFadden-Smith, 1995). Symptoms include elongated corky swellings or knots on shoots, spurs, branches, and trunks. Newly formed knots are greenish and soft; old knots are black and hard. Limbs may be stunted and eventually killed as knots expand. Control is based on a combination of cultural and chemical methods.

14.2.2 The bacterial diseases of cherries

Bacterial canker

Bacterial canker occurs wherever cherries are grown. It is particularly severe on sweet cherry, and less severe on sour cherry. The disease is also important on other stone-fruit crops and on some ornamental *Prunus* species.

Bacterial canker causes cankers on branches and twigs. Gum production may be evident on the surface of the cankers for part of the year. Dieback can occur any time during the growing season, when a canker girdles a scaffold limb or branch. Infection of leaves and fruits is relatively common in years with prolonged wet, cool weather during or shortly after bloom. Leaf spots (about 2 mm in diameter) are dark brown, circular to angular and sometimes surrounded with a yellow halo. The spots may coalesce to form large patches of dead tissue, especially at the margins of leaves. Lesions on immature cherry fruit are brown, with a margin of water-soaked tissue. Later, the affected tissues collapse, leaving deep, black depressions in the flesh of the developing fruit. On fruit pedicels, brown lesions with water-soaked margins are 2–3 mm long and extend a third to halfway around the pedicel. A dieback of fruit spurs may also be observed in years when leaf and fruit infections are common. Buds may be killed while dormant and fail to open in the spring.

Two closely related bacteria, *Pseudomonas syringae* pv. *syringae* van Hall and *P. s. morsprunorum* (Wormald) Young *et al.*, cause bacterial canker. The two pathovars are distinguished from each other primarily by some biochemical tests. Pathovar *morsprunorum* shows more host specialization than pathovar *syringae*. *P. s. morsprunorum* has been reported from sweet cherry, sour cherry and plum, while *P. s. syringae* has been reported from all stone-fruit crops and several ornamental *Prunus* spp. Also, the geographic distribution of *P. s. syringae* is worldwide while the distribution of *P. s. morsprunorum* is mainly in Europe, eastern North America and South Africa.

The bacteria are capable of surviving from one season to the next in cankers, in apparently healthy buds and in the vascular tissues of cherry trees (Sundin *et al.*, 1988). Bacteria from these overwintering sites colonize blossoms and leaf tissue as they unfold from buds in the spring. Growth and dissemination of the bacteria are favoured by cool, wet weather. Both pathovars are found as residents on symptomless blossoms and on leaf surfaces from anthesis through to leaf fall (Latorre and Jones, 1979). In autumn, the bacteria may be sucked into leaf scars, where they overwinter.

Outbreaks of bacterial canker are relatively rare, but they can be very serious when they occur. Disease outbreaks are often associated with late spring frosts or with severe storms that injure the emerging blossom and leaf tissues. Freezing can predispose the tissue to infection, but infection is dependent on the presence of wet weather during the thawing process (Sule and Seemuller, 1987). Free water on leaf surfaces and high relative humidity are required for at least 24 hours before significant infection to leaves following violent storms. Symptoms appear starting about 5 days later at temperatures of 20–25°C.

Copper-containing compounds are the only bactericides known to aid in the control of bacterial canker. But, because the leaves of sweet cherry are easily damaged by copper-containing compounds, treatment of sweet cherries is restricted to the autumn just before natural leaf fall and to the spring before bud break. Sour cherry is more tolerant of copper-containing compounds. On sour cherries, spraying trees weekly with a low rate of copper from bud break to about the shuck-split stage of bud development is used for disease control (Olson and Jones, 1983). However, strains of *P. s. syringae* resistant to copper were isolated in several orchards in Michigan, and the resistance was associated with a plasmid that may transfer to non-resistant strains (Sundin *et al.*, 1989). There are no cultivars that are highly resistant to bacterial canker, but some are affected less by the disease than others. High budding on scaffold limbs of Mazzard rootstocks, particularly Mazzard F.12/1, is used in areas of western North America to reduce the loss of whole trees. There is some evidence that some soil factors, particularly the presence of certain species of nematodes, favour the disease.

Crown gall

This disease develops on the roots of both sweet and sour cherry trees. It is a common problem in nurseries and may also occur on trees in the orchard. Crown gall is characterized by the formation of tumours or galls on the roots. Galls begin as small, smooth growths which enlarge to become dark, hard, woody tumours with gnarled irregular surfaces. Tumours range from 3.5 to 10 cm in diameter. Galls are typically globular but may be elongated or otherwise irregular. Galls are usually located on the roots or crown but, under certain conditions, they develop on aerial portions of trees. Several galls may occur on the same root or stem.

Crown gall is caused by the bacterium *Agrobacterium tumefaciens* (E.F. Smith & Townend) Conn., which carries a large DNA molecule outside of the chromosome referred to as the tumour-inducing (Ti) plasmid. When bacteria infect plants, a portion of the Ti plasmid is inserted into a plant cell chromosome and incorporated into the genome of the plant. Expression of this genetic material in the host results in overproduction of plant hormones and gall formation. After gall formation is initiated, galls can continue to develop in the absence of the bacteria.

When cherry trees are planted in infested soils, the bacteria enter the roots and crown through wounds produced when caring for and handling the nursery stock. Bacteria are also disseminated to new plants and planting sites by splashing rain or irrigation water.

Good sanitation in the nursery helps to control this disease. All nursery stock with symptomatic roots should be discarded. In many regions, nursery stock with visible galls cannot be certified for sale. Nursery planting sites should be rotated with sites that have not been used for fruit trees for at least 4–5 years.

Biological control has been achieved with a non-pathogenic isolate of *A. radiobacter* (Beijerinck & van Delden) Conn. strain K84 (Kerr, 1980). Root dips of nursery stock into a suspension containing strain K84 have proved very successful in preventing this disease. Some strains of *A. tumefaciens*, particularly those found on apple and grape, are not as well controlled by strain K84.

References

Bielenin, A. and Jones, A.L. (1988) Prevalence and pathogenicity of *Phytophthora* spp. from sour cherry trees in Michigan. *Plant Disease* 72, 473–476.

Brown, S.K. and Wilcox, W.F. (1989) Evaluation of sweet and sour cherry genotypes for sources of resistance to infection of fruits by *Monilinia fructicola* (Wint.) Honey. *HortScience* 2, 1013–1015.

Byrde, R.W.W. and Willetts, H.S. (1977) *The Brown Rot Fungi of Fruit. Their Biology and Control.* Pergamon Press, Oxford, 171 pp.

Eastwell, K.C. and Bernardy, M.G. (1993) The high molecular weight, double-stranded RNA associated with little cherry disease is highly divergent. *Abstracts of the IXth International Congress of Virology*, Glasgow, 8–13 August, p. 358.

Eisensmith, S.P. and Jones, A.L. (1981) A model for detecting infection periods of *Coccomyces hiemalis* on sour cherry. *Phytopathology* 71, 728–732.

Forer, L.B., Powell, C.A. and Stouffer, R.F. (1984) Transmission of tomato ringspot virus to apple rootstock cuttings and to cherry and peach seedlings by *Xiphinema riversi*. *Plant Disease* 68, 1052–1054.

Gilmer, R.M., Moore, J.D., Nyland, G., Welsh, M.F. and Pine, T.S. (1974) *Virus Diseases and Noninfectious Disorders of Stone Fruits in North America*. Agriculture Handbook No. 437. US Government Printing Office, Washington, DC, 433 pp.

Grove, G.G. and Boal, R.J. (1991) Overwinter survival of *Podosphaera clandestina* in eastern Washington. *Phytopathology* 81, 385–391.

Hansen, A.J. (1985) Canindex 1, a superior indicator cultivar for little cherry disease. *Plant Disease* 69, 11–12.

Jones, A.L., Ehret, G.R., Garcia, S.M., Kesner, C.D. and Klein, W.M. (1993) Control of cherry leaf spot and powdery mildew on sour cherry with alternate-side applications of fenarimol, myclobutanil, and tebuconazole. *Plant Disease* 77, 703–706.

Kable, P.F., Fliegel, P. and Parker, K.G. (1967) Cytospora canker on sweet cherry in New York state: association with winter injury and pathogenicity to other species. *Plant Disease Reporter* 51, 155–157.

Kerr, A. (1980) Biological control of crown gall through production of agrocin 84. *Plant Disease* 64, 25–30.

Latorre, B.A. and Jones, A.L. (1979) *Pseudomonas morsprunorum*: the cause of bacterial canker of sour cherry in Michigan, and its epiphytic association with *P. syringae*. *Phytopathology* 69, 335–339.

Mink, G.I. (1992) Prunus necrotic ringspot virus. In: Mukhopadhyay, A.N. (ed.) *Plant Diseases of International Importance*, Vol. 3. Prentice-Hall, Englewood Cliffs, New Jersey, pp. 335–356.

Mink, G.I. and Aichele, M.A. (1984) Use of enzyme-linked immunosorbent assay results in efforts to control orchard spread of cherry rugose mosaic disease in Washington. *Plant Disease* 68, 207–210.

Mircetich, S.M. and Matheron, M.E. (1976) Phytophthora root and crown rot of cherry trees. *Phytopathology* 66, 549–558.

Munkvold, G.P. and Marios, J.J. (1994) Eutypa dieback of sweet cherry and occurrence of *Eutypa lata* perithecia in the Central Valley of California. *Plant Disease* 78, 200–207.

Northover, J. and Briggs, A.R. (1990) Susceptibility of immature and mature sweet and sour cherries to *Monilinia fructicola*. *Plant Disease* 74, 280–284.

Northover, J. and McFadden-Smith, W. (1995) Control of epidemiology of *Apiosporina morbosa* of plum and sour cherry. *Canadian Journal of Plant Pathology* 17, 57–68.

Ogawa, J.M., Zehr, E.I., Bird, G.W., Richie, D.F., Uriu, K. and Uyemoto, J.K. (1995) *Compendium of Stone Fruit Diseases*. American Phytopathological Society, St. Paul, MN, 98pp.

Olson, B.D. and Jones, A.L. (1983) Reduction of *Pseudomonas syringae* pv. *morsprunorum* on Montmorency sour cherry with copper and dynamics of the copper residues. *Phytopathology* 73, 1520–1525.

Proffer, T.J., Jones, A.L. and Ehret, G.R. (1987) Biological species of *Armillaria* isolated from sour cherry orchards in Michigan. *Phytopathology* 77, 941–943.

Raine, J., McMullen, R.D. and Forbes, A.R. (1986) Transmission of the agent causing little cherry disease by the apple mealybug *Phenacoceus aceris* and the dodder, *Cuscuta lupuliformis*. *Canadian Journal of Plant Pathology* 8, 6–11.

Ragetli, H.W.J., Elder, M. and Schroeder, B.F. (1982) Isolation and properties of filamentous virus-like particles associated with little cherry disease in *Prunus avium*. *Canadian Journal of Botany* 60, 1235–1248.

Regner, K.M., Johnson, D.A. and Gross, D.C. (1990) Etiology of canker and dieback of sweet cherry trees in Washington State. *Plant Disease* 74, 430–433.

Sjulin, T.M., Jones, A.L. and Andersen, R.L. (1989) Expression of partial resistance to cherry leaf spot in cultivars of sweet, sour, duke, and European ground cherry. *Plant Disease* 73, 56–61.

Spotts, R.A., Facteau, T.J., Cervantes, L.A. and Chestnut, N.E. (1990) Incidence and control of Cytospora canker and bacterial canker in a young sweet cherry orchard in Oregon. *Plant Disease* 74, 577–580.

Süle, S. and Seemüller, E. (1987) The role of ice formation in the infection of sour cherry leaves by *Pseudomonas syringae* pv. *syringae*. *Phytopathology* 77, 173–177.

Sundin, G.W., Jones, A.L. and Olson, B.D. (1988) Overwintering and population dynamics of *Pseudomonas syringae* pv. *syringae* and *P. s.* pv. *morsprunorum* on sweet and sour cherry trees. *Canadian Journal of Plant Pathology* 10, 281–288.

Sundin, G.W., Jones, A.L. and Fulbright, D.W. (1989) Copper resistance in *Pseudomonas syringae* pv. *syringae* from cherry orchards and its associated transfer *in vitro* and *in planta* with a plasmid. *Phytopathology* 79, 861–865.

Wilcox, W.F. (1989) Influence of environment and inoculum density on the incidence of brown rot blossom blight of sour cherry. *Phytopathology* 79, 530–534.

Wilcox, W.F. and Mircetich, S.M. (1985) Pathogenicity and relative virulence of seven *Phytophthora* spp. on Mahaleb and Mazzard cherry. *Phytopathology* 75, 1451–1455.

Zagula, K.R., Aref, N.M. and Ramsdell, D.C. (1989) Purification, serology, and some properties of a mechanically transmissible virus associated with green ring mottle disease in peach and cherry. *Phytopathology* 79, 451–456.

15 Management and Control of Insect and Mite Pests of Cherry

J.F. Brunner

Washington State University, Tree Fruit Research and Extension Center, 1100 North Western Avenue, Wenatchee, WA 98801, USA

This chapter presents and discusses principles of arthropod pest management, using, where applicable, examples related to cherry production. It is not intended as a comprehensive review of all arthropod pest problems that occur on cherry. Nor is it intended as a detailed 'how to' manual for management of specific pests. There are several books on pest management, applied entomology and ecological aspects of entomology (Metcalf and Luckman, 1975; Huffaker and Rabb, 1984; Pedigo, 1989; Williams, 1991) that the reader is encouraged to explore for additional information and a more in-depth discussion of specific topics.

15.1 An Introduction to Pest Management

15.1.1 Pest management strategies

Some terms and definitions

Pest management (PM) or integrated pest management (IPM), as it is often called, is a philosophy of pest control founded on the principles of ecology. Simply stated, PM is the utilization of all pest control tactics in an appropriate manner so that crop loss is kept at an acceptable level. Other definitions include statements about environmental safety and social acceptability, but when any control tactic is used in an appropriate manner these considerations should be met.

PM is based on the understanding of a pest's biology, especially interactions with host plants and the impact of natural enemies. Knowledge and the ability to use it wisely will determine how successfully the principles discussed here are implemented. Eliminating all insects and mites from an orchard is not the objective of PM. PM assumes that pests can be tolerated at some level before unacceptable crop loss occurs. Direct pests, those which attack the fruit, are tolerated at very low levels and insecticides tend to be the primary tactic used for control. Indirect pests, those which attack the foliage, bark, or roots, can usually be tolerated at levels higher than direct pests and, as a result, tactics other than insecticides can be used to provide control.

The concepts of economic injury level (EIL) and treatment threshold (TT) are basic to PM. An EIL is the pest density that will result in crop loss equal to the cost of control. This concept is most often associated with the use of a chemical control (insecticides) but has application to other control tactics. The TT is the pest level at which a control tactic must be applied to prevent the pest from exceeding the EIL. With some insecticides, the TT is often the same as the EIL because their impact on pest density is immediate in stopping further crop loss. For other control tactics, such as the mass release of a biological control agent (e.g. a parasite) or use of a slower acting insecticide, the TT may be much lower than the EIL.

Figure 15.1 shows graphically the concept of an EIL and two TTs for a pest when using a chemical (A) or mass release of a biological control agent (B). The solid line shows the pest level that would result if no control was used. The shaded line shows the pest level that would result after a control was applied. Note that the release of the parasite must occur long before the pest reaches the EIL. Biological controls take time to affect pest populations – hence the lower

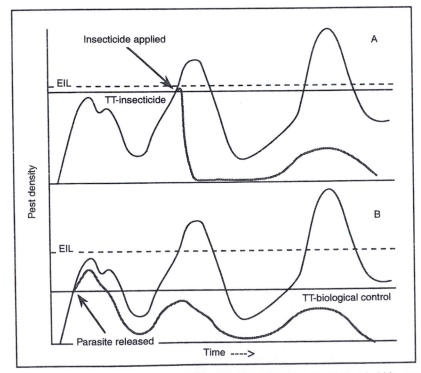

Fig. 15.1. The conceptualization of economic injury level and treatment threshold in graphic form. Treatment thresholds for different tactics, insecticides (A) versus augmentative release of a biological control agent (B), differ because of the response time required for them to have an impact on pest densities. ___ = Control; .... = treated.

TT. Their advantage is that suppression of the pest is often more lasting than when insecticides are used.

There are three general strategies for managing insect problems: eradication, containment and suppression. Eradication seeks to eliminate all members of a species from an area, thus driving it to extinction. This strategy is commonly employed against pests such as the Mediterranean fruit fly, *Ceratitis capitata* (Wiedemann), wherever it is detected in the USA. The tactics most often used to achieve eradication of this pest are aerial spraying of insecticides to reduce fly populations, followed by the release of sterilized males.

Containment seeks to prevent the establishment of a pest, usually foreign, in an area where it does not occur. This strategy is most often associated with regulatory control, the use of quarantines and inspections to prevent the shipment or distribution of infested products. Many countries use regulatory controls to prevent the introduction of new insect pests that could have a serious impact on their agriculture.

Suppression seeks to maintain pests at levels that will minimize crop loss and is the strategy used most often in PM. Several control tactics can be used in a suppression strategy, including chemical, cultural, behavioural and biological controls.

15.1.2 Methods for suppressing insect and mite populations

Chemical control

By far the most widely used tactic in the past has been chemical control. Following the development of modern insecticides, cherry growers have used these chemicals almost exclusively to control pests. Insecticides such as chlorinated hydrocarbons (e.g. dichlorodiphenyltrichloroethane (DDT)), organophosphates, carbamates and synthetic pyrethroids are neurotoxins and have activity against a broad range of organisms. Most chlorinated hydrocarbons can no longer be used because of their adverse environmental impact.

PM recognizes that insecticides are powerful tools for managing pests. Their great value stems from their rapid action, allowing them to be used close to the EIL. Too often, however, they have been used as the first line of defence, sometimes as a relatively cheap insurance against crop-loss. When chemical controls are used in this way they disrupt biological control and promote the development of insecticide resistance in pests. The more appropriate role of broad-spectrum neurotoxic insecticides is as a last line of defence against pests.

Insect growth regulators (IGRs) are a relatively new group of insecticides. They act by interfering with the normal growth and development of insects. Juvenile hormone and ecdysone are two insect hormones that regulate growth and the moulting process. Some IGRs mimic the action of these naturally occurring insect hormones and, when introduced at the wrong time in the insects' life cycle, they cause death or sterility. Other IGRs interfere with the formation or hardening of the insect's cuticle or the development of the egg. IGRs tend to have a narrower spectrum of activity, are less disruptive of biological control and are less toxic to mammals than most neurotoxins.

Other chemicals that fit into many PM programmes are the biological insecticides derived from the bacterium *Bacillus thuringiensis* (Bt). These Bt insecticides are stomach poisons so must be ingested to be effective. Bts have a short residue life and have been most effective against foliage-feeding caterpillars. They are low in toxicity to biological control agents and mammals.

Behavioural control

The most prominent method of behavioural control is called mating disruption or pheromone confusion. Pheromones are chemicals released by a member of a species that cause a specific behavioural response in members of the same species. For example, a sex pheromone released by a female attracts the male of the species from a long distance for mating. When a sex pheromone is used as a lure in a trap, the seasonal activity of an insect can be monitored and, in some cases, a relative estimate of an insect's density can be made. When applied to an orchard in much higher quantities, sex pheromones can be used to control insects. Dispensers containing pheromone are placed in the orchard, usually at a density of 500 to 1000 ha^{-1}. The exact mechanism by which mating disruption works is not clear but, in general, it works by preventing the male from locating the female. Males may follow false pheromone trails to their artificial source, they may be unable to locate the females' scent in a cloud of pheromone (masking), their sensory system may become overloaded by the large amount of pheromone present (adaptation) or some combination of these. Mating disruption offers a safe, easy-to-use means of controlling insect pests that does not disrupt natural controls. It is highly selective – usually only one pest is controlled by a single pheromone – but it tends to be more expensive than conventional insecticides.

Biological control

The attack of pests by natural enemies is well known. However, the use of this control tactic in cherry orchards has been limited by a reliance on broad-spectrum insecticides for pest control. The impact of insecticides on biological control agents has been well documented (Croft, 1990). While some biological control agents, especially several species of predatory mites, have developed resistance to certain organophosphate insecticides, most remain highly susceptible.

Predators and parasites are the most common forms of biological control agents; however, pathogens and nematodes also attack insects and, under the right conditions, can cause high levels of mortality. Biological control is most effective against indirect pests, such as aphids, leafminers, mites and scales, which can be tolerated at relatively high densities. A biological control agent cannot survive and reproduce in a crop where its host density is kept at too low a level.

Predatory insects are usually larger and more active than their prey and a predator consumes several prey during its life. Examples include beetles in the family Coccinellidae (ladybird beetles), predatory mites, true bugs, brown and green lacewings and larvae of the syrphid fly. A parasite is typically smaller than its host and usually only one parasite is produced per host attacked. Most parasites are small wasps, although some flies are also parasitic. Every effort should be made to take advantage of biological controls. PM programmes that minimize the use of broad-spectrum insecticides will, over time, increase the impact of natural controls in the orchard.

15.1.3 The basics: insect and mite development, identification, sampling and prediction

Arthropod pests can be placed in two groups: those that attack the fruit – direct pests; and those that attack the foliage, wood or roots – indirect pests. Direct pests are the target of most chemical controls applied to cherry because, even at low densities, they can dramatically affect crop yields or marketability. Examples of direct pests are the cherry fruit flies (CFF), *Rhagoletis* sp.; plum curculio (PC), *Conotrachelus nenuphar* (Herbst.); and leafrollers. Some direct pests do not require fruit as a part of their diet but may feed on fruit at some point during their life cycle.

Key pest refers to an insect whose control is 'key' to the successful production of a crop. Generally the key pest is a direct pest and in most cherry-growing areas of the world CFF is the recognized key pest. A secondary pest is one whose control is less critical. Secondary pests are usually indirect pests and often become problematic when chemicals used to control the key pest disrupt natural controls. In the absence of natural controls, the status of a secondary pest may be elevated, even to the point where it becomes a key pest if it also develops resistance to insecticides.

Indirect pests can usually be tolerated at higher densities than direct pests and therefore other control tactics, such as biological control, are often successful. Examples of indirect pests include mites, aphids, leafhoppers and leafminers, all of which attack leaves. Scales are also classified as indirect pests in cherry, since they are primarily found on bark. Another group of indirect pests is called borers. Examples include the peach tree borer (PTB), *Synanthedon exitiosa* (Say); lesser peach tree borer (LPTB), *Synanthedon pictipes* (Grote & Robinson); American plum borer, *Euzophera semifuneralis* (Walker); cherry bark tortrix (CBT), *Enarmonia formosana* (Scopoli); and shot-hole borer, *Scolytus rugulosus* (Miller). A few insects attack the roots of cherry trees but in most cherry-producing areas they are not a major problem and will not be considered here.

Life cycles and terminology

A basic knowledge of insect life cycles, development and the terms used to describe stages and events is necessary to gain a better understanding of later discussions in this chapter. The generalized life cycle of an insect is outlined in Fig. 15.2. Eggs are laid by the adult female following mating. After the egg hatches the insect passes through a series of immature stages called nymphs or larvae. The immature stages of aphids, leafhoppers and mites are called nymphs (although in Europe these are often referred to as larvae). Nymphs look like the adult, but are smaller and reproductively immature, and they transform directly to adults. The immature stages of moths, beetles, flies and bees or wasps are called larvae. Larvae do not resemble the adult and an additional stage, called the pupa, occurs between the larval and adult stages. Mating usually occurs

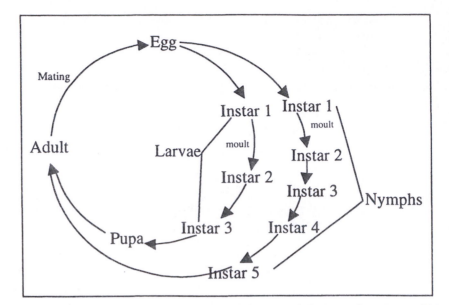

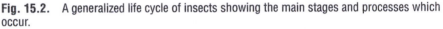

Fig. 15.2. A generalized life cycle of insects showing the main stages and processes which occur.

shortly after adult emergence and the cycle starts again with the laying of eggs. The adult is the reproductive stage and in most cases also selects the host plant and colonizes new areas.

The skin of insects and mites, called the exoskeleton, provides protection from desiccation and physical crushing and supports internal muscles and organs. It does not stretch. Thus, to allow for growth the old skin must be shed and a larger one produced in a process called moulting. The immature insect between each moult is called an instar and the first insect stage after hatch (larva or nymph) is referred to as a first instar (see Fig. 15.2). After moulting to the next stage, it is a second instar and so forth until the pupal or adult stage is reached. The number of instars varies from three to more than ten but is constant for a species.

Life histories

It is important that growers, or those who advise them, know the life history of the arthropod pests that attack cherry. The overwintering stage, the number of generations each year, the time when different stages are present, host plants used other than cherry and the timing of movement between host plants all comprise important life history information that can optimize the use of resources when making PM decisions.

Most arthropod pests of cherry complete one or two generations each year, although a few, such as aphids and mites, have several generations. In temperate growing regions, insects and mites usually have only one life stage especially adapted for winter survival. In areas with mild winters, more than one stage may successfully overwinter. Development begins as temperatures warm in spring and a life history pattern, consistent from year to year, is repeated. Table 15.1

lists some insect and mite pests of cherry, their overwintering stage and location, number of generations per year, stage(s) causing damage, the plant part attacked and type of damage and an approximate time when the damaging stage is present (in the northern hemisphere).

Memorizing the life histories of all cherry pests can be a formidable task. It is more practical to segregate the growing season into critical crop development stages or periods (e.g. bud break, bloom, shuck split, midsummer, etc.) and construct a table listing the pests active during these periods. In this way pest life histories can be compartmentalized and the orchard monitored only for pests likely to be present at any given time. Regardless of how the information on insect and mite life histories is mastered, its importance cannot be underrated. Life histories of selected pests are discussed later in this chapter, providing examples of the type of information needed to make good PM decisions.

Pest identification

Many decisions concerning insect and mite management are made in the field following detection of infestations. Correct identification of pests is essential to the successful implementation of controls. Many insects found in the orchard are not pests, only transitory visitors, and some are beneficial, acting as biological control agents.

There are several kinds of information to use when identifying an insect in the field. These include physical appearance, behaviour, and when and where it is found. An insect's colour, size and shape provide valuable clues when identifying it. A number of books and bulletins are available, some with colour plates of tree fruit pests and the damage they cause. Some also contain photographs of common natural enemies. Such reference materials (e.g. Alford, 1984; Beers *et al.*, 1993; Howitt, 1993) are very valuable when identifying insects and mites found in the orchard.

An insect usually restricts its feeding to certain plant parts. Knowing which insects are most likely to be found feeding on a particular plant part, such as a shoot tip or inside the fruit, helps eliminate many possibilities when making field identifications. Behavioural clues can also be used to identify a pest. For example, a caterpillar that wriggles violently backwards when disturbed is most likely to be a leafroller.

Sometimes the only evidence of an insect's presence is the damage left behind. Many insects leave characteristic feeding damage that provides a clue to their identity. It may be important to associate damage with the correct insect so that control of subsequent generations can be achieved.

Knowing the time when a life stage of an insect is likely to be present, i.e. its life history, can be useful when making identifications. For example, a green caterpillar found in a rolled leaf in August would probably be one of the two-generation leafrollers since the single-generation leafrollers would not be present at this time.

Experience is by far the best teacher when it comes to correctly identifying cherry pests. It may be necessary to seek an expert to identify a particular pest but, once identified, specimens should be kept for future reference.

Table 15.1. Life history information for several arthropod pests of cherry, with the time of year that stages are present in the northern hemisphere.

Insect or mite*	Stage overwintering and location	Number of generations on cherry	Stage(s) causing damage	Time of year stage(s) is present	Type of damage caused
Direct pests					
Plum curculio (ENA) Conotrachelus nenuphar	Adult in ground litter	1	Adults and larvae	May–early June, mid-May–July	Scars on fruit and internal fruit feeding
Cherry fruit fly (ENA) Rhagoletis sp.	Pupa in soil	1	Larvae	Mid-May–Aug.	Larvae feed in fruit
Pandemis leafroller (WNA) Pandemis pyrusana	Larva, 2nd or 3rd instar on tree	2	Larvae	Apr.–May, July–Aug., Sept.–Oct.	Consumes foliage and scars fruit surface
Oblique-banded leafroller (NA) Choristoneura rosaceana	Larva, 1st to 3rd instar on tree/leaf litter	2	Larvae	Apr.–May, July–Aug., Sept.–Oct.	Consumes foliage and scars fruit surface
Barred fruit tree tortrix (E) Pandemis cerasana	Larva and eggs on tree	1	Larvae	Apr.–early June	Consumes foliage and may damage fruit
Fruit tree tortrix (E) Archips podana	Larva, 3rd instar on tree	1	Larvae	Apr.–June	Consumes foliage and may scar fruit
Fruit tree leafroller (NA) Archips argyropilus	Eggs on tree	1	Larvae	May–June	Consumes foliage and scars fruit surface
Climbing cutworms (W) Agrotis sp.	Partially grown larva or pupa in soil	1–2	Larvae	Apr.–May	Feeds on buds and foliage, also scars fruit
Green fruitworms (W) Orthosia sp., Lithophane sp., Amphipyra pyramidoides	Pupa in soil, or egg on tree	1	Larvae	Apr.–May	Feeds on foliage and deep scars in fruit
Cherry fruitworm (NA) Grapholita packardi	Mature larva, in hibernaculum on tree	2	Larvae	June–July, Aug.–Sept.	Feeds on fruit

Indirect pests

Pest	Overwintering stage/location	Generations	Susceptible stages	Period of occurrence	Damage
San Jose scale (W) *Quadraspidiotus perniciosus*	Partially grown scale on tree	2–4	All stages	All year, first crawlers in May–June	Sucks plant sap, reducing of twigs and limbs
Lecanium scale, brown scale (W) *Parthenolecanium corni*	Partially grown scale on tree	1	All stages	All year, crawlers in June–July	Sucks plant sap, reducing tree vigour
Black cherry aphid (W) *Myzus cerasi*	Eggs on cherry tree	2–3	Nymphs and adults	Apr.–June	Sucks plant sap, reducing tree vigour
European red mite (W) *Panonychus ulmi*	Eggs on tree	6–8	All mobile forms	All year	Reduces photosynthetic ability of leaves
Two-spotted spider mite (W) *Tetranychus urticae*	Adult female on ground cover plants	Variable	All mobile forms	Variable	Reduces photosynthetic ability of leaves
Peach tree borer (ENA) *Synanthedon exitiosa*	Larva, all instars, beneath bark of tree	1	Larvae	July–Sept.	Girdles tree trunk, causing death
Lesser peach tree borer (NA) *Synanthedon pictipes*	Larva, all instars, beneath bark of tree	1	Larvae	June–Sept.	Girdles tree trunk, causing death
American plum borer (ENA) *Euzophera semifuneralis*	Larva beneath bark of tree	2	Larvae	May–June, Aug.–Oct.	Girdles tree trunk, causing loss of vigour
Shot-hole borer (W) *Scolytus rugulosus*	Larva, in galleries under bark	2	Larvae	Sept.–Mar.	Girdles twigs and limbs, causing death
Spotted tentiform leafminer (ENA) *Phyllonorycter blancardella*	Pupa, in leaf mine	3–4	Larvae	May, July, Aug.–Sept.	Mines leaf tissue, causing reduced vigour
Tent caterpillar (W) *Malacosoma* sp.	Egg masses on host plant	1	Larvae	Apr.–June	Consumes foliage
Fall webworm (NA) *Hyphantria cunea*	Pupa in cocoon on tree or sheltered location	2	Larvae	Apr.–May, June–Aug.	Consumes foliage

*Regions where pest occurs: W = worldwide, E = Europe, NA = North America, WNA = western North America, ENA = eastern North America.

Sampling

Sampling is an integral part of PM. Decisions on how to manage a pest need to be based on its density in the orchard. Since it is impossible to count all the insects, only a portion, the sample, is counted and this information is used to make inferences about an insect's density in the entire orchard. Samples are taken to: (i) determine if enough pests are present to justify a control treatment, i.e. if a pest population exceeds a TT; (ii) evaluate the impact of a control action; or (iii) follow the interaction between a pest and a natural enemy. Without good sample information, a manager can only guess at or make assumptions about pest levels and their threat to the crop. Management without a good sampling plan usually leads to an overuse of insecticides. Basic principles associated with sampling are presented by Southwood (1978) and Jones (1991).

Knowing when to sample and if there is a need to repeat a sample is as important as knowing how to sample. Figure 15.3 helps demonstrate how the timing of a sample can influence results. Generally, samples seek to estimate pest or natural enemy densities at their peak or at some critical time in their life history. Within a generation, the density of the insect stage to be sampled begins low, increases to a peak level and then declines, as shown in Fig. 15.3. If the objective of a sample is to estimate peak density, then only the sample at S_3 will be of value. Samples taken at S_1 and S_2 give the same estimate of pest density, but, if this is the TT, a decision to apply a chemical control based on S_1 will be good while one based on S_2 will be poor. In the latter case, most of the damage will have already occurred or it will be too late to control the most susceptible stage.

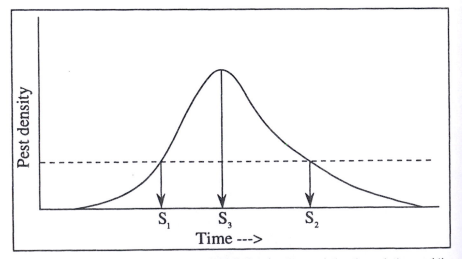

Fig. 15.3. Conceptualization of the rise and fall of an insect population through time and the influence of the timing of samples on the results. Samples taken at S_1 and S_2 give the same estimate of insect density, but one precedes the peak while the other follows. For most management decisions, samples which anticipate peak densities or target the peak density are most useful.

Predicting pest development

An important component of many predictive models is the tree itself. Cherry trees display easily recognizable stages of development early in the growing season, e.g. bud expansion, bloom and petal fall, and associating these plant growth stages with insect development is a reliable and useful means of timing sampling activities and control measures. After petal fall, however, there are few phenological events that can be used to predict the presence of an insect, so other methods are required. In today's PM programmes, degree-day or phenology models are widely used to predict insect development and to time management activities (Beers *et al.*, 1993; Croft and Hoyt, 1983). Degree-day models provide a simple means of relating temperature to insect development. Because these models are important and widely used in PM, it is helpful to understand their underlying principles.

Insects are 'cold-blooded' animals, that is, their rate of development is temperature-dependent. Figure 15.4 presents data from a laboratory experiment where an insect is reared at different temperatures. As the temperature increases it takes less time to complete the stage (part A). When development rate is plotted against temperature (part B), it increases linearly to 32°C and then declines precipitously. An estimate of the temperature below which growth no longer occurs, called the lower developmental threshold (LDT), is made by drawing a straight line through the points on the curve where the insect growth rate is increasing. The temperature where this line intersects the temperature

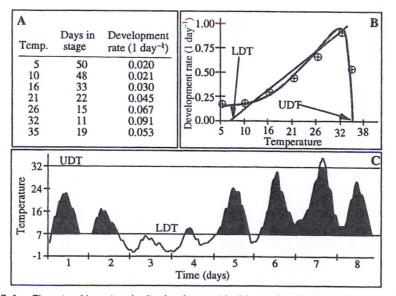

Fig. 15.4. The rate of insect and mite development is driven primarily by temperature. (A) As temperatures increase, the days required to complete a life stage decline and then increase again as the optimum for development is exceeded. (B) The rate of development can be plotted and these data used to estimate the lower developmental threshold (LDT). (C) Degree-days are units used to measure the rate of insect development from temperatures. The shaded area under the temperature line, above the LDT and below the upper developmental threshold represents the relative number of degree-days accumulated each calendar day.

axis is an estimate of the LDT, in this case 5°C. At temperatures higher than 32°C, where the development rate is maximized, growth is inhibited and eventually the insect dies. This upper limit to development is called the upper developmental threshold (UDT), and can be estimated by the point where the development rate curve again crosses the temperature axis, in this case 35°C.

How much an insect develops each day is visually expressed in part C of Fig. 15.4. Here the LDT and UDT are represented by the lines running horizontally across the graph. The shaded area below the temperature trace but above the LDT represents the relative amount of development an insect experiences each day. Degree-days are units used to measure insect development. The shaded area each day can be expressed as a number of degree-days.

A specific number of degree-days is required for an insect to complete a stage. The PM specialists in most production regions will have tabular information available for use by orchard managers using degree-days to make PM decisions (see appendix in Beers *et al.*, 1993). By knowing the number of degree days required to complete each life stage, it is possible to predict when that stage will be present in the field.

There are computerized instruments that sense temperature, calculate degree-days and automatically give predictions for some insects and diseases. However, anyone can calculate degree-days using a maximum–minimum thermometer and a degree-day look-up table.

15.2 Cherry Pest Biology and Management

Most arthropod pests of cherry do not discriminate between sweet and tart cherry, except possibly in the degree of attack. Thus, for the purposes of this chapter, when the term 'cherry' is used it will refer, unless specifically mentioned otherwise, to both species.

15.2.1 Direct pests

Cherry fruit flies

Several species of flies belonging to the family Tephritidae attack cherry and are among the most important pests worldwide. In North America three species of CFF are recognized as key pests. The eastern cherry fruit fly, *Rhagoletis cingulata* (Loew), and the black cherry fruit fly, *R. fausta* (Osten Sacken), are the species occurring in eastern USA and Canada. The western cherry fruit fly (WCFF), *R. indifferens* Curran, occurs in western USA and Canada (Fig. 15.5). The European cherry fruit fly, *R. cerasi* L., is a pest where cherries are grown in Europe. The apple maggot, *R. pomonella* (Walsh), although not considered a major pest of cherry in most areas, uses cherry as its primary host in Utah. In addition to commercially grown cherries, CFF infest fruits of native cherry plants, which, most probably, were their original hosts. CFF have also been reported from pear, plum and prune, but these are considered only incidental infestations, occurring where CFF populations are high in nearby cherry trees.

Fig. 15.5.　Adult of the western cherry fruit fly, *Rhagoletis indifferens* Curran.

Life stages. Eggs of the CFF are small, elongated and creamy white. They are seldom seen, since they are deposited in the fruit. The legless larvae are typically maggot-like, with a cylindrical body gradually tapering to a point at the head end. Dark mouth hooks can be seen at the head end of the creamy white larvae. They are about 5–7 mm long when fully grown. The pupa is brown and looks like a puffed kernel of wheat. Adults are about two-thirds the size of a common housefly and have distinctive dusky or dark banding patterns on their wings (Fig. 15.5).

Life history. The life histories of all CFF species are similar, so, while the WCFF, *R. indifferens*, is used as an example, it is a good generalization for the other species. The timing of life history events will vary slightly between species and from region to region, so it is important to follow local recommendations concerning these events. About 10 months are spent as a pupa in the soil. Adult emergence starts in mid-May and continues for about 8 weeks. Peak emergence occurs during sweet cherry harvest in late June or early July. Adults spend 7–10 days feeding before laying eggs and are active on warm, sunny days on leaf surfaces.

Eggs are inserted beneath the fruit skin, usually only one per fruit, and hatch in 3–5 days. The maggots feed around the pit and later in the cherry flesh. Maggots pass through three instars during a 2–3-week period and, when mature, drop to the ground and burrow into the soil. The maggot changes to a pupa in the soil and overwinters there. The WCFF has only one generation per year. A small percentage of pupae may remain in the soil for 2 or more years before emerging.

Damage. Damage occurs when larvae feed in the cherry. Fruit injury is almost impossible to detect, since infested fruits look normal until the maggot is nearly full grown, at which time the fruits appear sunken and flaccid. Due to quarantine restrictions, the detection of a single maggot can prohibit sales in certain markets.

Monitoring. Traps are the primary tool used to monitor the presence of CFF adults. Traps are of different colour, size and shape. Most traps are yellow and made of a single or double panel covered with trapping adhesive. Ammonium carbonate or ammonium acetate plus protein hydrolysate are added as lures to the trapping adhesive or held in a separate container on the trap to enhance fly capture. Traps should be placed in the orchard prior to the emergence of CFF adults in the spring. For maximum efficiency, traps should be placed within the fruiting canopy of the tree with foliage removed from around the trap for a distance of about 25 cm. Capture of flies in traps is used to determine the time to initiate protective chemical controls. Because CFF populations in commercial cherry orchards are usually low, a useful technique for detecting first emergence is to monitor abandoned orchards or areas that have historically had a high fly population.

Management. While traps are useful for monitoring WCFF activity when populations are relatively high, they are not considered reliable enough at low densities to make pest control decisions. Therefore, chemical controls are applied to cover the activity period of the flies. Degree-day models have proved reliable at predicting adult WCFF emergence and are used in Washington State to initiate the chemical control programme (Beers *et al.*, 1993). Consult a local or regional extension entomologist for detailed information for other areas.

Chemical controls are repeated as necessary, depending on the residual life of the product. Chemical controls are also applied in the postharvest period to prevent infestation of cherries left after harvest, since CFF activity can continue into late summer.

Alternatives to chemical control of CFF have been attempted but with limited success. A marking pheromone is used by the female after laying an egg to deter further egg laying in the same fruit. This pheromone was sprayed in orchards in the hope that it would reduce or prevent infestations. While some suppression of egg laying was reported, it was not sufficient to provide commercial control. Mass trapping of adult CFF has likewise not been an effective means of control. The release of sterile males was attempted in Europe. While the technique appeared to have some promise as an alternative control tactic, there were many difficulties encountered in rearing CFF and costs were high. CFF larvae are attacked by several species of parasite but levels of parasitism are too low to rely upon this tactic in commercial orchards.

Plum curculio

This beetle is native to North America and has been a serious pest of stone and pome fruits since early settlers began cultivating these crops. It is found throughout most of eastern North America. Principal native hosts include wild plum species, larger-fruited hawthorns (*Crataegus* spp.) and wild crab-apples.

The PC also attacks cultivated fruits, including nectarine, plum, cherry, peach, apricot, apple, pear and quince, with the same order suggested as host preference.

Life stages. The adult is 5 mm long and appears dark brown to grey-brown, although on closer examination it is a variegated grey, brown and black with white or silvery flecks. There are three pairs of humps on the back, the middle two being the most conspicuous. The most obvious characteristic is the long curved 'snout' protruding from the head. In the field, the adult PC usually feigns death and drops from foliage or fruit when disturbed. The egg is small, elliptical and creamy white. It is placed in a specially constructed crescent-shaped oviposition mark made in the fruit. The larva, which feeds internally on fruits, has a white to cream-coloured legless body with a well-defined brown head. When full-grown, the larva is about 6–7 mm long, curved or bow-shaped, cylindrical and tapered slightly at each end. The pupa is whitish or cream-coloured, about 5 mm long and found in the soil. The pupa resembles an adult with its appendages glued to its body.

Life history. Adults overwinter under ground debris, preferring fence rows or woodlot borders. They begin activity in the spring prior to bloom but are usually first seen as blossoms begin to open. Migration from outside sources may continue for 4–6 weeks but activity generally peaks 10–14 days after petal fall. Beetles feed on flower parts and foliage.

Egg laying starts once fruits are present. The female chews out a small cavity in the fruit where it deposits an egg. She then makes a crescent-shaped cut around and beneath the egg cavity. This cut relieves pressure from the growing fruit and possibly facilitates successful establishment of a newly hatched larva. Peak egg laying occurs 2–3 weeks after petal fall but may continue for 6 weeks. Eggs hatch in 6–7 days and the larva bores into the fruit, usually feeding near the pit. Larvae mature in 2–3 weeks and then exit fruit and enter the soil. The larva prepares a pupation chamber, where it remains for 10–12 days before pupating. The pupal stage lasts about 2–3 weeks. New adults appear in the orchard from mid-July through September. Adults may remain in the orchard for a short time after emergence, but by late September and October they seek overwintering sites.

Damage. Fruits are injured during egg laying when the adult punctures the skin to make the egg cavity and crescent-shaped oviposition mark. Even if the egg does not hatch, the fruit is scarred and unacceptable. Successful larval attack may initiate premature fruit drop, reducing yields. If fruits remain on the tree, larval feeding causes internal fruit breakdown.

Monitoring. There are no effective traps or selective monitoring tools for detecting PC. Thus, monitoring is difficult and more labour-intensive than for most cherry pests. Visual observation of adults or their crescent-shaped oviposition marks provides the only clue to their presence. A beating tray, a flat square or round panel covered with a cloth, is helpful for detecting adults. The beating tray is placed beneath a limb which is struck with a rubber mallet or hose. Beetles jarred from the tree fall to the tray. They may 'play dead' and look like bud scales or pieces of debris, so careful observation of material on the tray is required.

Management. Chemical control is the tactic most commonly used against this pest. Treatments are initiated when adults or damage is first detected and repeated at intervals based on the residual activity of the product used or as long as adults are detected. The PC moves into orchards from woodlots, fence rows or hedges, so trees bordering these areas should be checked frequently, beginning at this time. Monitoring activity in abandoned orchards or host trees nearby may help determine when to initiate observations in the orchard. Removal of host trees around the orchard or in old fence rows could help reduce local populations.

Leafrollers

There are several leafroller species reported as pests of cherry. In North America the leafrollers that have been reported from cherry include *Pandemis* leafroller, *Pandemis pyrusana* Kearfott, oblique-banded leafroller (OBLR), *Choristoneura rosaceana* (Harris), fruit tree leafroller (FTLR), *Archips argyrospilus* (Walker), and European leafroller (ELR), *Archips rosana* (L.). In Europe the leafroller species which are reported as pests of cherry include barred fruit tree tortrix, *Pandemis cerasana* (Hübner), and fruit tree tortrix, *Archips podana* (Scopoli). Since it is not possible to discuss each species, the OBLR has been selected as an example. Its life history is similar to that of several of the species mentioned and in many cases management tactics would be similar, although some leafroller species, e.g. FTLR, have only a single generation each year.

The OBLR is widely distributed throughout the USA and Canada. It is common in abandoned orchards or woodlots, preferring plants of the *Rosaceae* family, but feeding on many unrelated deciduous trees and shrubs. It has developed resistance to organophosphate insecticides in some regions, increasing its importance as a pest.

Life stages. Eggs are laid in flattened, overlapping masses of 100–300 on the upper surface of leaves. When first laid, the egg mass is a light green, turning a brownish colour as eggs near hatch. The fully grown larva has a green body, with the head capsule varying from brown to black (Fig. 15.6). The segment just behind the head, the thoracic shield, varies from black to greenish brown. The pupa, usually found inside a rolled leaf, is typical of leafrollers, being naked (without a silken cocoon). The pupa is greenish brown when first formed but turns brown after a few days; it is 11–18 mm long. The adult OBLR is tan or brownish, the forewing being crossed by alternating light and dark brown bands. The female moth is larger and more distinctively marked than the male.

Life history. The OBLR overwinters as a second or third instar larva within silken hibernacula. These usually occur in bark crevices, under bark scales or in debris at the base of the tree and are covered with faecal pellets, making them difficult to find. Larvae become active in spring as cherry buds open. They bore

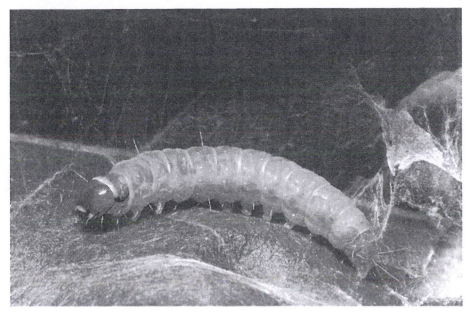

Fig. 15.6. Larva of the oblique-banded leafroller, *Choristoneura rosaceana* (Harris).

into the buds and begin feeding on leaves and flower parts. A larva often ties leaves and flower parts together, forming a mass of webbing and plant parts within which it feeds. Larvae are mature 2–3 weeks after petal fall. Pupation occurs within a rolled-up leaf and lasts 10–12 days. Moths of the overwintering generation appear in late May or early June, with peak activity in mid- to late June. Egg laying follows shortly after adult emergence. Eggs require 10–12 days to hatch. Summer generation larvae complete development by late July or early August, and a second moth flight occurs from early to mid-August through September. Larvae of the overwintering generation hatch from eggs laid by these moths. They feed for a short time on foliage and then seek overwintering sites and spin hibernacula.

Damage. Foliar feeding is usually of little concern except on young, non-bearing trees, where high populations can reduce leaf area and stunt growth. Larvae of the overwintering generation feeding within fruit clusters can severely damage fruit that either abort or are left deformed. Young larvae of the summer generation are present at or just before cherry harvest and may feed on fruit.

Monitoring. Examination of flower buds as they are opening in spring will reveal the presence of overwintering larvae. Larvae are often difficult to detect at this time of year without magnification. Summer generation larvae form tubular chambers from leaves at the tip of growing shoots and are easy to find. A pheromone trap can be used to monitor the activity of males. However, capture of moths in pheromone traps may not accurately reflect population densities within the orchard because males often move into the orchard from outside sources.

Management. Conventional insecticides remain the most common control tactic for this pest. Overwintering larvae are controlled by treatments made in the spring as fruit buds are opening. A careful examination of buds will reveal the presence of larvae and thus the need to treat. Examination of foliage and fruiting clusters after petal fall will reveal presence of larvae surviving spring controls. A degree-day model describing the development of OBLR is available to predict moth flight, egg hatch and larval development and can be useful to time insecticides.

IGRs which mimic the insects' juvenile hormone are effective against OBLR but are not registered for this purpose in North America. The bacterial insecticide, *Bacillus thuringiensis* (Bt), provides control of OBLR, although multiple treatments aimed at the larval stage are usually required. Mating disruption appears promising as a control and this tactic is being researched. Several parasites attack OBLR larvae; however, they are susceptible to conventional insecticides and have not played an important role in PM.

15.2.2 Foliage-feeding pests

Black cherry aphid (Myzus cerasi *(Fabricius)*)
The black cherry aphid (BCA) is a native European insect that is now found throughout the USA and Canada. This aphid prefers sweet cherry varieties but will also attack tart cherries.

Life stages. Eggs of the BCA are small, shiny and black. Except for size differences, adults and nymphs appear the same, black and shiny. They live together in colonies on the underside of leaves. Except when present in very high densities, aphid colonies are found on leaves at the tip of growing shoots.

Life history. The BCA overwinters as an egg on the bark of small branches. Hatch usually begins when buds start to open. The young aphids move to the foliage, pushing into opening buds. Within 3–4 weeks, wingless female aphids start producing live young. Two to three generations are produced on cherry. By early summer, most aphids have migrated to summer hosts, primarily plants of the mustard family. Several generations are produced on summer hosts and, in autumn, winged male and female aphids return to cherry. After mating, females lay overwintering eggs on bark.

Damage. The BCA causes most damage to young trees. Feeding causes a tight leaf curling and growth distortions in the stem. Heavy populations can stunt tree growth, and an accumulation of honeydew promotes a black, sooty fungus growth on fruit and foliage. High aphid populations have been known to limit fruit set in the following year.

Monitoring. The BCA is easy to detect. Colonies at the tips of actively growing shoots are usually first noted because infested leaves begin to curl. When densities of aphids are low, only the first few leaves will be infested. As populations increase, older leaves retain aphid colonies so that the proportion of infested leaves on a shoot increases.

Management. In most commercial orchards, the BCA is controlled by insecticides applied for other pests during the prebloom period. On young trees that have a limited pest control programme, it may be necessary to use chemical controls to limit the injury. There are many natural enemies of the black cherry aphid. These include ladybird beetles, lacewings, syrphid flies and several predatory true bugs. Because these predators tend to be generalists, that is, they feed on a wide range of plant-feeding insects, they do not always arrive in time to prevent aphid populations from reaching damaging levels. In addition, these natural enemies are usually susceptible to insecticides and may be eliminated if chemical controls for other pests become necessary. However, in many situations, if given an opportunity, predators will provide adequate control of the BCA.

Spider mites

Several species of mite are known as pests of cherry. These can be grouped into two classes: the spider mites and the eriophiids, or rust mites. The spider mites are often the most serious of the mite pests and the most common of these are the European red mite (ERM), *Panonychus ulmi* (Koch), and the two-spotted spider mite (TSSM), *Tetranychus urticae* Koch.

The European red mite. This is the most common and most important mite pest on deciduous fruit trees in many parts of the world. While greatest damage has been reported on apple, pear, plum and prune, it is also a pest of cherry, peach, grape and raspberry.

Life stages. The ERM egg is small, oval and bright red. Extending from the egg is a stalk or slender protuberance about as long as the egg is wide. There are three immature stages: the larva, protonymph and deutonymph. The larva can be distinguished from other immature stages because it has only three pairs of legs; others, including the adult, have four pairs. All the immature stages start as a light orange colour, which soon darkens to a greenish red as they feed. The adult female ERM appears as a larger version of the immature stages and is dark red or brownish red. The male ERM is smaller than the female, is straw-yellow to reddish yellow and has a pointed abdomen. All stages have bristles on top of their abdomen, but these are most evident in adults, in which white or lighter areas are visible at the bristle base.

Life history. The ERM stays on the tree all year in one stage or another. It passes the winter as an egg on smaller branches and twigs. Eggs begin to hatch about the time cherry blossoms open. Egg hatch is usually complete within 7–14 days. The larvae move to young leaves and begin feeding. Adults first appear shortly after petal fall. Female ERM live up to 20 days and lay 30–35 eggs. The life cycle usually takes 20–25 days, although under ideal hot, dry conditions it may take only 10–12 days. There are six to eight generations each year in most areas, although more could occur in regions with hot, dry summers. Populations usually peak in midsummer and females start laying overwintering eggs shortly thereafter.

Damage. Injury to cherry by the ERM is difficult to assess, as it is not easily measured and varies with varieties and growing conditions. Cherry, especially sweet cherry, can withstand large ERM populations without noticeable effects on the crop. Mites feed by inserting their mouth parts into leaf cells to withdraw the contents, including chlorophyll. Leaves can recover if mites are destroyed before the leaves are too badly damaged. Mite attack usually leaves foliage with a bronzed or brown appearance. Severe bronzing can result in reduced tree growth, small fruit, premature leaf drop and reduced fruit set in the year following a heavy infestation. Slight bronzing, especially if it occurs late in the year, seems to have little effect on cherry.

Monitoring. The small size and sometimes obscure habits of mites make them hard to detect. The ERM adults are usually visible on foliage, but a good hand lens is helpful for accurate identification. When populations are small, it may be difficult to locate any mites. The visual examination of leaves to determine the density of mites can be extremely time-consuming. A more sensitive monitoring technique commonly used to determine population densities of ERM or other mites is leaf brushing. Samples can also help determine if enough predatory mites are present to provide biological control.

Management. In well-managed orchards, the ERM and other pest mites can be controlled by natural enemies. In the eastern USA, the predatory mite *Amblyseius fallacis* (Garman) or predacious beetle, *Stethorus punctum* (LeConte), provide biological control of pest mites in apple as well as cherry orchards. In the western USA, the predatory mite *Typhlodromus occidentalis* (Nesbitt) is the most important biological control of pest mites. In other parts of the world, other biological control agents can completely or in part maintain ERM populations below damage levels. Where biological controls are not effective or are disrupted by chemicals applied to control other pests, miticides may have to be used.

The two-spotted spider mite. This is found worldwide. In many areas it is not considered a serious pest of cherry, since it usually remains on the cover crop feeding on many plant species. However, in hot, dry climates, when predators are destroyed by pesticides in the ground cover or when the ground cover becomes dried or killed by herbicides, the TSSM can cause severe damage when it migrates into the tree. In addition to cherry, the TSSM attacks apple, pear, plum, peach, many small fruit crops and numerous other crops and weeds.

Life stages. Eggs are spherical and appear clear when first laid, later becoming opaque and pearly white. After hatch, the six-legged larva of the TSSM appears pale to dark green with two black spots on the abdomen. The protonymph and deutonymph stages are progressively larger, have eight legs and more distinct dark spots. The adult mite is larger than the nymphs, but with the same general appearance. The male is smaller and has a more pointed abdomen than the female. While the TSSM has some bristles on its back, they are not as conspicuous as those of the ERM. The TSSM does not usually occur in the tree until late summer. They can often be seen on broad-leaf plants in orchard cover crops.

Life history. Full-grown females overwinter under bark scales on tree trunks or among fallen leaves and in other protected places on the ground. With the arrival of warm weather in the spring, the mites leave hibernation sites and wander about looking for food plants. Most of those on the tree trunk crawl down to the ground to feed on weeds in the cover crop. Eggs appear in early May and hatch in 5–8 days. A complete generation from egg to adult may require no more than 3 weeks. There are from five to nine generations each year; however, most do not occur on cherry trees. In mid- or late summer, when drought, herbicide use or other factors cause poor food conditions among host plants in the cover crop, the TSSM move into trees. This movement can be sudden and dramatic, often resulting in large increases in the mite population over a short period. After moving into trees they reproduce and remain until early autumn (September in Washington State), when they seek overwintering sites.

Monitoring. TSSM can be sampled as described above for ERM. In addition to monitoring foliage in the tree, it is wise to examine plants in the ground cover to determine if the TSSM is present. This mite is usually present on broad-leaved plant species in the cover crop and can be detected with the aid of a hand lens.

Management. In some areas the TSSM is an important component of an integrated mite management programme. This is the case in Michigan where the most important predatory mite, *Amblyseius fallacis*, overwinters in the cover crop. This predatory mite uses the TSSM as a food source early in the spring before it moves up the tree to feed on the pest mites. Miticides may be required when large numbers of mites migrate suddenly into trees from the ground cover or if pesticides that kill the predatory mites are used when controlling other pests.

15.2.3 Pests of the bark

Peach tree borers

The PTB, *Synanthedon exitiosa* (Say), and LPTB, *S. pictipes* (Grote & Robinson), were pests of great concern before modern organic insecticides were introduced. Now these moths are of less importance, although they can cause tree death, usually of young trees. Before the introduction of commercial stone fruits, these native North American insects used wild plum, cherry and other related plants as hosts. In addition to cherry, they will attack peach, prune, apricot, nectarine and plum. In Europe, the apple clearwing moth, *Synanthedon myopaeformis* (Borkhausen), and cherry bark tortrix (CBT), *Enarmonia formosana* (Scopoli), are pests of cherry, although they mainly attack older trees or trees in poor condition and are considered of secondary importance. The CBT is now known in North America, British Columbia, Canada and Washington State, although its distribution is restricted to non-commercial fruit-growing regions.

Life stages. Eggs of the PTB are about 0.7 mm long and chestnut or reddish brown. Eggs of the LPTB are slightly smaller and a cinnamon- or rust-brown. The egg surface has a finely netted appearance. The larvae of both species are white or cream-coloured with a yellowish brown to dark brown head. There are three pairs of segmented thoracic legs. Because larvae of both species appear similar, identification is difficult. Location on the tree can provide a clue as to the species; those larvae located higher than 15 cm above the ground are most likely to be LPTB. The light tan pupae vary in length from 10 to 17 mm and are elongated and cylindrical. Abdominal segments have rows of sharp spines which enable the pupa to push from the cocoon through the bark surface when ready to emerge as an adult. Cocoons are oblong, elongated and constructed from chips of bark and wood frass held together by silken strands. New cocoons are light yellowish brown; older ones are rust-coloured.

Both sexes have clear wings, females being slightly larger than males. The head, thorax, body, legs and antennae are metallic blue-black with some pale yellow markings on the abdominal segments. The fourth and fifth abdominal segments of the female PTB abdomen are covered by bright orange scales. The slender abdomen of the male PTB terminates in a wedge-shaped tuft of scales tipped with white. The finely tufted antennae of the male distinguish it from the female, whose antennae are long and slender. The behaviour of these moths differs from most other moth species since they fly during the daytime.

Life history. The PTB overwinters in various larval instars under tree bark, at or below ground level. When full-grown, the larva constructs a silken cocoon and pupates in early summer. Moth emergence usually begins in midsummer and continues into early autumn. Females live 6–7 days and most eggs are laid in the first few days of life. Eggs are deposited singly or in small bunches, most being found near the tree base, or higher in the case of the LPTB, especially where fresh wounds or cracks in the tree bark are evident. Eggs hatch in 7–10 days, and the young larvae bore through the bark to the cambium layer, where they feed.

The life history of the LPTB is essentially the same as that of the PTB. The time of emergence, female longevity and time of egg development are about the same as in the PTB. Eggs, however, are laid higher in the tree, from ground level to a height of about 2.5 m, although most are within 1–1.5 m of the ground.

Damage. The principal damage by larvae of both species is feeding on the inner bark and cambium tissue. Most larval activity of the PTB is confined from a few centimetres above to 15 cm below the ground line. Larvae of the LPTB are found in the trunk and scaffold limbs. Pruning wounds, areas damaged by disease or harvesting equipment, insect- or winter-injured areas or sun-scalded bark are frequent locations of infestation. Young trees may be completely girdled by larval feeding, resulting in death. Older trees are less likely to be completely girdled but are often injured enough to reduce vigour, making them susceptible to other insects, diseases and environmental conditions.

Monitoring. Adult males of both species can be monitored with a pheromone trap. Pheromone traps help indicate initial and peak adult activity periods. Larval feeding of the PTB is difficult to detect because much occurs below ground level. Borer-infested trees bleed or exude frass-infested gum from damaged areas. Active feeding sites are denoted by the presence of a light brown frass and a soft, sticky sap.

Management. Larvae feeding under the bark cannot be controlled. If insecticides are used for control, they must be present when eggs are hatching to kill the young larvae before they bore beneath the bark. Because emergence occurs over a long period it is difficult to maintain coverage with insecticides. Pheromone trap catch can be used to indicate periods of highest moth activity, thus providing valuable information on the optimum time to apply insecticides. Mating disruption using the pheromone for the PTB has shown promise as a control for both species.

San Jose scale (SJS)

Quadraspidiotus perniciosus (Comstock) is a native of China brought to California about 1870. By 1873, it was a serious pest in San Jose Valley – hence the name. By the late 1890s, it had reached all parts of the USA. It is known as a pest of fruit trees throughout the world.

SJS was the first insect in the USA known to exhibit resistance to a pesticide. Its resistance to lime-sulphur was reported in Washington in 1908. Tremendous damage, including tree kill by this pest, was reported before modern chemical controls were introduced. SJS prefers apple, pear and plum but also feeds on peach, apricot, cherry, quince, currants, gooseberry and other woody ornamentals.

Life stages. The female SJS produces live young in much the same manner as aphids. The newly born young ('crawlers') are extremely small, lemon-yellow and flattened. They have legs and move about bark and foliage before settling down and inserting their long sucking mouth parts into the tree. Once settled, the scale does not move again.

A day or two after settling down, the crawlers secrete white, waxy filaments from glands scattered over the body. These filaments soon form a waxy covering. At this stage of development, the insect is easily detected, but in a few days the waxy covering becomes dark and harder to see. When viewed with a hand lens the young scale looks like a miniature volcano. As the insect grows, the scale enlarges and the legs and antennae are lost.

The female scale is nearly circular with a nipple near the apex of the cone. The male scale is about twice as long as wide. When mature, the male, a delicate, light-coloured, winged insect, emerges from under the scale. The female remains under the scale covering her entire life.

Life history. In areas with cold winters the SJS passes the winter as a partially grown scale on tree bark. In areas, such as California, with milder winters other stages of the scale may survive winter. Growth continues in the spring and after the bloom period the males emerge and mate with females. In about 5–6 weeks crawlers can be detected. Each female produces between 150 and 500 crawlers over a 6-week period. Crawlers settle down after a few hours and insert their mouth parts into the bark or foliage. They mature in 5–7 weeks and produce a second generation. In the northern USA and Europe there are two generations of the SJS but in warmer regions four or five generations are possible.

Damage. After inserting its mouth parts into the plant, the tiny scale feeds on the plant's sap. It is difficult to realize fully the dangerous nature of the SJS without seeing its work. The species multiplies so rapidly that, from a few scattered parents, millions of progeny may be produced in a season or two, sufficient to completely cover the bark. Left unchecked, the SJS can kill a young tree or shrub in 2–3 years. Older trees withstand the attack longer but sooner or later are destroyed. SJS is usually found on the bark but also infests leaves and fruit.

Monitoring. It is difficult to detect this insect due to its colour and small size. Often the scale is concealed beneath bark scales and may be exposed by peeling these areas away. The base of suckers and spurs at the tree top, where insecticide coverage is poor, is another location where scale can often be found.

To monitor crawler activity, detect and tag live scale colonies and then periodically check those areas for crawler appearance. Detection of crawlers can be aided by placing a band of double-sided sticky tape around a limb near live scale. When the crawlers start to migrate they become trapped on the tape. Pheromone traps can be used to monitor the activity of male scale.

Management. Oil, sometimes accompanied by an organophosphate insecticide, applied to cherry trees in the dormant period or just as buds begin to show green tissue in the spring is the most common control for SJS. Control of scale at this time is critical to prevent populations from increasing. While chemical control of crawlers is possible, attacking this stage should be considered supplementary to good dormant control. Degree-day models are available to predict crawler activity and to time summer sprays.

Acknowledgements

My thanks to Dr Larry Hull, Entomologist, Pennsylvania State University, and Dr Larry Gut, Research Associate, Washington State University, for critical review of the manuscript for this chapter.

References

Alford, D.V. (1984) *A Colour Atlas of Fruit Pests*. Wolfe Publications, Glasgow, UK.

Beers, E.H., Brunner, J.F., Willett, M.J. and Warner, G.M. (1993) *Orchard Pest Management: A Resource Book for the Pacific Northwest*. Goodfruit Grower, Yakima, Washington State.

Croft, B.A. (1990) *Arthropod Biological Control Agents and Pesticides*. John Wiley & Sons, New York.

Croft, B.A. and Hoyt, S.C. (1983) *Integrated Management of Insect Pests of Pome and Stone Fruits*. John Wiley & Sons, New York.

Howitt, A.J. (1993) *Common Tree Fruit Pests*. Michigan State University Extension NCR 63.

Huffaker, C.B. and Rabb, R.L. (1984) *Ecological Entomology*. John Wiley & Sons, New York.

Jones, V.P. (1991) Sampling for integrated pest management. In: Williams, K.H. (ed.) *New Directions in Tree Fruit Pest Management*. Goodfruit Grower, Yakima, Washington State, pp. 61–73.

Metcalf, R.L. and Luckman, W.H. (1975) *Introduction to Insect Pest Management*. John Wiley & Sons, New York.

Pedigo, L.P. (1989) *Entomology and Pest Management*. Macmillan Publishing Co., New York, 646 pp.

Southwood, T.R.E. (1978) *Ecological Methods*. Chapman & Hall, London.

Williams, K.H. (1991) *New Directions in Tree Fruit Pest Management*. Goodfruit Grower, Yakima, Washington State.

16 Sweet Cherries: Protection of Fruit from Bird and Rain Damage

D. Pennell[1] and A.D. Webster[2]

[1]Brogdale Horticultural Trust, Faversham, Kent ME13 8XZ, UK; [2]Horticulture Research International, East Malling, West Malling, Kent ME19 6BJ, UK

16.1 Introduction

Throughout the world birds are significant pests of many agricultural crops at various stages of crop growth. Cherries, and in particular sweet cherries, are especially prone to damage wherever they are grown in the world. This is not a new problem, for in 1566 an English Act of Parliament placed a bounty on the heads of certain birds and also other animals considered to be pests (Wright and Brough, 1966). While bullfinches (*Pyrrhula pyrrhula* L.) were mentioned as a destroyer of fruit in bud and had a bounty of one penny per head, the same reward was offered for 12 starlings (*Sturnus vulgaris* L.). It is the starling which is the most common cause of damage to cherry fruit in the UK and many other parts of Europe. While starlings cause the major damage to ripening and ripe fruit in southern UK (Fig. 16.1), smaller amounts of damage can be experienced from blackbirds (*Turdus merula* L.) and jays (*Garrulus glandarius* L.). Wright and Brough (1966) showed that damage could also be caused to green cherry fruit, soon after stoning has occurred, and here the main culprit was the wood pigeon (*Columba palumbus* L.).

Starlings pose similar problems to cherry growers in Canada (Upshall, 1943), Belgium (Tahon, 1980) and the Netherlands (Bonnemayer and Dane, 1987). In New York State, USA, robins (*Turdus migratorius* L.) were also shown to be responsible for some damage, as well as the ubiquitous starlings (Way, 1968). Guarino (1972) reported damage to sweet and sour cherries in Michigan, USA, from common grackles (*Quiscalus quiscula*), in addition to that from robins and starlings. In Michigan, starling damage to the sour cherry Montmorency was a particular problem.

Economic losses resulting from bird damage to cherries arise in three ways. In addition to losses from birds eating fruit, fruit is also lost by the fouling of otherwise undamaged fruit. Also, because of the extreme likelihood of bird damage in most years, orchardists are often tempted to harvest fruit before it

Fig. 16.1.   Severe damage to the sweet cherry Starkrimson caused by starlings.

becomes fully ripe (Feare, 1980). In view of the significant increases in fruit size which occur in the few days prior to optimum harvest date, early picking can severely limit yield. Various estimates of losses to birds in cherry crops have been made, ranging from 0.5–10% of fruits in the week preceding optimum maturity (Upshall, 1943) to 90% (Way, 1968). In a US Department of Agriculture (USDA)-sponsored survey of sour cherry orchards in Michigan in 1972 (Guarino *et al.*, 1974), birds were estimated to have consumed up to 17.4% of the fruit. Losses can vary within the tree canopy, particularly in larger, mature trees. In a 16-ha orchard in the UK, losses of 14% were noted from the lower branches while 21% were lost from the upper tree canopy (Feare, 1980). Damage can also vary from season to season. Way (1968) observed that smaller orchards suffered a higher proportionate loss of fruit than larger orchards and that early-season cultivars and 'black' cultivars were more prone to damage than late or 'white' cultivars.

Records in the UK showed that most of the starlings feeding in commercial orchards were juveniles (Feare, 1980) except early in the day, before 0600 h, and later in the day, after 1800 h, when the number of adults increased. The predominance of juvenile birds in cherries has been noted elsewhere and limited data from the UK indicate that, while adults are mainly terrestrial feeders, juvenile starlings are mainly arboreal. Feare (1980) suggests that this behavioural difference avoids competition by naïve juveniles with more experienced adults for soil fauna.

In other studies, Summers (1985) has shown that captive starlings could not maintain their body weight on a diet of cherries alone and that the maximum daily intake of fruit was 25 cherries, some 1.5 times the bird's own weight. Furthermore, the average time birds were observed to spend in an orchard was

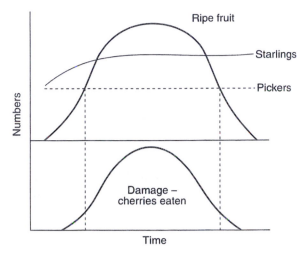

Fig. 16.2. The quantity of ripe fruit in a cherry orchard in relation to the number of cherry pickers and to the damage by starlings.

2.8 h, which was longer than the time required to consume 25 fruits.

Starlings form very large night-time roosts in woods which can attract birds from a radius of up to 30 km and may contain several million birds. As they assemble in the evening, they can feed intensively in very large flocks near the roost (Vaughan, 1982; Taylor, 1985). These observations help explain the damage caused by starlings by feeding and spoilage, compared with the more solitary bird species such as the English blackbird and the thrush.

Figure 16.2, which is based upon Feare's observations (1980), shows diagrammatically the availability of ripe fruit in relation to the abundance of starlings and the number of pickers in a traditional UK commercial orchard with large trees. Varieties ripen in sequence through the season, with most fruit ready for harvest in July.

During the early part of the season, bird numbers increase but the more than adequate number of pickers can harvest fruit sufficiently rapidly to minimize damage to fruit. If, later in the season, the number of pickers could be increased in relation to the abundance of the ripening fruit, it is possible that the losses to starlings could be reduced. This strategy could prove difficult in some countries, however, on account of high labour costs and lack of available labour as a result of competition from other sources of employment. Orchard management practices have an impact on the amount of damage sustained as a result of bird pressure. The single most important factor is tree size. Large trees provide many sites where birds can perch and feed, are difficult to pick sufficiently quickly to deny birds ripe fruit and can limit the effectiveness of other measures aimed at limiting bird damage.

The second major cause of crop loss, just prior to or at harvesting time, is fruit cracking following rain and uptake of water by the fruits. The causes of cracking and some of the possible remedies for its alleviation, particularly those dependent upon chemical sprays, are presented in Chapter 12. However, cherry growers in New Zealand, Norway and the UK, where rain-induced cracking is particularly severe, have recently tested protective canopies which are erected over the trees close to harvest time. Preliminary results with these rain shelters, which in some instances may also serve as protection against birds, are discussed in this chapter.

16.2 Protection of Cherries from Bird Damage

16.2.1 Chemical control measures

Chemical stimuli are mainly perceived through the specialist senses of smell and taste, and Wright (1980) concluded that, while the olfactory senses are functional in most birds, only in a few species is odour-related behaviour exhibited. This casts doubt, therefore, on the efficacy of the repellency of odours for use in bird control. Nevertheless, in the past, an anthropomorphic view of the likes and dislikes of birds has resulted in the use of a wide range of chemicals, many of which were derived as extracts from plants with pungent odours. Chemical repellents of various types have been used on crops for some 150 years.

Of the many compounds tested for avian repellent activity, thiram (tetramethylthiuram disulphide) was the most effective in a number of early studies and became a comparative standard. More recently, Guarino (1972) reviewed the development and use of methiocarb (4-(methylthio)-3.5-xyly-N-methylcarbonate), a non-systemic carbamate insecticide, as a bird repellent. In comparison with thiram and other chemicals, it was identified as an excellent potential broad-spectrum avian repellent. Methiocarb was applied to both sweet and sour cherry in Michigan at 451 g (1 lb) in 379 l (100 gallons) water and sprayed to the point of runoff. Highly significant differences in damage were found ($P < 0.001$), with controls receiving five times as much damage as treated sweet cherry trees. In sour cherries over 50% damage was noted to control trees compared with less than 20% in those treated with methiocarb. In subsequent trials using the same application rates, a 65.6% reduction in damage to sweet cherries and a 62.2% reduction in damage to sour cherries was obtained (Guarino *et al.*, 1974). There were also some indications from this work that different bird species reacted differently to methiocarb.

Unfortunately, in most trials only small blocks of fruit trees within an orchard are sprayed and the results are, consequently, difficult to interpret. It is probable that birds with an abundant and readily available supply of unsprayed fruits in adjacent control blocks within the experimental orchard were more likely to move on to these unsprayed trees than would birds which had much greater distances to travel to find unsprayed fruit, were the whole orchard to be sprayed.

Methiocarb, as the commercial product Mesurol, was registered in the USA for use on sweet and sour cherries (Rogers, 1980). Trials in the UK on sweet cherries gave similar encouraging results. Three applications of methiocarb were made at 2-week intervals, the last 7 days before predicted harvest date. A rate of 4.5 kg a.i. ha^{-1} applied at high volume was used and environmental impact assessed (Westlake, 1981; Price, 1982; Vaughan, 1982). Residues of methiocarb were low in fruit samples at harvest. When used on sour cherries for processing in Belgium, Hoyoux and Zenon-Roland (1984) found that a reduction of 70% in methiocarb residues occurred when fruit was transported from orchard to factory in water.

Research reported by Rogers (1980) indicates that methiocarb has an effect on birds via their sense of taste, as a conditioned aversion behavioural response. As such, a low level of damage must be expected, at least during the conditioning period, and 100% control of damage cannot be expected. Also, adequate alternative foods must be available to achieve the most effective repellent effect.

16.2.2 Physical control methods

These methods include trapping and shooting the bird pests, audio and visual devices for scaring the birds and nets or partial nets to keep the birds off the fruit.

Trapping or shooting birds

With the current climate of public opinion in many parts of the world favouring conservation of wildlife, including birds, shooting is unpopular and is in some areas illegal for many bird species. In Belgium (Tahon, 1980), even more draconian measures than shooting were assessed some years ago. Use of dynamite to blow up the evening roost sites of the starlings was tested as a possible method of control. Some 200–300 kg of dynamite was used per roost and detonated at 2300 h or 0200 h. The mean number of birds killed per explosion was 34,000, with the proportion of juveniles in roosts varying from 10 to 78%. Roosts destroyed in this way were not reoccupied for 2 years or more. Tahon comments, however, that the effectiveness of these methods was limited to the second half of the harvest period only and to only part of the starling population.

Scaring birds

The simplest scaring techniques, such as scarecrows or loud noises, have been widely used by orchardists for centuries. Slater (1980) and Inglis (1980) have both reviewed the use and effectiveness of sound and visual bird scarers. Figure 16.3 shows a typical propane-powered scaring gun used in many UK orchards. Broadcasting distress calls of birds initially proved successful in scaring a number of species, including starlings, from cherry orchards (Wright and Brough, 1966). Keil (1968) used distress calls in a 12-ha cherry orchard. A 10-s distress call, broadcast whenever starlings were about to settle in trees, played through five speakers distributed throughout the orchard, gave satisfactory results.

The rate at which bird pests may become habituated to any scaring device has stimulated much debate and a number of workers have tried combinations of scaring stimuli in an effort to delay any habituation and improve scaring efficiency. Zajac (1983) carried out trials over seven seasons in Poland using distress calls (biosonics) with or without gas bangers (pyroaccoustics) to deter starlings in both sweet and sour cherry orchards. Control equipment was operated manually from an observation tower in the centre of the orchards, which varied in size from 5 to 44 ha. Peak bird feeding occurred at 0730, 1530,

Fig. 16.3. A propane-powered 'gun' used for scaring birds in UK cherry orchards.

1800 and 2000 h, with its intensity depending upon weather conditions. Small flocks of birds were less susceptible to the distress calls than larger flocks. Optimum orchard size for this treatment was estimated to be 12 ha and distress calls gave 88% protection, which increased to 96% when used in conjunction with bangers. In England, radio-tagged starlings were used to monitor activity in commercial cherry orchards (Taylor, 1985). Gas bangers, audiovisual scarers, scarecrows and distress calls were used for 80 s every 10 min. Despite this array of devices, starlings still spent long enough in the orchard to obtain their daily food requirement.

Scarers did not prevent tagged birds from returning each day, although starlings did leave the orchard during the first 20 s of the distress calls and others were deterred from entering. However, birds did not respond to the scarers after 7–13 days (Summers, 1985). Dzhabbarov (1988) combined an optical scaring device with distress calls in cherries and grapes. The optical device consisted of a hanging sphere of mirrors 0.5–1 m above the crop and not less than 35 spheres ha^{-1} in cherry orchards. When used in combination with distress calls in an orchard of early cherry cultivars, 70% of fruit was saved, whereas 100% of fruit was lost in an unprotected orchard. A number of simple measures can be used to optimize the effectiveness of auditory and visual scaring devices (Table 16.1).

Protective netting structures

In view of the transitory effects of scaring devices, it is hard to escape the conclusion reached by Way (1968) that nothing except netting will fully control bird damage. Partial nets in the form of cotton (Lovelidge, 1976) or spun synthetic fibres draped to form a web in trees have also been used to some effect (Wright and Brough, 1966). Although observations in 1976 indicated some

Table 16.1 Guidelines for the effective use of auditory and visual scaring devices (after Inglis, 1984).

Use as infrequently as possible. Do not introduce into crop before it is at its most vulnerable stage
Use a variety of devices with different modes of action rather than rely upon one type
Immediately remove any scarer that has lost its effectiveness
Change the position of scarers as frequently as possible
Reinforce the deterrent effect by occasional shooting (this will be governed by local laws designed to protect some bird species)
Camouflage auditory scarers, but beware of any fire risk and alarm to people unaware of the position of equipment
Varying the height of scarers in the crop can be helpful
Control times of use and interval to take account of when feeding pressure is greatest
Attempt to reduce feeding pressure by being aware of alternative foods in the area

beneficial effect on yield (Table 16.2), these treatments do give rise to practical management problems, not least of which is access to the trees at harvest. Growers in the UK began considering nets for protection from birds in the early 1970s when a protected 2-acre orchard was reported to have yielded 50% more than a comparable area left unprotected (Lovelidge, 1973). Unfortunately, the difficulties and cost of protecting large trees in traditional orchards have deterred growers from fully developing the technique. However, with the introduction of semidwarfing and dwarfing rootstocks for cherry trees (Chapter 5), the use of netting is becoming a more economically viable solution to bird damage.

The nets most commonly used for protecting sweet cherries in the UK are made from ultraviolet light-stabilized polyethylene and have a mesh diameter of 20–30 mm. If birds smaller than starlings and blackbirds are a problem, a smaller-mesh diameter will be needed. Several basic systems of netting have been used (Pennell, 1981). Firstly, the net can be laid over and into the tree without any support other than that given by the tree structure (Bogdan *et al.*, 1981). Nets used in this way can inhibit the growth and shape of the tree crown and can be difficult to remove prior to picking without damaging leaves or crop. Also birds tend to perch on the exterior of the net and eat or spoil all the fruits

Table 16.2 Effect of cottoning as protection against birds on sweet cherries.

| Variety | Yield tree $^{-1}$ (kg) | |
	Cottoned	Control
Pointed Black	8.71	3.99
Vic	6.35	4.31
Vista	0.82	0.00
Venus	1.23	0.36
Belle Agathe	8.94	1.95
Merton Late	11.16	14.15
Mean of all cvs	6.2	4.13
Least significant difference	4.29	5.22

within their reach. With robust nets, the branches of the tree can damage the net as it is removed.

The use of degradable nets could facilitate removal from trees but a considerable amount of partially degraded debris is still left in the orchard. This debris has the potential to adversely affect use of machinery used in the orchard. Also, all netting systems have the potential to inadvertently trap wildlife in the mesh. This can be particularly evident at ground level when small mammals, such as hedgehogs, can become ensnared.

A second system of protection introduces some additional support, within the tree row, in order to remove pressure from the tree; this forms a tent, curtains, tunnel or draped effect. Stout poles (hop poles in the UK) are erected slightly higher than the trees and down the tree rows. The net is draped over wire(s) (10 SWG gauge) stretched between the tops of the poles (Fig. 16.4). This system requires more net than would a complete cage for covering the trees but because the net can be laid relatively quickly the labour requirement for erection is reduced compared with a cage.

A modification to this tent or tunnel strategy involves fixing a hoop of steel piping (1.0–1.5 m wide) to the tops of the support poles. This has the advantage of broadening the netted width at the apex of the tree and minimizes problems of upper branches becoming entangled with the net. More recently, a modification to this system has been tested. With this, two adjacent rows, trained as compact hedges using pruning and/or growth regulators and with a maximum height of 3 m, were covered using a single net (Fig. 16.5). Not only is this system very easy to erect but harvesting can be carried out from the central alley between the rows, without removal of the net.

Covering an orchard with a complete net cage has a number of advantages, not least of which is that once erected there is considerable freedom to work within the orchard for harvesting or routine husbandry operations. The side nets

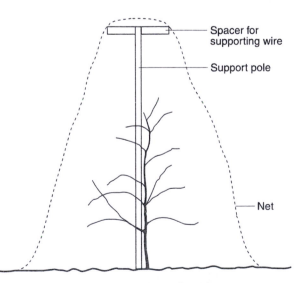

Fig. 16.4. Netting system for single rows of mature cherry trees.

Fig. 16.5. A double-row netting system for dwarf sweet cherry trees tested in the UK.

of complete cages can be in the form of a windbreak net, which may offer additional benefits at times of year when the overhead netting is not in place. The incorporation of a winch to move roof nets may reduce the annual labour requirement for erection and dismantling the nets (Fig. 16.6).

Nets will be degraded by ultraviolet light and damaged over the winter period if left outside. The life of nets can be prolonged by removal after harvest and careful storage. The system of netting adopted will be influenced by economics and the overall management of the orchard. In research situations, the ease of movement within a cage has much to commend it. Commercially, the netting of a two- or three-row bed planting may prove a suitable compromise between single-row tent-type nets and covering the whole orchard with horizontal overhead nets. It is essential to design any system of netting together with the planting system when an orchard is first planned.

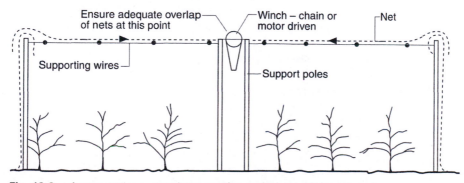

Fig. 16.6. A cage netting system incorporating a winch to withdraw the nets.

To summarize, given that bird damage is likely to remain a significant factor wherever sweet or sour cherries are grown, serious consideration must be given to protection measures. Traditional means of scaring by auditory and visual means are only partially effective. Chemical control is also only partially effective based on the limited evidence available and is under regulatory pressure in many countries. Netting offers the potential of full protection from birds but is costly.

16.3 Protection of Cherries from Rain Damage

The high value of sweet cherries in many world markets and the very substantial losses incurred following rain damage to fruit at harvest have prompted both researchers and cherry growers to seek methods of alleviating this problem. It is generally accepted that most damage is the direct result of rain deposited on the fruit being taken up and the fruit rupturing. Choice of appropriate sites with little or no rain incidence at or near harvest time is one of the best solutions to the problem. Unfortunately, this is an option in only a few cherry-producing countries. Other alternatives are to choose varieties which show some resistance to rain-induced cracking, or to spray the fruits with minerals, surfactants or plant growth regulators to improve their natural resistance to rain damage. These and other chemical preventive measures are fully discussed in Chapter 12.

Physically protecting sweet cherry fruit from rain damage beneath protective shelters or canopies is a strategy which, until recently, has always been considered too expensive and uneconomic. Nevertheless, with the high value of cherries on the world markets the use of such protective canopies is being reconsidered in some countries.

Trials began first in Switzerland, where rain at harvest is a frequent cause of severe crop loss (Meli, 1982). Stout posts (6.0–6.5 m in height) were erected within the tree rows, on top of which steel arcs up to 6.0 m wide were attached; a similar design of steel hoop used in Belgium is shown in Fig. 16.7. Three weeks prior to harvest, often about the time the fruit first begin to change colour and become sensitive to rain damage, ultraviolet-light-resistant polyethylene sheeting 6.0 m wide and 0.2 mm thick was draped over the metal frames along the length of the rows and firmly secured at its sides and ends. Despite problems of high temperatures and condensation beneath the covers and the increased risk of disease infection, they were considered economically worth while with early and midseason varieties grown in Switzerland. Picking efficiency and the quantity, size and quality of fruits harvested were greatly improved and the fruit from beneath covers commanded a 20–60% premium in years of persistent or heavy rainfall at harvest.

Similar systems of covering moderately vigorous sweet cherry trees under polythene draped over metal hoops on tall poles have been evaluated in Belgium and the UK. A higher-quality but more expensive system tested at East Malling in the UK utilizes frames made of metal tubes, on to which the polythene is spring-clipped. This sytem, manufactured in the Netherlands, is shown in Fig. 16.8.

Severe losses of fruits from rain-induced cracking in New Zealand have

Fig. 16.7. Metal hoops on stout poles used to support polythene rain shelters in a Belgian orchard.

stimulated the erection of much larger protective structures than those so far tested in Europe – structures capable of accommodating the largest mature sweet cherry trees (Gillespie, 1988). The supporting framework of these is built entirely of steel and they resemble Venlo-type glasshouses, with a form of ridge ventilation and very high eaves; they are clad in strong-gauge polyethylene.

Although these have proved successful in reducing, significantly, the degree of fruit cracking in protected orchards, the structures are extremely expensive and unlikely to be cost-effective in most areas of cherry production. Also, a small proportion of the fruits have still cracked, despite very effective shelter from the rain. This has stimulated researchers in both the UK and New Zealand to question the long-accepted hypothesis that sweet cherry fruit cracked only if water was taken up directly by the fruits themselves.

Sweet cherries are also protected from rain in Norway using polythene covers (Fig. 16.9) but here the canopy is erected with only a slight pitch to the roof. Securing these covers against wind damage has proved quite difficult.

Researchers in the UK are currently examining the effects of full or partial

Fig. 16.8.   End view of a cherry rain protection canopy mounted on a proprietary tubular steel support system (Rovero, the Netherlands).

polythene covers erected over sweet cherries from late May to harvest time in July. Effects on the microclimate surrounding the tree are being recorded, as well as those on shoot growth, yield and fruit quality. Preliminary results indicate that in seasons of cooler than average temperatures fruit size at harvest may be increased by as much as 1.0 g per fruit with no loss of soluble solids content.

In many areas of sweet cherry production, growing trees beneath polythene covers for periods longer than the few days just before and including harvest time may cause problems. Temperatures beneath the covers may become excessively hot, causing scorching and damage to the trees. To alleviate this potential problem of overheating, covering only the tops of the trees with umbrellas of polythene is being evaluated. These umbrella structures permit free air flow into the sides of the rows, so helping to ventilate the canopies. The requirement now is for covers which combine a polythene umbrella over the tops of the trees with netting at the sides to prevent entry by birds.

An alternative approach, previously examined in Washington State, USA (Opperman, 1988), and more recently the focus of research by scientists in Michigan, USA, is to use a covering material which is impermeable to rain but

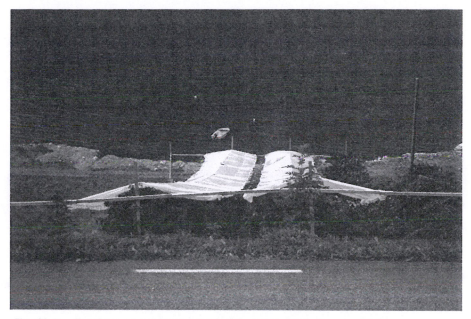

Fig. 16.9.   Polythene covers erected over a Norwegian sweet cherry orchard.

allows some air exchange. Unfortunately, most materials of this sort are far from perfect and in heavy rain some fine droplets invariably pass through the cover on to the trees and fruit.

If the rain canopy structures described above are to be cost-effective, they must be quick and simple to erect, secure against wind damage, easy to dismantle and relatively long-lasting. These objectives will only be fully achievable when the grower is able to limit sweet cherry tree height and width to maxima of approximately 3.0 m. New dwarfing rootstocks, recently released from Germany, Belgium and France (Chapter 5), may now make this possible. Also, in countries where plant growth-regulating chemicals, such as paclobutrazol (Cultar or PP333), are approved for use, these may provide a means of limiting excessive vigour so that the trees may be accommodated beneath relatively small rain shelters (see Fig. 11.3, p. 287). Finally, there is an urgent need for economic appraisals on the use of rain shelters over sweet cherry trees.

References

Bogdan, H.J., Dammann, H.J. and Zahn, F.G. (1981) Erste Erfahrung mit dem Tettmann Einwegnetz. *Mitteilungen des Obstbauversuchsringes des Alten Landes* 36, 184–188.

Bonnemayer, C. and Dane, J.H. (1987) De structuur van de kersenteelt in de Midden. *Betuwe Fruitteelt* 77 (28), 16–17.

Dzhabbarov, A.D. (1988) Means of repelling birds in orchards and vineyards. *Sadovedstvo-i-Vinogradarstvo* 11, 17–18 (CAB 1989 059–02729).

Feare, C.J. (1980) The economics of starling damage. In: Wright, E.N., Inglis, I.R. and Feare, C.J. (eds) *Bird Problems in Agriculture.* BCPC Monograph 23, pp. 39–55.

Gillespie, R.J. (1988) An evaluation of cherry production in New Zealand. *Acta Horticulturae* 223, 93–100.

Guarino, J.L. (1972) Methiocarb, a chemical bird repellent: a review of its effectiveness on crops. In: *Proceedings of the 5th Verbebrate Pest Control Conference*, Fresno, California, USA, 7–9 March 1972. pp. 108–111.

Guarino, J.L., Shake, W.F. and Schafer, E.W. (1974) Reducing bird damage to ripening cherries with methiocarb. *Journal of Wildlife Management* 38, 338–342.

Hoyoux, J.M. and Zenon-Roland, L. (1984) Sour cherry protection against starlings: can it be done with methiocarb? *Mededelingen van de Faculteet-Landbouwwet enschappen. Rijksuniversiteit-Gent* 49, 947–953.

Inglis, I.R. (1980) Visual bird scarers: an ethological approach. In: Wright, E.N., Inglis, I.R. and Feare, C.J. (eds) *Bird Problems in Agriculture*. BCPC Monograph 23, pp. 121–143.

Inglis, I.R. (1984) *Bird Scaring*. ADAS leaflet 903. *Ministry of Agriculture, Fisheries and Food*.

Keil, W. (1968) Elektroakustische Abwehr von Starenschaden im Kirschenanbau. In: *ISHS Symposium on Cherries and Cherry Growing*, Bonn, Germany, pp. 209–212.

Lovelidge, B. (1973) Netting cherry orchard increases yields by up to 50 per cent. *Grower* 80, 675–678.

Lovelidge, B. (1976) Cottoning could become practical measure for bullfinch control. *Grower* 85, 220, 222.

Meli, T. (1982) Abdecken von Susskirschen-hecken mit Polyathylenfolien: I. Voraussetzungen fur das Abdecken. II. Technik des Abdeckens. III. Betriebs- und arbeitswirtschaftliche Aspekte. [Covering sweet cherry hedges with polyethylene. I. Prerequisites of covering. II. Techniques of covering. III. Plantation and labour economy aspects.] *Schweizerische Zeitschrift für Obst- und Weinbau* 110 (25, 26, 27), 761–768, 778–794, 812–821.

Opperman, D. (1988) Cherry umbrella tried for protection against frost, wind and rain. *Goodfruit Grower* 39, 26–30.

Pennell, D. (1981) *Cherries*. ADAS Booklet 2372. Ministry of Agriculture, Fisheries and Food, UK.

Price, N.R. (1982) Methiocarb application to cherries. In: *Pesticide Science*, MAFF Reference Book 252, p. 36.

Rogers, J.G. (1980) Conditioned taste aversion: its role in bird damage control. In: Wright, E.N., Inglis, I.R. and Feare, C.J. (eds) *Bird Problems in Agriculture*. BCPC Monograph 23, pp. 173–179.

Slater, P.J.B. (1980) Bird behaviour and scaring by sounds. In: Wright, E.N., Inglis, I.R. and Feare, C.J. (eds) *Bird Problems in Agriculture*. BCPC Monograph 23, pp. 105–114.

Summers, R.W. (1985) The effect of scarers on the presence of starlings (*Sturnus vulgaris*) in cherry orchards. *Crop Protection* 14, 520–528.

Tahon, J. (1980) Attempts to control starlings at roosts using explosives. In: Wright, E.N., Inglis, I.R. and Feare, C.J. (eds) *Bird Problems in Agriculture*. BCPC Monograph 23, pp. 56–68.

Taylor, E.J. (1985) Starlings. In: *Mammal and Bird Pests 1983*. MAFF Reference Book 255 (83). HMSO, London, pp. 26–27.

Upshall, W.H. (1943) *Fruit Maturity and Quality*. Food Bulletin 447. Ontario Department of Agriculture.

Vaughan, J.A. (1982) Starlings. In: *Mammal and Bird Pests 1980*. MAFF Reference Book 255 (80). HMSO, London, pp. 55–58.

Way, R.D. (1968) Breeding for superior cherry cultivars in New York State. In: *ISHS Symposium on Cherries and Cherry Growing*, Bonn, Germany, pp. 121–137.

Westlake, G.E. (1981) Methiocarb application to protect ripening cherries. In: *Pesticide Science 1980*. MAFF Reference Book 252. HMSO, London, pp. 65–67.

Wright, E.N. (1980) Chemical repellents – a review. In: Wright, E.N., Inglis, I.R. and Feare, C.J. (eds) *Bird Problems in Agriculture*. BCPC Monograph 23, pp. 164–172.

Wright, E.N. and Brough, T. (1966) Bird damage to fruit. In: Synge, P.M. and Napier, E. (eds) *Fruit Present and Future*. RHS/Bles, pp. 168–180.

Zajac, R. (1983) Biosonics and pyroacoustics as protection means of large cherry orchards against starlings in Poland. *Fruit Science Reports* 10, 113–133.

V HARVESTING, HANDLING AND UTILIZATION

17 Harvest and Handling Sweet Cherries for the Fresh Market

N.E. Looney[1], A.D. Webster[2] and E.M. Kupferman[3]

[1]*Agriculture and Agri-Food Canada Research Centre, Summerland, BC V0H 1Z0, Canada;* [2]*Horticulture Research International, East Malling, West Malling, Kent ME19 6BJ, UK;* [3]*Washington State University, Tree Fruit Research and Extension Center, 1100 North Western Avenue, Wenatchee, WA 98801, USA*

This chapter describes and discusses the harvesting and handling of commercial volumes of sweet cherries, with stems attached, for the fresh market. It starts with a discussion about fruit maturation and ripening and the many components of fruit quality recognized by brokers, wholesalers and consumers. This is followed by a section dealing with the orchard environment and management practices that influence crop quality and uniformity. It covers the essential details of harvesting the crop and getting it to a packing-house without serious loss of quality.

A section on the theory and practice of removing field heat is followed by descriptions of various packing-house operations, such as sorting, grading and packaging. Finally, it deals with the protection of fruit quality during storage, shipping and marketing and discusses the postharvest diseases and disorders that afflict sweet cherry.

It is not meant to be a primer for direct marketing of cherries by individual producers, although many of the principles of preserving cherry quality are applicable. The commercial cherry packer faces unique time, distance and product quantity challenges that can only be met by careful planning and skilful execution, starting in the orchard and continuing to the moment of product consumption.

17.1 Sweet Cherry Maturation and Ripening

17.1.1 Fruit growth and the ripening process

Sweet cherry ripening occurs concomitantly with a rapid increase in fruit size and weight during the last few weeks before harvest (Table 17.1). As much as 25% of final fruit weight is added in the last week of growth prior to harvesting, and during this time there are dramatic changes in fruit colour, flavour and

Table 17.1 Relationship between fruit growth, softening and sugar accumulation in Royal Anne sweet cherries, Corvallis, Oregon, 1991. (Adapted from Barrett and Gonzalez, 1994.)

Weeks before commercial maturity	Mean fruit weight (g)	Flesh firmness (g)	Juice-soluble solids (%)
4	4.8	> 2000	5.2
3	6.9	1691	8.4
2	(8.9)*	675	9.0
1	10.5	506	10.6
0	12.3	301	14.3

* Estimated.

texture. Sugar concentrations increase as fruit ripen while acids, principally malic acid, remain relatively constant (Spayd *et al.*, 1986). Being a 'non-climacteric' fruit (Hartmann *et al.*, 1987), internal quality does not improve after harvest.

Sweet cherries may take varying times to ripen, i.e. change from first colour to full ripeness. Bulgarian research showed periods ranging from 15 to 24 days depending upon variety and season (Panova and Popov, 1983).

The importance of ethylene in sweet cherry development and ripening is still unclear but tissue ethylene levels and the levels of important ethylene precursors do rise during fruit ripening (Hartmann, 1989). This discovery has led to the suggestion that ethylene may be the natural trigger for the final fruit swell and ripening. The delay in fruit colouring and the improved firmness observed following gibberellic acid (GA₃) treatments (Looney and Lidster, 1980) are consistent with this suggestion. However, ethylene treatments *per se* do not greatly advance ripening of cherry, suggesting a low requirement for this bioregulator.

Research by Gucci *et al.* (1991) indicates that ripening cherries are strong sinks for the products of photosynthesis; fruit removal decreased leaf photosynthetic rate by 43% and increased leaf starch content by 59%, all within 24 h of harvest. Furthermore, the pattern of fruit sizing is such that about 25% of final fruit weight is accumulated during the last week before harvest. It would appear, therefore, that maintaining a healthy population of leaves and adequate soil moisture right through the harvest period may be especially important in relation to fruit growth and the development of good internal quality.

17.1.2 Colour development

Anthocyanin accumulation in the peel and flesh of black sweet cherries begins several weeks before harvest (Fig. 17.1). Colour intensity is the most commonly used indicator of ripeness. The colouring of black sweet cherries (e.g. Bing) progresses from straw colour to very light red, followed by red, which darkens to mahogany. Finally, the red/black colours brighten, at which time the colour loses its purity and anthocyanin synthesis slows down. This is thought to be associated with the beginning of anthocyanin breakdown, which is indicated by a progressive change to blue and then brown shades in the fruit.

The colour of black cherries is uniform on individual fruit but there can be quite a range of colour across the tree canopy. Black sweet cherries are at their

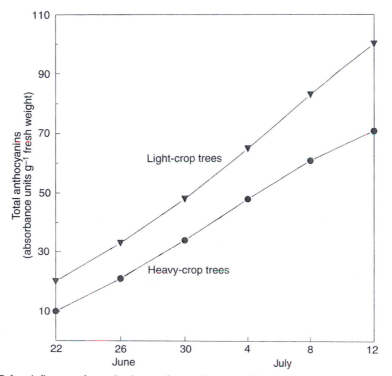

Fig. 17.1. Influence of crop load on anthocyanin accumulation in Bing sweet cherries approaching commercial harvest maturity. Redrawn from Spayd *et al.* (1986) with permission, averaging their data for 1982 and 1983 collected at Prosser, Washington State.

peak for commercial harvest when they turn a light to medium mahogany colour.

For a more in-depth discussion of cherry anthocyanins, the reader is referred to Chapter 19 in this volume.

Yellow sweet cherries (e.g. Royal Anne/Napoleon or Rainier) develop yellow flesh and skin as one of the first signs of maturity. Some cultivars are at their peak for commercial harvest when they develop a red blush on the cheek.

17.1.3 Flavour development

Fruit flavour is defined by sugar content (sweetness), acidity (tartness) and aroma. While the balance of sugars and acids is often as important as the actual amounts of either component of taste, Cliff *et al.* (1995) showed that sweetness was of primary importance to panellists ranking a number of sweet cherry cultivars for flavour. Drake and Fellman (1987) showed that stored sweet cherries lose acidity far faster than soluble sugars. The decline in flavour during modified-atmosphere packaging (MAP) observed by Meheriuk *et al.* (1995) was attributable primarily to acid loss.

Drake and Fellman (1987) also found that, within a single tree, yellow sweet cherries (cv. Rainier) containing 16% soluble solids were much firmer than those with 18%, but sensory tests showed that the former had poorer appearance and taste. Since soluble solids content was closely related to fruit colour and firmness throughout the season, the suggestion was that soluble solids content is a useful indicator of optimal harvest maturity for this cultivar.

The changes in sweetness and acidity after harvest are influenced by harvest maturity, with the decline in acidity faster in fruit picked at full maturity. Sugars also degrade more rapidly in fruit picked at full maturity, even though this fruit is initially the sweetest and most pleasant to eat.

Sweet cherries are rich in sugars, with 8–18 g 100 g^{-1} fresh weight being common (Table 17.2). The final value depends upon variety, stage of ripeness and growing conditions (Lichou *et al.*, 1990). The principal sugars are glucose and fructose with more of the former. The cherry is almost devoid of sucrose. The content of sugars in the juice is best measured using a refractometer and values of 10 to over 20% are common.

Total acidity is generally between 8 and 10 mequiv 100 g^{-1} of fresh material (Table 17.2) with a strong proportion of malic acid (around 85%). Citric acid is also important. The sweet cherry is, in fact, a moderately acidic fruit, with a juice pH of about 4.0.

Compared with many other fruit species, sweet cherries appear to be very low producers of aromatic compounds. Using headspace gas analysis techniques, Mattheis *et al.* (1992) detected more than 20 aromatic compounds from fresh Bing cherries but there was no clear quantitative relationship to fruit growth or colour development. Furthermore, none of the volatile compounds could be detected by sensory analysis (J.P. Mattheis, personal communication).

Table 17.2. Representative compositional values for sweet cherry fruit (amounts 100 g^{-1} of fresh fruit). (Adapted from Lichou *et al.*, 1990.)

Principal components		Mineral elements		Vitamins	
Water	80–86%	Potassium	200–250 mg	C	4–6 mg
Sugars	8–18 g	Calcium	8–19 mg	B_1	$50–65 \times 10^{-3}$ mg
Total acidity	8–10 mequiv	Magnesium	12 mg	B_2	$16–65 \times 10^{-3}$ mg
Lipids	0.2–0.8 g	Iron	0.4 mg	PP	$0.2–0.5 \times 10^{-3}$ mg
Proteins	0.9–1.0 g	Copper	0.14 mg	A	150–500 IU
Calories	60–80	Phosphorus	21–31 mg		

17.1.4 Accumulation of vitamins and minerals

The vitamins in mature sweet cherry fruit include vitamin C (around 5 mg 100 g^{-1} fresh weight) and very small quantities of vitamins B_1, B_2, PP and A (see Table 17.2). The cherry is rich in potassium and, among temperate fruits, is one of the richest in magnesium.

17.1.5 Fruit softening

The flesh of sweet cherries softens steadily over the month preceding commercial harvest (see Table 17.1), with the rate of change closely associated with changes in fruit volume (Barrett and Gonzalez, 1994). However, there appear to be great seasonal differences in rate of softening on the tree (Spayd *et al.*, 1986) and cultivars may differ considerably both in intrinsic texture (firmness, crispness) and in softening pattern (Brown and Bourne, 1988). Fruit approaching harvest maturity reach a more stable level of flesh firmness and, typical of a non-climacteric fruit, there is no precipitous loss of firmness after harvest so long as the fruit remain free of decay organisms.

17.2 Judging Fresh Fruit Quality

The sum of the processes described above determines fruit quality at the time of harvest. The following section discusses the various components used by the trade to evaluate the quality of harvested fruit, keeping in mind that careful attention to postharvest handling cannot improve fruit quality, only slow its decline.

It is also important to realize that fruit quality assessment is often highly subjective. Growers often consider all of their fruit to be of high quality and yet the packer who must remove cull fruit may refuse a shipment due to the cost of sorting. The expectations of wholesalers change over the course of a season (i.e. first early cherries are often judged less stringently). The retailer's view of quality may shift, depending on the amount of fruit in the market stream.

Judging cherry quality in the market-place is complicated by the fact that cherries are sold with stems attached. Thus the packer/shipper must try to preserve the quality of a fleshy fruit and a vegetative structure (the stem). Cherry fruit is evaluated visually for size, colour and lustre. Firmness and flavour are also very important. Fruit must be free from disorders and diseases such as brown discoloration, internal browning, decay, pitting and bruising. In addition, stems are evaluated for colour and turgor. In North America, long stems are valued by those markets accustomed to Bing as the quality standard among black sweet cherries.

17.2.1 Uniformity

Product uniformity is an important consideration when judging fresh fruit quality. For example, a box of cherries with a wide range of colour intensities will be less attractive than a more uniform box, even though all cherries are within the acceptable range for fruit colour.

17.2.2 Fruit size

Large fruit size is strongly preferred by most buyers. Larger fruit have greater visual appeal and often have better flavour. Since pit (stone) size is relatively constant, larger cherries have proportionally more flesh.

The commercial importance of this component of fruit quality is indicated by the fact that cherries are often packed and sold on the basis of 'row size'.

Table 17.3. Sizes of cherries commercially packed in Washington State.

Sales designation	Size (inches)	Size (mm)	Weight (g)*
9 Row	75/64	29.8	10.4+
10 Row	67/64	26.6	8.7–10.4
11 Row	61/64	24.2	7.1–8.6
12 Row	56/64	21.4	5.4–7.0
13 Row	52/64	20.6	4.2–5.3

* As cherries are sold by size, not weight, these figures are approximate.

This is the number of uniform cherries which, when packed side by side, fit across a 10.5-inch (276-mm) container (Table 17.3).

17.2.3 Fruit colour and lustre

The market expectation is that black sweet cherries are at their peak when they reach a light to medium mahogany colour and yellow cherries are at their peak when they develop a red blush on the cheek. The colour of black sweet cherries is judged commercially through the use of colour comparators or cards. Black cherries should be shiny, not dull in appearance. Unfortunately, this component of fruit quality, usually referred to as 'lustre', is very difficult to judge quantitatively.

17.2.4 Firmness

Firmness is one of the most important attributes of cherry quality, the market perception being that the best cherries are firm overall and the flesh is crisp. Unfortunately, the standard strain gauges tend to measure skin strength rather than flesh firmness and the data obtained are often of questionable value (Proebsting and Murphey, 1987). Furthermore, the close proximity of the pit to the epidermis makes the use of conventional penetrometers problematic at best. A constraint gauge developed in France measures the force necessary to crush the fruit a certain percentage of its equatorial width (Planton, 1992). This instrument shows promise as a tool for researchers but time will tell if it will find a place in industry.

In fact, it is still not fully established if any of these mechanical measures of firmness accurately predict the human sensory perception of what is good sweet cherry texture. However, it is likely that testing flesh firmness after removing the skin provides a better estimate of consumer acceptance than the standard penetrometer procedure. For example, Brown and Bourne (1988) found that Van sweet cherry, commonly considered to be a firm cherry, registered lower penetrometer values than other cultivars perceived to be softer. The explanation was that some soft-textured cherries register high skin strength. Flesh firmness as a proportion of total firmness proved to be high for Van.

Given these difficulties, the common approach to judging firmness is to eat a few cherries and squeeze a few others! Since 'in line' sorting of cherries for firmness has yet to be developed, it seems inevitable that some soft cherries will end up in the packaged product.

17.2.5 Flavour

Consumers strongly prefer cherries that taste sweet (Dever *et al.*, 1995), and sweetness is roughly equivalent to percentage soluble solids determined using a juice refractometer. Soluble solids content varies with fruit maturity, growing region, season and crop load. The cultivar is also important. For example, the soluble solids content of Bing cherries in an average Washington State crop may average 19–20%, whereas Lambert will be 18–19% and Van about 15%. The soluble solids level of Rainier cherries is often comparable to that of Bing.

It would be an advantage to the shipper if cherries could be sorted for sugar content. Research by Cho *et al.* (1993) suggests that low-resolution pulsed magnetic resonance could provide the theoretical basis for a non-destructive measure of juice-soluble solids.

Part of the characteristic flavour of black sweet cherries is determined by total acidity and the sugar–acid balance. Sugar–acid balance can be measured objectively but is usually assessed subjectively by industry. Total acidity of Bing cherries growing in western North America ranges from 0.25 to 0.75% acid and averages about 0.60% (8.9 mequiv 100 g^{-1} fresh weight).

17.2.6 Disorders

Unfortunately, sweet cherries are subject to an array of fruit and stem disorders that can seriously detract from their worth in the market-place. In this section we will discuss those defects usually apparent in freshly harvested fruit. A later section (17.7) will deal with those diseases and disorders, such as pitting and stem shrivel, that develop in or after storage.

External browning

External brown discoloration is a disorder which occurs following damage to the fruit cuticle by leaves moved by the wind or by the fruit rubbing against a branch. Cuticular damage can accelerate dehydration and softening of the fruit. This defect is apparent on freshly harvested fruit, especially yellow cherries, and afflicted fruit are removed during sorting.

Bruising

Cherries are very susceptible to injury by compression or impact or a combination of both. Compression injury, which mainly occurs on the tree and during harvest (Kupferman, 1994), results in a bruise that appears several days to several weeks after the actual injury. The evidence of bruising is a flat area on the fruit surface and soft tissue beneath that can be detected by feel. While bruises are quite apparent on yellow-skinned cherries, they may or may not be visible on black cherries. To the greatest extent possible, bruised fruit should be removed during sorting.

Cracking

Fruit cracking is usually associated with preharvest rain events. Splits and cracks in the fruit skin develop around the stem bowl, on the shoulder or at the stylar end of the fruit. Cracked cherries should be removed during the sorting process.

However, sorters may accept or overlook minor cracks and these cracks may increase in size during storage and shipment. Cracked fruit is very susceptible to fungal infection during storage, transport and marketing.

Stem damage

Problems with stems include shrivel and browning. Some of the deterioration in stem quality occurs even before the fruit reach the packing-house if cooling is excessively delayed or the fruit is subjected to very drying conditions. Mechanical damage to stems during picking results in rapid moisture loss and speeds browning.

In summary, cherry quality judged by those in the market chain is a function of 'initial' quality and a lack of damaging experiences thereafter. Quality is more than positive attributes – it is also the absence of diseases and disorders.

17.3 Orchard Factors Influencing Fruit Quality and Product Uniformity

Even under the best of circumstances the fruit on any given tree can vary quite a lot in stage of ripeness. In Europe it is not unusual for trees to be harvested two or even three times, but the common practice in North America is a single harvest. One notable exception is the multiple harvesting of Rainier cherries for speciality markets demanding a high degree of red blush on each fruit.

Since postharvest sorting and handling can go only so far toward achieving a uniform pack, it is important to consider the orchard management practices that can improve product quality and uniformity.

17.3.1 Within tree variability

The greatest contributor to poor product uniformity in black sweet cherry packs seems to be fruit colour. Proebsting and Murphey (1987) examined sweet cherry orchards at 15 locations in Washington State and used a multiple harvest strategy (weekly over 3 weeks) to assess the potential of 'colour picking' or delaying harvest to improve this situation. Interestingly, they learned that between the first and third harvest the coefficient of variation (CV) for fruit colour did decrease substantially, but the fruit was no less variable for soluble solids content and size and became even more variable for firmness. They concluded that, while product appearance might improve as a result of 'colour picking', overall quality would not improve substantially.

Fruit maturity and quality indicators such as soluble solids, colour, size and firmness are highly influenced by the position of the fruit on the tree (Drake and Fellman, 1987). That in the tree interior is less mature than that from the top or the periphery of the canopy. Furthermore, these locational differences persist throughout the harvesting period. This suggests that smaller tree canopies and pruning and training strategies that maximize light penetration into the canopy should reduce within-tree variability in fruit ripening.

Other Washington State research (Patten *et al.*, 1986) has demonstrated the importance of flower position on fruit size and quality. Flowers borne on spurs (found on 2-year-old or older branch sections) bloom later and produce fruit that is poorer in quality than fruit borne on 1-year wood. In work done at East Malling, fruit formed on axillary fruit buds on 1-year wood was higher in soluble solids than fruit on spurs on older wood (Webster and Shepherd, 1984). Since large tree canopies are more likely to have proportionally more spurs, this probably explains the canopy location effect noted by Drake and Fellman (1987).

17.3.2 Crop load

Fruit set influences crop load directly by determining fruit number and indirectly by its effect on fruit size at harvest. Crop load also influences internal fruit quality (see Fig. 17.1 for evidence of the effect on colour). Since fruit set can vary considerably from tree to tree and especially from season to season, it has been stated that fluctuating crop level is a major contributor to inconsistent fruit quality of sweet cherry.

The observation by Gucci *et al.* (1991) that ripening fruit is very demanding of photosynthate may explain why excessive cropping (i.e. a low leaf-to-fruit ratio) can have a serious effect on fruit size and sugar content (Facteau and Rowe, 1979; Roper and Loescher, 1987; Fig. 17.2).

Furthermore, fruit firmness and susceptibility to bruising may be influenced by crop load. For example, Spayd *et al.* (1986) showed that fruit from heavily cropping trees was less firm at any given tissue anthocyanin concentration than fruit from lightly cropping trees. It was also more susceptible to bruising at most harvest dates. On the other hand, Looney (1989) found that removing every second cluster 1 month after bloom of heavy-cropping Stella cherry trees (a self-fertile cultivar) improved average fruit weight, juice-soluble solids and titratable acidity but only marginally improved fruit 'hardness' estimated with a multi-purpose durometer. Furthermore, fruit thinning had no effect on impact-induced pitting in this experiment.

Unfortunately, achieving a moderate and consistent cropping level is a difficult task for sweet cherry producers and is seen as one of the industry's greatest challenges. Important components are tree age, frost protection, pollination management, mineral nutrition, irrigation, pruning, cultivars and rootstocks.

Growth regulators to enhance flowering and fruit set may be on the horizon but chemical thinning has yet to be successfully developed for sweet cherry. Interestingly, Looney (1989) found that the beneficial effects of GA_3 on fruit size, firmness and poststorage pitting appear to be independent of crop load.

17.3.3 Tree size and vigour

As is the case with all fruit trees, cherry trees become relatively less productive as they age and fruit quality declines because of internal shading and declining tree vigour. Fruit size and internal quality are seldom problems on young cherry trees.

Another potential adverse effect of large tree size on fruit quality is the

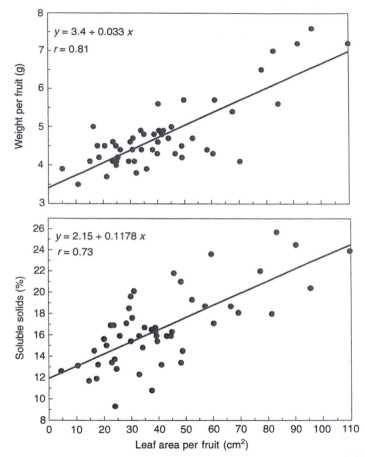

Fig. 17.2. Relationship between leaf area and fruit quality of Bing sweet cherry in Washington State. Redrawn, with permission, from Roper and Loescher (1987).

increased bruising that occurs when pickers, working from tall ladders, are unable to remove fruit as carefully as is desirable.

While tree size of some deciduous fruit tree species is controlled by use of dwarfing rootstocks and/or interstocks, suitable dwarfing rootstocks for sweet cherries have only recently been identified. Over the next few decades it will be necessary to determine the positive or negative effects of these new rootstocks on fruit quality. As an example of this, recent research in Norway (Cline *et al.*, 1995) showed that rootstock can influence the water and sugar content of several varieties of sweet cherry and A.D. Webster (East Malling, UK, unpublished) showed effects on fruit calcium content. Time will tell if there is reason to be concerned about adverse rootstock effects on fruit quality of sweet cherry.

Summer pruning as a technique to control tree growth and improve fruit quality has been explored by various researchers. Webster and Shepherd (1984) tipped the extension shoots of Van cherry when they reached 30 cm (mid- to late June) and observed increased juice-soluble solids in 2 years of a 3-year study. Fruit size was increased in one of those years. Earlier tipping affected neither fruit size nor sugar accumulation. Similar work by Looney (1989) showed that summer pruning could improve fruit firmness of Stella cherry without adverse effects on size, soluble solids, acidity or postharvest pitting. Overall, however, evidence to date suggests that the 'in season' benefits of summer pruning are rather modest. On the other hand, summer pruning may prove to be an important practice if the long-term effect is reduced tree size.

17.3.4 Climate and weather

Cherries are very sensitive to both rain and heat. Wind can also reduce fruit quality in the weeks leading up to harvest.

Rain occurring when the fruit is approaching harvest maturity is a major cause of crop loss and fruit quality deterioration in virtually all sweet cherry-growing regions. Cracking may occur in the stem bowl where free moisture remains, or in the case of heavy rains, on the cheek or calyx end of the fruit. Only the smallest of cracks are tolerated by grade inspectors.

Young trees with large leaves seem less prone to this problem, perhaps because the leaves intercept much of the rain and much of the crop is on 1-year wood. Although cultivars do vary quite a lot in relative resistance, true genetic resistance to rain-induced cracking has yet to be identified.

Various approaches have been used by cherry growers to reduce the risk of rain cracking. Where tree size can be controlled with growth regulators or with a custom-designed pruning and training system, tree covers are a practical possibility and are being evaluated in many countries (Opperman, 1988). Another common practice is to use helicopters or empty air-blast sprayers to 'dry' the fruit after a rain event, keeping in mind that fruit temperature and the length of contact with water interact to determine the injury level. Fortunately, postharvest application of cold water (i.e. hydrocooling) does not cause cherry cracking.

If rain cracking is prevalent in a block, the grower and shipper must decide whether to proceed with harvest. If rain is expected, it may be prudent to advance the harvest of unprotected blocks a day or two ahead of optimal maturity.

A second major climatic factor influencing fruit quality is heat. Temperatures above about 38°C, especially in desert climates, cause the cherry to 'wilt' as moisture is pulled from the cherry to the leaves by transpiration. Water relations within the tree and soil determine the extent of this phenomenon. In regions subject to this problem, growers are advised to confine harvesting to the early morning hours. They must also provide the trees with sufficient moisture so that the fruit can recover overnight.

The adverse effects of wind on fruit approaching harvest maturity include fruit abrasions that lead to moisture loss and skin discoloration. Windy conditions also intensify the temperature-related wilting described above. Windy conditions during bloom can reduce fruit set and thus contribute to the fruit quality problems associated with irregular cropping (see above). The best approach to reducing the wind hazard is to choose a protected site for the orchard. Windbreaks can also prove very beneficial.

17.4 Harvesting Sweet Cherries

17.4.1 Time of harvest

As discussed earlier (see Section 17.2), the various components of fruit quality develop and decline somewhat independently of one another. Consequently, deciding when to commence harvest always involves an element of compromise. Typically, time is of the essence if optimal quality and/or the greatest economic return is to be realized. Here are some examples of regional 'wisdom' relating to time of harvest of sweet cherries for fresh consumption.

Israeli experience is that fruit harvested early in the normal harvest period develops less decay in storage and transit. Washington growers are advised to harvest black sweet cherries at the mahogany stage of ripeness. Individual fruit darker than that are too subject to softening and the development of pitting after cold storage (Kupferman and Miller, 1990). Those harvested earlier are small and poor in flavour. Conversely, Rainier cherries intended for the 'carriage' trade are harvested several times to select those fruit very high in sugar content and showing the pink blush associated with advanced maturity. Such fruit must be handled very carefully to prevent marking.

17.4.2 Harvest and preharvest operations

Meli (1985) noted that, of the 1647 working hours needed in a season for the management of 1 ha of hedgerow sweet cherries in Switzerland, 90% was used to harvest the 10 tonne ha^{-1} crop. Average picking performance was 8 kg h^{-1}.

Clearly, harvesting sweet cherries for the fresh market is very labour- and management-intensive. It involves mobilizing pickers and equipment, arranging transportation, and coordination with the packing-house to ensure that fruit is received with the least possible delay.

Crop load must be considered when planning the harvest operation. Not only does it influence the labour requirement, but it can also, as already discussed, have a pronounced effect on fruit quality. Thus, while crop load is frequently beyond the control of the grower, recognition of this relationship to fruit quality will permit producers and fruit handlers to take steps to minimize potential problems.

One example would be to harvest, as quickly and carefully as possible, all fruit when it reaches the stage of maturity known to be optimal for the intended market.

Preharvest treatments

Cherries treated about 3 weeks before harvest with GA$_3$ (see Chapter 11) will be firmer, sweeter, larger and less likely to develop poststorage 'pitting' (Looney, 1989). GA$_3$ also delays fruit colouring, which means that harvest will usually commence a few days later than in untreated blocks. This bioregulator is widely used by Washington State and British Columbia growers, where earliness in the market-place is not a major consideration. Its effects on harvest time and product durability make it an important tool for the producer planning and executing the harvest operation.

Preharvest applications of fungicides are also effective for improving the quality of the crop and are especially important in those countries prohibiting postharvest applications of fungicides. Brown rot and powdery mildew are annual concerns in many districts. Where brown rot is particularly aggressive, a blossom-time spray is needed to reduce latent infections. A later spray (sometimes applied with GA$_3$) will prevent infection of cracked or otherwise injured fruit and thus prevent a build-up of inoculum in the orchard. In British Columbia, a preharvest spray with Benlate® and/or Captan® is recommended to protect the fruit during storage and marketing.

Fruit harvest

The most important component of a successful harvest operation is removal of the crop without excessive bruising or marking. Reliable pickers, appropriately trained and paid by the hour, will take the time to work carefully. They can also be asked to harvest and reject fruit with obvious defects, especially important in field-packing situations. Experienced pickers know they have to grasp stems rather than fruit and remove clusters from the limb with an upward movement. The aim is to leave the fruiting spur intact. Spur damage reduces future yields and may lead to disease entry where disease pressures are severe.

Cherries are damaged if dropped more than 15 cm. Pickers in Washington State use kidney-shaped buckets with shoulder straps. Picking buckets are made of light-weight aluminium or rigid plastic and the bottom of the bucket is padded to reduce bruising. Once the bucket is full, the cherries are gently poured into a plastic-lined wooden bin which holds about 200 kg of fruit when full. These bins are collected frequently and delivered to a point where the fruit is cooled. In Europe, fruit is usually picked into smaller containers to reduce bruising. It is increasingly common for fruit to be picked ('field-packed') directly into the baskets or boxes that will go to market.

Once picked, it is important to keep the fruit in the shade while awaiting transfer to the hydrocooler or packing-house. Studies have shown that fruit in bins or plastic punnets held in the shade lose far less moisture and gain temperature far less rapidly than those in direct sunlight (Fig. 17.3) (Micke and Mitchell, 1972; Wagner *et al.*, 1977). Black cherries rapidly absorb heat if directly exposed to the sun. It is also important to shelter the fruit from wind to help reduce desiccation. If shade is not available, insulated bin covers or wet foam can be used. Some growers apply cool well-water to reduce the temperature of harvested cherries awaiting removal to the packing-house or staging area.

17.5 Transport and Cooling

17.5.1 Transporting fruit to the packing-house

Depending on the distance to the packing-house, it may or may not be necessary to remove field heat in the orchard. If orchard cooling is unavailable, the fruit

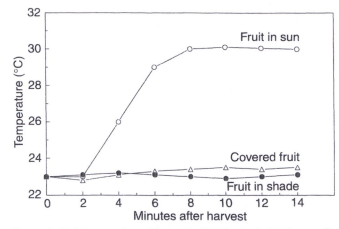

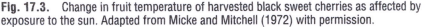

Fig. 17.3. Change in fruit temperature of harvested black sweet cherries as affected by exposure to the sun. Adapted from Micke and Mitchell (1972) with permission.

must be moved to where it can be cooled within 2–4 h of harvest. In any case, it is important to protect the fruit from excessive heating and desiccation during transport. This can be accomplished by using an enclosed truck bed (refrigerated if possible) or by covering the bins with wet burlap or other bin covers.

In Washington State, major cherry packing-houses have installed receiving areas in the more isolated growing districts to which growers bring binned cherries. These areas often have a hydrocooler and refrigerated truck-loading facility. The cherries are then brought to the packing-house by refrigerated truck.

Cherry fruit can be damaged during transport if the roadways are rough or the speed of travel is such that the bins are shaken excessively.

17.5.2 The importance of cooling

Many of the problems experienced by sweet cherry packers and shippers are quite profoundly influenced by the cooling practices adopted. These include softening and physiological breakdown; fungal rots; fruit dehydration and loss of 'lustre'; stem browning and loss of turgidity; and a progressive loss of flavour. Most of these processes are two to three times faster at 20°C than at 10°C but maximum product life is achieved when the fruit is held constantly near 0°C.

Depending somewhat on the intended disposal of the fruit, it is important that the delay between harvest and initial cooling should be as short as possible. Furthermore, the speed at which the cherries are cooled should be as fast as possible while taking care not to cause excessive desiccation or stem freezing (see below).

Decay organisms thrive at higher temperatures and it is inevitable that there will be abundant inocula and many fungal entry points in any large lot of unsorted cherries. See Section 17.7 for a discussion of the major fungal diseases of harvested sweet cherries.

Cherries, unlike apples or pears, have no starch reserves to provide a substrate for respiration. Thus, respiration relies primarily upon sugars and, perhaps as a consequence, cherries have a much higher rate of respiration than apples. The

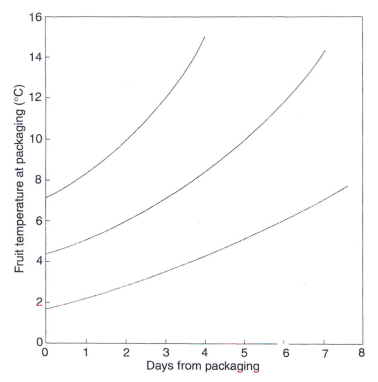

Fig. 17.4. Estimated temperature increase due to heat of respiration of packaged and palletized cherries in relation to initial fruit temperature. This figure assumes that the fruit is held in a cool room with negligible heat transfer between the fruit and the room air. Redrawn from Waelti (1986) with permission.

energy that is released by tissue respiration, the 'heat of respiration', must be removed to prevent a steady rise in fruit temperature. For example, if the fruit is packaged at 2°C, heat of respiration alone will cause the temperature to increase 2°C in a 5-day period (Fig. 17.4). If the initial temperature is 8°C, the heat of respiration will cause the temperature to rise to more than 16°C over the same time period. Shipping cherries at their lowest safe temperature ensures that the rate of respiration will be at its lowest.

European research indicates that cultivars may differ quite considerably in respiration rate. For example, cultivars with good storage performance, such as Oratovskii's Bigarreau and Bigarreau Francis, show a lower rate of respiration in storage than Drogan's Yellow, a poor-storing cultivar. Interestingly, respiration rate appears to be positively correlated with sugar levels in various cultivars (Sekse, 1988).

17.5.3 Temperature optima

Fruit that will be marketed and consumed immediately after harvesting may not require the same attention to cooling as fruit that must be shipped long

Table 17.4. Advice given to French shippers on shipping sweet cherries.

Days from harvest to sale	Holding temperature (0°C)
1–2	8–12
4–6	4–8
6+	0–4
8+	0

distances. For example, see Table 17.4 for the advice given to French shippers with respect to this time–temperature relationship.

In those producing regions serving distant markets, the generally accepted temperature management practice is to remove field heat as rapidly as possible and then hold the temperature near 0°C throughout the handling, storing and shipping operations. However, research examining the relationship between fruit temperature and impact injury during postharvest handling shows that warm fruit is less susceptible to the kind of injury that results in pitting during and after cold storage (Patten *et al.*, 1983; Crisosto *et al.*, 1993). However, it is yet to be determined if there is a critical fruit temperature above which such injury is substantially reduced and, if so, if cultivars differ with respect to this relationship.

When fruit can be sorted and sized immediately upon arrival at the packing-house (preferably having been partially cooled in the orchard) and then hydro-cooled just before packing, this may be the best possible approach to reducing impact-related disorders. However, holding warm fruit for even a few hours will probably negate any advantage of this approach due to the substantial reduction in product life.

17.5.4 Cooling technology

The following descriptions of the cooling technology used in cherry packing-houses are not sufficiently detailed for ready implementation. Thus, readers are referred to refrigeration texts such as *Industrial Refrigeration* (Stoecker, 1988) and *Forced Air Cooling* (Watkins and Ledger, 1990).

Hydrocooling

Hydrocooling is the most effective method available to the cherry packer to remove heat from the fruit. It involves total immersion or showering the commodity with cold water. In Washington State, a shower-type hydrocooler is used to remove the field heat from cherries in bins when they arrive at the packing-house. Bins are stacked three high and placed on a chain belt to be moved through the hydrocooler. Industry surveys indicate that a 11°C reduction in fruit temperature will occur over 7–10 min as the fruit move through a hydrocooling unit 6 m long. The current practice is to use either chlorine or chlorine dioxide to reduce the spore load in the water. Fungicides are not applied with the hydrocooling water.

A second hydrocooler is often located just ahead of the box filler. Such 'in-line' hydrocoolers are of two types, either drenching through a tunnel (immersion type) or showering the fruit on a conveyor belt with cold water. Both methods give very rapid cooling and no dehydration of the fruit. The immersion type uses conveyor belts to move the fruit under the cold water, where it remains for 7–8 min. In California, the cherries move into the hydrocooler on a porous belt and cold water showers on to the fruit at a rate of 40–60 l min^{-1} 0.1 m^{-2}. Shower-type hydrocoolers cool more rapidly than the immersion type, but may cause fruit damage if the drop height is too great.

It has been reported that hydrocooling can increase the rate of softening and the incidence of fruit rotting if the fruit absorb too much water. The assertion that hydrocooling is not yet sufficiently developed for some French cultivars (Lichou *et al.*, 1990) is consistent with the belief that varieties vary considerably in their suitability for hydrocooling. In Washington State, hydrocooling of Bing black cherry is much more common than with Van and Lambert, cultivars believed to be less firm. Rainier, a yellow cherry, is seldom hydrocooled in Washington State.

Overall, however, hydrocooling has clear advantages over forced-air cooling in that the rate of heat removal is faster and fruit and stem desiccation is minimized. Italian researchers compared hydrocooled and air-cooled cherries of four varieties and found the hydrocooled fruit of all varieties to be of much better quality following simulated transport of the fruit (Sozzi, 1979).

Forced-air cooling

Forced-air cooling (pressure cooling) is a rapid, low-cost method of cooling fruit (Waelti, 1986). Often cherries are sorted and packed on one day and then forced-air-cooled overnight and shipped the next morning. Figure 17.5 illustrates the temperature reduction that can be obtained with an effective forced-air cooling system compared with normal room cooling.

The principle is to create a pressure gradient across two sides of a stack of fruit which causes air to flow through the space in and around the boxes (Fig. 17.6). A fan pulls cold air (as cold as −3°C) from the room through the passages between and within the boxes into the inside of the stacks and out into the room. A number of engineering factors must be considered, including fan capacity, pressure gradient, air passages, channel length and temperature differential.

Forced-air cooling using humidified air is also an option. In this instance, the air is not cooled directly by an evaporator in direct contact with a refrigerant fluid but on an air/water exchanger, where it is humidified. The cold air is distributed from a central point containing iced water, often equipped with a store of ice. The system then distributes the cooled humidified air around the chamber(s). Performance of these systems is similar to that of the conventional forced-air system but has the advantage of reducing fruit and stem dehydration.

Some European cherry packers use prerefrigeration forced-air cooling chambers as an alternative to hydrocooling. These chambers are small in size

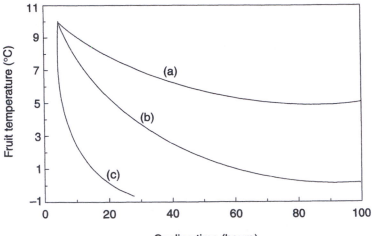

Fig. 17.5. Rate of cooling of packaged cherries as influenced by cooling method and stacking technique: (a) palletized boxes with liners, room cooling; (b) palletized boxes with liners, forced-air cooling; and (c) boxes stacked one box wide, forced-air cooling. Adapted from Waelti (1986) with permission.

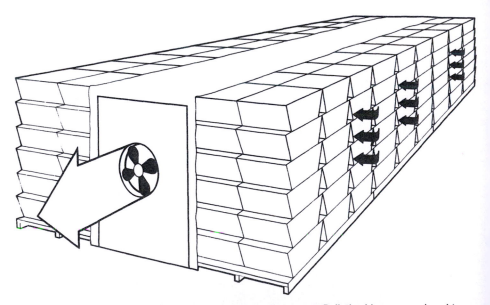

Fig. 17.6. Diagrammatic view of a forced-air cooling setup. Palletized boxes are placed to form a tunnel from which air is exhausted. Negative pressure pulls cold air through passages between stacked boxes. From Waelti (1986) with permission.

but have high refrigeration capacity and high air velocity. Speed of cooling in these chambers is proportional to air speed within the normal range of 2–5 m s^{-1}.

Room cooling

For cherries awaiting packing-house handling or for cold-packed fruit awaiting shipment, room cooling is often the operative technology. In California, for example, fruit arriving from the orchard, rather than being hydrocooled, is placed into such a room for cooling.

The holding room will often have water on the floor to increase humidity. The air temperature should be held under 2°C. Ideally, cherries should be held at a lower temperature, but ice on the floor causes accidents.

A refrigeration unit cools the air in relatively large storage rooms and fans effect the heat transfer. Room cooling is far less efficient at removing heat from the product than is forced-air cooling (see Fig. 17.5). None the less, high-quality room cooling is very important because even that fruit subjected to hydrocooling or forced-air cooling is subjected to this environment for most of its postharvest life.

'Cold chain' is the term sometimes used to describe constant refrigeration of a product. Cherries require the use of cold-chain systems if they are to remain a quality product. Cool-chain conditions must also be maintained throughout transport and distribution. One problem which can arise is that fruit may develop condensation when taken out of cold storage. Thus, it is advisable to warm fruit containers slowly in air with relatively low humidity.

17.5.5 Humidity management

Vapour pressure deficit is a term used to describe a difference in humidity across a membrane. It plays a very strong role in determining the rate of dehydration of cherry fruit and stems. This becomes apparent when one considers that cherries contain more than 80% water and, compared with many other fruits, stemmed cherries have a high surface-to-volume ratio.

The moisture-holding capacity of ambient air is very dependent on air temperature and sweet cherries are harvested and transported in midsummer when air temperatures are high and relative humidity (RH) values tend to be low. Thus, when cherries are warm and the surrounding air is warm and dry, water vapour rapidly moves from the fruit to the air. It is not surprising that fruit dehydration occurs most rapidly during the first 8 h after harvest, when both the fruit and air are warm (Micke and Mitchell, 1972).

However, as suggested above, dehydration also occurs during cooling and throughout the postharvest period of handling and storage. Thus, it is important to maintain RH as high as possible in the cooling air.

Placing warm fruit in cold rooms with very low coil-surface temperatures (e.g. -8 to -10°C) creates very favourable conditions for moisture loss by the fruit because under such conditions the air is excessively dry. The potential damage becomes even more acute as air velocity increases, such as in forced-air cooling.

Polyethylene liners in individual boxes and polyethylene wraps around pallets are used to prevent water loss. At the same time, however, they seriously interfere with heat transfer (see Fig. 17.5). Thus, it is very important that the fruit is fully chilled before being boxed or wrapped.

Australian research aimed at improving humidity management has shown that shelf-life and quality of Blackboy and Lambert cherries were greatly improved by storage at 95–99% RH (Sharkey and Peggie, 1984). Storage life at 0°C was extended 7–10 days and both fruit and stems remained fresher in appearance. This regime also resulted in a considerable improvement in firmness, flavour, juiciness and soluble solids. Where adequate ventilation was maintained, there was no increased loss due to decay.

17.6 Packing-house Operations

The role of the packing-house is to sort, grade, package and store the fruit so that the product sent to the market is the best that it can be, given the quality of the fruit received. If the producer and the packing-house have done their jobs well, the packaged fruit will remain in good condition throughout the marketing chain.

17.6.1 Handling, sorting and grading

The sorting, sizing and grading of sweet cherries usually consist of emptying the field containers and removing leaves, cutting the fruit stalks, eliminating under-sized fruit, hand-sorting, grading (may include sizing) and boxing. The boxes are then unitized on a pallet for cold storage or immediate shipping.

Transferring cherries on to the packing line
Recent improvements in cherry-handling practices have led to a greater use of water handling. Some packing-houses have hydraulic devices which grasp and invert the field container to gradually pour the fruit into a tub of chlorinated water. Some lines have a bin cover which pulls back to allow the cherries to slide into the water. A slanted link belt with cross-pieces (flighted conveyor) pulls the cherries out of the water up to the leaf eliminator.

Leaf eliminator
Cherries are dropped off the belt on to rods and carried over a fan, which blows up through the rods. The leaves which were carried in with the fruit are blown up and off the line.

Cluster cutter and small fruit eliminator
Fresh-market cherries are sold on the basis of size and, to size cherries mechanically, even just to eliminate small fruit, the clusters must be reduced to individual fruit. This is done by moving the cherries on a belt under revolving saws which cut the clusters apart. The fruit is lubricated with sprays of fresh water and protected from the saws by finger-like shields. Research has shown that this can be the most injurious part of the packing line. Increasing cherry volume and slowing belt speed can reduce damage. Damage shows up as bruising and eventual pitting.

The single cherries then move downwards through the small fruit eliminator, which consists of diverging rollers coated with hard rubber. The entire system is lubricated with fresh water. The smallest cherries drop through the top of the inclined rollers and are sold for processing. The remaining cherries pass over this device and are moved on to the sorting table in flumes containing chilled and chlorinated water. These flumes or pipes are commonly 100–150 cm in diameter. Chlorine concentration is usually within the range of 30–50 mg l^{-1} (Kupferman, 1992).

Sorting

Cherries sold on the fresh market are hand-sorted, each worker being responsible for the fruit passing by on a 60-cm-wide belt. Workers are instructed to remove fruit that is blemished, misshapen or lighter or darker than optimum. The remaining cherries fall off the end of the belt into a water flume which carries them to the in-line hydrocooler or they are placed directly into boxes.

Where rain-cracking is prevalent or other aspects of fruit condition or appearance are problematic, the sorting belts must be slowed. Since packing-house workers are paid by the hour, sorting can become very expensive.

Efficient sorting requires adequate light but does not become more efficient at light intensities greater than 50 foot-candles (4.6 lux) (Kupferman, 1991). Brown (1991) also deals with lighting systems and other important environmental considerations in the manual sorting of sweet cherries. He points out the benefits of sorting dark cherries on a black background, where browning of red or mahogany-coloured cherries is more easily detected.

A second hydrocooling

In many cases the major amount of cooling of cherries after harvest is accomplished by an 'in-line' hydrocooler located after the sorting operation. This second hydrocooler reduces fruit temperature by 10°C or more immediately prior to the fruit being sized and boxed. This is the most efficient place to lower the fruit to the desired shipping temperature.

Fruit sizing

If the warehouse policy is to sell fruit by size, rather than above a specific minimum size, the fruit is sized after the second hydrocooler. Belts or flumes receive the cherries as they fall through diverging rollers or sizing holes and carry them to the box fillers.

It is a challenge to size cherries because of the prominent stem and because they are irregular in shape. Most sweet cherries are heart-shaped with a flat side at the suture. Roller sizing sorts fruit according to its smallest dimension. The opposite situation occurs where fruit drop through holes of a given diameter.

Box filling and grading

Boxes are hand-filled by being held under a belt filled with cherries. In most packing lines the boxes then pass over a vibrating point in the belt to settle the cherries. Each box is weighed and the weight adjusted if necessary.

Washington State packers routinely use polyethylene liners and an absorbent pad in each box. The pad absorbs free moisture so that a pool of water cannot form in the bottom of the box.

It is at this point in the operation that the fruit is inspected and accorded a grade. Since standards vary from country to country, specific grade standards will not be addressed here. Suffice it to say that when assigning a grade the inspector considers freshness (fruit and stems), flavour and firmness (characteristic of the cultivar) and freedom from insects, diseases, defects and disorders. All of the fruit in the box must be of the same cultivar and origin and display adequate and reasonably uniform maturity. Fruit size above a minimum standard is an important consideration in the pricing of cherries but is not normally a grade consideration.

17.6.2 Packaging

Three types of fibre boxes are currently used by the Washington State cherry industry. Single-piece 'pyramid' boxes are the most common but double-sided boxes and two-piece shoe boxes are also important. The pyramid shape, an air space above the fruit and vent holes in the box are all designed to facilitate cooling. Expanded polystyrene boxes have been used for air transport to distant markets. Wooden boxes are also available and have the advantage of ease of cooling.

In the US Pacific Northwest cherries are sent to market in units of 12 or 20 lbs (5.4 or 9.0 kg). The California standard is an 18 lb (8.1 kg) box. An even wider choice of units can be found in Europe. In Italy, for example, cherries are packed into 5 kg boxes comprised of two 2.5 kg compartments, or into trays containing multiples of 500 g punnets. The North American equivalent of punnets is consumer packs using clam-shell plastic containers or poly bags. Irrespective of the package chosen, a unitizing step is necessary for efficient storage and shipping. In Washington State, 99 boxes of 9 kg are placed on each pallet and they are glued and strapped for stability. This operation normally occurs in a cold room.

Modified-atmosphere packaging

The development of plastic packaging materials with gas exchange properties that minimize the likelihood of anaerobic respiration promises to move MAP of sweet cherries to commercial prominence (Chapon and Bony, 1990). Research in British Columbia with Lapins sweet cherry has shown that, after 6 weeks of 0°C MAP storage, brightness and firmness started to decline but stems remained fresh and green for up to 10 weeks (Meheriuk et al., 1995). Sensory analysis indicated that overall appearance and flavour declined steadily with each week in storage but overall acceptability values remained satisfactory for at least 4 weeks. The decline in acceptability appeared to be related to a reduction in acidity during the storage period, since juice-soluble solids and texture did not change appreciably.

The atmosphere in the 38 μm low-density polyethylene bags equilibrated after about 2 weeks to 0.8% oxygen (O_2) and 4.5% carbon dioxide (CO_2). This was a much lower O_2 concentration than was expected but it was not until 8 weeks of storage that 'off' flavours were detected. These may have been attributable to decay organisms as well as anaerobic fermentation (Meheriuk *et al.*, 1995).

Commercial experience will quickly sort out the strengths and weaknesses of this technology. For example, shelf-life advantages (after the fruit is removed from MAP conditions) have yet to be demonstrated. Another concern is the possibility that the packaged fruit will be subjected to various temperatures through the market chain and that the atmosphere could become damaging at some point. Plastics that are more permeable to O_2 will lessen that risk.

17.7 Postharvest Diseases and Disorders

The perishable nature of ripe sweet cherries relates, in large measure, to their susceptibility to fungal diseases and to senescence-related disorders. Unfortunately, fruit arriving at the market to be judged by brokers, wholesalers, retailers and consumers may display diseases and disorders that were not apparent when the fruit left the packing-house. An assortment of 'poststorage' problems are dealt with in this section.

17.7.1 Postharvest diseases and their control

Disease problems are very different from location to location, so it is important that local advice is obtained relating to incidence and control. For this reason, no specific fungicides will be mentioned in the following paragraphs. For more detailed information about the postharvest diseases of cherry, the reader is referred to Hall and Scott (1989), Snowdon (1990), Ogawa and English (1991) and to Chapter 14 in this volume.

*Brown rot (*Monilia fructicola *and* Monilia laxa*)*

Brown rot is an important storage disease of sweet cherry in most production regions, Washington State being a notable exception. It seems to be more of a problem in wet seasons and there may be an association between the amount of brown rot in the orchard at bloom time and the seriousness of the disease in harvested fruit. Since bloom-time infections come from mummified fruit on the orchard floor, orchard sanitation is an important component of the control of this disease.

The symptoms appear as dark brown circular spots that spread rapidly over the fruit. The affected tissues remain relatively firm and dry. Buff-coloured masses of conidia are produced on the rotted area.

Fungicide sprays applied in the weeks leading up to harvest will protect the fruit from this disease and, where permitted, postharvest fungicide dips provide even better protection. Dipping fruit for 2 min in heated water (50°C) has also been used to reduce brown rot infections. Good temperature management and elevated CO_2 levels will further reduce the incidence of brown rot in stored fruit (De Vries-Paterson *et al.*, 1991; Brash *et al.*, 1992).

Grey mould (Botrytis cinera)

Grey mould is considered by many to be the worst disease problem affecting harvested cherries (Ceponis *et al.*, 1987). Although infection usually takes place in the orchard, damage during harvest and packing may also contribute to postharvest infections. Fruit harvested beyond optimal maturity is more susceptible to developing grey mould.

Grey mould rot first appears as a light brown spot on the skin. As the fungus grows, the underlying flesh becomes watery and dark brown. Under moist conditions an abundant white growth of fungus may cover affected fruit.

This disease has been controlled with fungicides applied as dip or drench treatments at the packing-house but preharvest sprays should also prove effective.

Blue mould (Penicillium expansum)

Opportunistic fungi such as *Penicillium* spp. may also enter fruit through rain cracks or other skin abrasions, eventually causing serious problems to packed fruit. It first appears as a circular, flat, light brown area. The affected tissue is soft and watery. As the rot develops, the skin cracks to reveal small white tufts of mould. Under humid conditions the mould grows, producing a crop of bluish green spores. Susceptibility to blue mould increases as the fruit ages.

Fungicides applied as postharvest dips have provided effective control of this disease.

Rhizopus *rot (Rhizopus spp.)*

This fungus can cause a serious postharvest disease of sweet cherry. Infection usually occurs after harvest and enters the fruit through cuts or cracks. It is recognized by the profuse development of coarse white mould strands which give rise to globular spore heads (sporangia) which eventually turn black. The fungus will not develop below about 8°C, so good refrigeration is an important part of its control, as is good packing-house sanitation. A fungicide spray applied during sorting has provided control of *Rhizopus* rot.

Alternaria rot (Alternaria spp.)

The affected area appears as a spot on the surface of the fruit and may be covered with olive-green spores and white strands of mycelium. The rotted tissue can be readily separated from the surrounding flesh. This fungus is widespread in nature and enters the fruit through cracks or cuts. Suppression has been reported with fungicide sprays or dips but there is no effective chemical control. Fruit held at or below 0°C in a high-CO_2 atmosphere is less likely to develop *Alternaria* rot.

Cladosporium rot (Cladosporium herbarium)

This fungus is also widespread in orchards but enters the fruit only through cracks or other wounds. The decayed tissue is hard, dry and grey to black and the rotted area is easily separated from the flesh. Careful handling, removal of damaged fruit and rapid cooling are the only effective tools for controlling this disease.

Other fungi have also been identified as causing rots of harvested cherries.

For instance, five species of *Aspergillus* were isolated from cherries in Indian fruit markets (Badyal and Sumbali, 1990).

17.7.2 Physiological disorders

Low-temperature injury

Cherries subjected to excessively low temperatures after harvest can develop generalized internal browning. Internal fruit temperatures of about $-2.0°C$ can result in freezing and tissue browning. Stems will freeze and brown at -1.0 to $-1.5°C$.

Cool storage breakdown

This term is used here to describe a range of symptoms that appear on fruit that has been held in cold storage for an excessive period of time. It differs from the symptoms of low-temperature injury described above (freeze damage). The predominant symptoms of cool storage breakdown of cherries are flesh softening and skin darkening, but shrivelling of whole fruit is also observed in overstored fruit (Hall and Scott, 1989). Some of the symptoms of this disorder may not be apparent while the fruit is cold, but once removed from storage the fruit deteriorates very quickly.

Pitting

Pitting is a dimpling of the cherry skin that becomes apparent after a period of low-temperature storage (Porritt *et al.*, 1971; Wade and Bain, 1980). Figure 17.7 illustrates this disorder, which results from impact injury during handling, mainly during various packing-house operations (Kupferman, 1994). It occurs when a relatively few damaged cells in the fruit cortex collapse and dehydrate. Because this injury cannot be detected before the fruit is boxed and stored, pitting is a very serious problem for shippers relying on cold storage to access distant markets. It is a cause for many claims and disappointments.

Cultivars differ substantially in susceptibility to pitting. Van, for example, is quite susceptible to this disorder and has been the cultivar used in many investigations.

The relationship between fruit temperature and susceptibility to impact injury has already been mentioned (Section 17.5.3). It seems that the temperature of the cherry fruit at the time of impact affects its ability to resist damage (Patten *et al.*, 1983; Crisosto *et al.*, 1993). Warmer cherries develop fewer pits than colder cherries when subjected to the same forces.

Australian work with Ron's Seedling and St Margaret showed that storage at temperatures near 0°C or the transfer of fruit from cool storage to room temperature worsened the disorder. Low-O_2, high-CO_2 and high-humidity atmospheres did not influence the development of pitting, although hypobaric (low-pressure) storage was beneficial (Wade and Bain, 1980).

The possibility that preharvest sprays or postharvest dips in calcium salts (e.g. $CaCl_2$) might reduce the incidence or seriousness of poststorage pitting has also been explored. Some results have been encouraging (e.g. Lidster *et al.*, 1979) but there has been little commercial interest in either approach. A prevalent

Fig. 17.7. Van sweet cherries showing the symptoms of impact-induced pitting.

concern that calcium sprays may reduce fruit size could explain the lack of interest in preharvest treatments.

Oregon work by Facteau and Rowe (1979) showed that the Lambert cherries least likely to develop pitting were those that were larger and higher in soluble solids. Firmer fruit was also less likely to develop the disorder.

17.8 Storage and Shipping

The importance of chilling and cold storage in determining the commercial life of harvested sweet cherries has been emphasized throughout. This section will deal with the practices used by packing-houses and shippers to keep packaged cherries in good condition until they reach the consumer.

17.8.1 Cold storage

Palletized cherries are placed in cold rooms for additional cooling. Forced-air cooling using makeshift cooling tunnels will help to bring down the fruit temperature (see Section 17.5.4). Some operators use high-velocity air flow rooms set at −1°C to rapidly reduce temperatures. See Section 17.5.4 for a discussion of room cooling of sweet cherries.

17.8.2 Controlled-atmosphere (CA) storage

There is now a considerable body of evidence showing that storage life of sweet cherries can be extended by the use of CA storage. Cherries are more tolerant of high CO_2 levels than most temperate-zone fruits. Levels as high as 40% are tolerated and 20% CO_2 has been used in commercial practice. When combined with good temperature management, high levels of CO_2 help to reduce decay and retain firmness, acidity, soluble solids and fruit colour (Patterson, 1982).

Chen *et al.* (1981) stored Bing cherries at CA atmospheres featuring either high CO_2 or low O_2. Fruit stored at low O_2 levels (0.5–2.0%) maintained a higher percentage of very green stems, brighter fruit colour and higher levels of titratable acids. High CO_2 (20% CO_2, 10% O_2) conserved fruit brightness and titratable acidity but did not prevent stem discoloration.

Overall, it appears that a storage life of about 8 weeks should be routinely achievable with CA storage technology. There are occasional reports of quality retention for as much as 3 months but it is unlikely that this fruit would retain its consumer appeal for more than a few hours when warmed to room temperature.

However, in commercial practice CA storage of sweet cherries appears to be less widely used than the encouraging research results would predict. One possible explanation for this is the desire to access the market as soon as possible after harvest. Another is the fact that it takes some time for sweet cherries to develop all of the symptoms of injury and disease infection that may have occurred during harvest and handling. These symptoms (e.g. pitting) appear during and after cold storage, even after high-quality storage. Thus, holding fruit for an extended period is always a gamble. One may succeed in preserving texture, flavour and product integrity, only to discover that overall appearance of the fruit is unacceptable because of pitting or shrivelled stems.

None the less, as cherry harvesting and handling practices improve, it is almost certain that there will be a steady increase in interest in the various storage and packaging techniques intended to extend the marketing season of sweet cherries.

17.8.3 Shipping sweet cherries

Most domestic cherry shipments in North America and many parts of Europe are by refrigerated truck. It is imperative that the truck be cooled when loaded and the refrigeration system be in perfect condition if the cherries are to arrive at their destination with good shelf-life.

International shipments often involve air freight. Refrigerated trucks deliver the fruit to an international airport, where they are loaded on to either passenger or cargo aircraft. This fruit is often subjected to less than optimum transit temperatures and may be delayed at the receiving airport awaiting inspection. Wrapping the pallets with insulating material helps to keep the fruit cool.

Intercontinental shipping of sweet cherries by boat is increasing in importance. Despite the long transit time, excellent on-board cold-storage facilities, coupled with good product management before and after shipping, have given very good results. CA containers and MAP will add another degree of sophistication to intercontinental shipping of fresh fruit.

17.9 Phytosanitary Issues in International Trade

One of the principal limitations on the export and import of cherries between various countries is the cherry fruit fly (*Rhagoletis cerasi*). Countries that do not have this pest are particularly keen to prevent its entry and enforce strict quarantine regulations over imported fruit. Japan is one of the strictest in enforcing such regulations.

New Zealand researcher Lay Yee (1989) compared six sweet cherry cultivars for their response to methyl bromide fumigation intended to destroy this pest (64 or 80 mg m^{-3} for 2 h at 12°C followed by time and temperature modulations designed to simulate transport to Japan). He found that cultivars differed significantly in their tolerance to this treatment, ranging from 80% (one outstanding experimental cultivar) to as low as 4% sound fruit following the full treatment regime. In all instances, methyl bromide fumigation significantly reduced fruit quality.

Codling moth (*Cydia pomonella*) is also of concern to some importing countries. Early research in the USA (Anthon *et al.*, 1975) indicated that fumigation of fruit with 32 mg m^{-3} methyl bromide for 2 h at 24°C guaranteed that shipments would be completely free of larvae of this pest. These treatments appeared to cause no deleterious effects on the quality or taste of the fruit.

References

Anthon, E.W., Moffitt, H.R., Couey, H.M. and Smith, L.O. (1975) Control of codling moth in harvested sweet cherries with methyl bromide and effects upon quality and taste of treated fruit. *Journal of Economic Entomology* 68, 524–526.

Badyal, K. and Sumbali, G. (1990) Aspergillus rot of cherry fruit. *Indian Journal of Mycology and Plant Pathology* 20, 280.

Barrett, D.M. and Gonzalez, C. (1994) Activity of softening enzymes during cherry maturation. *Journal of Food Science* 59, 574–577.

Brash, D.W., Cheah, L.H. and Hunt, A.W. (1992) Controlled atmospheres inhibit storage rot of cherries. In: *Proceedings of the 45th New Zealand Plant Protection Conference*, Wellington, New Zealand, August 1992. New Zealand Plant Protection Society, Wellington, pp. 136–137.

Brown, G.K. (1991) *Lighting for Manual Sorting of Apples and Sweet Cherries*. Paper No. 913553. American Society of Agricultural Engineers, St Joseph, Michigan, USA.

Brown, S.K. and Bourne, M.C. (1988) Assessments of components of fruit firmness in selected sweet cherry genotypes. *HortScience* 23, 902–904.

Ceponis, M.J., Cappellini, R.A. and Lightner, G.W. (1987) Disorders of sweet cherry and strawberry shipments to the New York market, 1972–1984. *Plant Disease* 71 (5), 472–475.

Chapon, J.F. and Bony, P. (1990) Cherries. The importance of modified atmospheres in maintaining quality. *Infos (Paris)* 62, 11–14.

Chen, P.M, Mellenthin, W.M., Kelly, S.B. and Facteau, T.J. (1981) Effects of low oxygen and temperature on quality retention of 'Bing' cherries during prolonged storage. *Journal of the American Society for Horticultural Science* 106, 533–535.

Cho, S.I., Stroshine, R.L., Baianu, I.C. and Krutz, G.W. (1993) Nondestructive sugar content measurements of intact fruit using spin–spin relaxation time (T_2) measurements by pulsed ^{1}H magnetic resonance. *Transactions of the ASAE* 36 (4), 1217–1221.

Cliff, M.A., Dever, M.C., Hall, J.W. and Girard, B. (1995) Development and evaluation of multiple regression models for prediction of cherry cultivar liking. *Food Research International* 28 (in press).

Cline, J.A., Sekse, L., Meland, M. and Webster, A.D. (1995) Rain-induced fruit cracking of sweet cherries. I. Influence of cultivar and rootstock on fruit water absorption. *Acta Agriculturae Scandinavica* Series B 45, 213–223.

Crisosto, C.H., Garner, D., Doyle, J. and Day, K.R. (1993) Relationship between respiration, bruising susceptibility and temperature in sweet cherries. *HortScience* 28 (2), 132–135.

De Vries-Paterson, R.M., Jones, A.L. and Cameron, A.C. (1991) Fungistatic effects of carbon dioxide in a package environment on the decay of Michigan sweet cherries by *Monilinia fructicola*. *Plant Disease* 75, 943–949.

Drake, S.R. and Fellman, J.K. (1987) Indicators of maturity and storage quality of 'Rainier' sweet cherry. *HortScience* 22, 283–285.

Facteau, T.J. and Rowe, K.E. (1979) Factors associated with surface pitting of sweet cherry. *Journal of the American Society for Horticultural Science* 104, 706–710.

Gucci, R., Petracek, P.D. and Flore, J.A. (1991) The effect of fruit harvest on the photosynthetic rate, starch content and chloroplast ultrastructure in leaves of *Prunus avium* L. *Advances in Horticultural Science* 5, 19–22.

Hall, E.G. and Scott, K.J. (1989) Stone fruit. In: Beattie, B.B., McGlasson, W.B. and Wade, N.L. (eds) *Postharvest Diseases of Horticultural Produce*, Vol. I. CSIRO and NSW Agriculture and Fisheries, Sydney, Australia, pp. 53–56.

Hartmann, C. (1989) Ethylene and ripening of a non-climacteric fruit: the cherry. *Acta Horticulturae* 258, 89–96.

Hartmann, C., Drovet, A. and Morin, F. (1987) Ethylene and ripening of apple, pear and cherry fruit. *Plant Physiology and Biochemistry France* 25, 505–512.

Kupferman, E.M. (1991) Cherry sorting table lighting. *American Society of Agricultural Engineers* Paper No. 913552, 9 pp.

Kupferman, E.M. (1992) Update on the use of chlorine. *Washington State University Tree Fruit Postharvest Journal* 3 (2), 12.

Kupferman, E.M. (1994) Cherry damage surveys in 1993. *Washington State University Tree Fruit Postharvest Journal* 5 (1), 17–19.

Kupferman, E.M. and Miller, K. (1990) Cherry warehouse survey shows room to improve. *Good Fruit Grower* 41 (6), 4, 45–47.

Lay Yee, M. (1989) New cherry variety outshines the rest. *Orchardist of New Zealand* 62, 30–31.

Lichou, J., Edin, M., Tronel, C. and Saunier, R. (1990) *Le Cerisier: la cerise de table*. CTIFL, Paris, France, 361 pp.

Lidster, P.D., Tung, M.A. and Yada, R.G. (1979) Effects of preharvest and postharvest calcium treatments on fruit calcium content and the susceptibility of 'Van' cherry to impact damage. *Journal of the American Society for Horticultural Science* 104, 790–793.

Looney, N.E. (1989) Effects of crop reduction, gibberellic acid sprays and summer pruning on vegetative growth, yield and quality of sweet cherries. In: Wright, C.J. (ed.) *Manipulation of Fruiting*. Butterworths, London, pp. 39–50.

Looney, N.E. and Lidster, P.D. (1980) Some growth regulator effects on fruit quality, mesocarp composition, and susceptibility to surface marking of sweet cherries. *Journal of the American Society for Horticultural Science* 105, 130–134.

Mattheis, J.P., Buchanan, D.A. and Fellman, J.K. (1992) Volatile compounds emitted by sweet cherries (*Prunus avium* cv. Bing) during fruit development and ripening. *Journal of Agriculture and Food Chemistry* 40, 471–474.

Meheriuk, M., Girard, B., Moyls, L., Beveridge, H.J.T., McKenzie, D.-L., Harrison, J., Weintraub, S. and Hocking, R. (1995) Modified atmosphere packaging of 'Lapins' sweet cherry. *Food Research International* 28, 239–244.

Meli, T. (1985) Ertrage und Kosten im Susskirschenanbau. [Yields and costs of sweet cherry cultivation.] *Schweizerische Zeitschrift für Obst und Weinbau* 10, 291–306; 12, 326–334; 13, 353–358.

Micke, W.C. and Mitchell, F.G. (1972) *Handling Sweet Cherries for the Fresh Market.* Circular 560. University of California, Davis, 18 pp.

Ogawa, J.M. and English, H. (1991) *Diseases of Temperate Zone Tree Fruit and Nut Crops.* Publication 3345. University of California, Davis, 461 pp.

Opperman, D. (1988) Cherry 'umbrella' tried for protection against frost, wind and rain. *Goodfruit Grower* 39 (4), 26–30.

Panova, R. and Popov, S. (1983) [Morphological and biochemical changes in the fruit of some sweet cherry varieties during ripening and storage under cold conditions]. *Nauchni Trudove Vissh Selskostopanski Institut Vasil Kolarov* 28 (2), 87–95.

Patten, K.D., Patterson, M.E and Kupferman, E.M. (1983) Reduction of surface pitting in sweet cherries. *Post Harvest Pomology Newsletter, Washington State University* 1 (2), 15–19.

Patten, K.D., Patterson, M.E and Proebsting, E.L. (1986) Factors accounting for the within-tree variation of fruit quality in sweet cherries. *Journal of the American Society for Horticultural Science* 111, 356–360.

Patterson, M.E. (1982) CA storage of cherries. In: Richardson, D.G. and Meheriuk, M. (eds) *Controlled Atmosphere for Storage and Transport of Perishable Agricultural Commodities.* Timber Press, Beaverton, Oregon, USA, pp. 149–154.

Planton, G. (1992) Fermenté des fruits et légumes. Des nouveaux outils de mesure. *Infos-CTIFL* 82, 27–28.

Porritt, S.W., Lopatecki, L.E. and Meheriuk, M. (1971) Surface pitting – a storage disorder of sweet cherries. *Candian Journal of Plant Science* 51, 409–414.

Proebsting, E.L. Jr and Murphey, A.S. (1987) Variability of fruit quality characteristics within sweet cherry trees in central Washington. *HortScience* 22, 227–230.

Roper, T.R and Loescher, W.H. (1987) Relationships between leaf area and fruit quality in Bing sweet cherry. *HortScience* 22, 1273–1278.

Sekse, L. (1988) Storage and storage potential of sweet cherries (*P. avium* L.) as related to respiration rate. *Acta Agriculturae Scandinavica* 38, 59–66.

Sharkey, P.J. and Peggie, I.D. (1984) Effects of high humidity storage on quality, decay and storage life of cherry, lemon and peach fruits. *Scientia Horticulturae* 23, 181–190.

Snowdon, A.L. (1990) *Post-Harvest Diseases and Disorders of Fruits and Vegetables,* Vol. 1. CRC Press, Boca Raton, 302 pp.

Sozzi, A. (1979) Precooling of cherries: influence of preservability and quality. International Institute of Refrigeration XVth International Congress of Refrigeration. *Bulletin de l'Institut International de Froid* 59, 1142–1143.

Spayd, S.E., Proebsting, E.L. and Hayrynen, L.D. (1986) Influence of crop load and maturity on quality and susceptibility to bruising of 'Bing' sweet cherries. *Journal of the American Society for Horticultural Science* 111, 678–682.

Stoecker, W.F. (1988) *Industrial Refridgeration.* Business News Publishing Co., Troy, Michigan.

Wade, N.L. and Bain, J.M. (1980) Physiological and anatomical studies of surface pitting of sweet cherry fruit in relation to bruising, chemical treatments and storage conditions. *Journal of Horticultural Science* 55, 375–384.

Waelti, H. (1986) Forced air cooling of cherries. *Postharvest Pomology Newsletter, Washington State University* 4 (1), 9–13.

Wagner, C., Schmid, P. and Hackel, H. (1977) Temperaturen und Qualitatsverluste von Susskirschen, die in Offenen und mit Folie verschlossenen schalchen der naturlichen Atmospharischen strahlung ausgesetzt Waren. *Erwerbsobstbau* 19 (11), 182–184.

Watkins, J.B. and Ledger, S. (1990) *Forced-air Cooling*, 2nd edn. Publication Q188027, Queensland Department of Primary Industries, Brisbane, 56 pp.

Webster, A.D. and Shepherd, U.M. (1984) The effects of summer shoot tipping and rootstock on the growth, floral bud production, yield and fruit quality of young sweet cherries. *Journal of Horticultural Science* 59, 175–182.

18 Harvesting and Handling Sour and Sweet Cherries for Processing

G.K. Brown[1] and G. Kollár[2]

[1]*United States Department of Agriculture, Agricultural Research Service, Fruit and Vegetable Harvesting Investigations, Agricultural Engineering Department, Michigan State University, East Lansing, MI 48824-1323, USA;* [2]*Department of Farm Management, University of Horticulture and Food Industry, PO Box 53, Budapest, Hungary*

18.1 Introduction

18.1.1 Sour cherries

The market for fresh sales of sour cherries is very limited in the USA and in western Europe. Nearly the entire production is utilized in some processed form (canned, frozen, juice, concentrate or dried). In middle Europe (Bulgaria, Czech and Slovak Republics, Hungary, Poland, Romania, Ukraine and some parts of the former Yugoslavia) most of the production is utilized in the local fresh market or is exported fresh to nearby countries (e.g. Germany, Austria).

The harvesting of sour cherries must be a timely operation in all production areas of the world. The fruit advance from a physiologically immature state to a fully mature state in just a few days. However, not all the fruit, either within the orchard or on a single tree, ripen on the same day. Ideally, the ripe fruit should be picked before it becomes overripe and the immature fruit should be left on the tree to increase in size. This approach would theoretically maximize the grower's yield per tree and provide the processor with large, firm, uniformly red, mature cherries for 'pitting' (removing the endocarp/stones) and packing. Unfortunately, today most growers cannot afford to approach their harvesting operation in this manner. Instead, the entire tree is harvested at one time when the fruit is considered to be mature.

There are about 18,000 ha bearing sour cherries in the USA, where the hand-picking labour requirements are high, in the range of 400 worker-hours ha^{-1}. To hand-harvest 1 ha day^{-1} would require about 40 workers, each devoting 10 h to picking. The average US grower has about 20 ha of sour cherries, and would, therefore, require nearly 50 workers, each picking 10 h day^{-1} for 16 days to complete the annual harvest. There are about 50,000 ha bearing sour cherries in middle Europe, and hand-picking labour requirements are in the range of 750 worker-hours ha^{-1}. In western Europe, much of the production is on dwarf-type trees, but the labour requirements are still in excess of 300 worker-hours ha^{-1}. Clearly, labour requirements are high if sour cherries are hand-picked.

Finding and managing large numbers of workers for only a few days each year is a task that most growers have now found to be impossible to accomplish. Added to this problem is the significant risk of crop loss due to unfavourable weather conditions during the harvest period, the high cost of supplying acceptable temporary housing for seasonal workers and the high cost of complying with all labour regulations, taxes, etc. These problems associated with hand-picking have become acute in all sour cherry-producing areas of the world.

US growers now rely completely on machines for harvesting their entire crop from trees that are older than 6 years. These machines use vibration or shaking to detach the fruit from the branches and stems, and various kinds of catching systems to collect the fruit as it falls from the tree. To accommodate these mechanical harvesting systems, certain horticultural, postharvest handling and processing plant practices must be followed so that both the grower and processor can experience profitable levels of income. Many of the mechanical harvesting systems used today require no more than 15 worker-hours ha^{-1}. Often, a grower can complete the entire 20 ha harvesting operation in only 7–10 days. Since one mechanical harvester is equal to many hand-pickers, its correct operation can reduce overall costs, maintain orchard productivity and maintain fruit quality. In contrast, incorrect operation of the mechanical harvester can result in financial disaster in any one season. Total costs kg^{-1} of fruit for mechanical harvesting typically ranged between 25% and 50% of the total hand-harvesting cost by the late 1960s (Bolen *et al.*, 1970) and are similar today. In Europe, mechanical harvesting typically costs only 10–30% as much as hand-harvesting (G. Kollár, unpublished).

Mechanical harvesting systems in the USA have been evolving since the late 1950s (Gaston *et al.*, 1959; Levin *et al.*, 1960). The first commercial mechanical harvesters were based upon limb shakers of various designs. To harvest a tree, the shaker was sequentially clamped on to each of the major scaffold limbs and shaken for 3–5 s at a frequency of about 800–1200 cycles min^{-1} and with a total movement of the limb (stroke) of 30–40 mm (Mitchell and Levin, 1969; Fridley, 1983). The recommended catching surfaces were soft so that bruising due to fruit impacting on hard surfaces was minimized.

The harvested cherries were immediately placed in a tank of cold water (< 15°C) on the harvester to begin the cooling and firming process and to minimize scald (browning) damage. Full tanks were moved to a cooling station near the orchard, where cold water was flushed through the tank for 30 min to reduce fruit temperature and clear the tank of debris. A soak time of 4–6 h, using a low flow rate of cold water, further reduced fruit temperature and caused some firming of the cherries so that the pitting process could proceed both efficiently and accurately. Pitting losses (decreases in product weight due to pit removal, plus some juice and flesh loss) usually averaged about 14–16%. However, excessive harvest and postharvest bruising, combined with a lack of adequate firming, could increase the pitting losses to as much as 28%. The fruits were kept in cold water until they were transferred on to the processing line, where, in less than 5 min, they were pitted and ready for freezing or canning. The aim was, and still is, to limit the total elapsed time from harvest to processing to a maximum of 9 h, so that fruit colour and scald, pitting loss and final fruit quality stayed within the desirable limits.

The trees had to be modified somewhat when changing from hand-harvesting to mechanical harvesting (Levin *et al.*, 1960). The number of main scaffold limbs was reduced to three or four, low-hanging branches in the way of the harvester and dead branches were removed, as were willowy 'hanger' growths or long slender shoot growths that would not shake well enough to detach the fruit. Most trees had been headed near the ground, a practice now discontinued. Such trees made harvester manoeuvring difficult, resulted in high fruit losses at the tree seal of the catching frame, and were not adaptable to trunk shakers.

The heavy pruning initially required in some orchards at the time of changing from hand- to mechanical harvesting, combined with chemical weed control under the trees and applications of nitrogen fertilizer, induced excessive tree vigour and resulted in soft, difficult-to-handle cherries (Mitchell and Levin, 1969). Leaf analysis in July was found to be a practical way of adjusting the fertilizer programme and thus the nutritional balance of the mature Montmorency trees so that the fruits were generally firm and fruit size was in the desirable range of 115–125 fruits 500 g^{-1}. Soft fruits still occur in some orchards in some years. The harvesters must be well cushioned to avoid severe impact bruising that will otherwise cause excessive losses for growers and processors.

In Europe, the adoption of mechanical harvesting systems occurred more slowly than in the USA. The problems of many small orchards (e.g. 2–8 ha per grower) in Western Europe or the lack of hard currency for the purchase of Western machines in Eastern Europe required special machinery development and management arrangements. In the 1960s a few large growers or cooperative farms bought Western harvesters. By the 1970s harvesters were being manufactured in Italy, Denmark, Germany and the Netherlands. By the mid-1980s the PSM, VSO and VUM systems were being produced in the Soviet Union, the Balkan systems in Bulgaria, the E-842 systems in Eastern Germany, and the TFH, VGF and VGR systems in Hungary. By the late 1980s more than 80 modern systems were in operation in Hungary. Experience has shown, as in the USA, that proper design, operation and maintenance of these machines are essential to maintain a healthy orchard, quality fruit and profitable conditions for the grower and processor. For more details on the European harvesting systems, the reader is referred to Kollár *et al.* (1981b), Andor *et al.* (1987) and Kollár (1987a).

In Europe, there are a number of sour cherry varieties that have very firm flesh which is strongly resistant to severe bruising by mechanical harvesting operations. These are described in Section 18.5. Since these varieties do not bruise or scald easily, they can be handled dry in bulk bins and held in cold storage, at 5°C, to maintain fruit quality prior to processing (Kollár and Kemenes, 1993). As in the USA, the trees must be trained and pruned to be compatible with the harvest systems (Pór and Faluba, 1982; Kollár, 1987b).

An industry-wide shift to the use of mechanical harvesting systems was necessary in the USA because the large hand-labour force from Mexico, supplied under Public Law 78 (the Bracero Program), was terminated in December 1964

Fig. 18.1. Two-half inclined-plane harvester. The left half carries the trunk shaker under the fabric inclined-plane catching surface, which diverts the fruit to the conveyor on the right half. The right half carries the water tank (rear), collector conveyor and a fabric inclined-plane catching surface (E.D. Kilby Manufacturing Inc., Gridley, California, USA).

Fig. 18.2. One-person full wrap-around harvester. The fabric catching surface (inverted umbrella) is wrapped under the tree by powered arms on the harvester. The trunk shaker is carried under the catching surface on the front of the harvester. The collector conveyors elevate the fruit from the low point of the catching surface to the water tanks on either side of the operator (OMC, Yuba City, California, USA).

(Levin *et al.*, 1969b). Since then, the harvesters, horticultural practices and postharvest handling practices have continued to evolve in response to the need for greater orchard productivity, higher fruit quality and reduced costs kg^{-1} of

Fig. 18.3. One-person half wrap-around harvester. The fabric catching surface (inverted umbrella) is wrapped under half of the tree by powered arms on the harvester. An inclined-plane catching surface is above the harvester. The scissors-type trunk shaker is carried midway at the side of the harvester, and extends out to clamp the trunk. A collector conveyor runs the full length of the harvester and delivers fruit to the water tank at the rear (E.D. Kilby Manufacturing Inc., Gridley, California, USA).

fruit. The mechanical harvesting systems now manufactured in the USA and Europe are of three basic types: (i) two-half inclined-plane harvesters (Fig. 18.1); (ii) one-person harvesters (either full wrap-around, half wrap-around or half roll-out) (Figs 18.2–18.4); and (iii) full roll-out harvesters (Figs 18.5 and 18.6). The present practices with respect to the use of modern mechanical harvesting and handling systems are summarized in the following sections.

18.1.2 Sweet cherries

In the 1960s, when the use of mechanical harvesting for sour cherries first took hold in the USA, approximately 100,000 tonnes of sweet cherries were also produced annually. Approximately 40% of these were sold fresh, while 40–50% were brined and the remainder canned. Hand-picking labour requirements ha^{-1} for sweet cherries are similar to those for sour cherries. Mechanical harvesting has never proved a viable option for cherries marketed fresh; too many bruised fruits and fruits without stems (pedicels) make the technique unprofitable. Mechanical harvesting of sweet cherries for brining or canning was, however, an industry goal. Except for fruit sold as cocktail cherries, which was and still is produced mainly in Oregon, stems were not required on brining or canning cherries, thus making chemically aided mechanical harvesting more feasible. Nevertheless, there were considerable initial problems experienced in mechanically harvesting sweet cherries. Not the least was the fact that traditionally

Fig 18.4. One-person half roll-out harvester. The roll-out fabric surfaces are extended from the harvester by powered arms. An inclined-plane catching surface is above the harvester. The tri-clamp trunk shaker is carried midway at the side of the harvester and extends out to clamp the trunk. A collector conveyor runs the full length of the harvester and delivers fruit to the water tank at the rear (note the safety chain) (Friday Tractor Co., Hartford, Michigan, USA).

Fig. 18.5. Full roll-out harvester (rear view). The shaker and carrier unit are on the right. The tractor pulled roll-out fabric catching surface, collector conveyor and water tank are on the left (OMC, Yuba City, California, USA).

Fig. 18.6. Full roll-out harvester (under-tree view). Scissors-type trunk shaker using round, particle-filled clamp pads. Cushioned ridges sewn into the fabric catching surface help convey the fruit up into the collector conveyor when the catching surface is pulled (by power) back on to the collector half (OMC, Yuba City, California, USA).

brining cherries were picked relatively immature. Successful mechanical harvesting of immature cherries was very difficult, needing high tree shaking forces and long shaking times to achieve fruit detachment, which result in severe fruit bruising. An abscission chemical is now applied preharvest when shake-harvesting most sweet cherry cultivars for processing. Also, varieties were found to differ greatly in their suitability for mechanical harvesting and, unlike sour cherries, sweet cherries handled in water frequently split before they could be processed. These problems needed considerable research effort before acceptable systems of mechanical harvesting were developed.

There are now about 20,000 ha bearing sweet cherries in the USA, producing about 150,000 tonnes. Approximately 45% of these are sold fresh, while 35% are brined and the remainder are canned. All of the fresh cherries are hand-picked, but the cherries for processing are harvested by shaking, using the same harvesters as for sour cherries (Gaston *et al.*, 1961; Norton *et al.*, 1962; Markwardt *et al.*, 1964; Stebbins *et al.*, 1967; Whittenberger *et al.*, 1968; Levin *et al.*, 1969a). The shaken sweet cherries are either handled dry in bulk bins for canning (Levin *et al.*, 1969a) or in SO_2 brine (e.g. 13,500 mg l^{-1}, pH 2.9) solution for brining (Whittenberger *et al.*, 1968). The cherries must be handled quickly from the orchard to the canning or brining facility so that oxidation in the bruised tissue is minimized. Brining immediately after harvest will maintain the highest quality.

The mechanically harvested sweet cherry trees were usually managed in ways similar to those used for sour cherry trees (Larson, 1969). Since sweet cherry trees are larger in size at full maturity, larger shakers and collecting surfaces are often required.

In Europe, approximately 170,000 tonnes of sweet cherries are produced annually. About half of this production is consumed fresh and half is processed (canned, frozen or distilled; brining is not used). The harvesting and handling practices are nearly identical to those used in the USA (Kollár *et al.*, 1981a). Appropriate tree training and pruning methods are essential to be compatible with the use of harvesters (Brunner, 1982; Mihályffy, 1982; Kollár, 1985).

In Italy and Hungary, experience has shown that consumers are willing to purchase certain kinds of carefully machine-harvested stemless cherries. These cherries are for local markets where they are both consumed fresh and processed at home. In these situations, the mechanical harvesting of sweet cherries for fresh market is fully profitable (Baldini, 1980; Kollár *et al.*, 1981a, b).

18.2 Commercial Harvesting Systems of the 1990s

18.2.1 Trunk and limb shakers

Most commercial harvesting systems now manufactured in the USA and other countries for sour cherry harvesting use a shaker that clamps to the tree trunk and shakes the entire tree at one time to accomplish fruit detachment (Figs 18.6 and 18.7). The shaker head assembly is heavy, typically weighing 600 kg. Most shakers rely on one or more unbalanced masses spinning about a vertical shaft

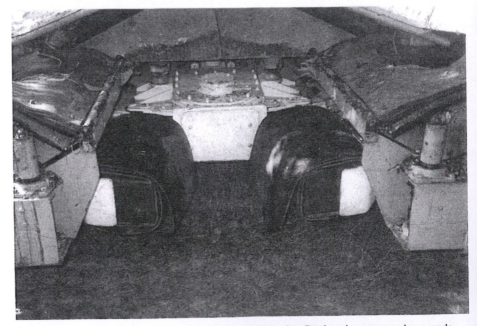

Fig. 18.7. Harvester-mounted scissors-type shaker using flat-faced neoprene clamp pads. Note the slip-belt assembly for holding the pads in place and limiting shear force transmission to the tree trunk (E.D. Kilby Manufacturing Inc., Gridley, California, USA).

to develop the shaking force (Fridley and Adrian, 1966). The shaking force can also be developed by an eccentric wheel (Andor *et al.*, 1987) or by a hydraulic piston. The shakers are hydraulically powered, so shaking frequency can easily be varied over a range that causes all of the fruit-bearing branches to shake enough to achieve fruit detachment. These shakers are usually operated over a frequency range of 700 to 1400 cycles min^{-1} (rpm) and move the trunk through a stroke of 16 to 12 mm, respectively, at the area of the shaker clamp (Fridley, 1983).

Some shakers have a single shaft about which two unbalanced masses spin at a fixed frequency ratio to develop the shaking force from a single hydraulic motor. Others have two shafts about which separate unbalanced masses spin at random ratios to develop the shaking force from separate hydraulic motors. In either case, the shaking action is multidirectional, so that the trunk is shaken in all directions to achieve quicker and more complete fruit detachment than is possible using only a single direction of shake (Fridley and Adrian, 1966; Esch *et al.*, 1989b). The amount of unbalanced mass can be increased or decreased in most shakers to accommodate the age of the trees or the differences between tree crops (i.e. cherry, apple, prune, citrus), but the eccentricity (distance between the centre of the unbalanced mass and the centre of the rotating shaft) may not be changeable. The peak input power to the shaker is typically in the range of 30–60 kW.

Many of the European harvesters are still designed using limb shakers instead of the trunk shaker. Older trees and orchards must be harvested in some areas, and limb shakers are often more effective as well as less damaging to these trees, even though fewer trees h^{-1} can be harvested. By manoeuvring the limb shaker, young trees can easily be trunk-shaken (trunk diameter up to 150 mm), and trees growing on very uneven ground conditions can be shaken. A limb shaker can also be attached at undamaged areas on the limbs or trunk. The input power to a cherry limb shaker typically ranges from 10 to 25 kW.

Trunk (bark) damage due to excessive shear, torsion and compression caused by the shaking action can be a problem if the shaker is operated too aggressively. A shake of only a few seconds (i.e. 2–5) is adequate to remove essentially all of the crop if the fruit is mature enough to harvest. All commercial trunk shakers must be clamped to the trunk, increased from zero to the desired shaking frequency and then decreased to zero before they can be unclamped and moved to the next tree. During the start-up and shut-down phases of shaker operation, resonance conditions occur between the shaker and the trunk causing large relative displacements (Affeldt *et al.*, 1989). These cause high shear and compressive stresses to be applied to the bark and cambium of the trunk, and the bark system can be damaged (Brown *et al.*, 1987). For this reason, the shaker should not be started and stopped more than once to complete the harvesting of each tree unless the tree is large, has a heavy crop and harvesting it in one 5-s shake would overload the conveying system of the fruit collector. Two designs for trunk shakers that use adjustable eccentricity to avoid the resonance problem were developed and are available (Affeldt *et al.*, 1989; Esch *et al.*, 1989a), but to date neither design has been commercialized.

18.2.2 The shaker clamp

The shaker is attached to the tree trunk by the use of a hydraulically operated clamp (see Figs 18.6 and 18.7). The clamps vary in design and use either two parallel clamping surfaces on opposite sides of the trunk or three surfaces that are arranged to surround the trunk in a triangle. The two-surface clamp is designed to operate like a scissors (scissors-type) or as a C (C-type) clamp. The three-surface clamp is designed as an equilateral triangle (tri-clamp), which can open at the point between the left and right sides (opposite the base) to receive the trunk and then close to clamp the trunk against the base of the triangle. The clamp pads (the media that provide the contact area on the trunk through which the shaking forces are transmitted from the shaker to the trunk) are made of either thick (~ 200 mm) neoprene rubber (see Fig. 18.7) or, as shown in Fig. 18.6, are made as a reinforced hollow tube filled with flowable particles (Frahm *et al.*, 1988). The pads should surround and grip the trunk so that it moves with the shaker during the shaking operation.

If the pads are too hard, the contact areas on the trunk will be small, causing excessive compressive stress to be applied to the bark and cambium by the clamping and shaking forces. The active phloem and cambium cells will be crushed. Excessive shear stress can also occur during shaking, which will split the bark and may tear it from the trunk. Such bark damage must be avoided in order to maintain a highly productive orchard with a normal lifetime (Schulte *et al.*, 1992). If the pads are too soft, the shaking action will not transmit effectively from the shaker to the tree. The neoprene pads should provide a flat surface about 200 mm high and 500 mm long to contact the trunk. The hardness of the neoprene should be 55 Durometer units for young trees (trees with 35–150 mm trunk diameter) and 65 Durometer units for older trees. The soft pad (55 Durometer) will avoid high compressive stress on small trunks that are easily damaged and yet provide good service life while efficiently transmitting the low shaking forces required by young trees (Brown *et al.*, 1988). The harder pad (65 Durometer) develops a large contact area on large trunks (compressive stress stays at a safe level) and yet will efficiently transmit the high shaking forces required for large trees and provide good service life (Frahm *et al.*, 1988).

The flowable-particle pad (diameter ~ 200 mm) cannot be measured for hardness by using a Durometer. They are 'soft' until tightly clamped to the trunk. The particles move in the tube during clamping to conform to the trunk. The tube must be nearly full for young trees and less full for old trees. It is important to follow the manufacturer's recommendations for particle fill, particle renewal and periodic pad rotation during use.

All pads are held in place on the shaker by a sling made of either smooth neoprene or nitrile-covered conveyor belting (Timm *et al.*, 1988), as shown in Figs 18.6 and 18.7. A second, loose, slip belt that contacts the trunk lies over the sling belt. The interface between these two belts is lubricated by either silicone spray or a food-grade grease. The lubricated interface provides a low-friction surface that can readily move when shear forces and displacements are developed by the shaker, and avoids the transmission of damaging shear stress to the bark and cambium. For young trees each belt should be thin (~ 3.5 mm thick), covered by nitrile and lubricated at the interface. For old trees each belt can be thicker (~7 mm thick), covered by neoprene and lubricated at the interface. The interface must be cleaned daily (to remove debris and cherry juice) and relubricated more frequently to ensure that the low-friction condition is maintained.

18.2.3 The shaker carrier

Only two methods are now used to carry, position, control and power the trunk shaker in the orchard. If a complete harvester and catching frame system is used, the shaker is mounted under the catching frame, as shown in Figs 18.1–18.4. In this case, the shaker is clamped on to the trunk after the catching frame is positioned at the tree, but before the catching surfaces are extended into collection position. If a full roll-out catching system is used, the shaker is carried on a separate vehicle and is clamped on to the trunk after the fabric catching surfaces (which often lie on the grass ground cover) are in collection position (see Figs 18.5 and 18.6).

When the shaker is mounted under the catching frame, it can be adjusted somewhat for desired shaking height and for horizontal distance between the trunk and the catching frame. Adjustment to make the angle of attachment 90° between the plane of the clamp and the vertical growing tree trunk (e.g. hillside alignment problems) is not possible. However, when the shaker is mounted on a separate vehicle, adjustments for distance, height and angle of attachment are readily made by the shaker operator. When cherries are grown on hilly or terraced sites the use of a two-half catching frame system may not be possible because neither the shaker nor collecting surfaces can be properly positioned at the tree. In such cases, either a one-person complete harvester or a full roll-out system should be used.

Limb shakers are always mounted above the catching frame on an inverted L-shaped pivot. Usually the mounting location will be at one end of the frame and on the outside corner (away from the tree). This provides room for the shaker operator to manoeuvre the shaker clamp on to each limb. When a pair of catching frames operate on two sides of the tree row, a limb shaker is often mounted on each frame (at opposite diagonal corners of the frames) to speed up the harvest operation. It is important that the number of main scaffold limbs be low (three to five) to maximize the trees harvested h^{-1} when using limb shakers. This also helps minimize bark damage on the scaffolds.

18.2.4 Fruit collection

The collecting surfaces used for catching the falling fruit must provide adequate cushioning for the fruit (Mitchell and Levin, 1969). Shaken fruit often falls 3 m or more to the collecting surface. Studies have shown that bruise damage from mechanical harvesting can be simulated by dropping the fruit twice from 2 m on to a hard surface. Cherries often split open from a single drop of 1.25 m on to a hard surface. A combination of washable fabric (e.g. neoprene on nylon) and foam rubber or plastic cushioning is used to provide cushioning for most harvesting systems. Cushioning should cover all structural parts of the harvester which the cherries would otherwise hit. Although the conveyors on most catching frames are not covered, they should be, because they are hard. The collecting surfaces on roll-out collector units should be made of light-weight flexible fabric to provide cushioning, and should be supported on grass or held in the air during fruit collection.

Bruise damage cannot be reversed. Bruised Montmorency sour cherries will scald if not quickly cooled in water (Levin *et al.*, 1960). Scald reduces the quality and value of the fruit. If bruising is so extensive that the flesh does not regain firmness during soaking in cold water, the pitting losses will be high and the accuracy of pit removal will be poor. In the USA, grade A processed pitted sour cherries should not contain more than one pit in each 1.134 kg of pitted fruit (USDA, 1964). However, a future industry goal is for not more than one pit in each 28.35 kg of pitted fruit. When good harvesting, handling and pitting practices are followed, the packed cherries now average about one pit in each 5.67 kg.

In Europe, several sour cherry varieties are more resistant to bruising and scald than is Montmorency. Cushioning is still required on the harvester, but the fruit can be handled dry in bins (see Sections 18.2.7 and 18.2.8).

The trunk seal design on the catching surface is important. It must be tight around the trunk to avoid losing fruit to the ground. If it sags, fruit will collect there, rather than roll to the conveyors, become damaged and contribute to low quality. An effective tree seal design is discussed by Peterson (1984) and Peterson and Kornecki (1990).

Cherries collected on catching frame or roll-out surfaces must be delivered to a conveyor and conveyed to the water-filled tank, dry bin or box. The proper selection of conveyor width and speed to handle the harvested fruit gently will help maintain fruit quality. Most conveyors are 400–600 mm wide. Some harvesters convey the collected fruit on a single conveyor to the handling container, while others must make one or more transfers involving a drop and 90° turn. The transfers must be gentle and avoid fruit pinching or compression; otherwise more bruising will occur.

Most harvesters use a fan to blow out leaves and light-weight debris as the fruit fall from the conveyor to the handling container. Remaining leaves and debris are skimmed from the top of the water-filled tank, using a screen or old tennis racket, before the full tank is sent to the cooling station. Debris in dry containers can be easily removed by hand.

18.2.5 Harvester operator training

The harvester operator is in control of an expensive machine that must operate in a timely, safe and efficient manner for a few days each year. If the operator has the proper knowledge of how the machine operates, how to harvest efficiently and maintain fruit quality, how to avoid trunk or limb damage, and how to recognize when something is wrong so that it can be corrected, the mechanical harvesting operation will be very successful. Proper operation will minimize cost for both the grower and the processor. Without proper operator knowledge, any mechanical harvesting operation can result in poor fruit recovery, poor fruit quality, trunk damage and reduced orchard life, equipment damage and high costs for both the grower and the processor. Proper operator training is critical to the success of mechanical harvesting operations, and is very important if the operator is only hired (does not own the orchard or harvester). In most cases, training can be obtained only by reading the operator's manual, talking with the dealers and talking with experienced owners or operators of similar harvesters.

18.2.6 Harvester maintenance

Although a mechanical harvester for sour cherries in the USA is used only a few days each year, its operation cannot be taken for granted or its maintenance ignored. The average grower with 20 ha of trees will need to cycle the harvester from tree to tree over 5000 times each year and operate 10–12 h day^{-1} for up to 10 days. Large growers may operate each harvester three times as long. Many US growers also harvest sweet cherries for processing using the same harvester. This spreads the investment cost over more tonnes of crops, but increases the need for timely maintenance. To avoid problems that will result in delays and higher costs at harvest time, the inspection and servicing of bearings, chain drives, hydraulic hoses, filters and fittings, hydraulic pumps, motors and cylinders, hydraulic valves and gauges, the collecting surface and fruit conveyors, the shaker clamp pads and slip belts, the power unit and the structural parts of the machine should be completed during the winter or several weeks before harvest.

In Europe, the same harvesters are used on several tree crops (e.g. sweet cherry, sour cherry, apricot, plum, apple, almond, chestnut) to improve profitability. The annual operation time may reach 350–500 h, so precise and regular maintenance is essential.

18.2.7 Handling containers

In the USA, to avoid scald damage resulting from bruising and to firm the cherries for efficient pitting at the processing plant, mechanically harvested cherries are placed into a tank of cold water on the harvester. These tanks, each weighing about 900 kg when full, are chained to the harvester for safety. A standardized and calibrated carbon-steel tank that can be handled by forklifts is used in the USA (Cargill *et al.*, 1970) and is shown in Fig. 18.8. The inside floor area of each tank is about 11,000 cm^2, so they hold 0.011 m^3 of cherries for each 1 cm of depth. Montmorency cherries in water have a density of about 754 kg m^{-3}, so each 1 cm of cherry depth represents 8.29 kg of dry cherries. A stainless-steel depth probe, graduated in mm units (with estimation to 0.1 mm) can be used to measure quickly the depth of cherries in water (Fig. 18.9) and convert this to the volume or weight of cherries for legal sale to processors (Tennes *et al.*, 1968; Levin *et al.*, 1970). Using this approach, the cherries do not need to be drained, transferred on to and handled across a scale – all operations that were found to be damaging, time-consuming and expensive. After the cherries have been flushed and hauled by truck for about 1 km, the probe can be used with accuracy.

The weight of sweet cherries in brine can also be measured using the depth

Fig. 18.8.   Line of full tanks being flushed on a cooling station. Note the drop hoses and cooling probes.

Fig. 18.9.   Cherry depth probe in a full tank of cherries. The bottom of the stainless-steel depth probe rests on the tank bottom, the adjustable circular plate rests on the top of the cherries. The depth of cherries is read from the vernier scale near the top of the plate extension (Tresco Inc., Spring Lake, Michigan, USA).

Fig. 18.10. Tank of cherries being dumped into the beginning of the processing operation by a rollover head on a forklift. The cherries are mechanically destemmed and sized (to remove small fruit and trash) before being elevated into the processing building.

probe, calibrated tanks and a bulk density of about 688 kg m^{-3} (Whittenberger *et al.*, 1969; Levin *et al.*, 1970).

The pallet tanks are usually emptied into water using a rollover head on a forklift, as shown in Fig. 18.10. The pallet base on the tank is designed to hold the tank on the forks in the inverted position. The base and top of the tank are designed to interlock for the safe stacking and transport of stacked tanks.

Many carbon-steel tanks are painted white to reflect solar energy. The paint must be food-grade and maintained periodically to prevent rust or paint flakes from becoming mixed in with the cherries. Paints that withstand prolonged submersion in water help minimize maintenance problems. Tank cleaning is a required operation each year prior to use.

In Europe, most shaken cherries are handled dry, either in bins that are up to 350 mm deep and hold 250 kg of fruit or in smaller lug boxes that hold 10 to 25 kg (used for the fresh market). These containers are set in cold storage within a few hours after harvest to cool the fruit prior to processing or packing for the fresh market.

18.2.8 Postharvest treatments

With mechanical harvesting of Montmorency sour cherries in the USA, a cooling station (see Fig. 18.8) is required on the farm. The station supplies clean cold well-water for each tank before it goes to the harvester and the cold water for flushing and cooling the cherries after harvest. The station has a sloping concrete floor that directs the flushing water away from the bins and provides a clean firm surface on which bins are set using orchard forklifts. The spent flushing water is used to irrigate the orchards, since it cannot be reused for cooling or be disposed of in nearby ditches or fields. The cooling station has an overhead set of water supply pipes that deliver water to each bin through a drop hose. The water flow rate to each tank should be 37.5 l min^{-1} for the first 30 min, followed by 7.5 l min^{-1} thereafter. The number of tanks on the cooling pad will determine the required flow rate needed from the well or other water supply. The cooling pad is usually designed to hold about two truck loads of tanks. This allows a full truck load to be cooled and ready for delivery to the processing plant, and additional space for continuing harvesting operations to place tanks that need to be cooled. A single mechanical harvester in a mature orchard will often fill a tank every 10 min. The timely pick-up of cooled tanks is a necessity to ensure good quality fruit, timely harvesting operations and efficient processing operations. A mutually agreed upon schedule for delivery of cherries to the processing plant helps to maintain high quality for the processor (Whittenberger and LaBelle, 1969).

The flushing and cooling water should be released into several locations on the tank floor so that hot spots in the fruit mass are avoided. A water probe delivery, much like the four legs on a square table, is used by many US growers (see Figs 18.8 and 18.9). It is made from standard plastic pipe and fittings. To minimize initial costs, the cooling station is usually located outdoors and near an existing well capable of producing at least 150–300 l min^{-1} of $\sim 10°C$ water harvester^{-1}.

When sour or sweet cherries are harvested into dry containers, they must be placed in humidified cold storage (0 to 6°C) as soon as possible. To maintain high quality, the holding time should not exceed 24 h, because after 48 h deterioration will be significant (Moser et al., 1977; Andor et al., 1987). At room temperature significant deterioration will begin after only 15 h. The cherries held dry in bins are usually intended for juice, purée or canned products. Cherries that are intended for brining are often handled dry from the orchards, but they are poured into vats or pits of brine solution (which removes the cherry colour and bruise colour) within a few hours of harvesting.

18.3 Optimum Time for Harvesting

Mechanical harvesting can recover nearly all of the crop from the trees when the force required to pull the average cherry from its stem (the fruit retention force, FRF) is less than 300–350 g (Mitchell and Levin, 1969).

To help the grower determine when to start harvesting, a random sample of 50–100 cherries should be hand-picked together with their stems (pedicels) from each orchard and tested with a 500 g pull-force gauge daily as the time for harvest approaches. The time required to complete the harvesting operation in any one orchard will depend primarily on orchard size (hectares and tree size) and the capacity of the harvesters (trees h^{-1}). Some parts of an orchard may mature a few days earlier than other parts. The harvesting operation should be planned so that the number of harvesting days fits the expected number of days from minimum to maximum fruit maturity.

In some countries, where the harvester is used on several types and varieties of fruit, determining when to schedule each harvest operation can become even more important. For example, in Hungary nearly 60 days can elapse between the first sweet cherry harvest and the last sour cherry harvest, and several varieties must be harvested during this period.

Research in the USA and Germany (Markwardt *et al.*, 1964; Stosser *et al.*, 1969; Stosser, 1971; Bukovac, 1979) first showed that there was a close relationship between the extent of abscission layer development between the stem (pedicel) and the fruit of sour cherry and the FRF. Cultivars that formed a complete abscission layer at maturity (e.g. Kelleris no. 14, Röhrigs Weichsel and Mörkes Bierkirsche) had lowest FRF, whereas Rote Maikirsche, which showed no evidence of abscission layer formation, had the highest FRF of the cultivars tested. Intermediate in response were the Heimanns types, North Star, Rexelle, Morelleufeuer and Diemitzer Amarelle. The abscission layer between stem and fruit is observed first in Montmorency fruits approximately 12–15 days before maturity and is composed of 5–8 rows of cells in the transition zone between the fruit and its pedicel. Initial cell separation occurs without rupture but later some collapse of cells is evident. The vascular bundles remain intact and no abscission layer is formed through them.

Sweet cherry cultivars with local but incomplete abscission layers have also been found and these cultivars exhibit lower FRFs than average for sweet cherries (Kollár and Scortichini, 1986). However, no abscission layer was observed in the transition zone between the fruit and stems of the sweet cherry cultivars Napoleon, Windsor and Schmidt.

The abscission chemical ethephon has been used to promote early and more uniform maturity on sour cherries for mechanical harvesting (Larson, 1969; Anon., 1977; Kollár *et al.*, 1981b; Kollár and Bukovac, 1993). Ethephon can be used to advance the maturity of part of a large orchard so that the harvest can extend over a greater number of days. Ethephon should not be applied when the weather is hot, however, nor should it be used on trees less than 5 years old. Once applied, ethephon will induce an abscission response (reduction in FRF and eventual fruit drop) in approximately 7–10 days. Fewer stems will be attached to the harvested cherries when ethephon is used. Always follow the recommendations on the label when applying ethephon to promote abscission.

The use of ethephon does not cause detrimental side-effects if the approved quantity and application method are used. Fruit quality, firmness and flesh consistency are maintained. Tree gummosis and leaf necrosis do not increase.

The optimum time for shake-harvesting can also be decided on the basis of fruit colour, flesh firmness, sugar content and optical density of the fruit, in addition to the measurement of FRF.

18.4 Cultural Practices that Aid Mechanical Harvesting

Uniform row widths and tree spacings within the row should be used when planting trees to standardize the 'fit' of the harvesting system to the orchard. In the USA, Montmorency orchards being planted today typically have a between-row spacing of 6.1–6.7 m and within-row spacing of 4.3–6.1 m. Until good size-controlling rootstocks are available, these spacings will continue to be popular. Severe mechanical side-hedging and topping has not proved a successful way to manage a sour cherry orchard for maximum income. In contrast, regular light mechanical hedging during the summer, plus additional hand pruning to maintain a smaller open tree, has proved successful.

In Europe, sour cherries are typically planted using a between-row spacing of 6–7 m and within-row spacings of 3–4 m. Sweet cherries are typically spaced 7–8 m and 4–5 m, respectively.

The orchard floor should be level and managed as a grass culture. This will speed up harvesting, especially during rainy weather, help avoid tree damage caused by equipment manoeuvring problems and avoid getting soil mixed in with the harvested cherries. Across the rows of trees, ridges are not tolerable. In the USA, a few orchards are successfully harvested where each tree row is planted on a ridge.

The trees should be planted with the trunks vertical and pruned to be free of side-branching up to 1.0 m above the soil level. Full roll-out harvesters can work with shorter trunks, but most other harvesters require taller trunks. The trees should be started as a whip and trained as a modified central-leader tree with strong side-branches (Anon., 1977; Pór and Faluba, 1982; Kesner and Nugent, 1984). After two growing seasons, winter pruning should consist of maintaining the central leader and several strong scaffolds, which are spaced 15–20 cm apart vertically and around the trunk and have a near 90° crotch angle. Scaffold branches high on the central leader should be cut to 15 cm stubs so that the central leader remains dominant. The maturing tree can have a natural structure. Low-hanging branches that block the harvester operator's view of the trunk or that drag on the collecting surfaces must be pruned back or removed completely. Weak hanger or upright branches that do not shake well must also be pruned back or removed. The mature tree must be open to light in order to promote high fruit set, uniform fruit size and uniform fruit maturity. Dead wood must also be removed so that it does not come out as debris during harvesting operations. Cultural practices that aided the initial successful adoption of mechanical harvesting were summarized by Larson (1969) and Andor *et al.* (1987).

On production sites that require irrigation, a trickle irrigation system should be installed and used as needed throughout the season, including during and after mechanical harvesting. This will promote tree growth and fruit production per unit area, and minimize tree stress following harvest.

When mechanically harvesting a young orchard it is essential that the correct shaker clamp system and clamping pressures are used so that trunk damage will be near zero (Schulte *et al.*, 1992). Ignoring correct harvesting practices in young orchards will result in serious trunk damage, and reduced orchard life.

18.5 Varieties Suitable for Mechanical Harvesting

One, if not the largest, sour cherry production area in the world is in Michigan,

USA. The cultivar grown in this area is, almost exclusively, Montmorency. Although evidence from Europe suggests that Montmorency is not necessarily the easiest variety to mechanically harvest, all of the US research effort into mechanical harvesting technology has been focused, with considerable success, on this one variety. However, not all markets for sour cherries will accept the clear-juice Amarelle types, such as Montmorency, and the suitability of other sour varieties for mechanical harvesting has been tested.

The factors to be considered in judging the suitability of a cherry variety for mechanical harvesting are:

- FRF and location of the favoured abscission layer (i.e. fruit : stem or stem : spur or branch);
- uniformity of ripening;
- tree habit;
- resistance of fruit to mechanical injury and juice loss;
- pit (stone endocarp) characteristics;
- yield.

Breeding objectives for new sour cultivars suited to mechanical harvesting include:

- increased fruit firmness;
- increased resistance to bruising;
- improved uniformity of ripening;
- lower FRF;
- uniform response to abscission-promoting chemicals.

Varieties with an FRF of less than 360 g (as measured using a Chatillon push–pull gauge) are harvested easily using shakers (Cain, 1967).

18.5.1 Sour cherries

Research by Blazek and Kloutvor (1985) in the Czech Republic showed that sour cherry cultivars differed considerably in their suitability for mechanical harvesting. In tests on 32 cultivars they recorded an average retention force of abscission between fruit stalk and spur of 7.4 N, while for abscission between stalk and fruits only 2.6 N was necessary. Generally, it was easier to get fruits of the Morello (Griotte) types to separate from stalks than the Amarelle types. Also, later-ripening cultivars seemed to abscise more easily than earlier ripeners. Ostheimer, Fanal and Schattenmorelle all abscised easily in tests. The greatest quantity of fruit removed by shaking was from the cultivar Köröser (Pándy). All four of these cultivars are recommended for mechanical harvesting in the Czech Republic. In these tests the cultivars Early Richmond, Krasnyjflag, Vackova, Favorit, Érdi Bôtermô, Schattenmorelle and, perhaps surprisingly, Montmorency all proved less suitable for mechanical harvesting.

Italian research (Anon., 1985) suggests that Montmorency and Schatten-morelle (the latter matures 3 days after the former) are both commercially viable and suited to mechanical harvesting. Nevertheless, not all sour cherry varieties are suited to mechanical harvesting and the Italian selection Visciola del Canalese is one which responds poorly.

Hungarian research (Brózik, 1989; Kollár, 1994) showed that several sour cherry varieties are well suited to mechanical harvesting and ripen both before and after Montmorency. Érdi Jubileum ripens about 14 days before Montmorency, Érdi Bôtermô ripen 8 days before, the Cigány Meggy clones (Hartai Meggy and Paraszt Meggy) ripen 8 and 4 days before, respectively, but can remain on the tree in good condition for another 10–14 days. The Ujfehértói Fürtös ripens 2–3 days after Montmorency, and the Pándy clones (Pándy 48, Pándy 279) ripen 4–5 days after.

Jakubowski *et al.* (1986) showed that Clone 761, a variety resulting from a cross between English Morello and Shirpotreb, was very suited to mechanical harvesting on account of its improved tree habit and low FRF. It ripens 5 days earlier than English Morello but has lower yields.

18.5.2 Sweet cherries

Generally, Bigarreaux types are easier to mechanically harvest than Heart types of sweet cherry. The principal sweet cherry cultivars mechanically harvested in the USA are Napoleon Bigarreau (Royal Ann), Windsor and Schmidt. Napoleon fruit allows more flexibility in harvesting, as it can be left on the tree longer to attain a more mature and easily harvestable condition and yet still be bleached effectively. This is not true with Windsor, which must be harvested in a less mature state for success. Most Schmidt cherries, which are very suited to mechanical harvesting, are allowed to ripen fully and then canned, usually as a syrup hot pack.

Research in Russia has shown that even varieties such as Hedelfingen may be mechanically harvested. FRF and the effectiveness of mechanical harvesting were improved if the trees were sprayed with ethephon shortly before harvest. Research in the Czech Republic by Blazek *et al.* (1981) showed that mechanical harvesting, not surprisingly, reduced the keeping quality of Van, Starking Hardy Giant and Kordia fruits. Loss of fruit pedicel also seriously reduced keeping quality and this effect was worst on those varieties with a poor abscission layer between fruit and pedicel.

Moore (1983), in suggesting breeding objectives for sweet cherries, lists shorter internodes, tougher trunks, stiffer branches, uniform harvest maturity, improved spur systems and lower FRFs as essentials for mechanical harvesting. Short pedicel length is thought to be a dominant character in breeding and is inherited readily from crosses involving the variety Van.

Experience in Hungary has shown that many of their sweet cherry varieties are suitable for mechanical harvesting (Kollár, 1994). Bigarreau Burlat ripens at the beginning of June, whereas Katalin ripens in the first half of July. Ripening between these two varieties are Margit, Linda, Germersdorf 1 (Van) and Germersdorf 3 (Stella).

Iezzoni *et al.* (1990) reviewed cultivars suited to mechanical harvesting and modified tables based on their review are presented in Tables 18.1 and 18.2.

Deckers (1974) noted that in Belgium the Morello types Griotte du Nord,

Table 18.1. Sour cherry cultivars suitable for mechanical harvesting.

Variety	Country	Reference
Fanal, Ujfehértói Fürtös, Oblacinska	Czech Republic	J. Blazek, personal communication
North Star, Osteimer, Late Morello	Czech Republic	Vondracek *et al.*, 1981
Stevensbaer	Denmark	Christensen, 1976
Oblacinska	Bulgaria	B'chvarov, 1976
Richmorency, Körözser Weichsel, Spanska, Conserven Weichsel, Amarena a Picciolo, Corto di Verona	Italy	Costa *et al.*, 1978
Luxardo, Schattenmorelle	Italy	Costa *et al.*, 1978
Tito Poggi, Fanal, Original Heiman's, Marasca di Zava, Petrodaravenka, Amarena P. Grilli	Italy	Liverani and Cobianchi, 1980
Montmorency	USA	Gaston *et al.*, 1959
Érdi Jubileum, Érdi Bôtermô, Harti Meggy, Parasyt Meggy, Pándy 48, Pándy 279	Hungary	Brozik, 1989; Kollár, 1994, unpublished data
Vladimir, Osteim Griotte, Moscow Griotte	Russia	Lukin, 1981

Gorsemkriek, Kelleris and Chenoharko could be harvested, with the aid of ethephon sprays, and transported to the processing plant dry without significant loss of fruit quality. In contrast the Amarelle types, Kleine Waalsci, Montmorency, Spanishe Glaskirsche, Ludwigs Fruhe and Diemitzer Amorelle all developed a brown discoloration (probably scald) soon after harvesting unless the fruit was immediately immersed in cold water.

18.6 Future Harvesting Systems

Today the sour cherry industry is using reliable mechanical harvesters that use a stop-and-go approach (i.e. the harvester must stop at each tree during the shaking operation and then go to the next tree and repeat the cycle) rather than a continuous-harvest approach (i.e. the harvester continues to travel along the row during the shaking operation). A harvester that operates over the row and continues to move during shaking has been developed (Fig. 18.11) and has been evaluated for the harvesting of sour cherries (Peterson, 1984). The sequence of harvesting operations (locate the tree trunk, clamp to the trunk and close the row seal, shake the tree, release the trunk and open the row seal, move the shaker to the next trunk and repeat the sequence) has been automated by using electronic sensors and time controllers. The operator only needs to control the harvester travel speed and alignment with the row, and observe the overall operation to ensure that the system is functioning properly. Future harvesters must be designed to avoid trunk damage (ensure long orchard life) and to maintain high fruit quality (minimize bruising and handling damage).

Table 18.2. Sweet cherry cultivars suitable for mechanical harvesting.

Variety	Country	Reference
Napoleon, Schmidt, Windsor	USA	Gaston *et al.*, 1961
Napoleon, Bing	USA	Norton *et al.*, 1962; Stebbins *et al.*, 1967
Van, Starking Hardy Giant, Kordia	Czech Republic	Blazek *et al.*, 1981
Kastanka e Techlovicka I (Ziklova), Ulster, Starking Hardy Giant, Van, Bing	Czech Republic	Blazek and Kucera, 1980
Jaboulay Fruke Mechenheumer, Rauna Lyaskovska, Flamentiner	Denmark	Ostergaard and Christensen, 1983
Iunskaya Ramyaya, Melitopolskaya Chernaya, Frantsuzskaya Chernaya, Izyumnaya	Ukraine	Malishevskaya *et al.*, 1979
Bryunetka, Frantsuzskaya Chernaya, Napoleon Chernyi	Ukraine	Gnezdilov and Aleinikova, 1975
Adriana, Vittoria, Durone di Cesena, Bianca di Verona	Italy	Bargioni, 1975, 1980; Baldini *et al.*, 1978, 1980; Baldini, 1980
Germersdorfer, Van, Hedelfingen	Hungary	Sárosi, 1978
Bigarreau Burlat, Margit, Linda, Germersdorf 1, Van, Germersdorf 3, Stella, Katalin	Hungary	G. Kollár, unpublished data
Venus, Starking Hardy Giant, Merton Heart	Germany	Stortzer *et al.*, 1977

The harvesting rates for stop-and-go systems are typically in the range of 60–120 trees h^{-1}. The harvesting rates for the continuous-harvest system were in the range of 200–400 trees h^{-1} on trees that are planted 2–3 m apart in the row and 5–6 m between rows. At these spacings, there are 600–1200 trees ha^{-1}, so only size-controlling rootstocks can be considered. The yield per hectare will be very high (B.F. Cargill, personal communication, 1982) and the high income per hectare should attract growers to such an orchard culture once the dwarf trees are available.

References

Affeldt, H.A. Jr, Brown, G.K. and Gerrish, J.B. (1989) A new shaker for fruit and nut trees. *Journal of Agricultural Engineering Research* 44, 53–66.

Andor, D., Kállay, E., Kemenes, M., Kollár, G. and Szenci, Gy. (1987) *Mechanical Harvesting of Stone Fruits* [in Hungarian]. Mezôgazdasági Kiadó, Budapest, 133 pp.

Anon. (1977) *Growing Cherries East of the Rocky Mountains*. Farmers Bulletin 2185. USDA, Agricultural Research Service, Washington, DC, USA, 31 pp.

Anon. (1985) Ciliegio acido [in Italian]. *Rivista di Fruticoltura e di Ortofloricoltura* 47, 18–22.

Baldini, E. (1980) Meccanizzazione della raccolta: aspeti tecnici [in Italian]. In: *Rin-*

Fig. 18.11. Continuous-moving over-the-row harvester during evaluation in high-density sour cherry orchard (B.F. Cargill, USDA, Agricultural Research Service, Appalachian Fruit Research Station, Kearneysville, West Virginia, USA).

novamento e Sviluppo della Coltura del Ciliegio. Asolo, pp. 67–102.

Baldini, E., Bargioni, G., Costa, G. and Intrieri, C. (1978) Ricerche preliminari sulla raccolta meccanica delle ciliegio dolci del veronese [Preliminary studies on the mechanical harvesting of sweet cherries in Verona]. *Informatore Agrario* 34, 2369–2372.

Baldini, E., Bargioni, G., Costa, G. and Miserocchi, O. (1980) The suitability of sweet cherries to mechanical harvesting for eating fresh: varietal differences [in Italian]. *Informatore Agrario* 36, 11595–11598.

Bargioni, G. (1975) Bianca di Verona, a new sweet cherry cultivar for mechanical harvesting [in Italian]. *Rivista della Ortoflorofrutticoltura Italiana* 59, 136–140.

Bargioni, G. (1980) A new cherry, Adriana [in Italian]. *Informatore Agrario* 36, 11617–11618.

B'chvarov, D. (1976) Oblacinska, a promising sour cherry cultivar [in Russian]. *Ovoshcharstvo* 55, 31–33.

Blazek, J. and Kloutvor, J. (1985) Evaluation of selected sour cherry cultivars in relation to some characteristics considered important for machine harvesting [in Czech]. *Vedecke Prace Ovocnaraske* 10, 87–101.

Blazek, J. and Kucera, J. (1980) Preliminary evaluation of mechanized harvesting of selected cherry cultivars by shaking [in Czech]. *Sbornik UVTIZ Zahradnictvi* 7, 177–188.

Blazek, J., Grossmann, G. and Zika, J. (1981) Storability of machine harvested cherries under normal conditions and in cold stores [in Czech]. *Sbornik UVTIZ Zahradnictvi* 8 (3), 173–180.

Bolen, J.S., Cargill, B.F. and Levin, J.H. (1970) *Mechanized Harvest Systems for Red Tart Cherries*. Michigan State University Extension Bulletin E-660, East Lansing, Michigan, USA, 6 pp.

Brown, G.K., Frahm, J.R., Segerling, L.J. and Cargill, B.F. (1987) Bark strengths and shaker pads vs. cherry bark damage during harvesting. *Transactions of the ASAE* 30, 1266–1271.

Brown, G.K., Rauch, M.H. and Timm, E.J. (1988) Improved clamp pad for trunk shakers. *Transactions of the ASAE* 31, 677–682.

Brózik, S. (1989) *Description of Most Important Hungarian Fruit Varieties* [in Hungarian]. Gyümölcs és Dísznövénytermesztési Kutató-Fejleszató Vállalat, Budapest.

Brunner, T. (1982) *Training of Dwarf Fruit Trees* [in Hungarian]. Mezôgazdasági Kiadó, Budapest, 337 pp.

Bukovac, M.J. (1979) Machine-harvest of sweet cherries: effect of Ethephon on fruit removal and quality of the processed fruit. *Journal of the American Society for Horticultural Science* 104, 289–294.

Cain, J.C. (1967) The relation of fruit retention force to the mechanical harvesting efficiency of Montmorency cherries. *HortScience* 2, 53–55.

Cargill, B.F., McManus, G. Jr, Bolen, J.S. and Wittenberger, R.T. (1970) *Cooling Stations and Handling Practices for Quality Production of Red Tart Cherries*. Michigan State University Extension Service Bulletin 659, East Lansing, Michigan, USA, 8pp.

Christensen, J.V. (1976) Description of the sour cherry cultivar Stevnsbaer [in Danish]. *Tidsskrift for Planteavl* 80, 911–914.

Costa, G., Intrieri, C., Baldini, E. and Grandi, M. (1978) Two years experiments on the mechanical harvesting of sour cherries [in Italian]. *Informatore Agrario* 34, 2355–2363.

Deckers, J. (1974) Ethrel and mechanical harvesting of fruits [in French]. *Le Fruit Belge* 42, 223–227.

Esch, T.A., Van Ee, G.R., Ledebuhr, R.L., Welch, D.P. and Brown, G.K. (1989a) *Design, Construction, and Testing of a Controllable Tree Shaker*. ASAE Paper No. 89–1072. ASAE, St Joseph, Michigan, USA.

Esch, T.A., Welch, D.P., Van Ee, G.R. and Ledebuhr, R.L. (1989b) *Analysis of a Controllable Trunk Shaker for Cherry Fruit Removal*. ASAE Paper No. 89–1647. ASAE, St Joseph, Michigan, USA.

Frahm, J.R., Brown, G.K. and Segerlind, L.J. (1988) Mechanical properties of trunk shaker pads. *Transactions of the ASAE* 31, 1674–1679.

Fridley, R.B. (1983) Vibration and vibratory mechanisms for the harvest of tree fruits. In: Brien, M.O., Cargill, B.F. and Fridley, R.B. (eds) *Principles and Practices for Harvesting and Handling Fruits and Nuts*, AVI Publishing Co., Florence, Kentucky, USA, pp. 157–188.

Fridley, R.B. and Adrian, P.A. (1966) *Mechanical Harvesting Equipment for Deciduous Tree Fruits*. California Agricultural Experiment Station Bulletin 825, Davis, California, USA.

Gaston, H.P., Levin, J.H. and Hedden, S. (1959) Experiments in harvesting cherries mechanically. *Michigan Agricultural Experiment Station Quarterly Bulletin* 41, 805–811.

Gaston, H.P., Hedden, S.L., Levin, J.H., Whittenberger, R.T. and Hamner, C.L. (1961) Sweet cherry harvest trials. *Western Fruit Grower* 15 (7), 23–24.

Gnezdilov, Yu.A. and Aleinikova, O.N. (1975) Mechanizing the cherry harvest [in Russian]. *Sadovodstvo* 8, 21–22.

Iezzoni, A., Schmidt, H. and Albertini, A. (1990) Cherries. In: Genetic resources of temperate fruit and nut crops. *Acta Horticulturae* 290 (1), 109–173.

Jakubowski, T., Dudziak, E., Zagaja, S.W. and Wojniakiewicz, A. (1986) Growth and fruiting of 15 tart cherry clones. *Fruit Science Reports* 13, 1–6.

Kesner, C.D. and Nugent, J.E. (1984) *Training and Pruning Young Cherry Trees*. Michigan State University Extension Bulletin E-1744, East Lansing, Michigan, USA, 4 pp.

Kollár, G. (1985) Connections between mechanical harvesting and cultivars, rootstocks, crown shape of industrial stone fruits [in Hungarian]. *Gyümölcsinform* 3, 98–100.

Kollár, G. (1987a) Hungarian experiences on mechanical harvesting and quality maintenance of stone fruit with special regards to processing [in Hungarian]. *Kertgazdaság* 2, 51–56.

Kollár, G. (1987b) Cultivars, pruning systems, general cultural practices for sour cherries in Hungary. In: *Proceedings, International Cherry Research Conference*, Traverse City, Michigan, USA, pp. 46–49.

Kollár, G. (1994) Mechanical harvest of sweet and sour cherries protected the quality [in Hungarian]. Dissertation for Hungarian Academy of Science, Budapest, 98 pp.

Kollár, G. and Bukovac, M.J. (1993) The effects of different fruit-removal-force decreasing treatments on sour cherry fruits and trees. *Acta Horticulturae* 410 (in press).

Kollár, G. and Kemenes, M. (1993) The effect of different fruit-removal-force regulating treatments on short term storage of sour cherry fruits. Lectures of ISHS Postharvest 1993 International Symposium, Kecskemét, 23.

Kollár, G. and Scortichini, M. (1986) Effeti di trattamenti chimi facilitanti I abscissione del frutto della cultivar di ciliegio dolce Germersdorfi [in Italian]. *Rivista Ortiflorofrutt. Italia* 70 (2), 85–95.

Kollár, G., Kemenes, M. and Szenci, GY. (1981a) Biological and technical aspects of mechanized harvesting of sweet cherries [in Hungarian]. *Kertgazdaság* 3, 169–180.

Kollár, G., Kemenes, M. and Szenci, GY. (1981b) Biological and technical aspects of mechanized harvesting of sour cherries [in Hungarian]. *Kertgazdaság* 4, 41–52.

Larson, R.P. (1969) Cultural practices for cherry mechanization. In: Cargill, B.F. and Rossmiller, G.E. (eds) *Fruit and Vegetable Harvest Mechanization – Technological Implications*. ASAE, St Joseph, Michigan, USA, pp. 687–697.

Levin, J.H., Gaston, H.P., Hedden, S.L. and Whittenberger, R.T. (1960) *Mechanizing the Harvest of Red Tart Cherries*. Michigan State Agricultural Experiment Station Quarterly Bulletin 42–60, East Lansing, Michigan, USA, 32 pp.

Levin, J.H., Whittenberger, R.T. and Gaston, H.P. (1969a) *When to Harvest Sweet Cherries Mechanically*. Michigan State University Extension Research Report 87, East Lansing, Michigan, USA, 5 pp.

Levin, J.H., Bruhn, H.D. and Markwardt, E.D. (1969b) Mechanical harvesting and handling of cherries. In: Cargill, B.F. and Rossmiller, G.E. (eds) *Fruit and Vegetable Harvest Mechanization – Technological Implications*. ASAE, St Joseph, Michigan, USA, pp. 677–685.

Levin, J.H., Tennes, B.R., Whittenberger, R.T. and Cargill, B.F. (1970) Weight–volume relationships of tart and sweet cherries. *Transactions of the ASAE* 13, 489–490.

Liverani, A. and Cobianchi, D. (1980) The varietal susceptibility of sour cherries to mechanical harvesting [in Italian]. *Informatore Agrario* 36, 11591–11593.

Lukin, E.S. (1981) Sour cherry for mechanical harvesting [in Russian]. *Sadovodstvo* 12, 33–34.

Malishevskaya, M.F., Dedova, I.M. and Gnevkovskaya, M.G. (1979) Sweet cherries for mechanical harvesting [in Russian]. *Sadovodstvo* 8, 24.

Markwardt, E.D., Guest, R.W., Cain, J.C. and Labelle, R.L. (1964) Mechanical cherry harvesting. *Transactions of the ASAE* 7, 70–74.

Mihályffy, J. (1982) *Crown Shape of Stone Fruit Trees Suitable for Mechanical Harvest* [in Hungarian]. MÉM Kutatási Eredmények, Budapest.

Mitchell, A.E. and Levin, J.H. (1969) *Tart Cherries – Growing, Harvesting and Processing for Good Quality*. Michigan State University Extension Bulletin E-654, East Lansing, Michigan, USA, 4 pp.

Moore, J.N. (1983) Mechanised harvest. In: Moore, J.N. and Janick, J. (eds) *Methods in Fruit Breeding*. Purdue University Press, USA, pp. 328–353.

Moser, E., Sinn, H., Bieler, E. and Vögtle, K. (1977) *Der Transport und die Zwieschenlagerung von Maschinell geernteten Steinobst* [in German]. Landtechnik, Lehrte, 10 pp.

Norton, R.A., Claypool, L.L., Fridley, R.B., Adrian, P.A., Leonard, S.J. and Charles, F.M. (1962) Mechanical harvesting of sweet cherries – 1961 tests show promise and problems. *California Agriculture* 16 (5), 8–10.

Ostergaard, C. and Christensen, J.V. (1983) Fruit retention force in cultivars of sweet cherries [in Danish]. *Tidsskrift for Planteavl* 87, 39–45.

Peterson, D.L. (1984) Mechanical harvester for high density orchards. In: *Fruit, Nut and Vegetable Harvesting Mechanization*, Special Publication 5–84. ASAE, St Joseph, Michigan, USA, pp. 46–51.

Peterson, D.L. and Kornecki, T. (1990) Catching-surface trunk seal for tree crop harvester. *Applied Engineering in Agriculture* 6 (2), 155–157.

Pór, J. and Faluba, Z. (1982) *Sweet and Sour Cherry* [in Hungarian]. Mezôgazdasági Kiadó, Budapest, 381 pp.

Sárosi, S. (1978) Suitability of sweet and sour cherry cultivars for processing with reference to mechanical harvesting [in Hungarian]. *Ùjab Kutatási Eredmények a Gyümölcstermesztésben* 75–81.

Schulte, N.L., Burton, C.L., Brown, G.K. and Timm, E.J. (1992) *How to Avoid Cherry Tree Trunk Damage Caused by Trunk Shakers*. Michigan State University Extension Bulletin E-2336, East Lansing, Michigan, USA, 12 pp.

Stebbins, R.L., Cain, R.F. and Watters, G. (1967) Mechanical harvesting of sweet cherries in Oregon. *Oregon State University Extension Tree Fruit Specialists' Newsletter* 6 (5), 1–11.

Stortzer, M., Blazek, R., Drobkova, R., Kuceva, J., Grossmann, G. and Unger, S. (1977) Untersuchungen des Süsskirschensorten in Holouvousy kinsichtlich der Sorteneigung für Maschinelle ernte. *Arch. Gartenbau, Berlin* 25, 397–441.

Stosser, R. (1971) Localization of RNA and protein synthesis in the developing abscission layer in fruit of *Prunus cerasus*, L. [in German]. *Zeitschrift für Pflanzenphysiol.* 64, 328–334.

Stosser, R., Rasmussen, H.P. and Bukovac, M.J. (1969) A histological study of abscission layer formation in cherry fruits during maturation. *Journal of the American Society for Horticultural Science* 94, 239–243.

Tennes, B.R., Anderson, R.L. and Levin, J.H. (1968) *Weight to Volume Relationship of Tart Cherries*. Michigan State University Farm Science Research Report 70, East Lansing, Michigan, USA, 4 pp.

Timm, E.J., Brown, G.K., Segerlind, L.J. and Van Ee, G.R. (1988) Slip-belt and lubrication systems for trunk shakers. *Transactions of the ASAE* 31 (1), 40–46, 51.

USDA (1964) *United States Standards for Grades of Canned Red Tart Cherries*. USDA, Agricultural Marketing Service, Washington, DC, USA.

Vondracek, J., Blazek, J. and Kloutvor, J. (1981) Evaluation of sour cherry cultivars [in Czech]. *Sbornik UVTIZ Zahradnicvi* 8 (4), 249–260.

Whittenberger, R.T. and LaBelle, R.L. (1969) Effects of mechanization and handling on cherry quality. In: Cargill, B.F. and Rossmiller, G.E. (eds) *Fruit and Vegetable Harvest*

Mechanization – Technological Implications. ASAE, St Joseph, Michigan, USA, pp. 699–711.

Whittenberger, R.T., Levin, J.H. and Gaston, H.P. (1968) *Maintaining Quality by Brining Sweet Cherries after Harvest*. Michigan State University Extension Research Report 73, East Lansing, Michigan, USA, 4 pp.

Whittenberger, R.T., Levin, J.H. and Cargill, B.F. (1969) *Weight to Volume Relationships of Sweet Cherries in brine*. Michigan State University Extension Research Report 89, East Lansing, Michigan, USA, 5 pp.

19 Cherry Processing

K. Kaack[1], S.E. Spayd[2] and S.R. Drake[3]

[1]*Danish Institute of Plant and Soil Science, Department of Food Science and Technology, Horticultural Research Centre, DK-5792, Aarslev, Denmark;* [2]*Department of Food Science and Human Nutrition, Washington State University, Prosser, WA 99350-9687, USA;* [3]*United States Department of Agriculture – Agricultural Research Service Tree Fruit Research Laboratory, Wenatchee, WA 98801, USA*

19.1 Introduction

Cherries have traditionally been used for a wide variety of food and beverage products in Europe, and with European migration this tradition has spread to North America and to many other parts of the world.

Sweet cherries (*Prunus avium* L.) are mainly grown for fresh consumption but both red/black and yellow sweet cherries are canned in syrup, individually quick frozen (IQF), brined for producing cocktail (Maraschino), glacé or candied cherries, and used in the production of jams and jellies. Sweet cherry juice is popular in some countries and is sometimes used to make wine after blending with sour cherry juice. Sweet cherry 'raisins' are a popular new product in North America.

Sour cherries (*Prunus cerasus* L.) are primarily grown for use in processing. They are used to produce juice and nectar, jams and jellies, and are canned and frozen whole (usually pitted) or further prepared as 'fillings' for use in a wide variety of bakery products. Small quantities of sour cherries are dehydrated and in some countries speciality wines are made from sour cherry juice.

North American cherry processing was reviewed by Marshall (1954) and most recently by Drake (1991). While there have been very few new products developed in recent years, the processes have evolved and the body of knowledge about the constituents and characteristics of fruit that influence product quality continues to grow.

This chapter provides an overview of present-day cherry processing in Europe and North America. A brief review of processing cultivars, raw product chemistry and other factors influencing processed product quality will be followed by a discussion of the major products and processes. In-depth descriptions of processing lines and machinery can be found elsewhere (Anon., 1974, 1979, 1982; Heuss, 1974).

19.2 The Importance of Cultivar

Many fruit and tree characteristics may prove important when selecting a cultivar for use in processing. Fruit size, appearance, colour, flavour, juiciness, flesh percentage and flesh firmness after processing are of central importance. Disease and cracking susceptibility, season of maturity, productivity and suitability for mechanical harvesting can be equally important. Some processed cherry products arise from a relatively small number of cultivars. Thus, while sweet and sour cherry cultivars are discussed elsewhere in this book, it is instructive to mention here some specific cultivar/product associations.

Across Europe, the most commonly grown sour cherry cultivars for industrial processing are Schattenmorelle, Cigany, Pándy and Marasca. However, many other cultivars are under test in various countries and many local cultivars are important for specific uses (Christensen, 1986, 1990). For example, in Denmark, Stevnsbär and Kelleris sour cherries are specifically recommended for juice processing. The cultivars Schattenmorelle and Schwaebische Weinweichsel can be used for processing of juice, compote and jams. Heimanns Rubin, Schwäbische Weinweichsel, Heimanns Konservenweichsel, Leitzkauer Presskirsche, Ostpreussische Bierkirsche, Mötzlich, Schattenmorelle, Vacecks Weinweichsel and Sämling von Braer are all suitable for juice processing.

Juices with a pleasant flavour can be processed from Fanal, Heimanns Rubine, Meteor and Schattenmorelle. Kelleris 16, Koroser and Nefris are found suitable for freezing while Nefris and Kelleris are suitable for processing of compotes (Lenartowicz *et al.*, 1985). Out of 147 cultivars tested in Italy, Montmorency and Schattenmorelle were found suitable for producing the widest range of processed commodities (Anon., 1985).

Two cultivars dominate North American processed products, Bing for sweet cherries and Montmorency for sour cherries. In addition to Bing, Lambert, Royal Ann (Napoleon), Windsor, Schmidt and Emperor Francis are used for brining.

The dominance of Montmorency sour cherry in North America may relate to its suitability for mechanical harvesting and ease of pit removal. However, the flavour of Montmorency is also quite characteristic and has become the industry standard in this part of the world.

19.3 Some General Fruit Quality and Maturity Considerations

The normal range (expressed as a percentage of fresh weight – excluding the pit) of several processing-related chemical constituents in sweet and sour cherries is shown in Table 19.1. Note that the essential difference between the two species is in juice acidity. Within each species, cultivar differences account for most of the variation but, even for a single cultivar, climate, soils and other growing conditions can result in quite different values from region to region. For example, sour cherries grown in southern Europe differ in composition from those grown at higher latitudes. They are lower in sugar, anthocyanins and aroma substances.

Table 19.1. Chemical composition of sweet and sour cherries expressed as percentage of fresh weight.

	Sweet cherry	Sour cherry
Soluble solids	12–16	12–17
Sugar	6–9	6–9
Non-soluble solids	1–4	1–4
Total acid	0.4–0.8	1.5–1.8
Minerals	0.5	0.5
Fibre	0.4	0.5

Many of the fruit quality characteristics important for fresh market fruit are also considered when selecting fruits for processing. For those products involving whole fruits, such as canned, frozen (i.e. IQF), candied or brined cherries, flesh firmness and uniform fruit size and appearance are especially important. Even fruit intended for jam, jelly, juice and wine must be disease- and insect-free, but the most important quality component might be uniformity of maturity, which in turn influences soluble solids, anthocyanins and other pigments, titratable acidity, stone percentage and content of amygdalin in the kernels (Christensen and Grauslund, 1979; Kaack, 1990b).

19.3.1 Assessing fruit maturity

Sweet and sour cherries grown especially for processing are harvested when the various flavour components reach a minimum threshold level and this level is commonly different from that used for fresh market fruit. For example, sweet cherries to be used for processing almost always have higher sugar and lower acid levels than the same cultivar grown for fresh consumption.

For nearly all cherry products, the balance between soluble solids (sugars) and free acids is very important and, partly because of ease of measurement, is the most commonly used indicator of harvest maturity. While it is recognized that for some products the content of aroma substances is even more important than taste, it is usually assumed that when the sugar/acid ratio is right the other components of flavour will also be present.

In some cases the most important consideration is sugar content. A processor buying cherries to dehydrate will seek high soluble solids concentrations since energy requirements are reduced. It costs less to dry a 21% soluble solids cherry to 50 or 68% soluble solids than a 15% soluble solids cherry.

19.3.2 Fruit size and other factors influencing product yield

Fruit size *per se* is only important for products where the fruit, either pitted or whole, remains intact. In such cases the desired size is usually specified by industry. However, uniformity of size is often more important than actual size. In the USA, specifications for US Department of Agriculture (USDA) grades generally include uniformity of size as an evaluation criterion for most 'whole-piece' processed products.

When comparing cultivars there is not a strong relationship between fruit size and stone size. Some cultivars with large fruit may have small stones and vice versa. Zielinski (1964) compared a large number of sweet cherry cultivars and found that the volume of the pit as a proportion of total fruit volume ranged from a low of 7.2% for Napoleon to 10.6% for Spanish Yellow.

In terms of product yield, cultivars with small stones should be advantageous in the manufacture of any processed product where the stone is removed. Furthermore, since smaller fruit of a given cultivar have a higher proportion of total fresh weight devoted to the stone (Marshall, 1954), it follows that horticultural practices that improve fruit size are especially beneficial in terms of yield of pitted fruit products.

Ease of pitting also influences product yield. More tissue adheres to the stone of difficult-to-pit cultivars, resulting in lower drained weights.

19.3.3 Flavour

The taste of cherry products mainly depends on the content of sugars (fructose, glucose) and organic acids (malic and citric acid). As already mentioned, fully mature sweet and sour cherries have similar levels of soluble sugars but differ dramatically in acidity (Table 19.1).

Flavour is defined by aroma as well as taste (flavour = aroma + taste) and the aroma of cherry products depends on the content of volatile organic compounds, such as hydrocarbons, alcohols, aldehydes, ketones, acids and esters (Stinson et al., 1969a, b; Schmid and Grosch, 1986). Cherry cultivars can differ significantly in intensity of cherry flavour (Table 19.2).

In creating or enhancing cherry flavour the flavourists use some of these naturally occurring aroma substances. However, chemicals obtained from other sources have also proved useful (Broderick, 1975).

Aroma development can be enhanced or suppressed during processing. With enzyme- or acid-catalysed conversion of amygdalins, occurring in the seeds and the fruit flesh, glucose, hydrocyanic acid and benzaldehyde are released. The latter is a very important aroma component in cherry.

Because of these reactions, a pleasant cherry flavour can be obtained if about 20% of the stones are crushed. This is a practice often followed in the production of juices and wine.

As the aroma substances are very volatile, processing has to be carried out in a closed system. Heating during processing causes new aroma substances to appear with different chemical reactions. An example is the degradation of amino acids, from which several aldehydes can arise. Enzyme-catalysed conversion of

Table 19.2. Average points for cherry flavour of juice processed from four cultivars grown in Denmark: 0 = no cherry flavour; 10 = very strong cherry flavour (Kaack, 1990a).

Variety	Cherry flavour points
Stevnsbär	6.7
Crisana	6.1
Fanal	5.4
Schattenmorelle	4.7
Least significant difference	0.6

esters to alcohols and acids can occur unless the enzymes are denatured by heating.

19.3.4 Colour

For red-fruited cultivars, skin colour is another commonly used indicator of fruit maturity and colour comparators are often used to choose the correct harvest date for cherries intended for processing. For yellow or blush sweet cherry cultivars (white-fleshed), loss of chlorophyll and increase in xanthophyll pigments are used to judge maturity.

Thus, both anthocyanins and xanthophylls increase during ripening (Dekazos, 1970a; Kaack, 1990a). The predominant anthocyanin pigments in sweet cherry are cyanidin-3-glucoside and cyanidin-3-rutinoside, but some cultivars contain lesser amounts of peonidin glucoside and rutinoside (Mazza and Miniati, 1993). The amount of anthocyanin in the peel of ripe sweet cherries is very dependent on cultivar and can range from as little as 4 to more than 400 mg 100 g^{-1} of tissue (Mazza and Miniati, 1993).

Cyanidin-3-rutinoside and cyanidin-3-glucoside are also important sour cherry anthocyanins but cyanidin-3-(2g-glucosylrutinoside) and cyanidin-3-sophoroside are also important in some cultivars. Morello sour cherries appear to be high in cyanidin-3-(2g-glucosylrutinoside) (Hong and Wrolstad, 1990). Dekazos (1970b) found each of these anthocyanins (and minor amounts of others) in Montmorency cherries as they matured and reported that, while the total amounts increased from 2 to 44 mg 100 g^{-1} of cherries, the relative amount of each anthocyanin remained quite constant.

From a fruit processing perspective, the quantity of anthocyanins and other pigments is less important than their stability. For example, during heating or storage of various products the anthocyanins are prone to degradation, resulting in product browning. The nature of this problem and ways to reduce it are reviewed in detail by Mazza and Miniati (1993).

19.3.5 Firmness

Fruit firmness is of particular importance for products that need whole cherries, such as canned and IQF fruit. Poor fruit firmness can lead to reduced drained weights and 'mushy' processed fruit. Addition of calcium chloride to syrups used in canning pitted sweet cherries sufficiently improved texture to justify its use (Drake and Proebsting, 1985), but choosing more suitable cultivars may be a better approach to avoid leaking and bruising of the fruits during processing or storage.

Fruit firmness is also an important determinant of processed product quality in the case of cocktail, glacé and candied cherries. Here, calcium is routinely added to the brining solution to increase firmness.

19.4 Preprocessing Operations

19.4.1 Harvesting and transport

Harvesting of cherries should be carried out at the optimum maturity stage for each commodity. Bruising, cracking and leaching from the cherries should be minimized because heavy growth of microorganisms in juices or bruised fruits may destroy the fruit.

If the temperature of the harvested cherries is above 25°C, cooling by addition of ice flakes or hydrocooling to 10–17°C can be an economical advantage because bruising and leaching are minimized (see Chapter 18). When transporting in water, sugars, acids and other water-soluble solids can be extracted into the water.

19.4.2 Washing and sorting

Washing and cleaning to remove dust, dirt, insects and plant parts is carried out by floating fruit in cold water and then spraying with cold water in cylindrical drums. Stemming is carried out by use of machines with parallel inclined rubber rollers revolving against each other and thereby pulling out the stems.

Mechanical sorting in connection with floating in water can be used for sorting by size. Sorting of washed or frozen cherries by maturity or content of anthocyanins can be carried out by use of electronic sorters (Anon., 1982).

Sorting machines for separation of cherries with and without stalk are available. This has special interest in connection with the processing of candied cherries with stalks (glacé cherries).

19.5 Cherry Processing

19.5.1 Canning

Sweet cherries are primarily canned for consumption with no further preparation, while sour cherries are canned for use as preprepared pie fillings. Figure 19.1 shows a diagrammatic sketch of canning sweet and sour cherries for the North American market.

After harvest, cherries are delivered to the processing plant in field containers, which are emptied into a water-filled receiving vat. The fruit is washed, graded and any remaining stems are removed. Fruit that is underripe, damaged or mouldy is removed. Fruit may also be sized to improve uniformity of the finished product.

Pits are removed via a punching action although some sweet cherries are canned with the pits intact. Cans are filled with cherries and then with hot syrup. Syrups range from heavy (up to 45° Brix) to light (less than 16° Brix). Use of neutrally flavoured fruit juices, such as white grape or pear, as 'natural' sweeteners is popular. A starch filling is added when canned sour cherries are to be used as pie or pastry filling.

After filling, cans are exhausted to remove air in the headspace, sealed and

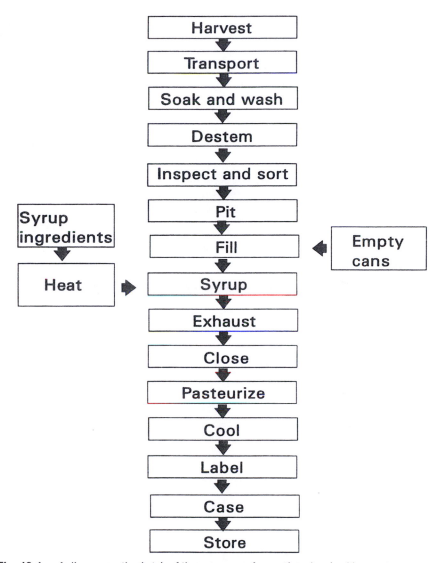

Fig. 19.1. A diagrammatic sketch of the sequence of operations involved in canning sweet and sour cherries for the North American market.

processed under atmospheric conditions. Process time depends on product, can size and fill temperature. Prior to storage, cans are immediately cooled to about 38°C to prevent heat damage to the finished product. The product may be labelled prior to casing and warehousing or labelled at a later date. It is important to note that canned cherries are subject to colour degradation if not stored in a temperature controlled warehouse. Even 20°C storage can result in colour degradation over time.

In Europe, Kelleris and other large-fruited cultivars are very suitable for processing of canned cherries. Washed, sorted and pitted frozen cherries are heated to about 0°C and then packed in glass jars with hot sugar brine in a concentration of 30% in the processed product. The jars are pasteurized by heating in water baths to a centre temperature of 80°C. After cooling to about 15°C, the jars must be stored for 5–6 weeks to obtain sugar equilibration.

An increased firmness can be obtained if the pitted cherries are heated (60°C, 5–20 min) in solutions of calcium chloride (LaBelle, 1971; Van Buren, 1974).

19.5.2 Freezing

Individually quick-frozen (IQF) is the term used to describe the very rapid freezing of single cherries. The advantage of IQF fruit is that they can later be removed individually from the container or package.

Initial processing steps are similar to those for canned cherries. However, rather than being placed in a can, a single layer of cherries is transported by stainless steel belt into a freezing unit. Here the fruit is subjected to a blast of nitrogen or air that virtually lifts it off the belt and immediately freezes the outer layer of every fruit. Complete freezing may be accomplished in the second stage in a thicker bed layer (Morrison and Ellis, 1984).

Steam-blanching of cherries at 100°C for 40–60 s before freezing can prevent degradation of anthocyanins by the enzyme polyphenol oxidase during slow thawing or storage after thawing (Siegel *et al.*, 1971; Pifferi and Cultrera, 1974).

Individually frozen cherries can be partially thawed, using microwave energy, pitted and then further processed (Weil *et al.*, 1970). By use of this method, the juice loss during pitting is considerably decreased.

Cherries are also frozen in bulk for institutional and industrial use. Fruit may be packed with or without sugar added prior to being placed in frozen storage. These cherries are later processed into such products as juice, wine, bakery fillings, jams and jellies. Some frozen cherries are dehydrated for cherry 'raisins' (see below).

19.5.3 Drying

Cherry 'raisins' have become a speciality product in North America, where a few small firms are producing limited quantities of dehydrated sweet and sour cherries. In Washington State, Rainier as well as Bing sweet cherries are used to produce cherry raisins.

Either fresh or frozen cherries may be used for drying. Pitted cherries can be dried either in a batch (tray) or on a belt fed into a continuous dehydrator. Care must be taken to prevent case hardening, where the outer portion of the cherry dries too rapidly and moisture is trapped in the centre of the cherry, preventing proper dehydration. J.N. Cash (Michigan State University, personal communication) recommends initial dehydration temperatures of 63–75°C with a gradual reduction in temperature to 52–65°C. Air velocity should be 150–180 m min^{-1}. Dehydrated cherries average about 20% moisture. Moisture levels above 22% can result in mould growth.

19.5.4 Maraschino, glacé or candied cherries

Marshall (1954) and Watters and Woodroof (1986) discuss cherry brining in detail. Brined cherries are eventually further processed into cocktail or Maraschino, candied or glacé cherries.

The fruit of cultivars known to produce firm fruits of uniform size is washed, size-graded, blanched and bleached. Primary bleaching is carried out by treatment of the cherries in a solution of sodium bisulphite (1% SO_2), calcium chloride (0.3% Ca) and pH adjusted to 3.5. The bleaching effect is due to sulphite and the firmness is increased by the calcium ions, which improve the binding of pectin molecules. Cherries sit in the brine for several months and can be stored in this condition. After the described treatment, the cherries are yellow and firm, but some discoloration may remain on cherries with skin blemishes, such as limb rubs, or on highly pigmented cultivars, such as Bing. This discoloration is removed and snow-white cherries without off flavour are obtained by a second bleaching with sodium hypochlorite (Beavers and Payne, 1969). After complete bleaching, the brine is drained off and the fruits are pitted and stemmed. The sulphite is leached to less than 200 mg l^{-1} by soaking the cherries several times in hot water and then in increasing concentrations of sucrose. In order to achieve the typical bright colour of the cocktail, candied or glacé cherry, artificial colouring is added to the sugar solution. Typical colours are red, green, blue or yellow. All colouring agents must be fully approved for food use.

This final step has been the subject of considerable research in recent years. Red dye no. 3 was traditionally used for dyeing cherries but fell into disfavour with the US Food and Drug Administration following reports of tumour formation in test animals. As an alternative, some processors are using carmine. Work is under way to develop a dye process using red carotenoid pigments as colourants for brined cherries (Sapers, 1994).

Maraschino cherries and candied cherries contain 48% and 72–74% sugar, respectively. Glacé cherries are candied cherries with stalks.

19.5.5 Jams and jellies

The processing of jam for manufacture of fruit yoghurt is quite typical of many similar processes. Clean, pitted, fresh or frozen cherries and water are heated to 30°C. pH is adjusted to 3.0 with a 50% (w/w) citric acid solution. Pectin is mixed with a small amount of sugar and then added to the thawed mixture of cherries and water. After heating to boiling, more sugar is added and heating to boiling is repeated. The jam is cooled to 50°C and tapped aseptically into glass jars. Preservation can be obtained by addition of sodium benzoate and potassium sorbate to levels of 0.1 and 0.05%, respectively. The cultivar Stevnsbär is used to produce a delicious jam for use in yoghurt, but variations in the colour of fruit yoghurt can be obtained by using other cultivars, such as Crisana 2 or Fanal (Kaack, 1990b).

Fruit yoghurt with 13% jam (6.5% cherries) is processed by addition of jam to natural yoghurt. Because the stones are normally removed before processing of jam, the content of the important aroma substance benzaldehyde is very low. A much better flavour can be obtained by extracting benzaldehyde from crushed stones and using the extract as processing water (Kaack, 1990b).

Jam intended for table use will have a higher content of sugar and more pectin.

Jelly can be made from whole fruits, juices or concentrates. When whole fruits are used, the cherries are boiled with water until the fruit flesh disintegrates. This pomace is pressed and cleared by filtering or centrifugation. A solution of commercial pectin is added and, if necessary, citric acid is used to lower pH to 3.3. After heating to boiling, sugar to a final concentration of 65% is added and solutions of sodium benzoate and potassium sorbate are added to a level of 0.1 and 0.05%, respectively. After cooling to about 60–70°C, the jelly is poured into glass jars which are cooled to about 25°C before storage.

Instead of juice processed directly from raw fruits, single-strength juice or cherry concentrates combined with essence and water can be used for processing of jellies.

19.5.6 Juices and nectars

Assessing the suitability of raw fruit for juice processing mainly involves determining juice-soluble solids, acidity and anthocyanin content. However, the incidence of cracking and decay, uniform maturity and the estimate of yield of juice by pressing are also very important considerations. Juice yield depends mainly on the percentage of fruit flesh.

Washing and sorting may not be required if juice processing is carried out within 24 h of harvest. If this is not possible, it may be necessary to sort, wash and freeze the fruit for later juice processing.

Marshall (1954) and Tressler *et al.* (1980) have reviewed North American cherry juice production in detail. In nearly all cases, cherries are pitted prior to juice extraction. Cherries may be either hot- or cold-pressed, using equipment used for pressing grapes and the juice is coarsely filtered to remove pulp. If cold-pressed, the juice is then heated to 75–80°C to inactivate enzymes that degrade colour and flavour. Juice is cooled and settled. Pectolytic enzymes are added to reduce flocculation and to ease filtration. Juice may be stored single-strength or concentrated to a 50–68° Brix product. Before bottling, the juice receives a final filtration, the sugar-to-acid ratio is adjusted to improve flavour and the juice is pasteurized. Cherry juice has a relatively short shelf-life due to colour and flavour degradation.

In Europe, industrial processing of cherry juice involves crushing about 20% of the stones to obtain the flavour desired by consumers. The macerated cherries are then heated to 90°C for 20 s, to inactivate enzymes and destroy microorganisms, cooled to 45°C, treated with pectolytic enzymes (2 h, 45°C) and pressed (200 atm) at the same temperature. The juice is then cleared by centrifugation and pasteurized (90°C, 25 s) before storage in tanks at 5°C (Schobinger and Dürr, 1980; Kaack, 1990a).

Because of the low content of pectins, enzyme treatment before pressing is not always necessary. However, it is possible to obtain better quality and higher juice yield by treatment with pectin-degrading enzymes at 40–50°C (Sulc and Vujicic, 1973; Baumann and Gierschner, 1974).

Table 19.3. Composition of raw fruit and industrial processed juice from sour cherries (Kaack, 1990a).

	Soluble solids (%)	Acid (%)	Anthocyanin (mg 100 g^{-1})	Benzaldehyde (mg kg^{-1})	Cyanide (mg kg^{-1})
Raw fruit	21	1.9	204	93	34
Juice	21	1.6	181	17	6

During industrial processing, a small amount of anthocyanin is degraded but, more importantly, the juice contains much less benzaldehyde and cyanide (from the crushed stones) than is found in the macerated fruit (Table 19.3). Only about 18% of the benzaldehyde and cyanide is extracted to the juice.

Furthermore, during storage of the juice, glucose cyanohydrin is formed by addition of hydrocyanic acid to glucose and cyanohydrin is converted to ammonium heptagluconate (Kröller and Krull, 1968). This process explains the low content of hydrogen cyanide in stored juices (Stadelmann, 1976; Kaack, 1990a).

For the production of cherry nectar, washed, stemmed and pitted cherries are milled (0.4–0.6 mm), heated to 85–90°C to inactivate enzymes and then cooled before enzymation or maceration with special enzyme mixtures. The aim of this process is to obtain complete degradation of protopectins, partial hydrolysis of pectins, hydrolysis of cellulose and de-esterification of larger pectin molecules. All plant material is either dissolved or suspended in the serum.

After addition of solutions of sugar and citric acid, nectar is pasteurized and bottled. Compared with juices, nectars have a higher content of colour substances, fibre and certain minerals. The cultivars suitable for processing juices are also used for processing nectars.

19.5.7 Wine

A limited quantity of cherry wines and liquors are produced in Europe and North America from both sweet and sour cherries. Bing sweet cherries produce a dark red wine, which can be fairly tannic. Generally, sweet cherry juice must be acidified to produce a suitable wine.

When sour cherries are used for wine production, about 10% of the stones are crushed for flavour. Yeast and some sulphur dioxide are added to the crushed fruit. The fermentation lasts for about 10–14 days. The fermented fruit is pressed and the wine finished as for a grape wine. As with juice, the shelf-life is relatively short, generally months, as compared with a red grape wine, which will last years.

Rather than relying on fermentation, alcohol is sometimes added to clarified juices. This is the case in Denmark, where ethyl alcohol added to the juice of Stevnsbär results in their famous Cherry Herring.

References

Anon. (1974) Wirtschafliche Herstellung von Pulpen, Nektaren und ihren Derivaten. *Industrielle Obst- und Gemüseverwertung* 59, 358–361.

Anon. (1979) Verarbeitungslinie für Tiefkühlobst. *Industrielle Obst- und Gemüseverwertung* 64, 563–567.

Anon. (1982) Compact unit sorts fruit according to ripeness. *Food Engineering* 53, 109.

Anon. (1985) Ciliegio acido. *Rivista di Frutticoltura e di Ortofloricoltura* 47, 18–22.

Baumann G. and Gierschner, K. (1974) Untersuchungen zur Herstellung von Fruchtsäftem aus Sauerkirschen (Weichseln) unter besonderer Berücksichtigung des Anteils zerkleinerte Samen in der Maische. *Flüssiges Obst* 41, 83–88.

Beavers, D.V. and Payne, C.H. (1969) Secondary bleaching of brined cherries with sodium chlorite. *Food Technology (Champaign)* 23, 573–575.

Broderick, J.T. (1975) Cherry common denominators. *Flavours* 6, 103–104.

Christensen, J.V. (1986) Evaluation of characteristics of 18 sour cherry cultivars. *Tidsskrift für Planteavl* 90, 339–347.

Christensen, J.V. (1990) A review of an evaluation of 95 cultivars of sour cherries. *Tidsskrift für Planteavl* 94, 51–64.

Christensen, P.E. and Grauslund, J. (1979) Changes in contents of important constituents during ripening of *Prunus cerasus* L., cv. 'Stevnsbär'. *Tidsskrift für Planteavl* 83, 95–99.

Dekazos E.D. (1970a) Anthocyanin pigments in red tart cherries. *Journal of Food Science* 35, 237–241.

Dekazos E.D. (1970b) Quantitative determination of anthocyanin pigments in the maturation and ripening of red tart cherries. *Journal of Food Science* 35, 242–244.

Drake, S.R. (1991) The cherry. In: Eskin, M. (ed.) *Quality and Preservation of Fruits.* CRC Press, Boca Raton, USA, pp. 169–180.

Drake, S.R. and Proebsting, E.L. Jr (1985) Effects of calcium, daminozide, and fruit maturity on canned 'Bing' sweet cherry quality. *Journal of the American Society for Horticultural Science* 110, 162–165.

Heuss, A. (1974) Wirtschaftliche Herstellung von Pulpen, Nektaren und ihren Derivaten. *Industrielle Obst- und Gemüseverwertung* 59, 358–361.

Hong, V. and Wrolstad, R.E. (1990) Characterization of anthocyanin-containing colorant and fruit juices by HPLC/photodiode array detection. *Journal of Agriculture and Food Chemistry* 38, 698–708.

Kaack, K. (1990a) Processing of juice from sour cherry (*Prunus cerasus* L.). *Tidsskrift für Planteavl* 94, 107–116.

Kaack, K. (1990b) Processing of jam and stewed fruit from sour cherry (*Prunus cerasus* L.). *Tidsskrift für Planteavl* 94, 117–126.

Kröller, E. and Krull, L. (1968) Uber die Möchlichkeiten der Metabolitenbildung bei der Begasung von Lebensmitteln mit Blausäure. *Zeitschrift für Lebensmittel Unterschung und-Forschung* 136, 321–324.

LaBelle, R.L. (1971) Heat and calcium treatments for firming red tart cherries in a hot-fill process. *Journal of Food Science* 36, 323–326.

Lenartowicz, W., Zbroszczyk, J. and Dzieciol, W. (1985) Technological evaluation of the fruits of tart cherry cultivars. I. Suitability of eight cherry cultivars and three clones of English Morello for compotes and for freezing. *Fruit Science Reports* 12, 5–17.

Marshall, R.E. (1954) *Cherries and Cherry Products.* Interscience Publishers, New York, 283 pp.

Mazza, G. and Miniati, E. (1993) *Anthocyanins in Fruits, Vegetables, and Grains.* CRC Press, Boca Raton, USA, 362 pp.

Morrison, P.C. Jr and Ellis, R.F. (1984) Rapid surface freezing of fruit preserves quality, prevents clumping. *Food Processing* 45, 60–62.

Pifferi, P.G. and Cultrera, R. (1974) Enzymatic degradation of anthocyanins: the role of sweet cherry polyphenol oxidase. *Journal of Food Science* 39, 786–791.

Sapers, G.M. (1994) Color characteristics and stability of nonbleeding cocktail cherries dyed with carotenoid pigments. *Journal of Food Science* 59, 135–138.

Schmid, W. and Grosch, W. (1986) Identifizierung flüchtiger Aromastoffe mit hohen Aromawerten in Sauerkirschen (*Prunus cerasus* L.). *Zeitschrift für Lebensmittel Untersuchung und-Forschung* 182, 407–412.

Schobinger, U. and Dürr, P. (1980) Werdegang eines Getrankes aus einheimischen Süsskirschen. *Flüssiges Obst* 47, 538–541.

Siegel, A., Markakis, P. and Bedford, L.D. (1971) Stabilization of anthocyanins in frozen tart cherries by blanching. *Journal of Food Science* 36, 962–963.

Stadelmann, W. (1976) Blausauregehalt von Steinobstsäften. *Flüssiges Obst* 43, 45–47.

Stinson, E.E., Dooley, C.J., Filipic, V.J. and Hills, C.H. (1969a) Composition of Montmorency cherry essence. 1. Low-boiling components. *Journal of Food Science* 34, 246–248.

Stinson, E.E., Dooley, C.J., Filipic, V.J. and Hills, C.H. (1969b) Composition of Montmorency cherry essence. 2. High-boiling components. *Journal of Food Science* 34, 544–546.

Sulc, D. and Vujicic, B. (1973) Untersuchungen der Wirksamkeit von Enzympräparaten auf Pektinsubstraten im Frucht- und Gemüsemaischen. *Flüssiges Obst* 40, 79–83.

Tressler, D.K., Charley, V.L.S. and Luh, B.S. (1980) Cherry, berry and other miscellaneous fruit juices. In: Nelson, P.E. and Tressler, D.K. (eds) *Fruit and Vegetable Juice Processing Technology*, 3rd edn. AVI Publishing Co., Westport, Connecticut.

Van Buren, J.P. (1974) Heat treatments and the texture and pectins of red tart cherries. *Journal of Food Science* 39, 1203–1205.

Watters, G.G. and Woodroof, J.G. (1986) Brining cherries and other fruits. In: Woodroof, J.G. and Luh, B.S. (eds) *Commercial Fruit Processing*. AVI Publishing Co., Westport, Connecticut.

Weil, K.O., Moller, T.W., Bedford, C.L. and Urbain, W.M. (1970) Microwave thawing of individually quick frozen red tart cherries prior to pitting. *Journal of Microwave Power* 5, 188–191.

Zielinski, Q.B. (1964) Resistance of sweet cherry varieties to fruit cracking in relation to fruit and pit size and fruit color. *Proceedings of the American Society for Horticultural Science* 84, 98–102.

Index